The History of Physics in Cuba

BOSTON STUDIES IN THE PHILOSOPHY AND HISTORY OF SCIENCE

VOLUME 304

For further volumes:
http://www.springer.com/series/5710

Angelo Baracca • Jürgen Renn • Helge Wendt
Editors

The History of Physics in Cuba

Springer

Editors
Angelo Baracca
Department of Physics
University of Florence
Florence, Italy

Jürgen Renn
Max Planck Institute for the History
of Science
Berlin, Germany

Helge Wendt
Max Planck Institute for the History
of Science
Berlin, Germany

ISSN 0068-0346 ISSN 2214-7942 (electronic)
ISBN 978-94-017-8040-7 ISBN 978-94-017-8041-4 (eBook)
DOI 10.1007/978-94-017-8041-4
Springer Dordrecht Heidelberg New York London

Library of Congress Control Number: 2014937804

Printed on acid-free paper

Springer is part of Springer Science+Business Media (www.springer.com)

Preface

Knowledge in the modern world is shared globally. Different knowledge communities in science, technology, economy, politics, and public opinion are becoming ever more entangled. This process has also affected the history of science, which is more and more becoming part of a global history of knowledge and ever less focused on specific disciplines or specific countries. Rather, it becomes clear that the history of scientific disciplines or national developments can only be understood as part of processes of knowledge exchanges and transformations, often over long time periods and large distances. Against this background, we consider this volume on the history of physics in Cuba, covering a period of more than 200 years, as a contribution to the study of the globalization of knowledge in history (Renn 2012).

With European colonial expansion, new forms of knowledge exchange were also created, partly destroying former regional networks. Cuba was a strategic center in the Caribbean and directly connected to all parts of the Spanish colonial empire. Havana became a colonial metropolis, communicating to all other cities in the Iberian world. The island of Cuba was also closely related to the non-Spanish colonies of the Caribbean. The relationships established during the colonial period persisted even after the independence of the Spanish-American nations. During the nineteenth century, commercial, cultural and political contacts to the United States increased considerably under Spanish colonial rule.

The early history of science in Cuba was closely connected to the history of its role as a colonial metropolis. The intellectual and political climate on the island was shaped by a continuous exchange with other parts of the Americas and with Europe. In Cuba, a rich and cosmopolitan aristocracy belonged to a worldwide exchange network transgressing the imperial frontiers.

The Cuban aristocracy relied on the so-called sugar-capitalism, a conjunction of capitalism, slavery and advanced technologies. But Cuba's impressive technological advance in the nineteenth century was not accompanied by an equally strong development in the educational and academic system, since such developments were opposed by Spanish colonial rule. Nevertheless, the worldwide diffusion of scientific knowledge in the eighteenth century and the ideals of the Enlightenment associated with it also affected Cuba. A constant migration of young Cubans to Europe and the United States, as well as the economic and social development of the island over the course of the nineteenth century also created new possibilities for the production and dissemination of knowledge.

These global connections persisted even after Cuba, beginning in the mid-nineteenth century and more strongly after its independence in 1898, became increasingly dependent, both politically and economically, on the United States. However, Cuba was never completely dominated by these influences and succeeded in maintaining intellectual networks outside their spheres. As a result, Cuba was open to the appropriation of global cultural and intellectual developments and in turn was able to disseminate its own achievements worldwide. The idea behind this volume on the history of physics in Cuba is to contribute to the reconstruction of this global entanglement of knowledge.

In this sense, the volume is a pioneering step toward providing a detailed account of global entanglements in the history of science by focusing on the global relatedness of one discipline in one country. Especially after the Cuban revolution of 1959, connections to the USSR and the Eastern European countries formed the basis for a co-evolutionary process that involved both local advances and the generous foreign support of physics research in Cuba.

During the last years, an anachronistic situation has evolved that weighs heavily on the future development of science in Cuba in a globalizing world. The United States' politics of blockade continues to represent a major obstacle: two major laws passed during the last two decades effectively restrain commerce and the exchange of information between Cuba and other countries. Meanwhile, the Internet and open access to scientific information have become an ever more crucial condition for the development of science. One of the starting points for our interest in the development of Cuban science was in fact the encounter between one of us (JR) and Professor Fidel Castro Díaz-Balart on the occasion of the conference on "The Role of Science in the Information Society" (RSIS) held in Geneva from 10–12 December 2003 in support of the role of open access. Consequently, the Cuban Society of Physics, headed at that time by Professor Oswaldo de Melo, became one of the first institutions on the American continent to sign the Berlin Declaration on Open Access to Knowledge in the Sciences and Humanities (http://openaccess.mpg.de/286432/Berlin-Declaration). The complex political and technological history of the establishment and usage of the new media in Cuba has itself now become an active subject of research, undertaken by scholars such as Bert Hoffmann (2004, 2012).

This volume appears in the context of a process of rethinking Cuban history in a global environment. Until very recently, the linear development from colonial and imperial subjugation to the *Revolución* competed with a rather marginal historiography that regarded these periods in their own right and not just as episodes in a one-dimensional historical account. New interpretations now commonly emphasize the autonomy of Cuba—at least for the time after the revolution. Here, the need for Soviet aid from the 1960s on, for instance, is contrasted with autonomy in both internal development and foreign policy. These studies emphasize the uniqueness of Cuba, supposedly impermeable to penetrative influences from the outside. Global history, in contrast, would demand the location of Cuba in a global environment that is defined neither by its hermetic confinement, nor by exclusive bilateral relations with Spain, the United States or the Soviet Union, but rather by multilateral entanglements. These different viewpoints promise to trigger interesting discussions about the relation between local and global epistemic traditions.

As the only participants in this project who are mere observers of the Cuban history of physics, we would like to express our gratitude and respect for all those who in past years have not only made this history, but also engaged in its writing. We consider it a privilege to have been able to work with our co-editor, Angelo Baracca, in bringing this volume together. He has not only initiated it and created the basis for its realization, but over many years has himself also been actively involved in Cuban physics, as a passionate participant-observer, so to speak. Since we do not share his first-hand experiences, our role was rather, together with Lindy Divarci, to act as midwives in helping to turn the available materials into a book. Nevertheless, also for us, the histories of Cuba and of Cuban physics have become a passion that we will continue to pursue.

Berlin, Germany

Jürgen Renn
Helge Wendt

References

Hoffmann, Bert. 2004. *The politics of the Internet in Third World development. Challenges in contrasting regimes with case studies of Costa Rica and Cuba*. New York: Routledge.

Hoffmann, Bert. 2012. Civil society in the digital age: How the Internet changes state-society relations in authoritarian regimes. The case of Cuba. In *Civil society activism under authoritarian rule. A comparative perspective*, ed. Francesco Cavatorta, 219–244. New York: Routledge.

Renn, Jürgen (ed.). 2012. *The globalization of knowledge in history*. Studies 1: Max Planck Research Library in the History and Development of Knowledge. Berlin: Edition Open Access.

Contents

About the Contributors

Diego de Jesús Alamino Ortega received a degree in physics in 1975 and a Ph.D. in physics in 1994 from the University of Havana. He was Professor of Physics and Professor of Philosophy at the Pedagogical University of Matanzas, a founding member of the Cuban Society of Physics, a member of the Cuban Society of Science History and Technology, the Cultural Society "José Martí" and the International History, Philosophy and Science Teaching Group. He has developed courses and attended events on the history of science in Cuba, Argentina, Mexico, Brazil and England. He was awarded with a Distinction for his services to Cuban education.

Marcelo Alonso graduated in physicomathematical sciences from the University of Havana in 1942; he later took postgraduate courses in theoretical and nuclear physics at Yale University. After working as a high school teacher on his return to Cuba, he became a member of the Department of Theoretical Physics of the University of Havana in 1949, and was appointed its chairman in 1957. In 1960, he moved abroad to serve as the director of science and technology for the Organization of American States, and took up residence in the United States serving as Professor at Georgetown University Graduate School and Florida Tech. Having held many significant posts in the academic world and at international organizations, he visited the University of Havana and other Cuban research institutions twice in 2000, and established friendly relations with local colleagues. Professor Alonso was author or coauthor of several physics textbooks, including one on atomic physics and another one which became a standard textbook used in many universities around the world. He passed away in Florida on November 11, 2005 aged 84.

Ernesto Altshuler was trained as a physicist at the University of Havana, where he also obtained his Ph.D. in 1994. He has worked scientifically on magnetic materials, transport in high temperature superconductors, vortex and granular avalanches, granular flows, collective behavior of social insects and bacterial dynamics in microfluidic devices – where the experimental approach has always dominated. While his scientific work has been strongly based in Cuba, he has collaborated with teams at the Texas Center for Superconductivity, The University of Oslo, and the

ESPCI (Paris), among others. His work in science popularization includes many papers in journals and magazines, and a science book on human vision for kids (*Through the Eyes*, 1993). He is currently Full Professor at the Physics Faculty, University of Havana, and Editor of the *Cuban Journal of Physics*.

José Altshuler received the electrical engineer's degree from the University of Havana in 1953, and the Dr.Sc. degree from the Czechoslovak Academy of Sciences in 1974. He is the author or coauthor of numerous papers and several books on subjects related mainly to the engineering sciences, higher education, and the history of science and technology. He served as both Professor of Electrical Engineering and as Vice-Rector at the University of Havana, and has also served as President of the National Space Commission and as Vice-President of the Cuban Academy of Sciences, of which he is an Honorary Academician. Professor Altshuler is currently President of the Cuban Society for the History of Science and Technology.

Oscar Arés Muzio was Professor at the Faculty of Physics of the Havana University for 35 years. He received his Ph.D. from the same university in 1985, in an issue related to permanent magnets. He has been a visiting scholar at several universities and research institutes in Russia, Italy, Germany and Mexico. His research interests have focused on material science (magnetics, superconductors, colossal magneto resistance, semiconductors, etc.) and devices for applications (magnetic, superconductors (SQUIDs), solar cells and solar thermal collectors). For many years he was head of the Laboratory of Magnetism and Superconductivity at the University of Havana and was co-author in obtaining the first high temperature superconductor made in Cuba. He was for many years a member of the Permanent Court of physical doctoral defenses in Cuba. For his scientific activities, he was awarded with the medal Carlos J. Finlay by the Cuban government.

Olimpia Arias de Fuentes is Senior Researcher at the Institute of Materials Science and Technology and Associate Professor at the Physics Faculty of the University of Havana. She graduated as a physicist at the University of Havana where she also obtained her PhD. Her current research area focuses on electrochemical sensors and biosensors and the necessary instrumentation for their environmental and biomedical application. She has also undertaken studies on the history of science and gender thematic in Cuba, published numerous papers in scientific journals and proceedings and presented many research contributions at diverse international congresses. She has given talks at universities in Italy, Germany, Chile, Mexico and Uruguay and has been a Visiting Professor at the University of Rome "La Sapienza." She is full member of the Third World Organization of Women in Science (TWOWS), the Cuban Commission of Women in Sciences, the Cuban Chemical Society and founder of the Cuban Physical Society where she was a staff member during the years 1978–1985. She was among the authors to be awarded with one of the Annual Prizes of the Cuban Academy of Sciences in 2001. She was also awarded with "The Distinction for the Cuban Education" given by the Cuban Minister of Higher Education and "The Medal José Tey" given by the Cuban Republic Government, both for her contribution to education in Cuba.

Angelo Baracca received his Ph.D. in physics in 1968 from the University of Florence, Italy, where he is currently Professor of Physics. He has collaborated with universities and research centres in France, the United Kingdom, Spain, and Germany, and is now collaborating closely with the University of Havana in Cuba. He is actively engaged in the movements for peace and disarmament and has devoted himself professionally to the study of nuclear technology – both civil and military – its structure and history. Besides several textbooks of physics, he has published many papers and books. Besides teaching and research activities in many fields such as high-energy physics, statistical mechanics, and the foundations of quantum mechanics, his interests concentrate on the history and social responsibility of science.

Duccio Basosi is Researcher in History of International Relations at the Ca' Foscari University of Venice since 2009. He obtained his Ph.D. in History of International Relations from the University of Florence in 2004. From 2002 to 2009, he has been a scholar at the Machiavelli center for Cold War Studies, a consortium of historians from seven Italian universities. He specializes in international relations from the 1960s to the 1990s, with a focus on US foreign policy, Cuban foreign policy, and international political economy. He co-edited the collection of essays (with A. Lorini), *Cuba in the World, the World in Cuba. Essays on Cuban History, Politics, and Culture* (Firenze, 2009). His monograph, *Il governo del dollaro. Interdipendenza economica e potere statunitense negli anni di Richard Nixon, 1969–1973* (Florence, 2006), was awarded the "Premio SISSCO Opera Prima" by the Italian Society for the Study of Contemporary History in 2007.

Carlos A. Cabal Mirabal was a co-founder in 1971 of Cuba's second school of physics at the University of Oriente, which he directed for several years. He was also the founder Dean of the Faculty of Mathematics and Physics (1980–1989) and Founder Director of the Medical Biophysics Centre (1989–2005) at the Oriente University in Santiago de Cuba. He defended his Ph.D. thesis in magnetic resonance at the University of Leningrad (1980). He has been Full Professor at the Oriente University for 40 years, Full Professor at Havana University for the last 5 years and is Honorary Professor at the La Plata National University, Argentina. Currently, he is Head of the Images Group at the Center for Genetic Engineering and Biotechnology and continues to work in the field of MR, in particular, in molecular imaging. In 2010 he received the Cuban National Prize for Physics. He is a member of the International Union of Pure and Applied Physics (Magnetism Commission) and merit member of the Cuban Academy of Sciences.

Ocatvio Calzadilla Amaya is Full Professor at Havana University, Department of General Physics. He received his Ph.D. in physics and mathematics at Havana University in 1994. Since 1972, he has served as Professor of general physics and training laboratory. He has authored books for physics education. His research interests are the materials science. His research first focused on the preparation of semiconductor materials by vapor phase transport in closed ampule using the melt method. At the present time, he is working on the growth of semiconductor thin films using the chemical bath deposition method and electrochemical method. He has published several scientific articles about the growth and characterization of materials.

Manuel Cardona is Founding Director Emeritus at the Max Planck Institute for Solid State Research in Stuttgart and Honorary Professor at the Universities of Stuttgart and Konstanz. He is also Adjunct Professor at Arizona State University. He received a *licenciado en ciencias* degree at the University of Barcelona in 1955, a Dr. of Sciences degree at the University of Madrid in 1958 and a Ph.D. at Harvard University in 1958. He has been awarded 11 honorary doctorates (Barcelona, Madrid, Valencia, La Laguna, Rome, Sherbrooke (Canada), Toulouse, Regensburg, Thessaloniki, CINVESTAV (Mexico), and Brno (Czech Republic)). He is a member of the National Academy of Sciences of the USA, that of Mexico and the Royal Society of Canada. He has received the 1988 Principe de Asturias Prize (Spain). He has authored or co-authored 1,300 refereed articles which have attained the *h* number of 98. He is the author or co-author of about 15 books and textbooks. He has supervised a large number of Latin-American students, having spent 3 months at the University of Buenos Aires (June–August 1965) shortly before the military revolution of General Onganía, whose effects in the academic life of the Continent he helped to palliate. His field of endeavor is solid state physics.

Jacqueline Cernogora graduated in physics at the Faculty of Sciences of Paris University. She then joined the GPS *Groupe de Physique des Solides* (Solid state physics group) in 1960, first studying the photoconductivity of germanium doped with deep-level dopant, (third level thesis in 1962), later studying disordered semiconductors by means of radiative recombination. The first study of highly doped germanium gave rise to her Ph.D. in 1968. Later studies were concerned with amorphous chalcogenides, semi-magnetic semi-conductors, porous silicon and amorphous carbon, each of which gave rise to various publications. In 1962, she joined the Centre National de la Recherche Scientifique (CNRS) as Research Attaché and went on to become Program Director there in 1968. She was part of the French-Cuban Committee during the early 1970s.

Víctor Luis Fajer Avila defended his doctoral thesis in physics in 1993. He was elected in 1999 President of the Cuban Physical Society and at present is Vice-President of the Society. He has led and conceived the developing works of seven versions of automatic and digital polarimeters, denominated LASERPOL, of economic and scientific-technical importance. He has presented 90 publications and 4 patents. He has published several papers about physics in Cuba. In 1999 he received the "Carlos J. Finlay Medal" and was nominated Academician of the Cuban Academy of Sciences. He is Titular Researcher and Head of Department of the Center for Technological Applications and Nuclear Development (CEADEN) and Titular Senior Professor of the High Polytechnic Institute (IPSJAE), Havana.

Federico Falcon is a researcher at the Institute of Science and Technology of Materials of Havana University. He completed his Ph.D. thesis in 1986. His research areas are crystallization and crystal growth. He has published papers on related subjects in various journals. In 2007, he obtained the medal "Serguei Koroliov" from the Cosmonautic Federation of Russia.

Roberto Fieschi obtained his degree in physics at the University of Pavia in Italy in 1950 and a Ph.D. in physics at the University of Leiden in the Netherlands in 1955. He was Professor of Physics at the University of Parma from 1965. From 1966 to 1972 he was Head of the Physics Institute, and from 1969 to 1976 Director of the National Research Council Laboratory MASPEC, overseeing its development. Professor Fieschi was a member of the CNR Physics Committee and Vice-President of the Technological Committee. In 1977, he was commended by the Ministry of Public Instruction and subsequently awarded a gold medal. He is also part of the National Council of the Union of Scientists for Disarmament (USPID). In 1997, Professor Fieschi was awarded the Laurea honoris causa in Materials Engineering by the University of Lecce. In the 1970s, he was a member of the Central Committee of the Partito Comunista Italiano. He is now Professor Emeritus at the University of Parma.

Juan E. Fuentes received his Ph.D. from Moscow State University, Solid State Faculty in 1975. Thereafter he has taught at the School of Physics and then at the Faculty of Physics of Havana University. At the same time he has been working in investigations in different branches of solid state physics. He is the author of articles on the growth and characterization of single crystals and films. He is currently working in the piezoceramics group and is interested in lead-free ceramics.

Federico García-Moliner founded the renowned Spanish research school in solid state physics, which reached a high international level. He was dedicated to teaching activities and played an important role in the formation of young scientists in many European countries. In addition, he played a key role in supporting scientists in developing countries, also in Latin America. He played an active role in the Pugwash Conferences on Science and World Matters. In 1992, he was awarded the *Premio Príncipe de Asturias de Investigación Científica y Técnica* for his contributions to solid-state physics.

Carlos R. Handy received his B.A., M.A., and Ph.D. degrees in theoretical physics from Columbia University in New York City. He was born in Havana, Cuba, in 1950, but raised in the United States. He is the grandson of the famous Blues composer, William C. Handy. Dr. Handy was a postdoctoral student at Los Alamos National Laboratory in the late 1970s. While there, he developed a strong interest in Moment Problem reformulations of Quantum operators, particularly for developing new computational methods in tackling singular perturbation/strong coupling problems. Based on his works in the mid to late 1980s, he is now recognized as a pioneer in the application of semi-definite programming related methods to quantum operators, anticipating these interests within the mathematics-optimization community by more than a decade. Simultaneously, through his appointments at Clark Atlanta University (1981–2005) and Texas Southern University (2005 to present), he was able to build internationally competitive research centers and departments in institutions dedicated to educating African Americans. He currently serves as the Chair of the Physics department at Texas Southern University in Houston, Texas.

Brian Hayes is Senior Writer for *American Scientist* magazine, where he writes a column called "Computing Science." He is also the author of *Infrastructure: A Field Guide to the Industrial Landscape* (W. W. Norton, 2005) and *Group Theory in the Bedroom, and Other Mathematical Diversions* (Hill and Wang, 2008). He is a former editor of both *American Scientist* and *Scientific American*.

Fabrizio Leccabue received a degree in chemistry in 1972 at the Chemistry Department of Parma University in Italy. From 1973 to 2007 he worked at the Institute MASPEC of Consiglio Nazionale delle Ricerche of Parma. He has coordinated national projects (Strategic Project/1997, Finalized Project/1997, ASI/2003), some European projects (Brite Euram Feasibility II/1992, HCM/1995, Alfa/1996, bilateral agreement between CNR and CSIC/2002, URSS Academy of Sciences/1984, MURST-British Council/1997) and international projects (CONACyT, Mexico/2002, University of La Habana, Cuba/2001). He has served on the management committee of European COST Action 514/1993 and 528. On materials of technological applications, he has developed collaborations between Italian companies (ST-Microelectronics, Alenia, Pirelli Pneumatici, Italcementi, Bormioli) and local companies (Vacuum Components & System, CAEN, ALSIM) devoted to the study of sensors, magnetically coupled mechanical systems, piezoelectrics, ferroelectrics, non-volatile memory (PZT, SBT), and hard (hexaferrites, intermetallic alloys) and soft (FeSiB based alloys) magnetic materials. In the last 10 years, he co-developed research on the preparation and characterization of ferroelectric thin film and oxides, obtained by pulsed laser ablation deposition and sol–gel processing, and amorphous, nano-/micro-crystalline magnetic materials using melt spinning. In this period, his scientific activities also included the design and construction of MOCVD reactors to prepare SiC film. He has co-authored more than 140 publications in international journals, given 185 scientific presentations at International Conferences and has supervised more than 35 degree theses, PhDs and fellowships.

Irving A. Lerch received his Ph.D. in biological and medical physics in 1969 at Chicago University. His research focused on radiation metrology, low-energy X-ray spectroscopy, radiation biophysics and biomedical statistics. In 1973, he was program officer at the Division of Life Sciences in the Department of Research and Isotopes at the International Atomic Energy Agency (IAEA) in Vienna, with developmental projects in East-Central Europe, the Middle East, Africa and Asia. From 1976 to 1992, he headed the Department of Radiation Oncology Physics of New York University. He has served on numerous national and international boards and committees. From 1992 through 2003 he served as Director of International Affairs for the American Physical Society at the American Center for Physics in College Park, Maryland. He has been consultant to the IAEA and UNESCO.

Jorge Lodos received his Ph.D. in Chemistry from Kiev Food Polytechnic Institute, Ukraine, in 1979 and became D.Sc. at Havana University in 1999. He has held different positions in the sugar industry. He acted as Director of Cuban Sugar Industry Research Institute (ICINAZ) for 15 years where he was responsible for and scientifically involved in experiments for the Cuba-USSR space flight in 1981. He is Associate Professor in the Faculty of Chemistry at Havana University. His research focuses on sugar and sugar by-products technology. He is the author of books, articles and patents on sugar technology and has received several national awards.

Werner Marx studied chemistry at the University of Bonn and prepared his diploma and doctoral thesis in physical chemistry, both dealing with atmospheric chemistry. In 1982, he joined the Central Information Service for the institutes of the Chemical Physical Technical Section of the Max Planck Society, located at the Max Planck Institute for Solid State Research in Stuttgart. This service provides assistance to access, explore and exploit the wealth of scientific information resources available worldwide. The target groups are researchers, librarians, administrators, and decision-makers of the Max Planck Society. Establishing impact data for research evaluation has become a major subject. His publications are focused on the possibilities and limitations when using and interpreting citation data. His special interests are the potential of databases for more comprehensive information demands and the application of bibliometrics to the history of science.

Osvaldo de Melo Pereira is Full Professor of the General Physics Department of the University of Havana. At this university, he received his Ph.D. in 1992 and also served as Dean of the Physics Faculty in the period 2001–2006. He was President of the Cuban Physical Society from 2005 to 2011. His research is mainly related to the preparation of semiconductor films and nanostructures as well as to optical and structural characterization techniques. He has published over 80 scientific papers, over 20 papers on the popularization of science, and has edited three books. He has received awards from the Cuban Academy of Science in 1999, 2004, 2007 and 2013. De Melo has been Visiting Professor of the Centro de Investigaciones y Estudios Avanzados (Mexico) and the Universidad Autónoma de Madrid (Spain), was IOP fellow (2006–2011), and a senior member of the IEEE. He has led a number of international projects and has chaired several international meetings.

Luis M. Méndez Pérez received a degree in physics in 1971 and M.Sc. in higher education in 2005 at Universidad de Oriente, Cuba. He took postgraduate courses on nuclear physics at San Petersburg State University, Russia, from 1972 to 1976. He is a Professor of General Physics, of History of Physics and also of History of Philosophy at Universidad de Oriente, Cuba. He is a member of the Cuban Society of physics, of the Cuban Society of History of Science and Technology. He has been a Visiting Professor at Eduardo Mondlane University, Mozambique, at the ISCED of Uige, Angola and at the Physical Institute of Sao Carlos, Sao Paulo University, Brazil. He was awarded with a number of distinctions and orders for his service to Cuban Education.

Jorge Núñez Jover has a Ph.D. in chemistry and is Director for Postgraduate Studies at the University of Havana (UH). He is the coordinator of the master's and doctoral programs on science, technology and society. He holds the chair in science, technology, society and innovation. He has published 10 books and over

100 papers. His research is in the areas of education, science, technology and society; postgraduate studies evaluation and accreditation, as well as university science policy. He is invited Professor in universities in Spain, Germany, Brazil, Mexico and other countries in Latin America and the Caribbean.

Isarelis Pérez Ones graduated in sociology in 1997, and also with a master's degree in science, technology and society (STS), at the University of Havana (UH). She is now an Assistant Professor there. She was a graduate student at Roskilde University, Denmark (2002–2003). In 2005, she trained in the Ph.D. program, 'Globelics Academy,' at the Technical University of Lisbon. She is finishing her Ph.D. on STS at the UH. Her main interests are related to the social studies of science and technology and gender and higher education. Her current work has focused on the role of universities in the national system of innovation.

Rosalind Reid is a journalist and university administrator. During her tenure as Editor of *American Scientist* magazine from 1992 to 2007, she visited several Latin American countries to promote international scientific networking and the visual communication of science. She serves on the Board of the Council for the Advancement of Science Writing and is currently Executive Director of the Institute for Applied Computational Science at Harvard University and Assistant Dean for External Programs at the Harvard School of Engineering and Applied Sciences.

Jürgen Renn is Director at the Max Planck Institute for the History of Science in Berlin, as well as Honorary Professor for History of Science at both the Humboldt-Universität zu Berlin and the Freie Universität Berlin. He has taught at Boston University, at the ETH in Zurich and at the University of Tel Aviv. He has held visiting positions in Vienna, Bergamo, Pavia and at CalTech. He is a member of the Leopoldina as well as of further national and international scientific and editorial boards. His research looks at the structural changes in systems of knowledge. The aim is to develop a theoretical understanding of knowledge evolution, taking into account its epistemic, social and material dimensions. As groundwork for such a theoretical approach to the history of knowledge, he has been studying some of the great transformations of systems of physical knowledge, such as the origin of theoretical science in antiquity, the emergence of classical mechanics in the early modern period and the revolutions of modern physics in the early twentieth century.

Carlos Rodríguez Castellanos received his *Licenciado en Física* (1970) and his Master of Science (1975) at Havana University, and his Ph.D. in theoretical and mathematical Physics at the Joint Institute for Nuclear Research, Dubna, USSR, in 1981. His research has focused on applications of quantum statistical mechanics to condensed matter theory and materials science. He has made contributions to polaron theory, dielectric relaxation, quantum Hall effect, adsorption and diffusion in porous materials. He was Professor of Theoretical Physics (1984) and Emeritus Professor (2013), Dean of the Faculty of Physics (1984–1991), Vice Rector for Research (1991–1996) and Director of the Institute of Materials Science and Technology (1999–2008) at the University of Havana. He was President of the

Cuban Physical Society (1991–1994); President of the Jury of the XXII International Physics Olympiades (1991); and Member of the Commission "Physics for Development" of IUPAP (1996–9). In 2012, he was Emeritus member and Vice President of the Cuban Academy of Sciences.

María Sánchez Colina is currently Dean of the Physics Faculty at the University of Havana where in 1996 she received her Ph.D. Her research focuses on semiconductor lasers. She has been Visiting Professor at the Centro de Investigaciones y Estudios Avanzados (Mexico) and at the Universidad Autónoma de Madrid (Spain). She has edited two conference proceeding volumes and was editor of the *Cuban Journal of Physics* from 2002 to 2010. From 1998 to 2001, she was Chair of the Cuban Chapter of the Electron Devices Society (EDS –IEEE).

Oscar Sotolongo-Costa received his Ph.D. from Havana University and is Professor for theoretical physics and history and methodology of physics there. He lectures on the physics of complex systems and on different branches of theoretical physics. His research interests include fluid mechanics, statistical physics, dynamical systems, and he has published widely on these subjects. He is currently President of the "Henri Poincare" group of Complex Systems at Havana University.

Maria C. Tamargo is Professor of Chemistry at the City College of the City University of New York (CUNY), where she has been since 1993. Prior to that, she was a Member of Technical Staff at Bellcore and AT&T Bell Labs. Her research interests focus on semiconductor materials growth by molecular beam epitaxy with emphasis on exploring novel materials properties and device applications of wide band gap II-VI compounds. She has authored over 250 research papers and is the editor of *II-VI Semiconductor Materials and Their Applications* (2002, Taylor and Francis). During the years 2001 through 2007, she served as Dean of the Division of Science at City College. She is a fellow of the American Physical Society and an editor of the research journals *Physica Status Solidi c* and *Semiconductor Science and Technology*.

Carlos Trallero-Giner is Professor at the Department of Theoretical Physics at the University of Havana. He received a Ph.D. in theoretical physics and mathematics in 1980, and a doctor's degree in physics and mathematics in 1990 from the Ioffe Institute, San Petersburg/Russia. He has been a visiting scholar at several universities and research institutes in Germany, USA, Russia, Spain, Brazil, and other countries. He was awarded a Fellowship of the American Physical Society and Distinguished Professor by the Cuban Ministry of Higher Education. His research interests focus on condensed matter theory, light scattering in solids, semiconductors and low-dimensional structures, phonons and electron phonon interactions in nanostructures, optical properties, mathematical physics and chaos, and Bose-Einstein condensation. He has published over 200 scientific research papers.

Theodore Veltfort (1915–2008) was an electrical engineer by profession and volunteered with the Abraham Lincoln Brigade during the Spanish Civil War. In Cuba, he worked first for JUCEPLAN, under the leadership of Ernesto Guevara, and then for 5 years at the University of Havana in the School of Physics, directed by Dr. José Altshuler.

Elena Vigil Santos received her B.Sc. and MAT degrees from Fairleigh Dickinson University, N.J. and her Ph.D. from Havana University (UH) in 1984. She teaches Solid State Physics as Professor or the Physics Faculty (UH). She engaged in III-V semiconductor physics and optoelectronic devices research for over 20 years. In the last 10 years she has studied materials and structures for new-concept solar cells. She designed and prepared the first experiment in which GaAs and AlGaAs epitaxial layers were grown in space during the Cuba-USSR space flight in 1981. She held different posts in Havana University: Research Vice-Director of the School of Physics (1970–1974) and Research Vice-Director of the Institute for Materials and Reagents (1985–1998). She is "member of merit" of the Cuban Academy of Science and member of the Latin American Regional Committee of the International Council of Science (ICSU).

Dina Waisman Argentinian, collaborated with Andrea Levialdi on solid-state physics in the Department of Physics of the University of Buenos Aires. Subsequently, she carried out teaching and research activities in the field of nuclear physics. At present she teaches physics outside of this specialized field.

Helge Wendt is a Research Fellow at the Max Planck Institute for the History of Science in Berlin, where he is associated with the project "Globalization of Knowledge." He received his Ph.D. from the University of Mannheim, where he taught early modern history. His research focuses on the history of Christian missions in different colonial contexts, the history and historiography of globalization and environmental history. He currently works on the analytic turn of mechanics between traditional models of thought and claims of autonomy, funded by the Collaborative Research Center "Transformations of Antiquity." He also works on coal as an object of knowledge in global history.

Abbreviations

ACC	Cuban Academy of Sciences
AENTA	Agency for Nuclear Energy and Advanced Technology
AID	Agency of Information for Development
ALASBIMN	Latin American Association of Biology and Nuclear Medicine Societies
APS	American Physical Society
BECICPA	Special Bureau for the Construction of Scientific Instruments with Adjunct Production (now CEDEIC)
BIOCEN	Center of Biopreparations
CAME	Argentina Confederation of Medium Enterprises
CBFM	Center for Biophysics and Medical Physics
CCE	"Ernesto Che Guevara" Electronic Component Complex
CEAC	Cuban Commission for Atomic Energy
CEADEN	Center for Studies Applied to Nuclear Development
CECT	State Committee for Science and Technology
CEDEIC	Development Center for Scientific Instruments and Equipment (originally BECICPA)
CENC	National Nuclear Energy Commission of Cuba
CENIAI	National Center of Automatic Exchange of Information
CENTIS	Isotope Center
CERN	European Organization for Nuclear Research
CID	Center for Digital Research
CIEN	Information Center for the Nuclear Sphere
CIES	The Solar Energy Research Center
CIGB	Center for Genetic Engineering and Biotechnology
CIME	Research Center for Microelectronics
CIMEQ	Center for Clinical and Surgical Research
CINVESTAV	Center of Investigation and Advanced Studies
CIPIMM	Institute for Research and Projects for the Mining-Metallurgical Industry

CITAMEL	Enterprise for Information Technologies and Advanced Telematic Services
CITMA	The Ministry of Science, Technology and Environment
CLAF	Latin American Center for Physics
CNEA	National Commission for the Civil Applications of Atomic Energy
CNIC	The National Center for Scientific Research
CNCT	National Council for Science and Technology
CNGC	National Commission for Scientific Degrees
CNRS	National Center for Scientific Research (France)
CNSN	National Center for Nuclear Safety
CPHR	Center for Radiation Protection and Hygiene
CSIC	The Spanish National Research Council
CUJAE	Ciudad Universitaria "José Antonio Echeverría"
CYTED	Ibero American Program of Science and Technology for Development
DNI	National Directory of Informatics
EF	School of Physics
ESI	Essential Science Indicators
FCT	The Science and Technology Forum
FCTN	Faculty of Nuclear Sciences and Technologies, University of Havana
FELASOFI	Latin-American Federation of Societies of Physics
FEU	University Students Federation
GET	Group for Electronics in Tourism
GDS	Semiconductor Devices Group
IAEA	International Atomic Energy Agency
ICID	Cuban Institute for Digital Research
ICIMAF	Institute for Cybernetics, Mathematics and Physics
ICM	Institute of Marine Sciences
ICINAZ	Cuban Institute for Sugar Research
ICO	International Commission for Optics
ICSU	International Council for Science
ICTP	International Center for Theoretical Physics, Trieste
IFN	Institute for Nuclear Physics
IGA	Institute for Geophysics and Astronomy
IILA	Italian Latin-American Institute
IMACC	The Mathematics, Cybernetics, and Computing Institute
IMRE	Institute of Materials Science and Technology
ININ	Institute for Nuclear Research
ININTEF	Institute for Fundamental Technical Research
INSAC	National Institute of Automated Systems and Computer Technologies
INSPEC	Web of Science, Essential Science Indicators
INST	Institute for Nuclear Science and Technology
INSTEC	Higher Institute for Applied Sciences and Technologies (until 2003, ISCTN)

IPHO	Organization for the International Physics Olympiads
IPVCE	Vocational Pre-university High School in Exact Sciences
ISCAB	Higher Institute of Intensive Livestock Farming Sciences
ISCAH	Higher Agricultural Sciences Institute of Havana
ISCTN	Higher Institute of Nuclear Sciences and Technologies (renamed INSTEC in 2003)
ISPEJV	Pedagogical Institute "Enrique José Varona"
ISPJAE	"José Antonio Echeverría" Higher Polytechnic Institute
ISPJAM	Higher Polytechnic Institute "Julio Antonio Mella"
ISCTN	Institute of Nuclear Sciences and Technologies, Havana
ITM	Military Technical Institute
IUCr	International Union of Crystallography
IUPAP	International Union of Pure and Applied Physics
JINR	Joint Institute for Nuclear Research
LACETEL	Central Laboratory of Telecommunications
LAGS	Laboratory of Synthetic Antigens
LIEES	Laboratory for Solid State Electronics Research
LRI	Industrial Radioisotope Laboratory
LTP	Planar Technology Laboratory
MASPEC	Special Materials for Electronics and Magnetism (now IMEM)
MES	Ministry of Higher Education
MIC	Ministry of Informatics and Communications
MINCOM	Ministry of Communications
MINED	Physics Department of the Ministry of Education
MINFAR	Ministry of the Revolutionary Armed Forces
MININT	Ministry of the Interior
MINJUS	Ministry of Justice
MINSAP	Institute of Oncology and Radiobiology of the Ministry of Public Health
MOCVD	Metalorganic Chemical Vapour Deposition
NED	National Endowment for Democracy
OTRI	Office for the Transfer of Research Results
PAHO	Pan-American Health Organization
PNUD	United Nations Development Programme
SCI	The Thomson Reuters Science Citation Index
SCUH	Superconductivity Laboratory, University of Havana
SEAN	Executive Secretariat for Nuclear Affairs
SIME	Ministry of Metallurgy and Electronics Industry
SLAFES	Latin-American Symposium of Solid State Physics
SUMA	Ultramicroanalytic System
SQUID	Superconducting Quantum Interference Devices
TWAS	Third World Academy of Sciences, Trieste
TWOWS	Third World Organization for Women in Science
UAI	International Astronomical Union
UCLV	"Marta Abreu" Las Villas University

UH	University of Havana
UJC	Union of Young Communists
UNAM	National and Autonomous University of Mexico
UNEAC	Cuban National Union of Writers and Artists
UNPD	United Nations Program for Development
UO	University of Oriente, in Santiago de Cuba
UPEC	Journalists' Union
USAID	United States Agency for International Development
UUCP	Unix-to-Unix Communication Protocol
VVER or WWER	Vodo-Vodyanoi Energetichesky Reactor (Water-Water Power Reactor)
WHO	World Health Organization
WoS	Web of Science

Introduction

Chapter 1
A Short Introduction to this Volume

Angelo Baracca, Jürgen Renn, and Helge Wendt

The history of physics in Cuba comprises two centuries of intensively interrelated processes in politics, society, science and worldwide communications. The evolution of physics as a discipline since the early nineteenth century may serve as a focal lens for studying the developments in Cuban society in a global context. At the eve of the 1959 Revolution, the new physics of the twentieth century was not widely known, nor taught, in Cuba and scientific research scarcely existed in its universities. It is only in the past half century that Cuba has built up an advanced scientific system.

To a broader public, the high level of Cuban science in biotechnologies, immunology and medicine is well known. But scientific development also took off very quickly in physics, achieving impressive results within the first 15 years after the Revolution. One may imagine that such a rapid take off and development was due exclusively to the far-reaching collaboration with—and support from—the USSR and other Socialist countries. It is true that this factor should not be underestimated. Hundreds of Cuban physicists graduated, specialized and worked in the prestigious scientific institutions of the USSR. Nevertheless, the contributions to this volume show that in the beginning many Western physicists also played a fundamental role. During various periods, they visited the newly founded School of Physics of the University of Havana and promoted initiatives in higher education and scientific research, such as the summer schools. Collaboration with the USSR and the Socialist countries grew gradually and assumed a significant role only from the 1970s on. Cuban physicists, however, never severed ties nor terminated their collaborations with Western colleagues and institutions, a circumstance that has played an important role.

A. Baracca (✉)
Department of Physics, University of Florence, Florence, Italy
e-mail: baracca@fi.infn.it

J. Renn • H. Wendt
Max Planck Institute for the History of Science, Berlin, Germany
e-mail: renn@mpiwg-berlin.mpg.de; hwendt@mpiwg-berlin.mpg.de

A. Baracca et al. (eds.), *The History of Physics in Cuba*, Boston Studies in the Philosophy and History of Science 304, DOI 10.1007/978-94-017-8041-4_1,

Another original aspect that must be considered in the scientific development in Cuba since the 1960s is that the strategies for this development were not simply copied or imported from more advanced countries. There was a lively debate within the Cuban scientific community (and with foreign Western advisers), inspired by the requirement that research should relate directly to the economic and social development of the country.

This volume presents the history of one scientific discipline in one country, from its origins in the early nineteenth century up to the present time. This task was facilitated by both the relatively short historical period of this development and the relatively limited size of the Cuban community of the physicists. Nevertheless, we hope to present a pioneering volume that depicts the development of a scientific community in an international context over a period of 200 years.

The Structure of the Volume

The contributions to the volume are organized in four parts. They are preceded with a personal introduction by Angelo Baracca to Cuban society with all of its complex and multifaceted aspects. This is an attempt to help the reader with little or no knowledge of Cuba to position the developments discussed in the volume within the lively historical and social reality. It is followed by a short critical bibliography compiled by Duccio Basosi that aims to suggest further reading for the non-specialist. A short biography of all of the authors who contributed to this volume is also included.

The first part of the volume includes several contributions that reconstruct the different stages of the history of physics in Cuba, from its beginnings in the late colonial era to the present. The first contribution by José Altshuler and Angelo Baracca gives an overview of the development from the late eighteenth century on. The second contribution by José Altshuler deals in more detail with the beginning of the systematic teaching of physics from the 1920s to the end of the 1950s. The third contribution by Angelo Baracca, Víctor Fajer and Carlos Rodríguez Castellanos treats the development of physics after 1959, covering the foundation of various institutions as well as international cooperation, mostly with countries of the former Eastern bloc. In their contribution, Carlos R. Handy and Carlos Trallero-Giner give a detailed overview of the achievements of Cuban physics after 1959, dealing with research as well as with education. The contribution by Luis M. Méndez Pérez and Carlos A. Cabal Mirabal describes how the University of Oriente in Santiago de Cuba, founded in 1947, became an important center of teaching and research in physics from the 1960s on. In his contribution, Diego de Jesús Alamino Ortega studies the training of physics teachers in Cuba from the colonial period on, with a focus on pedagogical efforts undertaken after the revolution. The paper by Isarelis Pérez Ones and Jorge Núñez Jover, reprinted from *Science and Public Policy,* discusses some of the political strategies of technological innovation and the development of physics in Cuba in the 1980s and 1990s.

The second part of the volume comprises testimonies of Cuban physicists, who offer lively insights from the perspective of the actors themselves. The first contribution is an interview by Angelo Baracca with Hugo Pérez Roja, focusing on the advancement of physics in the 1960s and 1970s. In a second interview, Melquíades de Dios Leyva reports on the changes after the revolution in the teaching of physics, on the international engagement of Cuban physicists and on the opening up of academic career opportunities for Cubans from the poor population who previously had no opportunities at all. In her contribution, Elena Vigil Santos discusses the development of semiconductor physics in the framework of Cuban engagement in the COMECON program Intercosmos. This program is the object of a more extended essay by José Altshuler, Ocatvio Calzadilla Amaya, Federico Falcon, Juan E. Fuentes, Jorge Lodos and Elena Vigil Santos. Oscar Arés Muzio and Ernesto Altshuler discuss the history of the synthesis of a superconductor at the University of Havana in the 1980s. Oscar Sotolongo-Costa in his contribution describes the emergence of the physics of complex systems in Cuba as a response to both an international trend and the need for ways of doing research in Cuba during the economic crisis of the 1990s. The contribution by Carlos A Cabral Mirabal discusses the development of the Magnetic Resonance Project during the 1990s. Carlos Rodgríguez Castellanos draws a conclusion for the present of the ongoing development of nanoscience and nanotechnology in the national and international framework. Osvaldo de Melo Pereira and María Sánchez Colina report on the development of physics education at the University of Havana and include a detailed overview over the topics and the number of students from the 1960s on. The contribution by Olimpia Arias de Fuentes is a gender study on the role of female students and researchers in Cuban physics.

The third part of the volume presents a series of testimonies by foreign physicists, some of whom were directly involved in developing Cuban physics, in particular in the development of teaching and research activities in the early years of the *Escuela de Física*. The first contribution is by Theodore Veltfort who passed away in 2008. A US engineer who took part in the Spanish Civil War, Veltfort spent many years at the *Escuela de Física* in Havana during the early 1960s, actively contributing to updating courses and developing the first research projects on solid-state apparatus before the first Cuban physicists returned from the Soviet Union to Cuba. The second contribution by the Argentine physicist Dina Waisman is a reprint of the obituary of the Italian solid-state physicist Andrea Levialdi, an anti-Fascist who spent many years in exile in Argentina. Levialdi supported an ambitious program of research at the *Escuela de Física* in Havana in the 1960s, but his projects were interrupted by his premature death. Roberto Fieschi's contribution reports on the close cooperation between the University of Parma and the University of Havana, which had been arranged by Levialdi. The subsequent articles focus on other partnerships between Cuban and European physicists and universities. The contribution by Jacqueline Cernogora, a member of the "Comité de Liaison Scientifique et Universitaire France-Cuba", deals with the *Esculeas de Verano* (summer schools) between the end of the 1960s and the early 1970s. Fabrizio

Leccabue, former researcher at the Italian *Consiglio Nazionale delle Ricerche* (National Research Council) in Parma, deals in his contribution with the collaboration of the Universities of Parma and Havana between 1970 and 2000. Helge Wendt depicts the first cooperation between the newly founded Academy of Science of Havana and the Academy of the German Democratic Republic (GDR) between 1965 and 1975. In his contribution, Federico García Moliner, a Spanish solid-state physicist, offers recollections of his collaborations at the University of Havana. The next two contributions are reprinted from the *American Physical Society News* (2002). The first is by Marcelo Alonso, a leading Cuban physicist in the 1940s. After the revolution, he left Cuba and did not return to the island for many decades. In his contribution, he reviews the development of the last decades of Cuban physics and endorsed a rapprochement between the US-American and Cuban physicist communities. The contribution by Irving A. Lerch discusses the first steps of collaboration between the American Physics Society and various Cuban physicists. The personal testimony by Carlos R. Handy, currently professor of physics at Texas Southern University in Houston, also deals with relations between US-American and Cuban physicists from the 1980s on. The contribution by Maria C. Tamargo discusses contacts between physicists from the US and from Cuba on the occasion of international conferences and workshops since 1995. The contribution by Rosalind Reid and Brian Hayes show the problems in doing research in different fields of physics at different institutions since the early 1990s.

The two contributions of the fourth part of the volume deal with some of the conditions for publishing scientific research in Cuba. The first contribution is a bibliometric study performed by Werner Marx and Manuel Cardona on Cuban publications from the late 1950s to the current time. The second contribution by Ernesto Altshuler reviews the last two decades of publishing activities of Cuban physicists in national and international journals.

Acknowledgements In 1996, the solid-state physicist and historian of science, Fernando Crespo Sigler of the University of Havana gave the decisive impulse for this project. The editors want to express their deep gratitude for his initiative and his considerable efforts for preparing the ground for this project. Unfortunately, his untimely death made it impossible to him to contribute as an author to this volume.

Angelo Baracca has spent long periods collaborating with members of the physics faculty of the University of Havana. Jointly with Cuban colleagues, he developed the research that has led to his contributions to this volume. He is deeply grateful for the openness, collaboration and friendly hospitality experienced during this time. The Department of Physics of the University of Florence has provided support for both the collaboration with the University of Havana and the with the Max Planck Institute for the History of Science (Berlin).

The reconstruction of the developments of physics in Cuba in the 1960s and 1970s had to rely, in large part, on oral history: it is impossible to mention and acknowledge all the Cuban colleagues who have kindly given interviews—not always formal—and who shared their personal recollections of the developments of Cuban physics and society.

Beyond the circle of the Cuban physicists, a specific acknowledgment goes to José Altshuler, President of the *Sociedad Cubana de Historia de la Ciencia y de la Técnica*, for the friendliness and professionalism throughout this collaboration as well as for his invaluable contribution to the translation into English of many of the papers. Work on the translations was also undertaken by Ernesto Altshuler, Birgit Kolboske, Roberto Días Martin, Brenda Porster and Elena Vigil Santos.

Alessandra Lorini and Duccio Basosi have provided important suggestions and advice concerning Cuban history and culture. Our thanks are also extended to Albert Presas y Puig who provided archive documents on the relation of Marcelo Alonso with Germany.

This volume would not have come into being without the ceaseless efforts of Lindy Divarci who coordinated the editorial work, revised the contributions and copyedited them with the help of Ross Fletcher, Sabine Kayser, Oona Leganovic, Barbara Lenk and Marius Schneider.

Chapter 2
The Cuban "Exception": The Development of an Advanced Scientific System in an Underdeveloped Country

Angelo Baracca

This island is a paradise. Cuba. If I am lost, look for me in Andalucia or in Cuba.

(Federico Garcia Lorca)

Cubanity *does not lie in showy touristic attractions, but in an ineffable underground tenderness, a being-not-being, the waving of the breeze, a certain lack of definition, a mixture of the earthly and the stellar. The most solid Cuban tradition may be looking forward to the future. Few peoples of America have been as determined to leap into the future so violently, with a shock of premonition. That is why there is a certain convergence of the generations. We are all marching towards a goal, somewhat distant and uncertain. This vagueness is convenient, it enriches us because it is limitless. Cuban means possibility, fantasy, fever for the future. We need to spread this character throughout the world.*

(José Lezama Lima)

Given that man has come in order to live, education has to prepare for living In the school it must be learned how to handle the powers of struggles in a lifetime. They should not be called schools, but workshops. And the quill should be used in the afternoons, while in the mornings, the hoe.

(José Martí)

A. Baracca (✉)
Department of Physics, University of Florence, Florence, Italy
e-mail: baracca@fi.infn.it

A. Baracca et al. (eds.), *The History of Physics in Cuba*, Boston Studies in the Philosophy and History of Science 304, DOI 10.1007/978-94-017-8041-4_2,

2.1 Physics in a Difficult Environment

2.1.1 Cuban Exceptionalism

Science, education, politics, social development and economics are today considered to be highly interdependent. Although none of these factors can exist alone, they have nevertheless often been considered in isolation from each other or studies of their interactions have been confined to the consideration of more or less local contexts. When it comes to studying the history of physics in Cuba, however, it is inconceivable to separate scientific developments from their social, political, and cultural contexts. As this volume shows, the history of physics in Cuba cannot just focus on local contexts since it is closely entangled with global history, from colonialism to the Cold War.

In this sense, also the history of Cuban physics displays the features of international entanglement, sometimes visible and sometimes latent, of other aspects of Cuban history, from music to politics. Cuba is a strip of land covering less than a thousandth of the earth's surface and counting roughly two-thousandths of the world's population. Yet few places or peoples in the world can boast a comparable capacity to excite strong passions and emotions and also to exert cultural influence in every field, permeating—at times explicitly, more often in hidden, indirect or even unconscious ways—styles, tastes and trends. Cuba is a melting pot of ethnic and cultural influences from three continents and civilizations. Over the course of the century this shaped and strengthened a peculiar kind of national and cultural consciousness. Starting with the *Grupo Renovación*, Cuban music was spread worldwide by musicians like Argeliers León (1918–1991) and Harold Gramatges (1918–2008) and by celebrated performers like Benny Moré (1919–1963), the Trío Matamoros and Enrique Jorrín (the inventor of the *cha-cha-cha*). It was in the 1930s that Joseíto Fernández (1908–1979) composed the world-famous song *Guantanamera* (which the North American composer Pete Seeger (born in 1919) adapted to verses by José Marti in the 1960s). In 1959 the victory of the Revolution on the small island of Cuba (which at the time had only six million inhabitants) introduced a clamorous rupture in the delicate world balance that had come into being at the end of the Korean War and that characterized the Cold War decades. Not only in the context of Latin American political emancipation but also in many other fields, Cuba played a disproportionally large role compared to its size.

In this volume only the history since the beginning of the nineteenth century will be considered, when Cuba was still under Spanish colonial dominion. Here some of the interdependence of science with economy and politics becomes obvious for the first time, when, for instance, the sugar industry exported more and more to the United States and, as a consequence, the *azucareros* sought for new scientific methods of sugar production. In this situation Cuba gradually detached itself from Spain and turned to the large and expansive northern neighbor. When by the end of the nineteenth century Cuba declared its independence from Spain it was rear-covered by Washington. This national liberation quickly entailed a new dependence—now it was the United States that dictated laws and economic measures. These first decades of the twentieth century were characterized by a radicalization that was not confined to the field of politics but also affected developments in education, science, economy and social politics. This radicalization, together with the consolidation of a middle class benefitting from new economic opportunities beyond the sugar industry, led to the birth of revolutionary movements. The outbreak of the revolution of 1958 was the result of economic stagnation, a deficit in education politics, intellectual criticism and political struggles.

Over the course of the last five centuries Cuba constituted an *exception*, an *atypical case* lying outside of traditional classifications taking national histories as their paradigm. We must underlie that our use of terms such as "exception," "exceptionalism" referred to Cuba does not imply any value judgment, nor reflect special sympathies, but correspond to concrete differences in his history and cultural and social reality with respect to any other (not only Latin American) country. The history of Cuba also constituted an exception in comparison with that of the other Spanish colonies in Latin America and the Caribbean. In a world that has become, since the Spanish expansion and colonial rule, increasingly globalized (Gruzinski 2004), Cuba has played a central role in many different aspects. It has served and continues to serve as a focal lens in particular for the cultural and intellectual processes accompanying globalization. In fact in Cuba people, customs, cultures, religious beliefs and even knowledge was mixed, giving way to synergisms that thus provided the basis for the concept of "transculturation" coined by Cuban sociologist Fernando Ortíz (1881–1969).

Ortíz developed the concept of 'transculturation' in 1941 to account for the interpretation of Spanish and African cultural influences in Cuban national identity that acknowledged the ongoing influence of the customs, traditions, and cultures of all those partaking in scenarios of cross-cultural contact and exchange. Ortíz insisted on the reciprocal influence that various groups have on each other in the creation of a new national identity (see e.g. Millington 2005). As Ortíz writes:

> because this process does not consist exclusively in acquiring another culture,[1] [...] rather, the process also necessarily implies the loss or uprooting of an original culture, which could be termed a partial deculturation, as well as the consequent creation of new cultural phenomena which could be described in terms of a neoculturation. (Ortìz 1995 [1947], 102–103)

This concept reflects the international situation of Cuba coming out of a long history of colonial and imperial domination and later inspired concepts such as Creolization (Otto 2007), *mestizaje* and Postcolonialism (Shih and Lionnet 2011), relevant also to current references in history of science (Anderson 2009).

The aim of the volume is to show that Cuba's *exceptionality* is not limited to such broader cultural questions, but instead comprises also knowledge, the natural sciences, and in particular physics. The articles collected in this volume show the rich history of Cuba's *exceptionality* regarding scientific developments. We shall limit ourselves to the field of physics, in which styles of work, methodological approaches and conceptual contents have undergone substantial homologation all over the world by now. Still, we feel that the reader will recognize some quite decidedly characteristic

[1] This is what the Anglo-American word 'acculturation' actually means.

features in Cuban physics, as part of an international science. Physics as practiced in Cuba has displayed an exceptional plasticity in its capability to adapt to an often improvised technological infrastructure, in its ability to respond to specific social and economic needs, but also in its openness towards diverse traditions and schools of research and, not the least, in its inventiveness of ways to integrate physics into the concert of a variety of disciplines without an agenda of hegemony.

Some examples may help to capture the spirit of this plasticity. At the end of the 1960s, shortly after the revolution, Cuban physicists combined advanced nuclear technology of Russian provenance with a new silicium technology introduced by French physicists in summer schools to advance in the field of microelectronics.[2] In more recent years, Cuban physicists improvised means for exploring the newly discovered phenomenon of superconductivity at high temperatures for ceramic materials and developed a new advanced research activity in this field.[3] The heavy burden imposed by the lack of funding and equipment due to the isolation of the country after the breakdown of the Soviet empire has been taken by Cuban physicists as a challenge to tackle new problems "under conditions of high tropicality" such as the collective behavior of ants or the movement of sand,[4] relying on modest material means and uncommon inventiveness.

There is another exceptional feature in the relation to science and society in Cuba. The percentage of university graduates and physicians in the Cuban population and the overall level of their scientific training have no equals among other underdeveloped countries (Hoffmann 2004, 166–168).[5] The country boasts considerable technical-scientific potential: there is a mid-level technician for every 8 workers, a university graduate for every 15; there are over 160 centers of scientific research, and 1.05 technologists and researchers for every 1000 economically active inhabitants (the latter amounts to 5.5 million): in 2011, 4.618 full-time researchers and 58.700 professors in higher education.[6] These indicators are comparable to those in many highly developed nations and there are branches of knowledge, such as biotechnology, whose development in Cuba is well known beyond its borders.

To sum up, the interest of Cuba for science lies in the fact that it represents an exceptional, and perhaps unique, case of *development of an advanced scientific system in a developing country*, and what is more one with an extremely small population and economy. This is all the more exceptional if we add that this development has taken place in a few decades, as the case of physics treated in this volume demonstrates. Physics in Cuba was for a long time a rather marginal discipline in the Cuban scientific community and still today physicists are only a small part of it. Nevertheless, physics always acted as a backbone of intellectual life and has provided many other sciences with scholars, methods and scientific approaches that were important for the further development of Cuban sciences such as medicine,

[2] See the chapters by Baracca, Fajer and Rodríguez, and by Cernogora in this volume.

[3] See the chapter by Arés Muzio and Altshuler in this volume.

[4] See the chapter by Sotolongo Costa in this volume. See also the homepage of Ernesto Altshuler: http://www.complexperiments.net/EAltshuler/HomeAlt.htm. Accessed October 18, 2013.

[5] See the chapter by de Melo Pereira and Sánchez Colina in this volume.

[6] http://www.one.cu

biotechnology, or the nanosciences. Physics is a good example to study this process also in an international context, as this science is highly dependent on the an international communication of research results, on well-equipped laboratories with instruments made in different countries and on the exchange of well-trained experts. We therefore consider the importance of the history of physics in Cuba as a case study of science in an international context. This volume illustrates ways in which international flows of scientific knowledge and an international political situations and the "local" scientific and political circumstances interacted.

Science in Cuba has represented, since the beginning of the nineteenth century, time and again, a field of emancipation from colonial and imperial submission. Even more, throughout the Cuban history of the late nineteenth and twentieth centuries important figures as Félix Varela (1788–1853), José Martí (1853–1895), Enrique José Varona (1849–1933) or Manuel Gran (1893–1962) were conscious of the fact that political independence had to be accompanied by an autonomous development of scientific education and research. While Simón Bolívar (1783–1830) still relied on Enlightenment ideas of education, considering it as an individual form of liberation (Briggs 2010, 99–106), José Martí was inspired by US-American and British models. He envisaged education as a crucial factor in the formation of the Cuban nation, independent from Spanish and US educational systems (Quiroz 2006). In Martí's perspective Cuba could achieve real independence only when the necessary skills were developed to overhaul the economic, political, social and technical underdevelopment inherited from the Spanish colonial regime.

The situation of the sciences in Cuba before the Revolution of 1959 depended on social and political conditions that inhibited the technological and scientific evolution of the country. The majority of the Cuban economic and political elites as well as foreign powers exploited the island and had no interest in any kind of autonomous development. This situation lasted over the times of Spanish colonial and US-American imperial domination, during which an elite of sugar producers impeded a real advancement of the society, especially that of science (Bravo 2001). However, the take-off of Cuban sciences after the Revolution would not have been possible without a minimal scientific infrastructure and trained personal. These achievements were the result of the development of a middle class and also a working class interested in the economic advancement of the country.

The social and economic circumstances fostered technological and scientific development, even against the will of the Spanish colonial regime. Looking for practical solutions, a cooperation between entrepreneurs and technicians established an international network and became a sort of a backbone of intellectual life on the island, as is shown by José Altshuler and Angelo Baracca,[7] before playing an important role in the industrial development from 1963 on.[8]

Studies of the history of physics in Cuba and its an international contexts are still in their infancy. This volume brings together, as a first step toward a more comprehensive account, some pioneering work also directly involving Cuban scholars, as well as interviews with protagonists, resulting in first attempts at a systematic

[7] See the chapter by Altshuler and Baracca in this volume.

[8] See the chapter by Baracca, Fajer and Rodríguez in this volume.

treatment of entire historical periods before and after the revolution. This material is complemented by recollections and reflections from both the inside and the outside, from Cuban physicists and their foreign collaborators, and also by glimpses into the conditions and impact of scientific work and communication in Cuba.

The political situation that still persists in Cuba makes it difficult to write a history that is free from political implications, as historical accounts not influenced by current political and ideological situations are hardly achievable. The result is unavoidably a kind of political history, and may be biased by a lack of distance to historical events that are still closely connected with the present.

2.2 Contradictions and Developments of Cuban Economy, Culture and Science in Late Colonial Times

Though poor in precious natural resources, especially metallic minerals, and with the exception of sugar lacking important processing industries, Cuba was a precious source of added value for goods thanks to its strategic geographical position, which gave it a trade advantage as juncture for commerce between Latin America and Europe and as time went on also with the United States. At the same time, there seemed to be little motivation for technical and practical innovations.

In the mid-eighteenth century the 250-year-old Spanish colonial domination in America was weakened by wars in Europe and North America. The political authority that before seemed to be prevailing was shattered by unforeseeable events. The situation in the Caribbean changed after independence of the 13 British colonies in 1776, the French Revolution of 1789 and the subsequent independence of Haiti as well as the Napoleonic occupation of Spain that led to the independence of Argentina in 1810–1816, Paraguay in 1811, Venezuela in 1811–1819 (in the context of the 'Gran Colombia', which in 1830 was divided into Ecuador, Venezuela and Colombia), Chile in 1818, Peru in 1821 (which Bolivia separated from in 1825) and Mexico in 1821–1823.

Cuba did not free itself from Spanish colonial dominion until 1898, when after the Spanish-American War the Iberian colonial power had to hand over Puerto Rico, Cuba and the Philippines to the new emerging imperial power. Thus, during the nineteenth century Cuba remained a Spanish colony and had to master the challenges to site itself between the colonial dependence and the newly emerged independent nations, mostly with United States.

2.2.1 Sugar and Tobacco in the Nineteenth Century

Cuba's economy relied mostly on a sugar monoculture. After the Haitian revolution of 1791 and the breakdown of its sugar production, the Cuban sugar economy was given a decisive boost. The threat of a similar social revolution (Ferrer 2005, 67–83) reinforced the determination of the Cuban Creole elite[9] to remain faithful to the Spanish crown. In this way, the colonial regime survived in Cuba, even when during the nineteenth century the dependence on the US-markets of sugar and from US-sugar companies increased.

The already mentioned Cuban sociologist Fernando Ortíz has considered tobacco to be the hero in Cuban society, while sugar is the villain (Ortíz 1940). Tobacco was grown mainly in the western part of the country, where it created a middle class. On the contrary, sugar was grown in the rest of the island and created two classes of people: masters and slaves (Reinert 2007). Tobacco farming developed a demand for specialized individual skills and the division of labor, since the market price depended on the pickers' and handlers' ability (an expert sorter can distinguish between 70 and 80 different types of tobacco). The sugar crop, instead, was based on a mass of slaves and on the brute strength needed to cut the cane with a machete. Tobacco, says Ortíz:

> is wealth and intelligence, and mainly domestic capital; sugar is poverty and ignorance, and foreign capital. In Cuba's history sugar represents Spanish absolutism, tobacco the liberation of the native population. Sugar has always been sustained by foreign intervention. [...] Sugar has always preferred slave hands, tobacco free men. Sugar imported blacks with force, tobacco fostered the voluntary immigration of white men. (Ortìz 1995 [1947], 65)

Cuba possessed a world advantage for both crops. Its tobacco was one of the few cases of "brand-name" products in the third world. It is interesting in this context to consider the thesis of Reinert, according to which the predominance of sugar in the Cuban economy was a cause of its underdevelopment. Indeed, as we shall see, though technological development was greater in sugar refining than in cigar production, cigar producers were consistently richer than sugar producers (here the author traces a parallel, in a certain sense reversed, with the situation of the United States after 1930, which boasted the highest level of efficiency in the world both in agriculture and industry, but farmers stayed poor while industrialists got rich).

In effect, in the first decades of the 1800s competition from other Caribbean islands that had increased their sugar production, along with the introduction of the sugar beet in Europe[10] and the British campaign for the elimination of the African slave trade (Adderley 2006; Brown 2006), that Spain was forced to cease in 1821 (Blackburn 1988), caused great changes in the structure of the Cuban sugar industry. Albeit the slave-economy, the owners of sugar mills invested in technological innovations (Scott 1985, 27–28; Zeuske 2012, 39, 104), including the introduction of steam-powered machinery and the first railroad. The number of *ingenios*[11]

[9] Creole is the term given to offspring of white immigrants born in Cuba.

[10] For the interrelation of the global sugar market and the technological change in Cuban sugar production, see Tomich (2005).

[11] The term *ingenio* indicates the former colonial companies for processing cane to extract sugar and its byproducts. Previously there had been the trapiche, which produced on a smaller scale. For an overview over the principal Cuban *ingenios*, see Cantero et al. (2005).

multiplied rapidly (from roughly 1,000 in 1827 to 2,000 in 1860). This economic boom of what has been called "sugar-capitalism" (McNeill 2010, 295) or "capitalism of humans," i.e. capitalism of slavery (Zeuske 2012, 35) constituted a factor that further delayed the formation of a national consciousness antagonistic to Spanish dominion, in contrast with what happened in the rest of Latin America.

Three developments polarized the Cuban situation even further, feeding a sort of vicious circle. Economic prosperity, linked to the single-crop sugar economy, fostered strong resistance to any possibility of eliminating slavery; on the contrary, the demand for slaves and the consequent illegal traffic from Africa grew. The result was that towards the middle of the 1800s the ratio between the white population and the population of African origin (made up of roughly 2/3 slaves and 1/3 free men) was reversed (de la Sagra 1842; Scott 1985). This circumstance worked to worsen fears of revolts, like in Haiti at the end of the eighteenth century or in Jamaica along the 1830s (Ferrer 1999). From the 1840s on, a high amount of Chinese came to Cuba by the coolie trade. Even if only a little number of them stayed in Cuba after in 1874 the trade was resolved, the Chinese impact in cultural life in Cuba was for a long time remarkable (López 2004).

Because of the unstable situation related to slavery and forced work, the Creole elite had an interest to strengthen the ties with Spanish colonial power. At the same time, the goal of many of those Cubans who wanted to put an end to Spanish domination was translated into a desire to be annexed to the United States, which embodied the ideal of a free modern country.

2.2.2 The Role of Sugar in Making Cuba Dependent on the US for Its Modernization

It is relevant to make some general observations at this point about how deeply the sugar economy conditioned and determined Cuba's entire history, economy and society. Sugar was not only the most important export good of Cuba, but also a motor for modernizing the country's technology, even when socially and politically this meant that Cuban social structure should remain backward as has been partly shown before. At the end of the nineteenth century the US-companies bought the majority of Cuban sugar production but invested in new forms of production. The Cuban economy was ever more oriented towards the United States, which absorbed the majority of Cuban exports (about 17 % in 1840, 86 % in 1895); US capital invested in Cuba also increased. This was also one reason why the *ingenios* sugar producers yielded to the greater productive capacity of the *centrales*.

Thus, US economic penetration of Cuba had been growing from the start of the century. Already in 1826 the volume of Cuban trade with the U.S. was almost three times greater than its trade with Spain. (Ramón de la Sagra 1831, 200–205). This was the reflection of a growing presence of North American capital in the Cuban economy, which led to the ownership and control of vast parts of its productive structure. As an authority like Fernando Ortíz emphatically claims:

> in 1850 this country's trade with the United States exceeded that with the Spanish metropolis and the United States definitively assumed its natural geographical condition of buyer for nearby Cuban products, but also its privilege as economic metropolis. Already in 1881 the United States Consul General in Havana officially wrote that Cuba was economically dependent on the United States, even if politically it was still held by Spain. (Ortíz 1995 [1947], 64)

2.2.3 *Technological Developments During the Nineteenth Century*

In this context it may come as a surprise that in the course of the nineteenth century Cuba experienced the introduction of advanced innovative technologies very early on, earlier not only than the mother country but also than other, more developed countries. This, in fact, constitutes another aspect of Cuban exceptionality. Examples are the introduction of the steam engine in the *trapiche*[12] as early as 1817, the first railroad for transporting goods between Havana and the village of Güines (46 km) in 1837, and the precocious development of advanced communication technology (Blaquier 2009; Baracca 2009).

The Cuban railway developed parallel to the sugar industry and telephone lines followed the train tracks (Blaquier 2009). The first railway was important to transport sugar and possibly also slaves (Zeuske 2012), as well as to explore remote parts of the island. Miquel Biada i Bunyol, who had a close relationship with the governor Miguel Tacón y Rosique (1755–1855), promoted the first railway in a Spanish territory that was inaugurated in 1837. He after this success returned to Catalonia, where he promoted the first European Spanish railroad from Mataró to Barcelona. Another migrant to Cuba was Antonio López y López, who promoted maritime steamship connections from Cuba to Spain (McDonogh 2009, 64). Such technical development on the island depended largely on Spanish investments. But its realization was more easily possible in the Cuban political environment, where the Creole elite was open to support such projects. They afterwards expanded the railway network in the Middle and Western parts of the island, while in the Eastern regions much less effort was made (Santamaría García and García Álvarez 2004, 170–172).

The telegraph played an important role during the Ten Years' War (1868–1878), when the interception of dispatches allowed the revolutionaries to avoid early arrest of the leaders of the revolt, while during its military intervention of 1898 the United

[12] The *trapiche* were colonial sugar plants for processing sugarcane.

States made early experiments with techniques for cutting underwater cables to isolate the Spanish army (Pruna Goodgall 2005, 227). It is therefore understandable that the Island felt the need for advanced scientific knowledge, modern higher education and new technical engineering competence only much later. Cuba's need for the development of railways, communication and information technologies, and the fight against tropical diseases is evident. It may also be supposed that Spain's scientific and technological backwardness and its pre-capitalistic, purely exploitative attitude towards its colonies created a further obstacle to Cuba's scientific development, at the same time as it contributed to the decline of Spain itself.

Also other infrastructures were renewed during the nineteenth century. The entire military defense system was reinforced in order to prevent an invasion as in 1762, when the British navy could have taken quite easily the harbor of Havana and occupied the town. Besides the buildings and the armaments the geographical knowledge of the Cuban island and the surrounding waters were improved, for instance by the cartography project of José de Valcourt undertook between 1821 and 1830. The map was the basis to build further defense fortification at the rim of the island as well as on surrounding islands (Nadal 1989, 329–356).

To these can be added the preliminary experiments on the telephone carried out in 1849 by the Italian Antonio Meucci in the *Teatro Tacón* of Havana (Ortíz 1940; Catania 1994, 2004; Altshuler and Díaz Martín 1999, 2004), the start-up of the first public telephone service in Havana in 1882 (González Royo 2004) and the first public illumination systems (Altshuler and González 1997). Cuba had also been a shipbuilding power; indeed, until the Battle of Trafalgar (1805) the most powerful Spanish ships were built in Havana.

2.2.4 Education and Intellectual Life in Cuba in the Nineteenth Century

In the nineteenth century, a contradictory situation arose which characterized Cuban society. Important technological advances were made, but were not reflected in the educational and the intellectual life of the country. The growth of more specific scientific disciplines, in particular chemistry, physics and engineering, remained slow, though they had made significant progress in other Central American countries. This contradiction can be explained on the basis of the peculiar Cuban colonial status we have discussed (Baracca 2009). The growth of the economy and commercial exchanges, in fact, boosted the need of advanced communication systems and production technologies. As we shall also see below, the progress made in Cuba in the field of medicine since the beginning of the nineteenth century, was greater than that of other colonies, and comparable to the level of medical knowledge and practice in more developed countries. In contrast to these noteworthy developments in technical progress, as well as in the medical and applied natural sciences, the

Spanish colonial rule and the fears of the Creole elite strongly opposed and prevented comparable progress in education and basic sciences.

Indeed, the most advanced sectors of Cuban intellectuality (and some progressive sector of the church) expressed the need of a scientific renovation: the priest Félix Varela (1787–1853), under the enlightened and progressive direction of Bishop Juan José Díaz de Espada (1756–1832), introduced as early as in 1817 modern contents of physics. However, the Spanish royal power in the motherland strongly opposed and prevented the further and full development of a modern education system, until the end of the colonial Spanish rule in 1900. During the second half of the nineteenth century, progressive and conservative circles could not agree on the question of how much education was required. While the Spanish conservatives rejected the initiatives of progressive Creoles, freed blacks and anarchist organizations, these latter groups tried to establish an educational infrastructure in partial opposition to the colonial institutions (Casanovas Codina 1998).

Earlier, at the beginning of the nineteenth century, the Catholic priest, Félix Varela who is said to have first taught Cubans to think, carried on educational work of fundamental importance, introducing the innovating spirit of the enlightenment in Cuba, as we shall see in the first contributions to this collection. Elected in 1822 as a representative to the Spanish Cortes, he voted in favor of partial autonomy from Spain and during his stay in Spain, wrote an influential treatise in favor of the abolition of slavery (Los restos del padre Varela 1955). With the restoration of absolutism under Fernando VII in 1823 he was forced to seek refuge in the United States and came to the conclusion that full independence was the only solution. Varela shared the destiny of exile with other intellectuals of this time, as José María Heredia (1803–1839), the first great Cuban poet.

After such promising innovation, Varela's fellows tried to develop his premises. All throughout the nineteenth century Creole elites in Cuba practiced a deeply rooted racism, which also had an impact on the educational system. For the entire first half of the nineteenth century there was no universal school system and where schools did exist they were limited to elementary education. The literacy rate was very low: only about 19 % of the white population knew how to read and write and hardly 2 % of colored people were literate (Melcón Beltrán 1989). In 1857 the first governmental *Escuela Normal*, or primary school, was founded, but in general education remained or was given back into the hands of religious orders (Melcón Beltrán 1989, 282). In the second half of the century provincial secondary schools (*de Segunda Enseñanza*) were created, but they had almost no funding. And while despite the commitment of its teaching staff the University of Havana languished in the indifference of the metropolis, several private colleges were born. In the countryside and among the colored population education was almost nonexistent. At the same time, the melting pot of ethnic groups and influences from all over the world (there were thousands of Chinese immigrants and French immigrants from Haiti and Louisiana) created fertile terrain for lively cultural manifestations. The periodical press flourished and a varied literature illustrated the features, figures and problems that characterized the island.

2.2.5 Academy of Science and Medicine

In contrast with the languishing situation of the education system and of the basic sciences, in the fields of medicine and natural sciences, during the course of the nineteenth century Cuba boasted renowned practitioners who made decisive contributions to the problems of a tropical country.[13] Here we will name only the most authoritative of these (García Blanco 2002; Pruna Goodgall 2006), some of them studied for some years in Europe. Many were pupils of Félix Varela at the San Carols Seminar and all belonged to the progressive part of Cuban society. In 1803 the physician Tomàs Romay (1764–1849) introduced the anti-smallpox vaccine. The naturalist Felipe Poey (1799–1891), who in the last years of his life abandoned his religious faith to embrace evolutionary theories, (Pruna Goodgall 1999), documented Cuban fauna and in 1877 founded the Sociedad Antropológica (Anthropological Society). After completing a PhD in Paris, Alvaro Reynoso (1827–1888) carried out fundamental research applying physiological chemistry to agriculture, and proposed a scientific system based on the physics and chemistry of soils for the cultivation of sugarcane. (Fernández Prieto 2005). In 1878 the researcher and physician Carlos J. Finlay (1833–1915) formulated the "theory of the mosquito" as carrier of yellow fever, coming up against the opposition of doctors internationally and in the United States (with highly unedifying outcomes, as we shall see later on). Carlos de la Torre (1858–1950) made important paleontological discoveries. In light of these scientific developments, in 1861 Queen Isabella II finally authorized the founding of the *Real Academia de Ciencias Médicas, Físicas y Naturales de La Habana* (The Royal Academy of Medical, Physical and Natural Sciences of Havana) (Pruna Goodgall 1994, 2001, 2003, 2006), which had in vain been proposed as early as 1826 by a series of scholars led by Tomás Romay and Nicolás José Gutiérrez (1800–1890) as we have already noted.

2.2.6 Liberation Movements

The international situation of the nineteenth century inspired the intellectual development in Cuba. Young Cubans who studied in France brought their acquired knowledge back to the Cuban sugar industry. The political ideals of Europe circulated among the progressive circles on the Island. In this way, Cuba became part of the international networks of intellectual life and could react immediately to scientific and political changes in Europe and the US. North American culture and lifestyle, the British abolitionist movement and gender equality in the North of the United States had its influences also on Cuban intellectual and political life. The more liberal developments led the Cuban Creole elite identify more and more with the slave-owning aristocracy in the US South.[14] Cuba remained a radically sexist

[13] Some cooperation with distinguished scientific centers was established, see García Gonzáles (2010).

[14] How useful a comparative view on Cuban and southern US slavery policies is revealed in Bergad et al. (1995).

and male chauvinist society, where women were not allowed to attend university, though women of color were objects of their white masters' sexual desire and the colored population entirely marginalized (Lamore 1980). The nineteenth-century masterpiece of Cuban literature, *Cecilia Valdés*, illustrates this situation very effectively. Its author, Cirilo Villaverde (1812–1894), accused of anti-Spanish conspiracy, spent most of his life in exile in New York. Here in 1869 his wife, Emilia Casanova (1832–1897), founded the first patriotic association for women. Freedom-loving and anti-racist sentiments were in fact deeply rooted among Cuban emigrants in the United States, whose most authoritative representative was undoubtedly José Martí.

The Cuban national liberation movement was shaped by both internal and external factors that determined its particular nature: the growing level of Spanish colonial exploitation, the ever growing problem of slavery (that lasted to 1886) and the patriotic sentiment that had grown up among some of the landholders, especially in the center-east of the country. Another part of the Creole social and economic elite remained faithful to Spain, although they occupied most important posts in the fields of administration and army (Kuethe 1986). After many months of preparation a revolt broke out in 1868, when a civil war in the mother country had temporarily overturned the monarchy, causing an earthquake in Cuba as well and giving the impression of a decline of the European powers (the defeat of the French in Mexico; the Spanish defeat in Santo Domingo in 1865; the rebellion in Puerto Rico in 1868). Though both slaves and free blacks joined the rebels in the Ten Years' War (1868–1878), the question of the abolition of slavery remained open, as their participation strengthened the rebels' fear of an uprising (Scott 1985).

The years following the war saw the expansion of strongly capitalist elements in Cuba, along with a process of concentration of production. The Cuban economy was ever more oriented towards the United States, which absorbed the majority of Cuban exports (about 17 % in 1840, 86 % in 1895); US capital invested in Cuba also increased. Meanwhile, patriotic sentiments and national consciousness continued to grow and, under the influence of Martí, prepared the decisive insurrection.

2.3 Cuba Between Independence, US-Interventions and Dictatorial Regimes in the First Half of the Twentieth Century

2.3.1 The War of Liberation and Independence and the US Rule

After the outbreak of the 1895 War of Liberation from Spain, which also awakened strong passions in Europe with many young men volunteering to fight for the Cubans, the twofold military intervention of the United States in 1898 against Spain in Cuba and in the Philippines brought an end to Spanish colonial rule, and actually marked the beginning of the foreign politics of intervention in the world. Inside the

US, while the black North American community held contradictory feelings towards Cuba—racial solidarity vied with benefits deriving from participation in the war as soldiers of the United States—the majority of the population was divided between those favoring annexation, "English-style" forms of government, and independence. The military intervention was perceived as a final step to reunite the country after the divisions caused by its Civil War; its expansion westwards was resumed and participation in the imperial race with the European powers began.

With regard to the influence, or the interference, of the United States with the countries of Latin America, we should remember that as far back as 1823 President Monroe had resolutely affirmed the principle of U.S. hegemony over the entire continent ("The Monroe Doctrine"), thereby excluding any further attempts on the part of European powers to colonize the Americas, or even merely to interfere in the politics of the newly independent states of central and south America (Foner 2004). In 1845 John O'Sullivan coined the expression "manifest destiny", which was to become an important part of US national ideology. In 1904 the Monroe Doctrine was reaffirmed and the "Roosevelt Corollary" was added, sanctioning the right of the United States to intervene militarily to restore order in the countries of the western hemisphere.

On January 1st 1899, the US occupation of Cuba officially began with General John Brooke (1838–1926); the peace treaty of Paris (December 10, 1898) between the USA and Spain excluded Cuban (as well as Philippine) representatives. The island was thus dominated by a neo-colonial regime, which had a profound influence on internal social developments. This situation, barely differing from Spanish domination, evidently frustrated the dynamics of development of a Cuban middle class, as well as of the popular movements that had led to the War of Liberation. This was true also for the intellectual and scientific milieus.

Brooke appointed some white Cubans, who had lived in exile in the US, to head the four main government departments (Interior; Finances; Justice and Education; Agriculture, Trade and Industry), while the economy of the island remained in US hands. To isolate the veterans of the Liberation Army (largely men of color), among other reasons, no Cuban army was created, but only a police force. This was to be a source of great difficulty for the future Republic.

The military government of Leonard Wood (1860–1927), which followed Brooke's government in December 1899, was favorable to annexation. It adopted the model of racial segregation that the US Supreme Court had legalized in 1896, effectively making Cuba a practicing laboratory for US hegemony by trying out new forms of political and economic neo-colonial dependence—the "informal" and "benevolent" empire—with both direct and indirect forms of interference (Zeuske and Zeuske 1998). The military government maintained Spanish law, administration and bureaucracy, while favoring the white elite and concentrating all economic affairs in US hands. The circulation of the US Dollar was promoted and investment of capital coming from the US was encouraged. In this way ties between the United States and sectors of the Creole bourgeoisie were strengthened, causing profound transformations and heightened tensions in the political, cultural and racial identity of the occupied country.

Foreseeing an all inclusive submission of the island's interests to those of the US, the first Cuban government promoted an education campaign, conceived as a way to make the Cuban population conform to US values and standards. The war had destroyed the majority of school buildings and the few teachers still employed were extremely poor. The school system was reorganized on the US model and the teaching of English and US history was promoted. In the framework of the effort to favor the values of the US system, a significant initiative was taken by Wood: in the summer of 1900, 1,273 Cuban teachers were sent to Harvard University (amounting to over half of the entire corps of elementary schoolteachers, a majority of them women). In realty, what the Cuban schoolteachers, especially the women, gained from the US tradition of liberty insights that reinforced their desire for independence.[15]

2.3.2 *The Period of Enrique José Varona*

The Ministry of Education and Fine Arts was assigned to the authoritative Cuban intellectual and educator, Enrique José Varona (1849–1933). He had been a deputy to the Spanish *Cortes* and a collaborator of José Martí and can be characterized as anticlerical, favoring gradual reforms and supporting a practical, objective and scientific approach to education. He, as José Martí, Andrés Poey Aguirre or José Varela,[16] was a positivist, supporting thus the development of science and technologies in order to industrialize the country (Pruna Goodgall 2005, 172–177). Varona introduced a reform of secular education founded on the supremacy of public over private schools and inspired by modern pedagogical ideas. Varona was well aware that without radical social change and the start of a process of industrialization, the objective of scientific-technical development was an illusion. In a letter of 15th October 1900 he wrote to the Cuban doctor and anthropologist Luis Montané:[17]

> You want to know the spirit that guided me when I undertook the reform of our education institutions. [...] I will be clear about this as you merit it to be, and open hearted. [...] I acted in the spirit of legal defense of the people of Cuba; a defense within its possibilities and in the field of the possible [...] When France was aware of its defeat against Germany, it renewed its military organization copying on the German example. We have to compete in the field of industries and in the field of sciences with the North Americans. And if we want to avoid to be completely erased of this field we have to educate us as the Americans do... [I] will transfer the fight to the only battlefield where we can fight. We are dealing with a social phenomenon and the consequences of an ineludible law. The only way to avoid the possible dangers of these consequences is to become part of the conditions producing this phenomenon.

This reveals Varona's great interest in the creation of technical careers in the University of Havana, though unfortunately he did not increase the teaching of basic

[15] For the different opinions on Cuban independence, see Zeuske and Zeuske (1998, 419–425).

[16] See the chapter by Altshuler and Baracca in this volume.

[17] Our thanks go to José Altshuler for bringing this quote to our attention.

sciences, such as mathematics and physics.[18] This view of technological progress being, to some degree, independent from support by an advanced scientific system, was, at the time, widely spread even among developed countries. Even in the US a culture of inventors (like Edison) still prevailed at the turn of the century, while the integration of science in technical innovation processes in Germany played an important role in fostering the "Second Industrial Revolution" (Baracca et al. 1979). In Britain this situation that was observed with concern, giving rise to a widespread warning for the backwardness of industry, science and higher education (Lloyd-Jones and Lewis 1998). Crucial turning points in this situation were the First World War, the Great Depression, the New Deal and the Second World War.

2.3.3 US-Exploitation of Cuban Sciences

The US goal to gain and maintain control dominated the scientific climate of Cuba of the early twentieth century. As a specific example, in the question of whether mosquitoes—as the Cuban physician Carlos J. Finlay had discovered—or bacteria were the cause of yellow fever, the head of the US Army medical services, George M. Sternberg (1838–1915), who strenuously opposed the "mosquito theory," in 1899 sent a medical commission to Cuba.[19] A member of this commission, Jesse Lazear (1866–1900), in agreement with his superior, Walter Reed (1851–1902), tested Finlay's thesis by allowing himself to be bitten by a mosquito provided by Finlay and died as a result of the experiment. Using Lazear's observations, Reed then published a work in which he credited himself with the discovery of the transmission of yellow fever. After some years the scientific community finally recognized the expropriation of the scientific discovery and Finlay was nominated for the Nobel Prize (Access to the Archives of the Nobel Foundation has revealed that Finlay was proposed for the Nobel Prize seven times before his death in 1915; the world congresses of medical history of 1935, 1954 and 1956 recognized Finlay's absolute, unquestionable priority). This episode illustrates how US governmental institutions influenced control over Cuban scientific research.

2.3.4 Between Republic, Autocratic Rule and Scientific Advancement

The range of positions that developed inside the country, with a majority in favor of independence and a growing awareness of the repressive nature of United States "humanitarian" domination, made it clear that the solution of annexation would be possible only if imposed by force. Washington therefore handled in order to leave

[18] See the chapter by Altshuler and Baracca in this volume.

[19] On this controversy, see Cirillo (2004).

substantial instruments of control in its own hands, imposing in 1901 the insertion into the new republican constitution of notorious *Enmienda Platt* (contemporaneously with the Spooner Amendment for the Philippines), insuring the United States a veto on Cuba's foreign, military and economic policy, as well as the right of military intervention to preserve the Island's independence and government stability. This Amendment had a devastating effect on Cuba's political development and set the stage for three US military interventions (in 1906–1909, 1912 and 1917–1923), but it was also used by various Cuban authoritarian governments as a scarecrow against adversaries (the interventions were partly demanded by the Cuban governments themselves, amongst internal uprisings). Its influence was still felt even after its abrogation in 1934.

In this framework the Republic was inaugurated on May 20, 1902, opening an extremely tormented and contradictory period which was to last almost half century. Despite the extremely serious problems the newborn Republic had to face, the United States gave priority to strengthening its system of dominion, the "Cuban experiment," which was to serve as a model for other countries. The immigration of whites from Spain was encouraged in order to reproduce the model of racial segregation that existed in the US South[20] (blacks were not allowed into fashionable public places), exasperating social contradictions (the living conditions of the majority of farm workers were miserable). At the same time, thousands of US citizens invaded the Island, flooding it with dollars, the country was obliged to open up to US products, thereby thwarting the possibility of development of other industrial sectors, and Cuban economy was ever more monopolized by the production and exportation of this single crop, sugar (36 % in 1900, 92 % in 1920: in the end Cuba was producing 28 % of the world's cane sugar; the US purchased over 50 % of Cuban sugar). The penetration of US financial capital (in the 1930s Cuba received more US capital than any other Latin America country) nourished a Cuban sugar-producing oligarchy which served as an internal prop for neo-colonial domination and which, referring to the *Enmienda Platt,* used this domination as a weapon of blackmail to solve its own contradictions.

While the political situation of the Republic continued to worsen, the presidency of Woodrow Wilson practiced between 1911 and 1916, behind the screen of a formally non-expansionist program, a kind of neo-colonialism different from European-style occupation, stepping up the politics of intervention in the Caribbean (Mexico, 1914; Nicaragua, 1912–1925; Haiti, 1915–1934; the Dominican Republic, 1916–1930). In 1917 the marines once again landed in Santiago de Cuba, mainly to preventively protect the US plantations from the spread of liberal insurrections.

A fleeting economic miracle during the First World War, due the rising to stars of the price of sugar, was followed in 1920 by an economic and financial collapse, while Cuba become a paradise of lavish business affairs for North American bankers,

[20] Actually, the profound changes in the island's social fabric created by US neo-colonial policy led to a form of discrimination different from the segregation existing in the United States, whose formal inspiration was a concept of "racial democracy" sanctioned by the Constitution in which discrimination derived from the "color line."

entrepreneurs and speculators. The growth of social discontent and turmoil offered Wilson the opportunity for a political rather than a military intervention, dispatching in the island General Crowder as "proconsul" to insure the interests of US bankers.

In this climate, discontent and protests reflected the growing intolerance in respect to US interference and led to the maturation of a new national consciousness. Even if Cuba remained an essentially agricultural country (in 1931 nearly 50 % of its four Million inhabitants worked in agronomy) torn by social contradictions and with a high illiteracy rate of nearly 40 % of the population and higher still among the colored population in 1919, the country's physiognomy and social composition was undergoing profound transformations. A sector of working class in production and services was growing and acquiring consciousness. The transformations comprised urbanization, the spread of the telephone (Altshuler and Díaz Martín 2004), completion of railway lines, public electric lighting (Altshuler and González 1997), tramways, radio broadcasting and the first airborne communications. The automobile, the radio, the cinema and advertising increased the penetration of United States culture and values into every aspect of Cuban life and reinforced racial discrimination.

The presidency of Gerardo Machado (1871–1939), elected in 1924, as it suffocated revolts and attempted coups, degenerated (with Crowder's favor) into a ferocious dictatorship inspired by Mussolini's fascism. The dictatorship also led to radicalization of the opposition movement. The University of Havana was the seat of the *Directorio Estudiantil,* a radical association directed by Antonio Guiteras (1906–1935). Banned by Machado in 1927, the *Directorio* reformed as a secret association, but then splintered into groups with different tendencies. Right-wing associations also arose, while in 1925 a group of anarchists, socialists and communists founded the *Confederación Nacional Obrera de Cuba* (CNOC) and, inspired by the Russian Revolution, the Communist party was born in 1925 led by Carlos Baliño (1848–1926) and Julio Antonio Mella (1903–1929). Many of its exponents were Spaniards and Jews of East European origin (a novelty in Cuba's ethnic mix). In 1929 Mella, who was in exile in Mexico, was assassinated according to Machado's orders.

The depression that followed the 1929 crack struck Cuba violently. Economic crisis provoked a radicalization of grassroots protest movements and in 1931 there was an attempted general strike with insurrectional overtones, in which Antonio Giuteras took a prominent part. In 1933, when his hope for a new United States intervention was not met, Machado was forced to resign.

2.3.5 The Intertwinement of Social, Intellectual and Political Growth

In these decades, mentality and the intellectual life in Cuban society underwent deep changes. A new generation of young people took the place of the veterans of the Liberation Army who until then had been the animating force behind social unrest. Sections of the working and middle classes, colored people (Bronfman 2004, 106) and above all students and intellectuals introduced a more radical interpretation of the national problem, based on a reformulation of the inheritance of José Martí and of the *Mambises*,[21] which was heavily influenced by the revolutions in Russia and Mexico. The University of Havana became the center of student protest, which grew more radical under the leadership of Julio Antonio Mella. This combination of nationalistic and revolutionary goals led to demonstrations and rebellion. Furthermore, between 1923 and 1925 a number of influential groups were founded, such as the *Federación de los Estudiantes Universitarios* (FEU), the *Asociación de Veteranos y Patriotas*, the first feminist groups and the Cuban Communist Party. The *Junta de Renovación Nacional*, whose head was Fernando Ortíz, launched a Manifesto of Intellectuals for the reform of civil society and the fulfillment of the ideals of the Revolution.

However, the tormented social climate hindered the kind of commitment required by scientific work, and at universities political struggles overshadowed even educational activity. Intellectuals mainly focused on the construction of a national identity and became organized in groups (Zeuske and Zeuske 1998, 426–427), many of whom were soon banned by the authorities (Bronfman 2004). At the same time the new nationalism boosted the foundation of national scientific associations such as the Cuban Society of engineers, to mention only one closely related to the subject of physics (Funes Monzote 2004, 286–292).

The 1920s were years of heavily social turmoil, but also intellectually vital for the years to come. The social ferment gave birth to an intellectual renaissance in politics, in cultural production and scientific research. In 1921 Varona gave his speech on "*El imperialismo yanquí en Cuba*" at the *Academia de Ciencias*. The great writer and musicologist Alejo Carpentier (1904–1980), who had been jailed by Machado in 1927, fled to France (a hazardous escape, using the passport of the French poet, Robert Denos); he returned to Cuba in 1939, emigrated to Venezuela in 1945, and returned to the Island in 1959, the year of the triumph of the Revolution. After his early training in the school of Lombroso, Fernando Ortíz (1881–1969) set out to search for the originality and essential traits of Afro-Cuban culture (Los Negros Brujos, 1916; Catauro de Cubanismos, 1923; Glosario de Afronegrismos, 1924). He founded and directed several reviews and invited Federico García Lorca (1898–1936) to lecture in Cuba. At the same time, science teaching at the University of Havana was modernized by the mathematician Pablo Miquel (1887–1944) and the physicist Manuel Gran.[22] In 1927 another important scientific institution was founded called the Finlay Institute, whose first task it was to instruct the future clerks of sanitary administration and later on developed departments treating tropical diseases with vaccination (Pruna Goodgall 2005, 224–227).

[21] The term *mambises* refers to the guerrilla Cuban independence soldiers who fought against Spain in the Ten Years War (1868–1878) and War of Independence (1895–1898).

[22] See the chapters by Altshuler and Baracca and by Altshuler in this volume.

2.3.6 The Establishment of Batista's Regime and the Consolidation of the Revolutionary Movement

The times when revolutionary movements worked mostly underground followed a phase of uncontrollable revolutionary furor, which was exploited by a group of low-level non-commissioned army officers—unusual in Latin American *golpes*—to stage a mutiny, the "Sergeants' Revolt," on September 4, 1933. The radical movement saw this as an opportunity to intervene, in particular the student organizations, which were divided between moderates and revolutionaries but were in fact dominated by Guiteras, who became the most influential figure in a short-lived radical government that nationalized the electricity company and several United States-owned sugar companies.

It was in the context of this increasing chaos that Fulgencio Batista (1901–1973), the most astute of the military officers, was able to exploit the situation: in 1934, maneuvering with the US ambassador, Summer Welles (1892–1961), he took advantage of the worsening of the crisis to gain popular favor and place the liberal Mendiata as head of government. The United States strengthened the position of the new president by abolishing the *Enmienda Platt* and signing a new treaty, which kept the military base at Guantanamo in its possession. Yet the protest movement did not die down, culminating in the bloody repression of the premature general strike of March and the elimination in an ambush of Guiteras's group, which was preparing an expedition to launch a revolution in the east of the country.

The masses and workers' movement (foundation of the *Unión Revolucionaria Comunista* and of the *Confederación de Trabajadores de Cuba*, CTC in 1939) renewed with a combativeness brought new challenges to an already enfeebled political and social system.

As a response, the military organization was strengthened, obviously with the approval of the United States, and became a parallel power directed at guaranteeing internal stability. The same process was happening in other countries of the region, reflecting an international climate favorable to the rise of fascist regimes in Europe. Batista ably manipulated the seven short-lived civil governments that followed one another from 1934 to 1940, and in the end won the 1940 elections.

The economic boom that occurred during the Second World War gave Batista's social-democratic government of 1940–1944 a certain degree of popularity. The war consolidated the integration of the American continent and its economic mechanisms under the North American shield: between 1942 and 1947, Cuba sold the whole of the *zafras* to the United States.

As it was the case after World War I the crisis had only been postponed. The workers' movement maintained a position of strength in the post-war situation, when the electoral victory of the *Partido Auténtico* (*Partido Revolucionario Cubano*, not to be confused with the Communist party) awoke great hope in 1944. However, the wartime economic recovery had not eliminated contradictions existing between the dependent position of the sugar companies, the various sectors of the internal bourgeoisie and pressures coming from the lower classes.

In Cuba as elsewhere, the development of the Cold War translated into an offensive against the Communist and workers' movement. The moderate spirit, the subservience to oligarchic and imperialistic interests, the rampant corruption, the nepotism and the proliferation of gangster bands aggravated the institutional crisis. Popular pressure demanded greater national sovereignty and the restitution of the US military bases of San Antonio de los Baños and San Julián. But the 1948 law that lowered the sugar quotas included a clause that instituted United States retaliation against countries that denied "fair and impartial" treatment to its citizens, trade, navigation or industry. Between 1950 and 1952 military agreements were signed that placed Cuban armed forces under the control of North American missions.

In the 1950s once again the Cuban bourgeoisie showed its inability to carry out a process of national independence autonomously, a limit it was to pay dearly for in the choices imposed after 1959 with the victory of the Revolution. Only the FEU made some feeble attempts at resistance to Batista's coup, but they were destined to failure because of a lack of arms. In fact, at first the *golpe* was rather favorably welcomed, even by the Communist Party. But Batista's politics became more and more repressive and anti-working class, and showed to be completely dependent on United States interests. Batista enlarged the rights of North American military personnel, broke off diplomatic relations with the Soviet Union and, with the support of the CIA, created the *Buro Represivo de Actividades Comunistas* (BRAC). Meanwhile, the balance of trade continued to worsen and the policy of public and urban works increased public debt without improving economic conditions. In fact, the traditional parties became Batista's accomplices, taking part in the governments or the elections of 1954 and 1958.

The revolutionary movement created new organizations, such as *Acción Revolucionaria Oriental*, directed by Frank País, and García Bárcena's *Movimiento Nacional Revolucionario* (Ibarra 1998, 157–160). In Havana the young lawyer Fidel Castro (born in 1926 presented a denunciation of Batista and published a manifesto calling for mobilization; he attracted a group of young people identified as the *Juventud del Centenario de Marti,* who prepared themselves for armed struggle. In this way the assault on the *Cuartel Moncada* in Santiago de Cuba of 26th July 1953 was conceived. Though it was a failure in military terms, it represented a point of no return. Fidel transformed the trial that followed into a denunciation of the regime (*La historia me absolverá*). From prison he launched the *Movimiento Revolucionario 26 de Julio,* and after he was released thanks to popular pressure and exiled to Mexico he prepared for the invasion of the Island. In 1954 the FEU also took up the struggle, under the presidency of José Antonio Echeverría, creating the *Directorio Revolucionario* as its armed branch. On 31st August 1956, Fidel Castro and Echeverría signed the *Carta de México* and on 2nd December, 82 combatants led by Castro landed from an overloaded *Granma* yacht in the Eastern Province.

2.3.7 Social Conditions, Cultural Ferments and Modernization in Science

Meanwhile, Cuba was becoming a more complex society. In 1953 the population reached 5,829,000 inhabitants, of whom 23.6 % were still illiterate. Urban concentration grew. In the towns 51.6 % of the population lived in masonry houses and

37.2 % in palm-leaf houses; in the countryside the relative proportions were 2.5 % and 78.2 %. Agriculture remained the main occupation for men, and services for women, who represented only 17.1 % of the economically active population. Nevertheless, in spite of its explosive contradictions and social inequalities and the discrimination against blacks, Cuba was not properly underdeveloped: in the 1950s the average pro-capita income was the second highest in Latin America and the country ranked among the first five on the basis of other socio-economic indicators. The standards of healthcare were among the best on the continent, not very far behind those of the United States and Canada. As regards the number of doctors in proportion to the population, Cuba ranked 11th world-over and third in Latin America, although the situation was decidedly worse in the rural areas and especially in the Eastern Province.

The younger generations gave birth to a lively cultural revival: the Grupo Origenes produced writers of the caliber of Lezama Lima (1910–1976), Cintio Vitier (1921–2009) and Eliseo Diego (1920–1994), and of the playwright Virgilio Piñera (1912–1979). In the visual arts, Wilfredo Lam (1902–1982), of mixed Chinese and Afro–Cuban origin, was a world-renowned interpreter of symbolism and surrealism, also showing Cubist influence. Cuban music was spread by musicians like Argeliers León (1918–1991) and Harold Gramatges (1918–2008) and by celebrated performers like Benny Moré, the Trío Matamoros and Enrique Jorrín (the inventor of the cha-cha-cha). It was in the 1930s that Joseíto Fernández (1908–1979) composed the world-famous song Guantanamera (which the North American composer Pete Seeger adapted to verses by José Marti in the 1960s).

A group of intellectuals that included Rubén Martínez Villena (1899–1934), Mirta Aguirre (1912–1980), Juan Marinello (1898–1977), José Antonio Portuondo (1911–1996), Julio le Reverend and Carlos Rafael Rodríguez (1913–1997) spread Marxist ideas in Cuba. A fresh approach to historical studies was given by Ramiro Guerra (1880–1970), whose Manual de Historia de Cuba was published in 1938, and by Emilio Roig de Leuchsenring (1889–1964).

Fernando Ortíz's production flourished: besides the already cited Contrapunteo Cubano del Tabaco y del Azúcar (Ortíz 1940) he published La Africanía de la Música Folklórica de Cuba (1940), and Los Instrumentos de la Musica Afrocubana (1952–55). Nor can we ignore the birth of the writer Italo Calvino, whose parents were Italians working at the *Estación Experimental Agronómica* of Santiago de las Vegas.

When in 1950 the government approved a law on private universities, the Universidad de Oriente (Eastern University) (Méndez Pérez et.al. 2012, 33–36) had already been founded in Santiago de Cuba[23] and the Università Marta Abreu in Santa Clara. In the 1950s scientific studies were updated, even if the only employment possible for graduates that remained was secondary school teaching.[24]

The Batista regime promoted sectors of investigation in physics and searched after international cooperation, as in nuclear physics. In the framework of the Atoms for Peace campaign, a program to foster the civil use of nuclear power and to build

[23] See the chapter by Méndez Pérez and Cabal in this volume.

[24] See the chapter by Altshuler in this volume.

a nuclear power plant was initiated by the US-government. The only physicist who knew nuclear physics in Cuba, Marcelo Alonso (1921–2005) contacted colleagues in the Netherlands and in Germany in order to evaluate potential cooperation with European institutions and firms (Politisches Archiv Auswärtiges Amt). In this field, the pre-revolutionary years laid a foundation on which later research and cooperation with Soviet and Eastern Europe research units could be built on.

Meanwhile, the crisis of the Cuban system deepened on every level. United States foreign investments were gradually redirected towards oil and industry, with the result that Cuba fell from its place as first investment market for North American capital to second in 1940 and third in 1956, after Venezuela and Brazil. And though sugar exportations were declining, Cuban industry did not take off in other sectors. In 1954, 63.2 % were quite small enterprises employing less than 10 workers, while the rest of these enterprises hired 25 employees.

While sugar was in the late nineteenth century an economic motor for innovation and technical advancement at the island, it now retarded any introduction of new technologies and further mechanization in sugar production processes. The azucarero-oligarchy, in straight collaboration with the changing dictatorships and bearing in mind the US-interests in Cuba, preferred to employ the cheap main d'oeuvre of impoverished land-workers instead of investing in costly machines. By doing so, unemployment rates were quite low—at least in times of harvest—and no social riots were to be feared. The producers of sugar did not have the capital to buy any machines. Consequently, hardly any technical innovations were known in Cuba for this period, neither by importing machines nor by developing them inside the country (Edquist 1985, 79–82).

2.4 Revolution, Modernization and Political and Economic Changes Between 1960 and 1990

2.4.1 A Revolution That Broke All Moulds

One can be critical of or radically contrary to the Cuban Revolution and its ardent supporters but one can hardly deny the originality of this event and its developments. For half a century it has defied all analyses and predictions, representing yet again Cuba's "exceptionality," for better or worse, to remain impartial (if for not other, the disproportion between the smallness of this country, and its worldwide influence). As Mark T. Berger underlines, the particularity of post-revolutionary Cuba was not that it would have been a Soviet satellite state close to the US, but that despite being a close ally of the USSR it continued to pursue its own politics, as can be seen in Nicaragua, Guatemala or even Angola (Berger 2004). Another example would be Cuba's relation to China after the revolution that troubled the Cuban-Soviet relationship for some years (Ying-Hsiang 1973). The limited independence to act on its own count in international politics was also a schema recognizable in the field of physics, as will be shown later.

Legions of journalists, intellectuals, writers and revolutionaries immediately precipitated to this small island (Gott 2004). Among them were Jean-Paul Sartre (Sartre 1961), Simone de Beauvoir (1908–1986), Leo Huberman (1903–1968) and Paul Sweezy (1910–2004) (Huberman and Sweezy 1960) and Charles Wright Mills (1916–1962). Indeed, Mills' last work was on Cuba (Mills 1960); in it, while denouncing the United States' ignorance of history, including the history of its own imperialism, he stressed that one of the challenges the Revolution would have to face was the lack of well-qualified people. The consequences of this problem on the choices made as regards physics are discussed in some of the essays included in this *volume*.

Ernest Hemingway (1899–1961) lived in Cuba for 20 years (he wrote *The Old Man and the Sea* there), where he met both Fidel and Che. From its very dawning, the Cuban Revolution challenged all interpretations and attempts to trace (or reduce) it to existing ideologies or movements. This evasive character in all likelihood was neither a special gift of the Cuban revolutionaries nor a sign of sophisticated political analysis. On the contrary, the Revolution's first steps seem rather uncertain, pragmatic and at times unclear: one should recall the very young age of the leaders at the beginning of the *guerrilla*: Castro 29, Ché 28, Raul 25, Camilo Cienfuegos 24 (3 years more in 1960). But it is probably these very uncertainties, this character so difficult to fit into traditional classifications, which contributed to disorient the great powers and above all the imperial logic of the United States, still under the influence of McCarthyism. It is likely that the United States made some of the most serious mistakes in its history towards this "object of desire," moved as it was by arrogance and a sense of pride that had been hurt by this small Caribbean David.

The leaders of early post-revolutionary tried to position Cuba internationally into a context that would not determine a priori the ideological ally. But it became clear very quickly that the new regime was refuted by the US-government and that Moscow was akin willing to spend support. Although, it is difficult (and is neither the job nor the purpose of this presentation) to discern clearly how much there was of shrewdness or deliberate calculation in the events of the Cuban Revolution, how much improvisation or more probably pragmatism and intuition, how much luck or destiny, and how much in the end depended on mistakes made by others, as Richard Gott (2004) has stated, from whom we borrow some of the following observations. This is not a merely theoretical question, because the playing-out of the various positions and judgments, with all their consequent misunderstandings and rigidities, in only a few months decided Cuba's choices regarding internal affairs and its international position for the next half century. In this context it is interesting to learn the ideas professed by Fidel Castro at the dawn of the Revolution, as received by an educated United States population when, in April 1959, he accepted an invitation to speak at a conference in Washington. On purpose to avoid any occasion of a meeting, President Eisenhower had left the capital for a game of golf in South Carolina.

Castro gave a speech at Princeton University on the subject "The United States and the Revolutionary Spirit":

> He first addressed existing theories of revolution, noting that the Cuban experience had shattered three myths. First, that 'a revolution could be successful even if the mass of people are not starving.' Second, that revolutionary forces 'could defeat regular military forces;' and third, that a revolution 'could succeed against modern arms.' Guerrilla tactics aside, Castro attributed the success of the Cuban Revolution to the widespread 'fear and hatred of Batista's secret police,' and to the fact that the rebels 'had not preached class war,' thereby gaining '95 per cent support' from the population.

A student journalist noted that Castro promised "he would lead the country to economic and cultural progress without sacrificing individual freedoms." Castro also was reported to have said that democracy was "the most beautiful political and social idea." The latter remark was greeted with a standing ovation. *Long Island Newsday* reported that the Premier had claimed in his address that he

> expected and would allow minority parties to develop 'in opposition to his regime', and that while there were 'no plans to nationalize any lands,' his government would 'expropriate legally' any 'idle or unproductive lands.' (Bogenschild 1998)

The first years after the revolution were owing to favorable circumstances where decisions made by the new government, as a minimum wage for sugarcane cutters, a decrease in electricity and telephone rates, rents and the cost of medicines, the creation of new jobs and an increase in the salaries of low-level public workers, while the pay of high-level bureaucrats, judges and members of the government was lowered. This met with the inaptitude by the old pro-US-bourgeois to reinstall its powers.

What proved to be the decisive turning point in Cuba-United States relations was the agrarian reform of 17th May 1959 (Benjamin 1990, 179–181). It was not in fact a radical reform, but the essential point is that it was impossible for Cuba to carry out a real reform without encroaching on the interests of the large imperialist corporations. Economists arrived from other Latin American countries and from East Europe, as Michal Kalecki (1899–1970) from Poland, or the French Marxist economist, Charles Bettelheim (1913–2006) (Hamilton 1992, 36–54). The final goal was ambitious indeed: to develop an industrial economy. It was Ernesto Guevara who promoted the ambitious project of turning Cuba from a prevalently agricultural country into an industrialized one through a process of accelerated industrialization that would start with heavy industry came up against objective difficulties, such as the country's economic isolation and its backwardness (Boughton 1974).

The revolutionary government was strongly committed to developing basic and scientific-technical education. The whole country was swept by a wind of change and enthusiasm. In 1961 a widespread campaign mobilized 100,000 teachers to eliminate illiteracy (some of them were brutally murdered by counter-revolutionary bands still present in the island) (Abendroth 2009). A campaign of on-going adult education was also undertaken, along with a program to develop an advanced school system open to one and all, which had no equal in Latin America. On 20th December

1959 the first *Reforma Integral de la Enseñanza* was promulgated (Wylie 2010, 82). Sixty-nine army barracks were transformed into schools, over three thousand new schools were built in the first year and about seven thousand teachers were trained, so that three hundred thousand children could attend school. The doors to secondary and university education were opened to workers, including farm workers.[25] In 1961 the *Consejo Nacional de Cultura* was founded to foster cultural development, along with the *Escuela Nacional de Arte*, ENA (its buildings in west Havana, audaciously designed by outstanding architects, are well worth a visit),[26] while a widespread network of art schools was built all over the country. That same year the *Ley de Nacionalización de la Enseñanza* was promulgated, putting an end to private profit-making schools. Immediate measures to improve and extend healthcare to the entire population were also taken (Wylie 2010, 83).

In 1960 Fidel Castro explicitly affirmed that "The future of our country must necessarily be a future of men of science",[27] while Ernesto "Ché" Guevara even foresaw the future importance of solid-state electronic devices. As regards university, the urgent need to update the old system and introduce a modern spirit of scientific research fostered probing discussions involving directors, professors and students. The result was the University Reform of 1962, which was to prove a fundamental stimulus. But even as early as 1960 there were spontaneous experiments to reform the Engineering School of the University of Havana by introducing new programs of study, despite the difficulties caused by the lack of well-prepared staff. The students themselves made up for this lack by giving lessons to students in lower years; this work on the part of the *alumnos ayudantes* went on for years. These processes are described in detail in this volume as regards the field of physics. In 1960 the first six students left to study engineering in Moscow, thanks to Ché Guevara's personal involvement, to be followed by physics students soon after. Their return as graduates in 1966 made it possible to establish a stable professional teaching staff and the curriculum of the School of Physics of the University of Havana.

2.4.2 In Search of New Allies

The reactions inside the United States against the Cuban Revolution grew at all levels, independently from the nature of the measures it adopted, but based on the

[25] See the interview with Melquades de Dios Leyva in this volume.

[26] The project for the Escuelas, conceived of by the Ministry of Culture to host 1,500 students from Latin America, Asia and Africa, was entrusted to the Cuban architect, Ricardo Porro, who called his colleague Roberto Gottardi and his Italian friend Vittorio Garatti to collaborate. It soon became clear that the plan was beyond the island's concrete possibilities, and in fact in the following years there was a gradual waning of the extraordinary enthusiasm that had taken hold of architects, workers, teachers and students in the initial phases of planning and construction. See Loomis (1999), Daley (2000), Craven (2002), chap. 2 and Giani (2007).

[27] Speech given on January, 15 1960 by Fidel Castro. See Torres Yribar (2011).

real or presumed Communist leanings. The first counter-revolutionary group (*La Rosa Blanca*) inside US territory was formed with the complicity of the CIA and the FBI, and the sabotage actions grew (through US air raids in October 1959 which sabotaged the sugar mills) up to the unsuccessful military invasion on 17 April 1961 at the Bay of Pigs.

The obstinate intensification of clandestine actions to create unrest in Cuba and overturn the regime ("Operation Mangoose"), with the enforcement in January 1962 of the expulsion of Cuba from the Organization of American States (a decision that was not revoked until June 2009, see below), objectively pushed the country to increase links with the Soviet Union and converged with Khrushchev's plan to deploy nuclear missiles (Bain 2008). Nikita S. Khrushchev (1894–1971) obtained from US President John F. Kennedy (1917–1963) the promise not to attack Cuba and to withdraw US-American nuclear missiles from Turkey and Italy, before the head of the Kremlin agreed to abandon his plans (George 2013).

The already big problems the country had to face were further aggravated by the fact that many entrepreneurs, technicians, economists professors or intellectuals left the Island after 1959 (Gott 2004). The international crisis had nearly no influence on an animated intellectual and artistic life: the Cuban cinema was born, new writers became worldwide known and the new Cuban music met internationally with success. Efforts were also focused on publishing and cultural diffusion and growth. In 1962 the *Editorial Nacional Cubana* was created, and in 1967 the *Instituto Cubano del Libro*. The result was that while in 1958 the country produced only about a million books, in 1967 production had grown to eight million and to 35 million in 1975. Meanwhile, innovative activity in the fields of healthcare and education went on apace. The 1962 University Reform greatly stimulated higher education, the level improved enormously and study materials were brought up to date. The start-up of scientific research necessitated the creation of scientific institutions and the formation of researchers. In fact, many of the most important research institutes that the country has today were founded in the 1960s. And most science, engineering and research programs were introduced in universities in those years (Núñez Jover et al. 2008).

Cuba's cultural ferment continued to exercise its fascination all over the world. The articles in this volume reconstruct the singular presence in the University of Havana's School of Physics of physicists and technicians from many western countries, who held courses in modern physics, equipped workshops and laboratories and started the first scientific projects. During the same decade some of Cuba's other main research institutes were born. Among them figures the newly founded *Accademia Cubana de Ciencias*, under whose aegis a series of projects and new institutions were developed, as it was the norm in Soviet Union, where the Academy hold the monopole for research (Domínguez 1979, 397). The universities were able to consolidate stable teaching staff and curricula. By the end of the decade there was a critical mass of graduates prepared to carry out scientific research, thanks also to the return of Cubans who had left to study in the Soviet Union and other socialist countries, while collaboration with these countries was consolidated.

2.4.3 The Sharp Turn of Cuba's Economy and Politics at the End of the 1960s

As we have remarked, a change of the early revolutionary Cuban (Guevara's) economic and political program was underway since 1963, but the critical turn was accomplished in the last 1960s. Guevara had recently disappeared, and the echoes of his denouncement of the Soviet and the Chinese models burnt out, when Moscow's tanks crushed the "Prague Spring", on 21th August 1968. Castro's denunciation of the Czech reformers took everybody by surprise.

The rigid bipolar equilibrium of the Cold War drastically limited space for autonomy in the countries that chose to remain in the Soviet Block. It seems like a twist of fate that Cuba debunked its non-aligned international role while the world was undergoing shocks that seemed to confirm those previous perspectives: the North Vietnamese "Tet Offensive" of January 1968 fed the crisis that was sweeping the United States, with the massive movement against the War in Vietnam, and the assassinations of Martin Luther King (1929–1968) in April and Robert Kennedy (1925–1968) in June, and in September Mexico City was bathed in the blood of *Plaza de la Tres Culturas* massacre; while the "French May", and its international extensions, challenged the hegemony of the orthodox Communist parties in the western left—flying as its own icon the image of "Ché", next to that of Mao Tse-tung.

The peculiarity of Cuba's position can probably be better appreciated taking into consideration the endemic instability that was to run over Latin America, characterized by coups and bloody military regimes, which increased segregation and social inequalities. Anyway, even if Cuba had entered the Soviet orbit, it still showed the originality of its choices in the international arena. Castro kept a space of autonomous action, taking decisions and positions that disoriented the big powers and interfered with their plans. In 1975 he surprised both Moscow and Washington by returning into the African arena, which Ché and his guerrillas had left suddenly 10 years before. Castro's decision to intervene militarily to support the MPLA in the defense of Luanda in 1975 was so resolute that he sidestepped the reservations of Brezhnev, who at that time was aiming at a substantial rapprochement with the United States. The venturesome (reckless?) Cuban expedition, without any Soviet logistical support, allowed Agostinho Neto (1922–1979) to route the assault on the capital and proclaim the independence of Angola. Further Cuban reinforcements convinced South Africa to withdraw from Angola, and the symbolic value of the defeat by a largely black army was one of the triggers of the Soweto revolt in June 1976. Once more images of Ché were hoisted next to those of Malcom X and Mao Tse-tung.

Cuba's intervention in Africa continued in the form of aid, medicines and teachers to many countries, besides another military intervention in Ethiopia. Cuba regained great prestige in the anti-colonialist milieus, and in September 1979 hosted Conference of the Non-Aligned Movement. In the 1970s, Cuba continued to be a sort of Mecca of Afro-Cuban culture. Performers like Sidney Poitier (born in 1927),

Harry Belafonte (born in 1927), and the South African singer Miriam Makeba (1932–2008) arrived.

Following the terrorist attack of October 6, 1976, in which 73 persons lost their lives, including Cuba's entire fencing team, in 1977 Jimmy Carter's administration opened a phase of (although contrasted) rapprochement between the two countries. Some concrete initiatives were taken (opening *Interest's Sections*, travel and remittances), but this opening tottered after the success of the Sandinists in Nicaragua in 1979, and was closed when Reagan's victory in the 1981 elections led to a strengthening of the embargo against Cuba and intensification of acts of sabotage.

In the years following the "Prague Spring" Cuba brought its economy into line with reassuring Soviet standards, thereby condemning itself for 30 years to dependency on sugar. Despite the failure in 1970 of the clearly overestimated objective of a sugar harvest (*zafra*) of 10 million tons), while the ideological and revolutionary ferment of the 1960s was reduced. Fidel Castro decided to nationalize all private companies in 1968 and a national party organization was installed that broadened its impact in any social activity. In 1975, the first Congress of the Cuban Communist Party was held.

In fact, thanks to substantial Soviet support the Cuban economy overcame the difficulties of the past. Between 1975 and 1985 it enjoyed an over-4 % growth rate, and for the population there began a period of prosperity. Some original aspects remained in the Cuban system, confirming it as an exception even within the Soviet system. Available resources were used not only to strengthen the armed forces, but also to develop schools and healthcare (Bain 2007), science and technology (Bethell 1993) and to give Cubans a higher standard of living and more available consumer goods. However, we cannot ignore critics like the writer Heriberto Padilla (1932–2000), who was segregated and then arrested in 1971, which provoked a position statement by legions of authoritative international intellectuals, including Sartre, Simone de Beauvoir, Octavio Paz (1914–1998), Carlos Fuentes (1928–2012), Vargas Llosa (born in 1936), García Márquez (born in 1927) and Julio Cortázar (1914–1984) (the last two subsequently withdrew their support of the letter).

Adhesion to the Communist Block entailed a growing reliance of Cuban agriculture on sugar cane, which was to have deep implications in the 1990s (Lorini and Basosi 2009). Nevertheless, although Cuba counted heavily on the exchange of "sugar-for-oil" and other goods with the Soviet Union, it never cut off trade with non-socialist countries. Though carried on at lower market prices for sugar, this supplied a certain quantity of strong currency and some margin of autonomous maneuver. The percentage of exportations towards non-socialist countries was 30.5 % in 1970, and the situation did not sensibly change even after 1972 when Cuba joined the COMECON market; in 1980 this percentage was 29 % (60 % of the 71 % of exportations towards socialist countries was with the Soviet Union).

With the Soviet Union's economic support Cuban society progressed greatly and enjoyed a period of prosperity, as the state was able to satisfy the population's basic needs and also to supply consumer goods. In the field of healthcare, while in 1958 there was only one rural hospital and 161 first-aid stations with 6,000 doctors (of whom half left the country in 1959), in 1975 the country had 56 clinics, 118 dispensaries, 396 polyclinics and 10,000 doctors. Over the same period, average life expectancy rose from 55 to 70 years.

2.4.4 The Crucial Leap in Education and Science

Following the great commitment to schools and universities, the total number of students passed from 811,345 in 1958 to 3,051,060 in 1975 (on roughly 7 and 9.5 million inhabitants respectively). The spread of primary education grew by a factor of 2.7, middle and higher education by 6.1 and university education by 5.5. In the same period the number of students in higher education passed from 15,000 to 83,000.

In December 1968 the *Congreso Cultural de La Habana* brought together a numerous and vivacious international public that also included authoritative scientists, (mainly French and Italian physicists), who expressed relevant advices for the basic choices in the development of Cuban physics.[28] From a proposal launched on that occasion by French and Italian scientists was born the initiative of the *Escuelas de Verano* (summer schools), which from 1968 to 1973 brought to Cuba international specialists in every discipline, who introduced equipment, techniques and materials that gave a strong impetus to scientific research in Cuba.[29]

The 1970s saw a first readjustment of the Cuban scientific system after the preparatory phase and the take-off in the 1960s. This represents record time for a small third-world country, an absolute exception. Cuba by now could count on a critical mass of scientists with a solid up-to-date formation and a network of research centers. This is certainly true in the field of physics. As the articles in this volume, advanced scientific sectors such as microelectronics were developed. Forms of cooperation and exchange with the most important centers of excellence in the Soviet Union and other socialist countries were consolidated (after first degrees were taken in these countries, advanced PhD studies began), but this did not mean that important instances of collaboration with the western world were interrupted (Reid-Henry 2010). The historical analysis developed in this book mentions several collaborations developed, mainly in the field of solid state devices and physics, with Canada, Spain and France, while the article by Leccabue documents the lasting and fruitful collaboration with Parma, Italy, which is still alive: it is relevant for the appreciation of Cuban peculiarity Leccabue's remark that an analysis with the Cuban colleagues showed that reliance on Soviet technology appeared too restrictive for the needs and perspectives of subsequent development of research in this field.

Anyway, the relevance of the strengthening of the collaboration with, and support by the Soviet Union and the Socialist countries could hardly be underestimated, such as it was the case for big international cooperation treaties, as for instance the

[28] See the chapter by Baracca, Fajer and Rodríguez in this volume.

[29] See the chapter by Cernogora in this volume.

participation of Cuba in the INTERKOSMOS program. This cooperation existed already since the mid-1960s but only in 1976 a formal institution having for aim the exploration of the outer space, depending on the Soviet Academy of Science was founded in cooperation of most of the Soviet allies states in Eastern Europe.[30]

In general terms, as Jorge Núñez Jover states:

> The existence of the science and technology policy has long come hand in hand with the notion that scientific advances fail to fully translate into the practical use of its results. So in the mid-1970s, evidence started to mount indicating that the practical use of scientific results to solve the problem of production and services was a matter of greater complexity. This brought about a number of changes in the science and technology policy establishing what has been called the 'model of centralized direction' (1977–1989), whose objective was to complete the efforts from the supply with a deliberate strategy to use scientific and technological results; this was called the 'introduction of results.' It was supposed to be achieved through a very centralized model relying on the identification of 'research problems' to direct research towards issues of high priority and the use of results in the fields of production and services. Although the use of results was emphasized, this staged was based on the same lineal concept towards scientific research as a triggering element in the relationship between science, technology and production.
>
> Problems of concept were compounded by a very relevant practical circumstance. In parallel with the emphasis on science and the expectations that it would increase its contribution to development ran an implicit technological policy which was characterized by generalized imports of technology, quite frequently from the European socialist countries. They were characterized, *inter alia*, by being moderately modern technologies with low energy efficiency and environmentally aggressive. The tendency of assimilating, rather than producing, technologies and the frequent lack of interest to innovate shown by the entrepreneurial segment of the agents of technological change explain why the scientific development and the human potential achieved were not expressed in the expected practical results. This situation justifies the critical view that was formed around this issue throughout the 1980s, a discussion that was included in a broader debate on the praxis of the socialist transition in Cuba and, particularly, the economic efficiency of the country. (Núñez Jover et al. 2008)

2.5 Politico-economic Crisis and New Cooperations (1985–1999)

2.5.1 Toward the Breakdown of the USSR

The collapse of the Soviet Union without doubt caused traumatic change in Cuba. There had been a forewarning in 1983, when President Andropov had let the Cuban leaders know that the Soviet strategic guarantee, established in 1962, had not been operative for some time (Gott 2004; Purcell and Rothkopf 2000, 84). Cuba changed its military strategy, instituting a system of defense based on a "people's war," which was to play an important role in the following two decades. But Castro had more or less sensed what was coming and preceded Gorbachev by presenting a new

[30] For the history of Interkosmos and the role of the GDR, cf. Katharina Hein-Weingarten (2000), esp. 153–165.

economic program in February, 1986, that was called *Politica de rectificación de errors* (politics correcting errors.). This program was far from the Soviet model and represented an explicit refusal of the reforms attempted by some socialist economies of Eastern Europe. The economic situation in Cuba at the mid–1980s became increasingly difficult as the fall in the international price of sugar had created an increasing dependence of Cuba on socialist countries' markets (Lorini and Basosi 2009). In contrast to previous years, in 1986 this market accounted for 85 % of Cuban exports, and that same year 98 % of the fuel. The nutrition was in huge parts assured by COMECON-countries from where 50 % of the calories and 57 % of the protein consumed in Cuba were imported (Tablada 2001, 39).

> Also in its science and technology policy since the mid-1980s the country introduced relevant changes. The more pointed changes include the re-launching of university research, now with a more applied orientation, the definition of new priorities for scientific and technological development, comprising biosciences, biotechnology, pharmaceutical industry, high-tech medical equipment. The newly created productive scientific parks became true networks of integrated cooperation where research, creation of technologies, production and commercialization of products are part of a continuous process led by unique strategies, the enhancement of the Science and Technology Forum, a unique Cuban experience geared towards increasing social participation in the scientific and technological development and its applications. (Núñez Jover et al. 2008)[31]

These changes had important consequences in physics as well, leading to a probing process of critical revision of prior choices and serving as a stimulus for new applied sectors such as nuclear physics, including the project to build a nuclear power plant at Juraguà, near the bay of Cienfuegos, with Soviet aid (Blasier 1993).[32]

In international affairs Cuba became involved in some political conflicts in Africa in 1988, after the attack launched by South Africa the previous year, sending 9,000 soldiers to complete a contingent of over 50,000 men (George 2005). As Castro observed, in relation to total population, this was equivalent to over a million US soldiers. Not only did Cuba save Dos Santos' government in Angola (which had succeeded that of Agostinho Neto) but it literally changed African history: as Nelson Mandela declared upon his visit to Havana in 1991, the defeat of the racist army at Cuito Cuanavale destroyed the myth of its invincibility and greatly contributed to the subsequent end of apartheid.

Even before this decisive intervention in Africa Castro had made another able move (Lorini and Basosi 2009) that restored Cuba's credibility in Latin America, despite the fact of being partly isolated by most of the military juntas governing in the Continent. In 1982 Mexico opened the "Latin American foreign debt crisis", which then spread to Brazil.[33] Loans from international financial organizations were subject to the "Washington consensus" and they imposed plans of adjustment based on privatization, economic deregulation, reductions in public expenditure and liber-

[31] For development in biotechnology, see also Reid-Henry (2010, 28).

[32] See the chapter by Baracca, Fajer and Rodríguez in this volume.

[33] Though the situation varied from one country to another, Latin American foreign debt had swollen from $30 billion in 1970 to $331 billion in 1982 and would reach $410 billion in 1987.

alization of monetary flow. Castro made an astute move in March 1985, once again disorienting the politics of the great powers, by maintaining that the debt was too high to be repaid and had in fact already been abundantly paid to the rich North through neo-colonial exploitation (Castro 1989). This move had been carefully prepared by the Cuban leadership, which had already found an agreement for its own debt of $2.9 billion (Lowenstein 1985). In August 1985 Havana hosted a large continent-wide meeting “against the debt”; though the ministerial delegations were limited to Ecuador, Argentina, Bolivia, Nicaragua and Panama, there were over a thousand delegates from NGOs, unions, religious groups, political parties and even businessmen. The headline in the *New York Times* was, “Cuba’s emergence, America’s myopia” (Szulc 1987).

After the fall of most dictatorships in the early 1980s political, diplomatic and trade relations were established first with Ecuador, then with Argentina, Uruguay and Brazil. As of 1989 there were only three Latin American countries that had not resumed relations with Cuba: Chile, Paraguay and Colombia.

Despite internal economic problems, Cuba has continued to send doctors and teachers to Africa and to Latin America, where it has undoubtedly played an active role, participating intelligently in the reawakening of this continent in the last 20 years and thereby continuing to represent an indispensable interlocutor, also as regards the United States. But, the Cuban diplomatic offensive in Africa and Latin America lost energy over the years because of the worsening of the economic crisis. As well, the trial and execution in 1989 of the Soviet-affiliated General Arnaldo Ochoa Sánchez, a hero of the wars in Africa, shows that the political situation inside Cuba became more and more difficult in regard to the influence the Soviet Union was allowed to take.

2.5.2 *The Deep Troubles of the ‘Periodo Especial.’ New Cooperations in a Changing World*

In January 1990, faced with the global retrenchment of the Soviet Union, Fidel Castro spoke of the need for Cuba to enter a “special period in peacetime” (Gott 2004, 335–349). It was the official beginning of the most dramatic period for the Cuban economy since the *Revolución*. Technological supplies from the USSR came to a halt between 1989 and 1992, and the sudden lack of strategic raw materials and spare parts caused further economic problems in industry and transport. In 3 years imports plummeted by 72 % and exports by 67 %, the investment rate was 7 % down from 26 %, gross capital formation fell by 60 % and oil imports dropped by over 50 %. In 1993 the GDP was a startling 35 % lower than in 1989 (Núñez Jover et al. 2008). Annual sugar production sank from over 8 million tons to little more than 4, while the end of Soviet subsidies forced direct sales on international markets at very low prices. Even tobacco production halved (Lorini and Basosi 2009). Official Cuban statistics reported that in 1996 daily caloric intake was down 27 % compared to 1990 levels.

To the negative effects of the changes in east Europe was aggravated by the consequences of a process that from this time on was called globalization.[34] In a situation when the entanglement of the world's political and economic processes became obvious by the downfall of the seemingly clear cutting frontier between the West and the East, the United States strengthened the so-called *bloqueo* against Cuba by the Torricelli (1992) and Helms-Burton (1996) Acts. International financial institutions denied loans to Cuba, one of the few remaining communist regimes in the world. In these years, the CIA dedicated at least two detailed studies to Cuba, in 1991and 1993:

> the first one advanced the hypothesis that "the deterioration of the Cuban economy (would have) further undermined Castro's legitimacy", while the second spelled out clearly that there was "a better than even chance that Fidel Castro's government would fall within the next few years.
>
> Given these conditions, with a display of Cuban sense of humor, an economist of the Havana-based *Centro de Investigaciones de la Economia Mundial* observed some years later that, with virtually no access to foreign credit, the socialist island had undergone, during the *periodo especial*, the only true 'classical' economic adjustment: international prices had grown, the external deficit had soared, domestic consumption had been curtailed to offset the loss in purchasing power. (Lorini and Basosi 2009)

Certainly, this new global and internal situation required some important adjustments in Cuban politics. In 1993 the government reacted rapidly to take more radical economic measures. As regards international trade, possession and use of the US dollar were legalized, creating the rather paradoxical situation of the existence of a double currency, in order to try and stimulate foreign investment in public-private joint ventures, tourism and selected industries in which the Cuban state kept at least 51 % ownership. Instead of the traditional low technology sectors for export (sugar and tobacco), non-traditional sectors in which capacity had been developed in the past and which could attract foreign currency were promoted: high value-added service products such as international tourism, healthcare and education; and high-tech industrial sectors, especially biotech products, pharmaceuticals and medical equipment. At the same time, the export potential of traditional sectors was sustained, with a restructuration of the sugar agribusiness and promotion of the nickel industry (Moreno 1998).

As regards the domestic economy, there was a partial liberalization of small-scale enterprises, in particular restaurants, called *paladares* and rented rooms for foreign tourists, along with more limited businesses that do catering for the population in houses and streets (Corrales 2004). A program for the production of food for the population was developed, allowing farm cooperatives to sell surplus products to urban markets, the so called *agromercados* (Babb 2011). Taxes were introduced for the first time in over 40 years and they were heavy for private enterprises. Cuba continued its transformation to a service economy oriented towards activities aimed at fostering foreign trade, saving and reducing its dependency on fuel, and fully utilizing the Island's skilled human resources. The outcome was a marked growth in the tertiary sector at the expense of the primary and secondary sectors.

---

[34] Some at this time spoke even of a reversed global situation, see Jatar-Hausman (1998/1999).

The transformations were profound: greater diversification of the economy was promoted, with more cultivable land farmed for crops other than sugar, like vegetables and soy. Small-scale organic farming was developed in urban gardens, the *organoponicos*, which covered a significant percentage of the country's food requirements (Paponnet-Cantat 2003). Half of the existing sugar refineries were closed, while nickel and cobalt mining doubled from 1993 to 1997, mainly in the form of public-private joint ventures with foreign countries (Navarro Vega 2013). Despite the economic bottleneck, huge investments were made in medical and pharmacological research 120 million dollars between 2005 and 2008 (Núñez Jover et al. 2008), allowing Cuba to become one of the world's most advanced countries in the field of biotechnology. The sudden lack of the energy supplies that had come from the Soviet Union created very serious problems, especially the interruption of electricity, with frequent long blackouts (*apagones*) (Pérez-López 1992). Deep drilling led to the discovery of substantial (though low-quality) oil deposits, while the production of natural gas grew from virtually zero to over 450 billion cubic meters from 1990 to 1999.

In 1994 the worst phase of the *periodo especial* was over, economic decline slowed and recovery began. (Núñez Jover et al. 2008). By 2004 the sum of products and services had registered a 46 % recovery from 1994 and the country enjoyed a balance of payments surplus. In 2005 the economy boasted a growth rate of 11.8 %, the highest since 1959, and it continued to grow in 2006. And this despite the hurricanes that devastated the Island in 2008, when the evacuation of a million people and the other measures that are always taken in Cuba avoided the hundreds of victims that such cataclysms regularly claim in the other countries of the region (including the United States).

This period can be considered to exemplify again the exceptionality of Cuba, as the increasing diversification of Cuban economy has not let to a collapse of the country. Although the egalitarian society of the 1960s and 1970s was deeply shaken by the creation of tangible economic differences between those who could access foreign currencies (through tourism and foreign remittances) and those who could not, the government has made an effort to keep up support the key sectors of healthcare, education and social security. Even if the persistent shortage of medicines, products and basic equipment, paper, books and other materials for consumption undoubtedly lowered the general level of these services, Cuba has determinedly worked at preserving and in some cases even improving the features that distinguished its social model from the one prevailing in the rest of the world, especially in Latin America.

> Midway through the 1990s, a report of the United Nations Economic Commission for Latin America and the Caribbean (CEPAL, from the Spanish acronym) stated that 'given the size of the external shock, the policies of adjustment (in Cuba) has been relatively low [...] in comparison with other Latin American economies,' attributing such an outcome to the 'orientation toward solidarity' of the Cuban social context. (Lorini and Basosi 2009)

UNICEF confirms that Cuba is the only country in Latin American or the Caribbean to have eliminated childhood malnutrition (UNICEF 2004). The infant mortality rate at birth is among the lowest in the world and continues to improve despite the economic crisis.

As far as international politics are concerned, Cuba has enormously strengthened its influence on the Latin American continent. In 1987 eight Latin American countries for the first time proposed Cuba's readmission to the Organization of American States (after its expulsion in 1962, cf. *supra*). The United States opposed the motion, but 2 years later was humiliated when the Latin American countries elected Cuba as non-permanent member of the U.N. Security Council. In the U.N. General Assembly's votes against the US blockade against Cuba the Latin American countries passed from abstaining to voting in favor, with the result that the total number of votes against the US jumped from 59 to 101 between 1992 and 1994. Left-leaning Latin American leaders have shown renewed interest in the Cuban Revolution. Venezuela, above all, has become Cuba's most important political and economic partner, while China has also given significant support to Cuba.

Finally, after 47 years the Organization of American States unanimously cancelled the U.S.-imposed resolution excluding Cuba from the continent's community structures during the 2009 summit in Honduras. While Hillary Clinton was forced to grin and bear it, the Honduran President Manuel Zelaya (who soon after, as if by historical nemesis, was overthrown in a coup not without United States' manipulation!) turned directly to Fidel Castro and, referring to his historic defense speech of 1953, pronounced: "I say to Commandeer Fidel Castro, today you have been absolved by history."

A time of reckoning has come as concerns Cuba's internal problems as well. In part these have become more critical because of the repercussions the world crisis has had on the domestic economy and the population's living conditions. But the passage of political leadership from Fidel, the historic leader now afflicted with health problems, to his brother Raul, at the same time as it granted substantial political continuity also awakened great hopes in the population. The commitment to carry out significant changes can no longer be postponed, but it is evidently difficult to carry out because of bureaucratic fossilization, international isolation and internal taboos in political debate in relation to repressive responses by the government.

2.5.3 Cultural Vitality, Higher Education and Low-Funded Universities

Few words are sufficient to recall the ongoing of Cuban cultural vitality, in spite of the crisis: the international success of Cuba's popular music (Pablo Milanés, Silvio Rodríguez, the Van Van, Charanga Habanera, etc.), Cuban ballet (Alicia Alonso, the *Escuela Cubana de Ballet*), the plastic and figurative arts, the *Havana Biennale d'Arte*, uninterrupted Cuban film production, the annual *Festival del Nuevo Cine Latinoamericano*, the good health of Cuban literature (both at home and abroad: besides by now historical figures like Dulce María Loynaz, Cintio Vitier, Eliseo Diego, Dora Alonso, and Lisandro Otero, is testified by names that include

Guillermo Cabrera Infante, Miguel Barnet, Roberto Fernández Retamar, Senel Paz, Pablo Armando Fernández, Reynaldo González, Leonardo Padura).

The effect of the crisis on scientific research has been particularly heavy. However, Cuba has managed to resist in this field as well, though paying a high price. The government has done everything possible to support the most important activities. On 18th January 2002 the than president Fidel Castro stated that "this country will live on intelligence and intellectual property", which means that the political commitment to bet on knowledge, education and science remains firm. There has been far-reaching reorganization in many areas; inevitably, experimental research has been the most affected, while computer-driven research has been significantly boosted. The relations with non-socialist countries cultivated by Cuba in the first decade of the Revolution, which had never been completely suspended, have turned out to be very useful in this phase: many Cuban scientists (even if a certain number have left the country) take advantage of collaboration and exchanges with Spain, Mexico, Brazil and other countries. This reconfirms the high level reached by Cuban science on the international scene as well as the solidity of the results achieved before 1990.

Though the necessary reorganization has produced sacrifices and penalized important activities, it was done intelligently and in the end has even had some useful outcomes. Since the country has made every effort not to abandon its free education and healthcare systems, interdisciplinary projects have been intensified. The excellence of Cuba's biotechnological and medical research and the importance of the results it has reached are known worldwide.[35] Naturally, the interruption of relations with the socialist countries has led to a worrying crisis in employment opportunities for university and polytechnic school graduates. Still, thousands of students from Latin American and other third-world countries study medicine or art in Cuba, something that would be impossible for them in their own native countries.

The new global situation of Cuban science alters thus between the excellence in some fields of study and research and difficulties that are not very different of those in other countries.

2.5.4 The Challenge of the Future in the Context of the World Crisis

This is not the place to discuss the present world economic crisis, and speculate on the place and the future of Cuba. Nevertheless at least a short remark seems necessary. Indeed, starting around 2010 the Cuban government has turned into reforms of the economic system, which are difficult to evaluate, but are changing at least the external aspects of Cuban life, but are deeply influencing also the mentality of the citizens. Anyone who visits Cuba now will remark at first glance the change in the

[35] See the chapter by Pérez Ones and Núñez Jover, republished in this volume.

urban panorama itself, people selling goods everywhere, small trades, but inexistent before 2010. On the one hand, the world economic crisis strikes also this country, even if in different ways from others. The American Administration does not show any sign of loosening the blockade to the Island, in contrast with the initial perspectives of the Obama presidency. On the other hand, the Cuban government has acknowledged (all the more in these conditions) the inefficiency of the "universal" State economy, and in particular the necessity of dramatically reducing the number of the State workers and bureaucracy personnel, an extremely painful choice. Among other economic reforms (commerce of apartments and cars, permission to expatriate, etc.), the permission of the *trabajo por cuenta propia* is the most visible at glance.

We think that it is impossible for everyone to give a well-advised evaluation of these economic measures, and even more of the general and political perspectives: too many times in the past quarter of century specialist have predicted nothing less than the collapse of the Cuban regime. Nevertheless we felt obliged at least to point out these changes, which could give a further turn to the situation that we have tried to discuss: Cuba could set aside one more surprise.

References

Abendroth, M. 2009. *Rebel literacy: Cuba's national literacy campaign and critical global citizenship*. Duluth: Litwin Books.

Adderley, R.M. 2006. *New negroes from Africa. Slave trade abolition and free African settlement in the nineteenth-century Caribbean*. Bloomington: Indiana University Press.

Altshuler, J., and R. Díaz Martín. 1999. Primeros experimentos telefónicos de Antonio Meucci: La Habana 1849. In *Aniversario 150*. La Habana: Sociedad Cubana de Historia de la Ciencia y la Tecnología.

Altshuler, J., and R. Díaz Martín. 2004. *El Teléfono En Cuba, 1849–1959*. La Havana: Sociedad Cubana de Históría de la Ciencia y la Tecnología and Empresa de Telecommunicacióñes de Cuba.

Altshuler, J., and M. González. 1997. *Una luz que llegó para quedarse: Comienzos del alumbrado eléctrico y su introducción en Cuba*. La Habana: Editorial Científico-Técnica.

Anderson, W. 2009. From subjugated knowledge to conjugated subjects: Science and globalisation, or postcolonial studies of science? *Postcolonial Studies* 12(4): 389–400.

Babb, F.E. 2011. Che, Chevys, and Hemingway's daiquiris: Cuban tourism in a time of globalisation. *Bulletin of Latin American Research* 30(1): 50–63.

Bain, M.J. 2007. *Soviet-Cuban relations 1985 to 1991: Changing perceptions in Moscow and Havana*. Lanham: Lexington Books.

Bain, M.J. 2008. *Russian-Cuban relations since 1992*. Lanham: Rowman/Littlefield.

Baracca, A. 2009. Science (physics) in Cuba: A lag between technological and scientific development? In *Cuba in the world, the world in Cuba: Essays on Cuban history, politics and culture*, ed. Alessandra Lorini and Duccio Basosi, 81–93. Florence: Florence University Press.

Baracca, A., S. Ruffo, and A. Russo. 1979. *Scienza e Industria, 1848–1915*. Bari: Laterza.

Benjamin, J.R. 1990. *The United States and the origins of the Cuban revolution: An empire of liberty in an age of national liberation*. Princeton: Princeton University Press.

Bergad, Laird W., Fe. Iglesias García, and María del Carmen Barcia. 1995. *The Cuban slave market 1790–1880*. Cambridge: Cambridge University Press.

Berger, M.T. 2004. After the Third World? History, destiny and the fate of Third Worldism. *Third World Quarterly* 25(1): 9–19.

Bethell, L. 1993. *Cuba: A short history*. Cambridge: Cambridge University Press.

Blackburn, R. 1988. *The overthrow of colonial slavery (1776–1848)*. London/New York: Verso.

Blaquier, M. 2009. Cuba and Latin America at the end of the cold war: From the "Foreign Debt Diplomacy" to the soft power of the Período Especial, 1985–1995. In *Cuba in the world, the world in Cuba: Essays on Cuban history, politics and culture*, ed. A. Lorini and D. Basosi. Florence: Florence University Press.

Blasier, C. 1993. The end of the Soviet-Cuban partnership. In *Cuba after the cold war*, ed. C. Mesa-Lago, 59–97. Pittsburgh: University of Pittsburgh Press.

Bogenschild, T.E. 1998. Dr. Castro's Princeton visit: April, 1959. *Boletín*, *Princeton Program in Latin American Studies* 21–26.

Boughton, G.J. 1974. Soviet-Cuban relations 1956–1960. *Journal of Inter-American Studies and World Affairs* 16(4): 436–453.

Bravo, O.M. 2001. *Undesirable neighbors: The U.S. Naval Base at Guantánamo*. Havana: Editorial José Martí.

Briggs, R. 2010. *Tropes of enlightenment in the age of Bolívar. Simón Rodríguez and the American essay at revolution*. Nashville: Vanderbildt University Press.

Bronfman, A. 2004. *Measures of equality. Social science, citizenship, and race in Cuba 1902–1940*. Chapel Hill/London: The University of North Carolina Press.

Brown, C.L. 2006. *Moral capital. Foundations of British abolitionism*. Williamsburg/Chapel Hill: Omohundro Institute of Early American History: University of North Carolina Press.

Casanovas Codina, J. 1998. *Bread or bullets! Urban labor and Spanish colonialism in Cuba, 1850–1898*. Pittsburgh: University of Pittsburgh Press.

Castro, F. 1989. *Fidel Castro y la deuda externa*, ed. Rosa Alfonso. La Habana: Editorial Política.

Catania, B. 1994. *Antonio Meucci: L'inventore e il suo tempo, da Firenze a l'Havana*. Rome: Seat.

Catania, B. 2004. La labor precursora de Antonio Meucci sobre el teléfono, desde La Habana hasta Clifton. In *El teléfono en Cuba 1849–1959*, ed. Altshuler José and Díaz Roberto, 9–24. La Habana: Sociedad Cubana de Historia de la Ciencia y la Tecnología: Empresa de Telecomunicaciones de Cuba.

Cantero, J.G., E. Laplante, L.M. García Mora, and A. Santamaría García. 2005. *Los ingenios de la isla de Cuba: colección de vistas de los principales ingenios de azúcar*. Madrid: Edición Doce Calles.

Cirillo, V.J. 2004. *Bullets and Bacilli. The Spanish-American war and military medicine*. New Brunswick: Rutgers University Press.

Corrales, J. 2004. The Gatekeeper State. Limited economic reforms and regime survival in Cuba, 1989–2002. *Latin American Research Review* 39(2): 35–65.

Craven, D. 2002. *Art and revolution in Latin America: 1910–1990*. New Haven: Yale University Press.

Daley, Yvonne. 2000. Cuba's lost art schools: An American Unearths Truly Revolutionary Architecture. *Stanford Magazine*. September/October: http://alumni.stanford.edu/get/page/magazine/article/?article_id=39904

de la Sagra, R. 1831. *Historia economico-politica y estadistica de la Isla de Cuba*. Habana: Impr. de las viudas de Arazoza y Soler.

de la Sagra, R. 1842. *Historia física, política y Natural de la Isla de Cuba*. Paris: Arthus Bertrand.

Díaz Martín, R. 2004. El Primer Servicio Telefónico En Cuba Entre 1888 Y 1915. In *El Teléfono En Cuba 1849–1959*, ed. J. Altshuler and R. Díaz, 45–57. La Habana: Sociedad Cubana de Historia de la Ciencia y la Tecnología: Empresa de Telecomunicaciones de Cuba.

Domínguez, J.I. 1979. *Cuba: Order and revolution*. Cambridge: Belknap.

Edquist, C. 1985. *Capitalism, socialism and technology: A comparative study of Cuba and Jamaica*. London: Zed Books.

Fernández Prieto, L. 2005. *Cuba agrícola: Mito y tradición*. Madrid: Consejo Superior de Investigaciones Científicas.

Ferrer, López A. 1999. *Insurgent Cuba: Race, nation, and revolution, 1868–1898*. Chapel Hill: University of North Carolina Press.
Ferrer, A. 2005. Temor, poder y esclavitud en Cuba en la época de la revolución haitiana. In *Las Antillas en la era de las luces y la revolución*, ed. José A. Piqueras, 67–83. Madrid: Siglo XXI.
Foner, E. 2004. *Give me liberty!: An American history*. New York: W.W. Norton.
Funes Monzote, R. 2004. *El despertar del asociacionismo científico en Cuba (1876–1920)*. Madrid: CSIC.
García Blanco, R. (ed.). 2002. *Cien figuras de la ciencia en Cuba*. Havana: Editorial Científico-Técnica.
García González, A. 2010. *Cuerpo abierto. Ciencia, enseñanza y coleccionismo andaluces en Cuba en el siglo XIX*. Madrid: CSIC/Universidad de Sevilla/Diputación de Sevilla.
George, E. 2005. *The Cuban intervention in Angola, 1965–1991*. New York: Frank Cass.
George, A.L. 2013. *The Cuban missile crisis: The threshold of nuclear war*. New York: Routledge.
Giani, Esther. 2007. *Il ricarro del progetto: Vittorino Garatti e l'Ena dell'Avana. Occasioni di architettura*. Rome: Officina.
González, R.M. 2004. El primer servicio telefónico en Cuba. In *El teléfono en Cuba 1849–1959*, ed. J. Altschuler and R. Díaz Martín, 19–44. Havana: Sociedad Cubana de Historia de la Ciencia y la Tecnología: Empresa de Telecomunicaciones de Cuba.
Gott, R. 2004. *Cuba: A new history*. New Haven: Yale University Press.
Gruzinski, S. 2004. *Les quatre parties du monde: Histoire d'une mondialisation*. Paris: Editions de la Martinière.
Hamilton, N. 1992. The Cuban economy. Dilemmas of socialist construction. In *Cuba. A different America*, ed. W.A. Chaffee and G. Prevost, 36–54. Boston: Rowman/Littlefield.
Hein-Weingarten, K. 2000. *Das Institut für Kosmosforschung der Akademie der Wissenschaften der DDR*. Berlin: Duncker/Humblot.
Hoffmann, B. 2004. *The politics of the Internet in Third World development. Challenges in contrasting regimes with case studies of Costa Rica and Cuba*. New York: Routledge.
Huberman, L., and P.M. Sweezy. 1960. *Cuba: Anatomy of a revolution*. New York: Monthly Review Press.
Ibarra, J. 1998. *Prologue to revolution: Cuba 1898–1958*. Boulder: Rienner.
Jatar-Hausman, Ana Julia. 1998/1999. What Cuba can teach Russia. *Foreign Policy* 113: 87–103.
Kuethe, A.J. 1986. *Cuba 1753–1815. Crown, military, and society*. Knoxville: University of Tennessee Press.
Lamore, J. 1980. 'Criolllismo blanco' et conscience nationaleà Cuba (1820–1868). In *Esprit créole et conscience nationale. Essais sue la formation des sciences nationales en Amérique Latine*, 97–122. Paris: CNRS.
Lloyd-Jones, R., and M.J. Lewis. 1998. *British industrial capitalism since the Industrial Revolution*. London: UCL Press.
López, K. 2004. 'One brings another': The formation of early-twentieth-century Chinese migrant communities in Cuba. In *The Chinese in the Caribbean*, ed. Andrew Wilson. Princeton: Markus Wiener Publishers.
Loomis, John A. 1999. *Revolution of forms: Cuba's forgotten art schools*. New York: Princeton Architectural Press.
Lorini, A., and D. Basosi (eds.). 2009. *Cuba in the world, the world in Cuba: Essays on Cuban history, politics and culture*. Florence: Florence University Press.
Los Restos Del Padre Varela En La Universidad Del La Habana. 1955. Havana: Comisión Técnica para Investigar la Autenticidad de los Restos del P. Félix Varela.
Lowenstein, M.Z. 1985. Energy applications of biomass: [The proceedings of the National Meeting on Biomass R&D for Energy Applications held 1–3 October 1984 at Arlington, Virginia, USA]/ [National Meeting on Biomass R&D For Energy Applications.] London: Elsevier.
McDonogh, G.W. 2009. *Iberian worlds*. New York: Routledge.
McNeill, J.R. 2010. *Mosquito empires: Ecology and war in the greater Caribbean, 1620–1914*. Cambridge: Cambridge University Press.

Melcón Beltrán, J. 1989. La geografía en el sistema de instrucción primaria en España, Cuba, Puerto Rico y Filipinas (1838–1898). In *Ciencia, vida y espacio en Iberoamèrica*, ed. J.L. Peset, 267–292. Madrid: CSIC. III.

Méndez Pérez, L.M., E.J. Roca Oria, et al. 2012. Roberto Soto del Rey, Fundador de la Universidad de Oriente y de su departamento de Física. *Rev. Cub. Fis.* 29(1): 33–36.

Millington, Mark. 2005. Transculturation: Taking stock. In *Transculturation: Cities, spaces and architectures in Latin America*, ed. Felipe Hernández, Mark Millington, and Iain Borden, 204–233. Amsterdam/New York: Rodopi.

Mills, C.W. 1960. *Listen, Yankee: The revolution in Cuba.* New York: Ballantine Books.

Moreno, J.A. 1998. *Cuba: Período especial. Perspectivas.* Havana: Ediciones de Ciencias Sociales.

Nadal, F. 1989. La formación de la "Carta grográfo-topográfica" de Valcourt y los trabajos geográficos de las comisiones de estadística y divisón del territorio de Cuba (1821–1868). In *Ciencia, vida y espacio en Iberoamèrica*, ed. J.L. Peset, 329–356. Madrid: CSIC. III.

Navarro Vega, A. 2013. *Cuba, el socialismo y su exodos.* Bloomington: Palibrio.

Núñez Jover, J., L.F. Montalvo Arriete et al. 2008. *University, innovation and society: Cuban higher education in the national innovation system.* Paper presented at the VI Globelics Conference in Mexico City, September 22–24, 2008.

Ortíz, F. 1940. *Contrapunteo cubano del tabaco y el azúcar: Advertencia de sus contrastes agrarios, económicos, históricos y sociales, su etnografía y su transculturación.* Havana: Jesús Montero.

Ortíz, F. 1995 [1947]. *Cuban counterpoint: Tobacco and sugar.* Durham: Duke University Press.

Otto, M. 2007. The Caribbean. In *The Routledge companion to postcolonial studies*, ed. John McLeod, 95–107. Milton Park/New York: Routledge.

Paponnet-Cantat, C. 2003. The joy of eating. Food and identity in contemporary Cuba. *Caribbean Quarterly* 49(3): 11–29.

Pérez-López, J.F. 1992. Cuba's transition to market-based energy prices. *The Energy Journal* 13(4): 17–40.

Pruna Goodgall, P.M. 1994. National science in a colonial context: The Royal Academy of Sciences of Havana, 1861–1898. *Isis* 85(3): 412–426.

Pruna Goodgall, P.M. 1999. El evolucionismo biologico en Cuba a fines del siglo XIX. In *El darwinismo en España e Iberoamérica*, ed. T.F. Glick, M.A. Puig-Samper, and R. Rosaura. Madrid: Universidad Nacional Autónoma de México Consejo Superior de Investigaciones Científicas.

Pruna Goodgall, P.M. 2001. *Ciencia y científicos en Cuba colonial: La Real Academia de Ciencias de la Habana, 1861–1898.* Havana: Editorial Academia.

Pruna Goodgall, P.M. 2003. *La Real Academia de Ciencias de la Habana, 1861–1898.* Madrid: Consejo Superior de Investigaciones Científicas.

Pruna Goodgall, P.M. 2005. *Historia de la ciencia y la tecnología en Cuba.* La Havana: Editorial Científico Técnica.

Pruna Goodgall, P.M. 2006. *Historia de la ciencia y la tecnología en Cuba.* Havana: Editorial Cientifico Técnica.

Purcell, S.K., and D.J. Rothkopf (eds.). 2000. *Cuba: The contours of change.* Boulder: Rienner Publishers.

Quiroz, A.W. 2006. Martí in Cuban schools. In *The Cuban Republic and José Martí*, ed. Mauricio A. Font, 71–81. Lanham: Lexington Books.

Reid-Henry, S. 2010. *The Cuban cure: Reason and resistance in global science.* Chicago: University of Chicago Press.

Reinert, E.S. 2007. *How rich countries got rich ... and why poor countries stay poor.* New York: Carroll & Graf.

Santamaría García, A., and A. García Álvarez. 2004. *Economía y colonia: la economía cubana y la relación con España (1765–1902).* Madrid: Editorial CSIC.

Sartre, J.-P. 1961. *Sartre on Cuba.* New York: Ballantine Books.

Scott, R.J. 1985. *Slave emancipation Cuba: The transition to free labor 1860–1899.* Princeton: Princeton University Press.

Shih, S., and F. Lionnet. 2011. Introduction. The creolization of theory. In *The creolization of theory*, ed. F. Lionnet and S. Shih. Durham: Duke University Press.
Szulc, T. 1987. *Fidel: A critical portrait,* reprint ed. New York: Harper Collins.
Tablada, C. 2001. Les nouveaux agents économiques dans une societé socialiste (Cuba). In *Cuba quelle transition?* ed. A.A. Tejada, 27–49. Paris: L'Harmattan.
Tomich, D. 2005. Material process and industrial architecture. Innovation on the Cuban sugar frontier, 1818–1857. In *Nature raw materials and political economy. Research in rural sociology and development*, ed. P.S. Ciccantell et al., 287–307. Amsterdam: Elsevier Jai.
Torres Yribar, W. 2011. Ciencia cubana en tiempo de Revolución. In *Salud pública* 37. http://bvs.sld.cu/revistas/spu/vol37_05_11/spu06511.htm
UNICEF. 2004. *Progress for children*. New York: UNICEF Strategic Information Section, Division of Policy and Planning.
Wylie, L. 2010. *Perceptions of Cuba. Canadian and American policies in comparative perspective*. Toronto: University of Toronto Press.
Ying Hsiang, Cheng. 1973. *Idylle sino-cubaine. Brouille sino-sovietique*. Paris: Presses de la Fondation nationale des Science politiques.
Zeuske, M. 2012. *Kuba im 21 Jahrhundert: Revolution und Reform auf der Insel der Extreme*. Berlin: Rotbuch Verlag.
Zeuske, M., and M. Zeuske. 1998. *Kuba 1492–1902: Kolonialgeschichte, Unabhängigkeitskriege und erste Okkupation durch die USA*. Leipzig: Leipziger University Verlag.

Chapter 3
Cuba: A Short Critical Bibliographic Guide

Duccio Basosi

An island with a population of approximately eleven million citizens, Cuba has been the topic of a huge amount of books and articles by scholars, politicians, artists, tourists and—why not?—foreign undercover agents. A random search in a well-known on-line bookshop gives some 118,000 results for the island's name. In brief, to present a selection of basic works on Cuba is a very harsh task that necessarily leads to difficult choices.

Cuba's general history from the pre-Columbian era to (almost) the present is told in two widely celebrated and reliable works: Richard Gott, *Cuba: A New History* (New Haven: Yale University Press, 2004); and Hugh Thomas, *Cuba: The Pursuit of Freedom* (New York: Da Capo, 1998). A good introduction is: Leslie Bethell (ed.), *Cuba. A Short History* (Cambridge: Cambridge University Press 1993). An extremely useful collection of documents and short essays on Cuba's history, politics and culture (with some 100 short contributions by Cuban and foreign "authors," including Christopher Columbus, Fernando Ortiz, Father Félix Varela, Carlos Manuel de Céspedes, José Martí, Alejo Carpentier, Ernesto "Che" Guevara, John F. Kennedy and Fidel Castro) is the recent Aviva Chomsky, Barry Carr, Pamela Maria Smorkaloff (eds.), *The Cuba Reader: History, Culture, Politics* (Durham: Duke University Press, 2003).

The nineteenth century was a crucial period in the making of Cuban modern identity. Two indispensable general works are: Ramiro Guerra, *Manual de historia de Cuba desde su descubrimiento hasta 1868* (Havana: Ciencias Sociales, 1971); and Eduardo Torres-Cuevas, Oscar Loyola Vega, *Historia de Cuba, 1492–1898. Formación y liberación de la Nación* (Havana: Pueblo y Educación, 1998). The volume by Louis A. Pérez Jr., *Cuba Between Empires* (Pittsburgh: Pittsburgh University Press, 1983), is a fundamental work on Cuba's relationship with the two "empires" (Spain and the United States) that competed for it. In a way, part II of the book, is the

D. Basosi (✉)
Department of Comparative Linguistics and Cultural Studies, Ca' Foscari University of Venice, Venice, Italy
e-mail: duccio.basosi@unive.it

A. Baracca et al. (eds.), *The History of Physics in Cuba*, Boston Studies in the Philosophy and History of Science 304, DOI 10.1007/978-94-017-8041-4_3,

author's follow-up work Cuba. *Between Reform and Revolution* (New York/ Oxford: Oxford University Press, 2005) The more recent book by Louis A. Pérez Jr., *On Becoming Cuban: Identity, Nationality, and Culture* (Chapel Hill: North Carolina University Press, 2007), is an investigation in Cuban-US cultural relations from 1850 to the 1959 Cuban revolution. The history of slavery and anti-slavery (and anti-Spanish) rebellions on the island is narrated in detail by Ada Ferrer, *Insurgent Cuba: Race, Nation, and Revolution, 1868–1898* (Chapel Hill: University of North Carolina Press, 1999). Cuba's rise to formal independence from Spain, and its subsequent fall into protectorate status under the US in 1898–1902 is analyzed in Marial Iglesias, *Las metáforas del cambio en la vida cotidiana: Cuba, 1898–1902* (Havana: Ediciones Unión, 2003). Cuba's multifaceted relationship with the United States from the early nineteenth to the early twentieth century, including an in-depth analysis of the island's semi-colonial status under the so-called "Platt Amendment" to the Cuban constitution of 1901, is the subject of the work by Alessandra Lorini, *L'impero della libertà e l'isola strategica. Gli Stati Uniti e Cuba tra Otto e Novecento* (Naples: Liguori, 2007). Last but not least, speaking of nineteenth-century Cuba, the life and works of Cuba's national hero, José Martí, are the subject of hundreds of articles and books. The Havana-based Centro de Estudios Martianos promotes the study of Martí's writings and offers a wide catalog of works about and by Martí, including a digital version on CD-Rom of J. Martí, *Obras Completas* in 27 volumes (Havana: Ciencias Sociales, 1975). The complete catalog can be downloaded from the Centro's website at http://www.ecured.cu/index.php/Centro_de_Estudios_Martianos. An interesting selection of recent essays on Martí's political thought, mainly by US scholars, is in Jeffrey Belnap, Raúl Fernandez (eds.), *José Martí's "Our America": From National to Hemispheric Cultural Studies* (Durham: Duke University Press, 1998).

Compared with the flood of works covering the post-1959 period, there is far less scholarship dealing specifically with Cuba's difficult times in the first six decades of the twentieth century, marked by repeated US military interventions, authoritarian governments (with scant and unlucky openings to constitutional rule) and powerful mob presence. Two important contributions, respectively on the Machado (1925–1933) and Batista (1952–1959) dictatorships are Francisca López Civeira, *Cuba entre Reforma y la Revolución 1925–1935* (Havana: Editorial Félix Varela, 2007); and Samuel Farber, *The Origins of the Cuban Revolution Reconsidered* (Chapel Hill: University of North Carolina Press, 2006).

The Cuban revolution in 1959 changed the shape of the country and projected it onto the world stage. A useful work covering the revolutionary years and the socialist transformation of Cuba's economy, polity and society after the triumphal entry of the *barbudos* into Havana in January 1959 is Geraldine Lievesley, *The Cuban Revolution: Past, Present and Future Perspectives* (New York: Palgrave MacMillan, 2004). Abundant reflections on Cuba's post-revolutionary history can be found in the many biographies of some of the revolution's main actors. Unfortunately, it is simply impossible here to list but a few among the—literally—thousands of available references about the life and complex personality of Cuba's *comandante en jefe* Fidel Castro: the recent Volcker Skierka, *Fidel Castro: A Biography* (Cambridge, UK:

Polity, 2006) has been translated into several languages from the German original edition; less up-to-date is the "classic" biography by Tad Szulc, *Fidel: A Critical Portrait* (New York: Harper Collins, 2000). Castro's autobiographic interview with the former director of *Le Monde Diplomatique*, Ignacio Ramonet, is another valuable reading on Cuba's 50-year leader and is easily available in several languages: Fidel Castro, Ignacio Ramonet, *Fidel Castro: My Life. A Spoken Autobiography* (New York: Scribner, 2009). The life and political thought of the Argentine-born *guerrillero heroico* Ernesto "Che" Guevara, which partially overlap with Cuba's attempts to export revolution in Latin America during the 1960s, have been the subject of another flurry of works. Among the most valuable are: John Lee Anderson, *Che Guevara: A Revolutionary Life* (New York: Grove Press, 1997); and Paco Ignacio Taibo II, *Guevara, Also Known as Che* (New York: St. Martin's Press, 1999), translated into various languages from the Spanish original.

The Cuban revolution greatly affected world politics in the second half of the twentieth century. It is not surprising that Cuba's foreign relations have become the subject of intensive research in recent years. In particular, in October 1962 Cuba came to be the stage of one of the most dramatic episodes of the confrontation between the United States and the Soviet Union. While the "Cuban missiles crisis" is recollected in virtually any book dealing with the political history of the twentieth century, the Cuban viewpoint on those events and their aftermath is recollected in two valuable works based on Cuban oral and written sources: Bruce J. Allyn, James G. Blight, David A. Welch, *Cuba on the Brink: Castro, the Missile Crisis, and the Soviet Collapse* (Lanham, MD: Rowman and Littlefield, 2002); and James G. Blight, *Sad and Luminous Days: Cuba's Struggle with the Superpowers after the Missile Crisis* (Lanham, MD: Rowman and Littlefield, 2007). More generally, Cuba's thorny relationship with its powerful northern neighbor is the subject of the recent 500 page work by Lars Schoultz, *That Infernal Little Cuban Republic. The United States and the Cuban Revolution* (Chapel Hill: North Carolina University Press, 2009), based on extensive research in US archives and Cuban public sources. The indispensable reading on Cuba's involvement in African revolutionary and national liberation struggles, based on Cuban sources, is Piero Gleijeses, *Conflicting Missions: Havana, Washington, and Africa, 1959–1976* (Chapel Hill: University of North Carolina Press, 2002).

The collapse of the Soviet Union in 1991 provoked a dramatic shock in Cuba's economy and society. However, despite bookshops everywhere in the world being filled with tenths of variations on the theme of Cuba's transition out of socialism and into capitalism, such a transition did not actually occur. The adaptation of the island's economy to the new conditions is analyzed, with a prescriptive attitude and some contempt at Cuba's "incomplete transition", in Carmelo Mesa-Lago, Jorge F. Pérez López, *Cuba's Aborted Reform: Socioeconomic Effects, International Comparisons, and Transition Policies* (Gainesville, FL: University Press of Florida, 2005). A more descriptive attitude, focusing on both the difficulties and the resilience of the post-1991 Cuban economy, is instead that taken by Jorge Domínguez, Omar Everleny Pérez Villanueva, and Lorena Barberia (eds.), *The Cuban Economy at the Start of the Twenty-First Century*

(Cambridge, US: Harvard University Press, 2005), which collects the works of economists based both in Cuba and the US. Finally, while virtually any book or newspaper article dealing with Cuba will touch on controversial questions regarding the island's political system, there has been very little scientific work done on the latter's actual working: one of the few notable exceptions, the fruit of years of in-the-field research with Cuba's municipal, provincial and national assemblies in the 1990s, is the work by the political scientist Peter Roman, *People's Power: Cuba's Experience with Representative Government* (Lanham, MD: Rowman and Littlefield, 2003).

Part I
Historical Surveys

Chapter 4
The Teaching of Physics in Cuba from Colonial Times to 1959

José Altshuler and Angelo Baracca

4.1 General Introduction

The history of science has experienced remarkable progress in recent decades, especially thanks to detailed studies referring to the most scientifically advanced countries. This approach has led to fundamental contributions to the understanding of the mechanisms responsible for the advances of science in general, and their interaction with social, economic, cultural and human development. Such contributions pave the way for a complementary subject of general interest: the study of the advancement of science in the so-called Third World countries. As far as these are concerned, generally speaking it can be said that each one presents itself as a separate case, for which complete histories of individual scientific disciplines have been published only rarely.

In this paper, we examine the teaching of physics from its origins up to the 1959 Revolution.[1] While this choice originated in the intrinsic interest of the subject to national historiography and culture, the result may have a wider significance. It shows how dealing with a small country with a relatively recent history, whose scientific development is concentrated essentially in the last two centuries, can be of practical advantage for the historian. To this, it may be added that Cuba represents the very special case of a country that, after a long period of scientific underdevelopment, has been able to create in the last decades a considerably

This paper is based on an earlier Spanish version which was published in 2004 in *Llull: Revista de la Sociedad Española de Historia de las Ciencias y de las Técnicas* (vol. 27, n°60, 557–608).

[1] The subsequent period will be dealt with in the chapter by Baracca, Fajer and Rodríguez.

J. Altshuler (✉)
Cuban Society for the History of Science and Technology, Havana, Cuba
e-mail: jea@infomed.sld.cu

A. Baracca
Department of Physics, University of Florence, Florence, Italy
e-mail: baracca@fi.infn.it

A. Baracca et al. (eds.), *The History of Physics in Cuba*, Boston Studies in the Philosophy and History of Science 304, DOI 10.1007/978-94-017-8041-4_4,

advanced national scientific system where some specialties have reached genuine international levels, in spite of the country's limited natural resources and its involvement in a particularly complex international situation.

The present paper will concentrate on the study of the first one and a half centuries of the teaching of physics in Cuba, starting from the early interest or activities in this. To provide the appropriate context, we will begin by going back in time to the early local signs of a modern scientific sensitivity, scornful of scholastic teaching and fond of experimental approaches, under the belated but stimulating influence of the European Enlightenment. We will go over individual cases during the initial phase of the country's academic physics in some detail, firstly during the initial period of colonial rule, and then from the beginning of the twentieth century to the triumph of the 1959 Revolution. We thus hope to make the fundamental aspects of our subject reasonably consistent and abiding, even though further historical research may eventually throw new light on relevant details, personalities and events.

4.2 General Survey

To appreciate the circumstances and the character of the period of the first cultural and scientific interests in Cuba that are described in the following, it seems important to take into account the specific conditions under which the island was ruled. For almost three centuries, Spanish colonial rule prevented by all possible means the formation of a culture, self-consciousness and an identity rooted in the country. One of these means was the systematic prohibition of the circulation of ideas, and such other restrictions which transformed the island into a large-scale military domain. A young person wanting to pursue higher studies in Latin America had to enroll in one of the three universities existing at that time in the region: the University of Santo Tomás de Aquino in Santo Domingo (1538), the University of San Marcos in Lima (1551), and the Real y Pontificia Universidad de México (1551). It was not until 1728 that the king of Spain Phillip V approved the foundation of the Real y Pontificia Universidad de San Jerónimo in Havana and the first printing shop opened in Havana a little earlier. But the climate for intellectual debate and scientific investigation remained hostile, as can be gathered from the fact that the "Consejo de Indias" banned the printing of the monthly newspapers, the Gazeta and the Mercurio. As late as 1778, a law prohibited Americans and Spanish residents in Cuba from "studying, observing or writing on subjects related to the colonies." Due to those restrictions the scientific and intellectual situation in Cuba showed a significant delay in comparison with other Spanish colonies, as we will see.

The economic and social situation in the colony underwent some remarkable developments in the last decades of the eighteenth century. A part of the white population of the island formed a small bourgeoisie, while a restricted creole middle class enriched itself by participating in contraband, the slave trade and the sugar industry. In this environment, the early ideas of technical modernization arose, driven also by the wind of new ideas arriving from abroad. The stimulating and subversive ideas of the century of enlightenment began to infiltrate Cuba, also through the young people who returned from their studies abroad, in particular, from the United States.

This wind of renewal, however, entered with considerable delay the sphere of the university which in Havana was under the rule of the Dominican order; the developments were faster in other Spanish colonies where the academic milieu was under the control of the much more progressive order of Jesuits.

The real adoption of the spirit of enlightenment and the introduction of modern teaching of physics was due to Father Félix Varela in the Seminar of San Carlos in Havana, around 1814 (Sect. 4.5) while at the same time the University of Havana continued to cultivate an obsolete scholastic culture.

Specific courses on mathematics, physics and chemistry were formerly offered only in the modern secular courses by the San Carlos Seminar, and later introduced at the University of Havana, when it was secularized in 1842. Several reforms were incorporated in the subsequent decades, following major institutional changes in Spain. Despite progress, throughout the nineteenth century and in the republican period of the first two decades of the twentieth century (national sovereignty being grossly limited by the government of the United States), the scientific activity in Cuba was mostly limited to the Academy of Medical, Physical and Natural Sciences (established in 1861) and the University of Havana. Higher-level science courses were basically aimed at the training of high-school teachers in mathematics, physics, chemistry and biology, and remained at an essentially qualitative and conceptual level. The use of mathematics as a tool was generally marginal and research activity was largely absent.

In spite of the great political changes following the war for independence (1895–1898), the domain of scientific teaching was not affected. The situation began to change only from the 1920s. Up to the 1940s, for the first time a modernization and extension of the physics courses took place at the University of Havana with the creation of courses worthy of that name, with rigorous scientific criteria and substantial use of mathematics and of laboratory work serving as an effective means of teaching. These improvements depended moreover on a very limited number of essentially self-taught professors with outstanding scientific sensitivity (both in physics and in mathematics) and were driven by the great scientific progress in the industrialized countries, and by goals of reaching a higher scientific rigor. They took place in the context of the growing aspirations of the new middle-class milieus following the modernization of the country.

A very important aspect is that this wind of renovation was reflected in the improvement of the teaching of scientific subjects in secondary schools: several high-school physics textbooks in the early 1940s and the 1950s show an excellent scientific level, while their pedagogical approach was as good as that of the standard textbooks used in Europe and the United States at that time.

In spite of objective limitations, one must acknowledge that when the 1959 Revolution broke out there was among the Cuban population a small stratum of individuals with a good basic training in physics, who could provisionally make up for the due lack of trained university personnel, since the above mentioned professors either retired, or left the country, as will be seen in Chap. 5 below.

4.3 Experimental Physics vs. Scholasticism

The innovatory movement of the scientific Renaissance that was intensifying in sixteenth-century Europe could not penetrate the closed environments of the first universities in Latin America since Spain rejected the germination of this progress. Therefore, peripatetic scholasticism shaped the content of the universities in Santo Domingo, in Lima and in Mexico.[2]

The Enlightenment in Europe included the generalized spreading of the ideas of scientific physics and the definitive incorporation of these to the Western cultural heritage. Throughout the eighteenth century the prestige of the Newtonian viewpoint was consolidated after it was successfully applied to the solution of a large number of physical problems, while experiment in itself became the subject of real cult, heavy with philosophical implications. With obvious exaggeration, Voltaire wrote: "We learn more from the single experiments of the Abbé Nollet than from all the philosophical works of antiquity."

In particular, accepting the Copernican astronomical model or not, and more generally, the scholastic approach, which was purely discursive and hostile to the practice of observation and experimentation, became time after time a motive for ideological quarrel between the most enlightened elements and the most conservative ones within the Catholic Church.

It was not until the second half of the eighteenth century that experimental science began to awaken intellectual interest and generated a movement in Spain as well as in America (as observed by Humboldt on his travels). The universities of Spanish America became a preferential stage for the struggle between the old school and the renovating ideas. But this intellectual panorama of the second half of the eighteenth century did not reach the University of Havana, whose statutes were inspired by the original Dominican university. Therefore, in absence of the scientific spirit, the direct study of nature, of experimentation, the disengagement from dogmas, made a late appearance. Apart from medicine, no scientific matters played a role in the university. What was taught was the physics of Aristotle.

Thus, for example, while at the Xaverian University of Bogotá, governed by the Jesuits, the Copernican model was already taught in 1755, following a manuscript entitled *Physica specialis et curiosa*.[3] Not until July 1797 was the doctrine of Copernicus referred to and publicly defended at the *Real y Pontificia Universidad de San Jerónimo* (Royal and Pontifical Saint Jerome University) of Havana, which was under the aegis of the Dominican friars since its founding in 1728.[1] It was the student Manuel Calves who did it while he was engaged in the defense of two Bachelor of Arts theses which favored that doctrine. It is quite telling that the following year he denounced head-on "the universal grudge [toward] the experimental

[2] See, for example, Vitier (1938), 21–33 and Gruzinski (2004).

[3] Arboleda and Soto Arango (1991), 13. Betancourt–Serna specifies how the Jesuit author of *Physica specialis et curiosa* Francisco Javier Trías obtained his information about actual physical debates in Europe. It seems that Trías read books written by Spanish Jesuit professors of mathematics at the Imperial College in Madrid, see Betancourt–Serna (2007), 733–734.

doctrine" held by the Dean of the Medical School of the same institution (López Sánchez 1973, 8–10).

The rector of the university, the Dominican priest Juan Chacón, asked the "enlightened" Spanish King Charles III to create the chair of "Philosophia experimentalis" for setting up an Academy similar to that of Spain. But in 1767 (the year of the expulsion of the Jesuits) the King denied the request. Instead the King ordered to dedicate one of the two previously existing mathematics chairs to the teaching of legal matters. Again, a quarter of a century later, another modernizing rector, priest José Ignacio Calderón, did not succeed in introducing a study plan that included "experimental philosophy, Geometry and Calculus, with everything related to this transcendental part of Mathematics" (Simpson 1984, 43–45, 69–70). In spite of those attempts to renew the study of science, the University of Havana was consequently condemned to remain for a very long time a rigid bastion of the most obsolete scholasticism, as was the case with other universities in Spanish America also under the aegis of the Dominican order.

That obstacle considerably delayed the spreading of the physical sciences in Cuba with respect to progress made in their teaching in some of the more enlightened institutions in Spanish America. Universities of other capitals of the colonial Vice-Royalties were less constrained by royal rule as was Havana, where the Royal and Pontifical Saint Jerome University was the only institution of this kind. It was possible, for instance that in one of them—the St Nicholas of Bari University in Bogotá, governed by the Augustinian order—a general reform was instituted in 1773, that demanded the

> … banishment or total elimination from schools of all that peripatetic philosophy and theology full of useless and apprehensive subjects, which are good for nothing save for wasting time, and teaching instead a useful and worthwhile philosophy, capable of preparing the students for fruitful study at the other faculties and finding the truth, which is the end to which all aim at. (Arboleda and Soto Arango 1991, 30–31)

Another example that illustrates the point in a very concrete way is the publication in Mexico, in 1774, of a textbook "for teaching modern physics at the chair of philosophy of the St Francis of Sales College," written by Juan Benito Díaz de Gamarra (Ramos Lara 1994, 35). A similar text came to light in Cuba only four decades later, as will be seen below.

The introduction of modern science to Cuba had the general support of the most influential native sugar barons. One of them, Nicolás Calvo y O'Farrill, the owner of one of the country's largest sugar factories, was interested in experimental physics, and more generally in science. This could have responded to the need for development of the island's productive forces. He excelled with the introduction of "various mechanical improvements he had invented" in his model sugar factory, and for having actively championed the creation by the end of the eighteenth century of a school for chemistry and botany. This he did with a view to increase the profitability of the country's sugar agroindustry.[4] At that time, he had built up a collection

[4] See Ely (2001), 95 and Moreno Fraginals (1978), 128; on the Cuban sugar industry, see also Guerra y Sánchez (1964).

of physics instruments and equipment in the European style, whose contents are mentioned by Father Caballero in his March 1801 panegyric, dedicated to Calvo, who he says was

> … surrounded by machines, books, plans, and instruments. The harpsichord, the *camera obscura*, the electric machine, the pneumatic machine, the lodestone, the celestial and terrestrial spheres, the barometer, the thermometer, the hydrometer, a whole chemical apparatus, a collection of botanical and pictorial wonders, Newton's prism, the telescope, a solar microscope, and who knows what other thousand contrivances related to the exact sciences, were the jewels that embellished his tireless devotion. (Caballero 1801, 301)

Nothing was as fitting as giving priority to the development of chemistry and botany over physics as a support for the plantation economy of Cuba that existed at the time.[5] On the other hand, mining development—especially that of silver—, which was the primary economic interest of Spain in a country like Mexico in the second half of the eighteenth century, urgently required greater scientific support for the training of the corresponding branch specialists. This urgency is reflected in a document dated 1774, which was addressed to the King "in the name of the mining industry of this New Spain." The document contains a proposal made to create a school where, in addition to chemistry, mineralogy and metallurgy, an experimental physics course would be taught that included "mechanical machinery, hydrostatics and hydraulics, aerometry and pyrotechnics." The high-level institution that was demanded—The Royal Mining Seminar—was inaugurated on January 2, 1792, with a European teaching staff (Ramos Lara 1994, 57–64, 70).

4.4 The *Papel Periódico* and the Patriotic Society for Modern Science

On January 9, 1793 the *Real Sociedad Económica de Amigos del País* (Royal Economic Society of the Country's Friends or Royal Patriotic Society) was inaugurated at the Governor's Palace of the island of Cuba.[6] Its creation was the result of the

> … insistence of some twenty rich Cubans of distinguished families, enlightened through their trips abroad and their contact with foreign civilizations; who backing the initiatives of an illustrious governor—don Luis de las Casas—, demanded the foundation chart of the Economic Society for identical purposes as those of the similar institutions that had been created in the mother country years before.[7]

[5] This does not mean that knowledge of some of the modern industrial elements that began to be introduced in the Cuban sugar industry, particularly steam engines, were totally insignificant: see Guerra y Sánchez (1964), 53–54.

[6] Societies of this kind were part of a movement within the enlightenment to renew the economical organization of a country. The Spanish societies took the example of French precursors.

[7] Raimundo Cabrera, president of the *Sociedad Económica de Amigos del País* (Economical Society of the Country's Friends) from 1910 to 1923 in 1952, quoted by Fernando Ortiz (2001, 122).

The enlightened forces that gathered around the Economic Society had begun to show themselves publicly in the *Papel Periódico de la Havana* newspaper, founded in 1790, where it was not long before the first foreign news appeared about scientific subjects, especially in the realm of experimental physics (Díaz Molina 1991, 64–71). The spread of information of this kind no doubt had the effect of making the scientific backwardness of the island more and more obvious to the more enlightened strata of the population, because years went by and nothing was done to remedy the situation from an institutional viewpoint. Accordingly, it was no surprise that a moment arrived in which the situation seemed untenable to some, to the point that in the March and May 1798 issues of the *Papel Periódico* articles were published that launched an unusually harsh attack against the fossilized scholastics still prevailing at the university, which blocked the introduction of modern science teaching. The author was Father José Agustín Caballero who, under a pen name, wrote the following:

> Scholasticism, terrifying as it is, has died forever in Europe [...] The black shadows that darkened delicate minds disappeared with it. Its place was occupied by the torch of truth: experiment. These were repeated. Their effects were collated. Experience was formed and from the dirty diapers of syllogistic arguments, rational philosophy, experimental physics, methodic chemistry and all the other natural sciences, emerged luminous and brilliant, shaking out the dust of chimerical entities. [...] Thus it is in all of wise Europe, and it should be so all around the world. But, is it so in Havana? [...] I do not want to affirmatively say no, because there are some people that by getting over immense worries have succeeded in touching by themselves the light of true philosophy. [...] I can see progress (and anyone will see them) in music, painting, writing, poetry. But where are the lights that should have be introduced after those ill-praised Aristotelian darknesses were eradicated? What is our chemistry? What is our experimental physics? What is our mathematics? What is...? Perhaps I want too much. But will anyone want to deprive me of my yearning for the desire that my wish be known? (Caballero 1801, 73, 164–165)

However justified Caballero's lament was, almost two decades would pass before modern physics—i.e., essentially experimental, as it was then understood—began to be systematically taught in the country, though there were some attempts at advancing its introduction. Some individual initiatives illustrate how new physical knowledge began to spread at the island. In two advertisements published in the August 1804 issue of the *Papel Periódico*, a Spanish-born physician living in Havana, Eugenio de la Plaza, offered to teach mathematics and physics

> ... with all the paraphernalia of machines and instruments needed for the instruction of the students to whom [he would explain] the subjects in Spanish language according to the method of the Madrid Imperial College. (Plaza 1804)

We do not know who Eugenio de la Plaza's students were—if he ever had any—nor whether he really got hold of the material media required to support the corresponding experimental demonstrations in the announced courses. In any case, we have already seen that some of the main physics apparatus of the time were known by certain members of the country's elite. It should be remembered, by the way, that Volta's battery—the device that opened the door to electrodynamics in 1800—did not appear in the collection of Nicolás Calvo as described by father Caballero, which is quite natural, for it was a very recent invention at the time. We do not know when the first one arrived in

Cuba, but it is quite probable that it came as part of the luggage of a certain Nicolás Niderburg, a self-styled physician who proclaimed that he was a disciple of Galvani and Volta, since in 1807 he declared that he had demonstrated in Havana applications of the "galvanic fluid" to people suffering from various illnesses.[8]

Whatever the reality might have been, it should be noted that while the knowledge of the advances of modern physics was spread in the island of Cuba, thanks perhaps to some public show, the news and comments supplied by the *Papel Periódico* or at most by the extremely limited gatherings of some enlightened member of the upper classes,[9] the teaching of the matter had already been institutionalized quite seriously in some regions of Spanish America of more economic importance for the mother country. This was perceived by Alexander von Humboldt when he arrived in New Spain (México) in 1803, where he found that its Mining School had a chemical laboratory, a geological collection, "and a physics department which included not only precious [European] instruments, but also models locally made with the greatest accuracy, and of the best woods of the country." Humboldt, who had already visited Cuba in 1801 and saw nothing of the like worth mentioning, could perceive something radically different in New Spain: "No city in the New Continent, including the United States—he wrote years later—presents scientific establishments as big and solid as the capital of México" (Ramos Lara 1994, 59–64, 70).

Quite the opposite, at the beginning of the nineteenth century the outlook for science in Cuba could not be less devastating, with the possible exception of the scientific work of the Havana born chemist José Estévez y Cantal, who had been a student of the outstanding French chemist Joseph Louis Proust in Madrid (Leroy y Gálvez 1954b). Because of its obvious interest for economics and medical practice, after some time chemistry was to receive the wholehearted support of the Patriotic Society, in which most of the *azucareros* were organized in order to increase production rates (Sáenz and Capote 1989, 23–24). This kind of support was much less direct and clear in the case of physics. The introduction of this science in Cuba took place mainly for ideological reasons, as a part of the struggle against scholastics which was waged by the island's learned, led by the archbishop of Havana, Juan José Díaz de Espada y Fernández de Landa, who had taken up his diocesan post at the beginning of 1802 (see in the following).

[8] This information comes from the summary of a pamphlet by Niderburg (1807) which appears in Catalog #36 of *Hesperia*, an antique bookshop in Saragossa. It is described as "a very curious advertising pamphlet on the Cuban stay of a showy itinerant physician, self–styled pupil of Galvani and Volta, who installs in Havana his "machines" for the application of the galvanic fluid, and also sophisticated ones invented by himself: "fumigatory (*azote del gálico*)" and "inhalatory" (for the "'inhalation of oxygen air, fixed air, hydrogen gas and nitrous gas," ideal for the "illnesses to which the female sex is subject to, applied to the corresponding part of the body)."

In the minutes of the Ordinary Meeting of the *Real Sociedad Económica de Amigos del País* dated February 20, 1807, it is stated that Niderburg submitted to the institution a dedicated "report on Galvanism," no doubt with the purpose that the Society might endorse and print it, but it was understood that "the *Sociedad* had nothing to gain from accepting such a dedication [and] it decided to return it to its author."

[9] See also Zeuske and Zeuske (1998), 171–188.

In several viceroyalties, non-university teaching centers had played an important role since the sixteenth century. These colleges were under the control of religious orders and aimed to form select social groups. In Cuba, for example, the *Real y Conciliar Colegio Seminario de San Carlos y San Ambrosio* was established in Havana in 1773 as an institution to influence Cuban society. It was not primarily a religious seminar dedicated to ecclesiastic issues, but also covered other subjects, such as physics, chemistry and even offered a short-lived course on the constitutional law. No other teaching center in Cuba succeeded in encouraging the progressive minorities as did the San Carlos college: its influence was particularly strong in the period in which the colonial society developed its cultural identity and underlined its *cubanidad* ("Cubanity").

4.5 The First Regular Courses in Physics in the Style of Félix Varela (1814–1841)[10]

In 1800, Juan José Díaz de Espada y Fernández de Landa[11] (1756–1832) was appointed Bishop of Havana and arrived in the city in February 1802. A peculiar priest endowed with a deep intelligence and uncommon energy, Bishop Espada undertook a thorough overhaul of the church and of the social institutions. His activities ranged from fields such as culture, education, health, and moralization of the diocese, to public works. He was a member and later president of the *Sociedad Patriótica*. His writings express his views against slavery and were in favor of the development of agriculture in Cuba beyond the sugar economy. He also joined Tomàs Romay in his campaign for vaccination against smallpox. Espada made use of the control he had over the *Real y Conciliar Colegio Seminario de San Carlos y San Ambrosio* (abbreviated to the San Carlos Seminar in the following) to disseminate from their new ideas that had no place at the University of Havana and thus provided an institutional opportunity to support the most gifted individuals.[12] In order to appreciate his attitude, it can be remarked that frequently he burst into the seminar in order to personally ascertain how the pupils behaved.

In the seminar, Bishop Espada decided to "gather together in a single direction the studies of philosophy and prompt a substantial change in their contents." In April 1811, to take care of the chair of philosophy he chose Félix Varela y Morales, a 23-year old Cuban born deacon whom, by the end of the year, he also made a presbyter exempted from the normal age requirement (Torres-Cuevas 1995, 124–125). Since the fourth volume of Varela's work *Instituciones de filosofía ecléctica para uso de la juventud*, published in 1814, probably embodies a course of

[10] For additional details to those given in this section, see Díaz Molina (1991).

[11] Extended books on Espada are for instance those by Garcia Pons (1951), Figueroa and Miranda (1975) and Torres-Cuevas (1995).

[12] For generally accepted viewpoints and ways of teaching, See, for instance, Garcia Pons (1951), chapter XVI, 136–153, and passim.

the elements of mathematics and physics that he taught his students at the time, it is quite reasonable to consider it to be the first physics textbook written in Cuba.[13]

We might appreciate once again the direct interest of the Cuban sugar industry owners in the development of scientific disciplines, which could promote technical improvements:

> Mill owners were hard-working members of the Economic Society and the Development Board; mill owners persuaded Father Varela and Saco to initiate the teaching of physics and chemistry in Cuba in order to meet the needs of the sugar industry; mill owners brought over the chemist, Casaseca, and founded our first botanical garden; mill owners imported steam-driven machinery to run sugar mill rollers and … also organized our first public service railway company. (Guerra y Sánchez 1964, 56)

Here, the role of Alejandro Ramírez must underlined: He established an important link between different Spanish domains in the Americas, fostering the ideal of enlightened institutions. In his first administrative position in Guatemala, he became a member of the Sociedad Economica de Amigos del País and the Consulado de Comercio, both of which promoted liberalization in Guatemala. The aim of these groups of enlightened men was to reform agriculture, industry and mining. The reformers focused on the introduction of new plants and techniques to agriculture. In 1816, Alejandro Ramírez was appointed supervisor of finances (*Intendente de Hacienda*) in Havana, after having spent some years reforming the economy in Puerto Rico. In Havana, he immediately became a member of the Sociedad Económica and was put in charge of the education section of this organization. He then proposed the creation of chairs for botany and chemistry at the University of Havana, in addition to founding a botanical garden (Puig–Samper and Maldonado 2005).

4.5.1 Félix Varela's *Lecciones de Filosofía*

There is, however, good reason to think that it was not until 1816 when Varela's physics classes began to be supported by a solid experimental base, as can be observed in volumes 2 and 4 of his *Lecciones de filosofía* (philosophy lessons), published in 1819 and 1820, respectively. For there is every indication that only in 1816 and afterward could the philosophy chair of the Seminar acquire quite a good collection of instruments and devices appropriate for work on some important branches of physics.[14] As explained by Varela himself in 1822:

---

[13] Meaningfully enough, the first two volumes of this work were written in Latin and were entitled *Institutiones philosophiae eclecticae ad usum studiosae iuventutis.*

[14] That 1816 was the year when the physics laboratory of the San Carlos Seminar was created can be inferred from a letter by José de la Luz y Caballero, dated May 4, 1832, where, when referring to what was then kept from that laboratory, he states that "after sixteen years of continued use, and in the hands of the students, a part of the apparatus that were made to come from London at that time at the request of Mr. Don Félix Varela, still subsist serviceable and in very good shape after sixteen years of continuous use, and in the hands of students."

> [the collection of instruments] we owe to the generosity of the bishop of that diocese [Espada], but it is only stocked up in the branches of pneumatics, electricity, galvanism and astronomy, though in this last one much of the absolutely necessary is lacking. All of the instruments are the most modern ones and from the best English factories, since for the greater part they are from the well-known Adams firm.[15] Those for electricity and galvanism are identical to the ones depicted in the plates of my lessons of philosophy [...] (González del Valle 1942, 201)[16]

According to the authoritative testimony given in 1832 by José de la Luz y Caballero, who had been a student of Varela and later succeeded him in the chair, his former teacher

> was the first on this soil who put the philosophy class of the Colegio de San Carlos, spiritually and corporally as well, allow me the expression, at the level still preserved, which the advances of the experimental sciences required already at the time. (Luz y Caballero 1832, 97)

Manuel Gran, author of a detailed critical study of the contents of volumes 3 and 4 of Varela's *Philosophy Lessons*, points out that in order to assess the merits of his work,

> it is important to realize that when he [...] wrote his physics he had been a teacher for only a few years, and was teaching many other things at the same time [...] while organizing the laboratory work, paternally training his students and, at the same time, studying and writing. From what remains from him he doesn't seem to have gone into the notions of higher mathematics; [to which it should be added that] he lived at a time in which it was very difficult to be updated on the latest developments of the sciences. (Gran 1945)

Still, according to his former students, Varela "received books and periodicals published in Europe on the physical sciences, and [...] kept himself and his pupils up to date with the latest discoveries and novelties" (Rodríguez 1878, 37). While on the subject, it should be pointed out that throughout his textbook, reference is made to some 50 works (by Brisson, Biot, Chaveneau, Nollet, Haüy, Davy, Almeyda, Gamarra...), and that the third volume of each of the successive editions (printed in 1824, 1828, 1832, and 1841) of the *Lecciones*, where the physics topics are concentrated, was systematically updated.

Judging by the contents of the *Lecciones*, the physics course which was taught at the San Carlos Seminar corresponds to what we would call today "high school

---

According to the same source, a refractor telescope and three artifacts for the demonstration of astronomical phenomena, acquired in London by Bishop Espada, were incorporated to the same collection the following year. On the other hand, it cannot be ruled out that previous to all those acquisitions from abroad, Varela might have used some locally made instruments—perhaps with his own hands—, for his experimental demonstrations, as one is led to imagine from the explicit mention by Luz of the existence in the instrument collection referred to, of "two concave silver plated mirrors [...] made here in Havana to demonstrate the law of the reflection of the caloric" Luz y Caballero (1832), 96–97, 118, 131.

[15] For the catalogue of Adams, see Damerow und Lefèvre (1985).

[16] These lines belong to the "Brief account of the state of the studies in Havana," that Varela presented to the *Dirección General de Estudios* in Madrid, on May 14, 1822, while he was in the Spanish capital as a Cuban *diputado a Cortes* (member of Parliament).

level," explained with plenty of experimental work and demonstrations. Varela's textbook is clear and concise, peppered with personal opinions and observations, which lend it a very *sui generis* hallmark. The following paragraph, taken from the original volume 4, can be taken to illustrate this:

> Last year [1819] during the heaviest downpours, at seven o'clock in the morning, and in the room used for the classes of philosophy in this [San Carlos] seminar, which is quite humid because of its location, it happened repeatedly that when a Leyden jar was hung on the conductor of an electrical machine so that it would be perfectly isolated, in spite of that, after a few turns of the machine it was charged to the point that it gave a strong detonation with a spark when the exciter was applied. This experiment did not come out as well when the atmosphere was clearer, but I must confess that I have never been able to absolutely avoid charging an isolated bottle, and I conclude that our atmosphere is never capable of providing such an insulation as obtained in cold places, for which reason electrical machines in this country do not give the same effects as in Europe. (Varela 1820, 307)

Varela continued teaching physics at the San Carlos Seminar (and teaching lessons on Constitution, since 1820) until, after being elected as a Cuban representative to the Cortes (Spanish Chamber of Deputies), he left for Spain in April 1821. The teaching of philosophy was entrusted to his former student, José Antonio Saco, who would later on shine with his own light in public and intellectual life but whose command of the subject does not reach the level of his teacher. Saco published in 1823 a brief treatise entitled *Esplicación de algunos tratados de física* (Explanation of some physics treatises), where he deals with the subject "not breaking the barriers that isolate physics from other sciences, chemistry in particular," but at the same time clearly stating:

> I do not write for those who have a vast knowledge of these subjects, I write only for those who wanting to acquire some ideas and in the absence of authors who may teach them, come to me to find a summary of what they have said. (Saco 1823)

Saco's text is basically qualitative and descriptive. It explains, in a very elementary form, some (mainly chemical) characteristics of gases, the properties of water, and some minimal notions of cosmography, chronology, geography, meteorology, and geometry. In a volume he published at a later date, Saco reprinted three reviews that he considered "samples that [would] tell the Cuban posterity what was the state of the teaching of the physical sciences in Havana in 1823 and 1824" (Saco 1824), this last year being that when he resigned his professorship at the San Carlos Seminar to travel abroad to meet his former teacher in the United States. Varela had been living there since the end of 1823, having established himself in that country to escape the death sentence imposed by king Ferdinand VII on the members of the Cortes who, like Varela, had opposed the return of absolutism to Spain. To replace Saco, Bishop Espada appointed José de la Luz y Caballero to take care of the teaching of philosophy and physics. Luz y Caballero, who was another brilliant disciple of Varela, remained active in that post until 1826, when he resigned for health reasons.

In this phase of Cuban history, therefore, the peculiar figure of Bishop Espada played a fundamental role in the introduction of an enlightened attitude and the first rigorous scientific conceptions, by compensating to a certain extent for the inactivity of monarchical institutions, in spite of further attempts by himself, as we will see in the next sub-paragraph. Two gifted individuals, namely Varela and Saco, benefitted from his policies at the San Carlos Seminar to help lay the foundation of modern physics in Cuba.

4.5.2 *The Backwardness of the University: Arango's Reform Proposal*

In the first decades of the nineteenth century, the *Colegio Seminario de San Carlos y San Ambrosio* enjoyed a solid reputation as the leading teaching institution in the country, in contrast to the situation then prevailing at the University of Havana:

> Public opinion has duly recognized the teaching quality of the Colegio Seminario—wrote its director, Justo Vélez, in 1826—, for while the university is nearly deserted (except in the medicine classes [the only ones taught at the university that the Colegio did not duplicate]), the Colegio has an average of five hundred external [non-religious] students attending the chairs of grammar, philosophy, mathematics, and jurisprudence. (Simpson 1976)

Besides the backwardness of the teaching system at the University of Havana, the entrenchment of corrupt practices had considerably contributed to erode the prestige of the institution in the public eye. The mere fact that the rector and the secretary of the university received an additional income for each degree awarded, contributed to promote corruption at the institution, to the point that it ended up awarding "an amazing number of major degrees [...] in the most informal, disgusting and indecent way," as literally expressed by a distinguished university professor in a letter to Francisco de Arango y Parreño (Simpson 1976).

Arango was a top leader of the sugar mill owners in Cuba. In 1825, he was entrusted by the colonial authorities with the reform of higher education. In October 1826, he met with the staff of the university with the goal to obtain from them opinions and suggestions for reform. This was in the spirit of the reform intended for the Spanish universities, which, according to its original aims, should guarantee "solid monarchical and Christian education and instruction" with the purpose of "forming new men and new customs, and close forever the abyss of all revolutions," though "without ignoring [...] the real progresses of the useful sciences" (Simpson 1984, 117).

In the end, the dialogue with the university led nowhere. Between 1826 and 1828, Arango and Bishop Espada also exchanged views on this subject which were much closer to each other. Espada's viewpoint, and that of Varela as well, was that in order to produce a substantial reform of higher education in Cuba, it was imperative that the University of Havana get rid of the tutelage by the Dominicans and be reconstructed in the style of the San Carlos and San Ambrosio Seminar College. This seems to be the basic idea behind the first article of the first title of the final version of the plan elaborated by Arango: "The San Jerónimo [Havana] University will be eliminated and in its place another one will be established." It was further specified that, among other subjects, "experimental physics, chemistry, elements of natural history, principles of mathematics, of nautics, of surveying, and of geometry applied to the arts" would be taught. But the proposal never prospered, since it was shelved in Spain, where it probably arrived in September 1828 (Simpson 1976). The colonial authorities confirmed once more their enduring unenlightened attitude.

4.5.3 Luz y Caballero and the Gabinete de Física del San Carlos

Luz y Caballero, meanwhile, who in 1826 had resigned his post at the San Carlos Seminar, had recovered considerably from his ailments to the extent that on March 1828, accompanied by José Antonio Saco, he set sail for the United States, where both would establish a close relation with their former teacher, Félix Varela. One year later, Luz extended his trip to Europe, where he met quite a few scientific and cultural personalities, among them the eminent physicists and chemists physicists and chemists Joseph Louis Gay Lussac, Louis Jacques Thénard and Jean Baptiste André Dumas, and the famous geographer and naturalist Alexander von Humboldt, who had visited Cuba in the early 1800s. Humboldt offered to establish in Havana "a regular service of hourly magnetic observations," a project that Luz would promote after his return to Cuba, but without success because, once again, of lack of official support (Chávez Rodríguez 1992).

By the end of 1830, while visiting Venice, Luz was asked by Justo Vélez, then director of the San Carlos Seminar College, to buy equipment and enrich the institution's physics, chemistry and astronomy laboratory and lecture demonstration hardware, which he accomplished with outstanding rationality and efficiency. In 1832 he complemented it with a detailed information on the items bought in an extensive letter addressed to Vélez, who found it "so full with exact ideas and enlightening data" that he sent it for publication in the prestigious journal *Revista Bimestre Cubana*, the organ of the Patriotic Society (Vélez 1832). The document listed the instruments and apparatus the College had at its disposal in 1832 for the teaching of physics, chemistry and astronomy, among them those inherited from Varela's time. Luz differentiated "research instruments" from "instruments for pure demonstration or teaching," and remarked that occasionally he did not buy the best offered in the European market, for he thought that "not always the *best* was *the best for the case*." And he added:

> According to the principle of economy, that I have adhered to whenever possible without detriment to teaching, I have endeavored to modify various apparatus, asking the manufacturers to make them not according to their description in the treatises, but with some alterations that tend to make them simpler and even usable for assembling other ones similar to them, and also endeavored every now and then to make the demonstrations easier. Plenty of examples of the first are offered by the instruments of mechanics, hydrostatics, etc., which I omit to avoid excessive detail, and there will suffice to illustrate the second the addition of an opaque glass to Biot's apparatus for light polarization, so that each student doesn't have to look for himself one at a time, as is the case with the standard instrument, but all of them can observe at the same time, as in a *camera obscura*, the phenomena of double refraction, and all the rest [...] In this way, once accustomed to save money and apparatus, the student is taught to find a way to do much with meager means [...] (Luz y Caballero 1832)

As stated by Justo Vélez in May 1832, the San Carlos and San Ambrosio Seminar College owned by then "a collection [of instruments and equipment] so numerous and brilliant [acquired] at so little expense" that it possessed "everything required for the teaching of natural sciences" thanks to Luz y Caballero's enlightened collaboration (Vélez 1832). At that time, physics was taught at the College by Francisco Ruiz, who benefited from Varela's personal advice by mail (Torres-Cuevas 1995, 352, 359).

Still, Varela's basic approach, already adopted by the "enlightened youth" or "liberal youth" led by Luz y Caballero—which, of course, transcended the teaching of physics—, met with strong opposition from the supporters of the status quo. Justo Vélez himself, though for different reasons, tried to discard Varela's textbook and replace it with another one that followed the so-called "Scottish school"[17] (Torres-Cuevas 1995, 32), against the opinion of Luz, which by the end of 1833 he expressed as follows:

> Physics can be advantageously taught from volumes 2 and 3 of Mr. Varela's philosophy lessons, by adding to it a treatise on physical astronomy [...] There are several data that recommend the work by Mr. Varela for teaching. It is brief, it is at the level of the recent discoveries, written under an excellent plan; and as far as style goes, suffice it to say that [...] no writer has given among us a better demonstration of what a truly didactic language should be. (Díaz Molina 1991)

The very fact that the fourth edition of Varela's *Lecciones de filosofía* appeared in 1832, and the fifth and last one 9 years later, is a clear indication that this textbook continued to be widely used in the country for a long time.

The introduction of modern concepts and method in the teaching of physics at the San Carlos Seminar therefore played a fundamental role in spreading the ideas of the enlightenment and thereby circumventing the persisting resistance of the Dominicans, which prevented such ideas from entering the *Real y Pontificia Universidad.*

4.5.4 In Search of Alternatives for the University Crisis

With the relegation in Spain of the educational reform proposed by Arango y Parreño and the passing of bishop Espada in August 1832, the attempts at a higher education reform in Cuba came to a halt. On the other hand, the need for the island to train not only future physicians, lawyers, teachers and churchmen, but also technical personnel in various branches (i.e., creating some form of what we would now call a public "polytechnic institute"), became more and more obvious.

Already by the end of 1826, the Navy commander Angel Laborde had made the proposal to move the Nautical School that had been in operation since 1812 in the town of Regla, on the opposite shore of the bay of Havana, so as to use that School

[17] Traditionally, the Scottish system has emphasized breadth across a range of subjects, while the English, Welsh and Northern Irish systems have emphasized greater depth of education over a smaller range of subjects at secondary school level.

as the foundation of an institution where "useful" matters would be taught such as pilotage, mechanics, civil architecture and languages. By the end of 1832, the island's Agriculture and Trade Royal Board decided to support the idea and ordered the drafting of a report on the matter to a committee where José de la Luz y Caballero would play the leading role (Simpson 1976).

With inspiration drawn from the Asturian Institute created by Gaspar Melchor de Jovellanos in Spain, and also from the information he had obtained during his previous trip to Europe, mainly Germany, in December 1833 Luz presented to the Board a report on the creation of an "essentially practical and experimental [Cuban Institute] similar to that established in Gijón, for the benefit of all the Asturian people, as that of Havana would be for all those born in Cuba." Since quite obviously this project competed fundamentally with the university reform proposed by Arango y Parreño, this caused some friction between the two authors. In the end, however, both proposals came to nothing, except for the belated creation of a much needed chemistry chair and laboratory which were inaugurated in Havana around the middle of 1837 by the *Real Junta de Fomento* (led by sugar plantation owners) and headed by the remarkable Spanish chemist José Luis Casaseca, who was mentioned earlier (Simpson 1976).

In any event, it must be said that as with the case at the San Carlos and San Ambrosio Seminar College, at the Havana San Jerónimo University physics teaching (intermingled with that of chemistry) was a part of the philosophy lessons that were taken during 3 years by the students who were working for their Bachelor of Arts degree. But while at San Carlos the subject was explained in a "modern" style, in the Spanish language and emphasized experimental work—especially since 1816, with the help of the physics laboratory created that year as already said—, at San Jerónimo the subject was taught and scholastically discussed in Latin, in an abstract and speculative way, with the predominance of Aristotelian-Thomistic matters, though at times some "modern" physical ideas cropped up, as can be gathered from the following propositions defended by university degree candidates:

> Horror to vacuum as an attribute of Nature should be totally discarded, and should be put among the inventions of the Cartesians. [Year 1815]
>
> Astronomy declares that the light of the Sun and that of the planets is the same, and Fraunhofer's experiments totally confirm it. [Year 1829]
>
> Water is decomposed by galvanic electricity. [Year 1840] (Leroy y Gálvez 1976)

Still, physics teaching remained extremely poor and outdated at the university, compared with that given at the San Carlos Seminar College and at some private educational institutions dedicated to train students for taking the Bachelor of Arts (or philosophy) degree at the university.[18] An important contribution to the

[18] These schools, which existed for several decades in Havana with spacious premises, libraries, and laboratories, were the ones named *San Cristóbal* or *Carraguao*, founded in 1829, *San Fernando*, founded in 1832, and the *Colegio de Humanidades*. In particular, José de la Luz y Caballero, appointed *Carraguao*'s Literary Headmaster in 1832, was in charge of the teaching of physics from 1834 to 1835. To enter those schools, the *San Carlos*, or the university, a "purity of

indefinite continuance of such a situation was the fact that it was the rule at San Jerónimo that the subject was to be taught not by a permanent professor but by different religious readers who changed each year and had to share this task with the teaching of the other subjects in the Bachelor of Arts study plan (Leroy y Gálvez 1976).

The backwardness and general deterioration of teaching at the university became more evident as time went by. By the mid-1830s, a group of doctors who were professors there—among them, the highly respected doctor Tomás Romay—openly asked the colonial governor, Miguel Tacón, to tidy up the university by taking it out from the convent of the Dominicans and relieving it from their control. If it is taken into consideration that at the time the great period of nationalization of the idle property owned by the Church had begun in Spain under minister Mendizábal, it will be understood that the action demanded by Romay was not unrealistic. On the other hand, reports continued to arrive that provided evidence of the backwardness and corruption at the university and demanded from the colonial authorities some quick action to improve the situation. In view of all this, in August 1840 governor general Pedro Téllez Girón created a commission to study the matter and make the appropriate recommendations. In June of the following year, the commission presented to the newly appointed governor, Jerónimo Valdés, the results of their work, which included a new study plan and new regulations for the university. While this was happening in Cuba, in Madrid a commission specially created by the *Dirección General de Estudios* (Head Office for Education) completed the drafting of its report on the state of public education in the Spanish colonies, which was handed over to the Overseas Minister on July 31, 1840. As stated in the report, from the data at the disposal of the commission (many of them directly arrived from the island)

> a truth that was already known to the Dirección General now shows itself totally confirmed; and it is that the reform of higher education in the Island of Cuba should not be delayed, the emancipation of the university from the dependence on the Dominican Friars [being urgent…] The university must experience no less than a complete reform: its management, its teachings, everything demands modifications and neither one or the other can be delayed any longer. (Simpson 1984, 164–165)

The Spanish monarchy was very much in favor of introducing such a university reform as the one brought forward, not so much because of the progress it meant from an academic point of view, as for the fact that it put in the hands of the Government the total control of public education in the colonies, which gave it the opportunity to check on the development of national and anti-colonial conscience that was growing at the time in overseas educational centers. All of this, supposedly

---

blood" certificate was required from prospective students, which blocked the entrance of blacks and mulattoes. For economic reasons, only the descendants of the well–to–do classes had the possibility of entering these schools, except for the few offspring of the underprivileged who were lucky enough to be awarded free tuition. About the same time, other quality private schools were founded, such as those known as *Buenavista* in Havana, *La Empresa* in Matanzas, and *Santiago* in Santiago de Cuba, respectively run by José Antonio Saco (in 1832), the Guiteras brothers, and Juan Bautista Sagarra, see Buenavilla et al. (1995), 36–38.

did not require new expenditures, since it was anticipated that any additional expenses derived from the introduction of the reform would be paid for by an increase in the registration fees and other surcharges.

In summary, these developments show how institutions in Cuba were organized and how they intermingled in the development of sciences. The conservatism of Dominican higher education that was reflected in the San Jerónimo University contrasted with the relatively low-funded Seminar of San Carlos. Two forces helped to overthrow this situation: Cubans who were highly aware of international standards in sciences and who aspired to participate in an international scientific or economic community formed one of the pressure groups. The second force was the development of colonial rule itself which guided even the Spanish governments in reworking the academic landscape left after the independence of so many Latin American colonies between 1810 and 1825.

4.6 The Secularized Colonial University Takes Over (1842–1898)

On December 29, 1841, the Spanish Crown decreed a Royal Order establishing the new regulations that thereafter would become the basic tenets of public education in Cuba in general, and of higher education in particular. It also included a stipulation that unfortunately was not carried out: the creation in Havana of an institute as the one formerly proposed by Luz y Caballero, where "mathematics, living languages, physics, elements of chemistry and other principles useful for industry, agriculture and commerce" would be taught. Soon after the arrival of the Royal Order on February 1842 a *Junta General de Inspección de Estudios* (General Board for the Inspection of Studies) was created, in charge of specifying the details for the application of the reform. On April 24, the Junta handed out its proposals pertaining to secondary and higher education, which were officially approved on August 24 (Simpson 1984, 170–171). Complying with an order issued by Governor Jerónimo Valdés on October 31 the Dominicans left the building where their convent of Santo Domingo or San Juan de Letrán had been lodged, just behind the Governor's palace, and established themselves in the town of Guanabacoa. Two days later they formally handed the building over to the San Jerónimo University of Havana. The first course of the now secularized university was inaugurated on November 19, 1842 (Armas et al. 1984, 119–120).

The now secularized University of Havana was created with the faculties of law, pharmacy, and medicine and surgery, respectively dedicated to the training of lawyers, pharmacists and physicians in the degrees of bachelor, *licenciado* (graduate), and doctor. New students had to first acquire a degree of bachelor of arts or bachelor of sciences before entering into these so-called "major faculties." This was awarded after studying for 4 years at the Faculty of Philosophy, known as a "minor faculty." Theology was no longer included as a degree course in the study plans of the University of Havana.

Bachelor studies at the Faculty of Philosophy had an encyclopedic character. It included

> ... a wide range of subjects: mathematics, physics, chemistry, natural history in all its branches, Greek language, oratory, and literature, especially Spanish literature; geography, chronology, and history, especially national history, philosophy (logic, metaphysics, ethics, and moral philosophy and history of philosophy), religion and natural law. [... Only] in the final degree examinations special emphasis would be put either on arts or on sciences, according to the kind of bachelor degree aimed at. Obviously, the course was excessively wide-ranging, even for students of the same age of those taking junior high school these days [...] (Simpson 1984, 174)

No additional physics courses were included in the "major faculties," nor in the science courses required to obtain a *licenciado* degree in the Faculty of Philosophy. The difficulties in creating a chair of physics at the University of Havana were not simply due to colonial issues. Sáenz and Capote (1989, 22) point out that also in the Spanish universities of Alcalá de Henares and Salamanca, the teaching of Newton and Galileo was refused.

The first professor appointed to teach physics at the newly secularized university was the director of the botanical garden of Havana, Pedro Alejandro Auber (1786–1843), a French-born botanist who had become a naturalized Spaniard. He had acquired some formal training in experimental physics while studying medicine in Madrid (Puig-Samper and Valero 2000, 169). He died some 5 weeks after the inauguration of the course, which was then taken over by the Canarian-born lawyer Domingo León, an extraordinary professor at the Faculty of Philosophy (Leroy y Gálvez 1979, 4).

4.6.1 The 1840s and 1850s: The First Physics Laboratory at the University

Appointed by Governor Jerónimo Valdés, the Asturian born extraordinary professor of the Faculty of Philosophy, Feliciano Carreño (1810–1847), was entrusted in September 1843 with the professorship of physics at the University of Havana. During the three and a half years he taught physics at the University, he published, for the first time in the institution, a formal study program for the subject and promoted the buying of instruments and related hardware to equip the physics laboratory. According to the annual report for 1869–1870,

> The physics laboratory did not start to materialize until November 27, 1843, when at the request of Mr. Feliciano Carreño, professor of the subject that he was at the time, some apparatus worth 600 pesos were bought, whose number has been repeatedly increased whenever allowed by the allotted budget. (Universidad de la Habana 1870, 22)

After the passing of Carreño in March 1847 his place was occupied by the Havana born lawyer and doctor of philosophy, José Zacarías González del Valle (1820–1851), who became an extraordinary professor of the Faculty of Philosophy after taking a competitive examination, and published in 1849 his *Lecciones elementales de meteorología* (elementary lessons in meteorology) for use of his physics students. Since González del Valle had become seriously ill in early 1851, Ramón Zambrana (1817–1866), a bachelor of philosophy and doctor of medicine and surgery, also born in Havana, substituted him until the end of the year (Leroy y Gálvez 1979, 10).

It is almost superfluous to remark that none of the professors in charge of the physics courses at the University of Havana had received specific training in this scientific field. On the other hand, interest in this matter of modern teaching was spreading in Cuban society. It is interesting in this respect that in 1850 a very good quality translation from French into Spanish of François Marcet's *Cours de physique expérimentale* (Marcet 1850), published in Havana, was announced as a work accessible to "beginners and readers of all kinds," which followed the lessons taught by the author at the Geneva Industrial School. The translator, José Manuel Mestre, who was a prominent cultural figure in Cuba (Pruna Goodgall and González 1987, 57), made clear in the first pages that he had undertaken the translation "at the request of the physics professor of the Royal University" González del Valle and with the help of his advice. This textbook had, as compared with that of Varela, the appeal that it dealt quite fully with steam engines and their applications, but its treatment of electromagnetism and optics was less extensive. The growing interest in scientific matters is testified by another initiative, that is, the foundation in 1861 of the *Real Academia de Ciencias Mèdicas, Fìsicas y Naturales de La Habana*, which gathered the most prominent scientific personalities of the country.[19] The already mentioned Ramón Zambrana y Valdés was appointed as the first General Secretary of the Academy, until his death.

Another scientific institution created was the Physical-Meteorological Observatory, founded in 1856 by Andrés Poey, though it had to be closed in 1869 for political reasons. All this shows that in the mid-nineteenth century, different initiatives promoted modern science in Cuba, as the introduction of the decimal metric system in 1849 also indicates (Sáenz and Capote 1989, 20, 22).

Nevertheless, these ferments were late in arriving to the University of Havana. From the above, it can be gathered that although the first 10 years of the presence of "modern" physics at the University of Havana were characterized by the instability of the teaching staff in charge of it. However, during the same time, a physics laboratory was created and the first program for the subject and some complementary teaching material were published. From the fact that the fifth and last edition of Félix Varela's *Lecciones de filosofía* was published in 1841, it is reasonable to assume that the third volume of this work—the one that dealt with physics and chemistry in an excellent pedagogical way—substantially influenced the fashion in which the corresponding subjects were taught during the period referred to. Actually, although the

[19] See Pruna Goodgall (1986, 2010) and Pruna Goodgall and Ortega (1985). Actually, the first proposal for the creation of an academy of sciences had been sent to the Governor of Cuba in 1826 by Nicolás J. Gutierrez and Francisco Alonso, whom Tomás Romay joined, sending his proposal to the king of Spain, who did not answer. After further meetings, proposals and petitions, on November 6, 1860, the Royal Decree from Queen Isabel II authorized the creation of the Academy.

teaching of physics at the University of Havana stabilized during the following four decades, the course continued to be delivered by professors with no specific training in physics. Correspondingly, the textbook used up until the end of the century was the Spanish translation of the 1851 edition of Adolphe Ganot's *Traité elementaire de physique expérimentale et appliquée et de météorologie*, whose contents were essentially descriptive.[20] The professor who took charge of the course from the end of 1851 until his death was the Cuban born Antonio Caro y Cerecio (1826–1891), a medical doctor by training, who had learnt some (elementary) physics himself[21]: from 1851 on, save the period 1876–1879,[22] he was in charge of the all the teaching of physics at the University of Havana until 1881, when, in addition to a subject called expanded physics, the subjects named higher physics 1st and 2nd courses were introduced (Leroy y Gálvez 1979). In this period modest measures were taken, such as allocating larger and independent premises to the physics laboratory, and providing it—as stated by rector Antonio Zambrana in 1861—with

> ... a sizable assortment of machines and instruments which, while not making up a fully equipped laboratory, they do cover the essential teaching needs [while] at the same time [...] the walls and the floor have been repaired to the extent that the humidity that is so detrimental to the success of experiments and to the best preservation of apparatus has disappeared. (Zambrana 1861, 10)

As will be seen in the following, though several attempts were made later to elevate the teaching of the subject by the introduction of higher-level courses, the end results left much to be desired.

4.6.2 The 1863 Study Plan: The Creation of Secondary Education Institutes and of the Faculty of Sciences

In accordance with a Royal Decree by Isabella II dated July 15, 1863, the Spanish Crown established a public education plan for Cuba, a plan originally proposed by the former governor of the island, José Gutiérrez de la Concha. With a view to

---

[20] Ganot's textbook was wholly reworked several times by Georges Maneuvrier so that it incorporated a certain mathematical level in dealing with some topics and was very much used as a textbook or reference work on general physics in Spain, France and other countries until well into the twentieth century. The fact that in its twentieth French edition, published in 1887, it is stated that the work has been rewritten "according to the most recent university programs" is quite illustrative. However, in the Spanish translation of the 24th edition, published in 1909—whose mathematical level is lower—, it is pointed out that the book has been written "according the official programs for secondary teaching."

[21] A student of the outstanding Spanish chemist José Luis Casaseca from 1845 to 1848, Antonio Caro was awarded the *licenciado* degree in sciences (1850) and a doctoral degree in medicine and surgery (1857), and became an expert embalmer.

[22] In these years, Caro was temporarily put in charge of the chair of therapeutics, medical matter and art of prescribing, which was part of the Faculty of Medicine: in his place, the chair for experimental physics was covered by Manuel J. Cañizares.

effectively implement it, on September 28, governor Domingo Dulce issued a decree which eliminated the former Faculty of Philosophy and transferred the more elementary teaching in their charge to newly created high schools called Institutes for Secondary Education, where pupils who finished their studies would be awarded the bachelor of arts degree. The first institutes for secondary education were established in Havana, Santiago de Cuba, Matanzas and Puerto Príncipe (Camagüey), the island's most populated cities at the time.

The old Faculty of Philosophy was replaced by two newly created faculties, that of Philosophy and Humanities (*Filosofía y Letras*), and that of Sciences, but only nominally until 1871 (when they were granted permission to teach courses up to the bachelor degree (Simpson 1984, 224)). In practice they engaged in teaching subjects corresponding to a so-called "preliminary period" with a duration of 1 academic year, which was implemented so as to provide the new incoming university students with a reinforcement and extension of certain subjects whose notions they had been taught in high school.

In the preliminary or "extension" period "certain subjects [were given] appropriate to the faculty or degree course that the students would have to follow," such as law, pharmacy, and medicine and surgery (Universidad de la Habana 1870, 10). The students who graduated from high school and wanted to enter the first degree courses had to previously pass the subjects of the "arts section" given in the Faculty of Philosophy and Humanities: world history, geography, and Latin literature, while those who wanted to follow the degree courses in pharmacy or medicine and surgery had to pass the subjects corresponding to the "sciences section," taught at the Faculty of Sciences: mineralogy, botany, zoology and geology (in the charge of Felipe Poey), general chemistry (in the charge of Cayetano Aguilera), and experimental physics (in the charge of Antonio Caro, as already mentioned) (Universidad de la Habana 1866).

The chair of zoology, botany and mineralogy (with its natural history museum), the chair of general chemistry (with its laboratory), and the chair of experimental physics (with its laboratory) all belonged to the Faculty of Sciences, which comprised three sections: exact sciences, physical sciences, and natural sciences.

It should be noted that the physics laboratory then belonging to the university left much to be desired, as suggested by the fact that in the university report for 1864–1865 it is stated that the said laboratory "still needs many means for the students to be able to do their extension studies in a convenient form" (Universidad de la Habana 1866, 17). In later reports some 120 apparatus for mechanics, thermology, optics and electricity—some of them said to be in bad state or not used at all—are listed as part of the collection of items belonging to the physics laboratory.

While this was happening at the University of Havana, in 1864 at the Institute for Secondary Education of Havana, thanks to the initiative of its first headmaster, Antonio Bachiller y Morales, a physics laboratory was created "so that the teaching of physics could have an experimental basis." According to a rare pamphlet published at the beginning of the twentieth century and preserved as an item of the physics apparatus collection of the Institute, "with a lot of effort Bachiller y Morales managed to provide it with 154 apparatus, which soon became useless because of careless handling."[23]

Within the context of the educational reforms of 1863, high school teaching at schools run by Jesuit priests was put on a level with that of the governmental Institutes for Secondary Education. As far as the San Carlos Seminar College is concerned, it was associated with the university in the sense that it was entrusted with the task of preparing its students to take the philosophy or Theology degree exams. The San Ildefonso Seminar College of the island of Porto Rico was associated with the University of Havana in a similar fashion (Universidad de la Habana 1870, 10).

4.6.3 The Academic Restrictions of 1871–1878

The outbreak of the first war for the independence of Cuba on October 10, 1868, had an impact on the academic life of the country, as the university became a source of conspiracy against colonial rule. With the explicit wish to control the teaching staff so that "it would not [instill] pernicious doctrines, or [turn] science into a revolutionary platform," hard line governor Blas Villate (Count Valmaseda), decreed on October 10, 1871, a draconian "reform" that drastically diminished the content taught at the university.[24] This decree eliminated doctoral studies from the Law, Pharmacy, and Medicine and Surgery faculties. From then on those studies had to be done in Spain, as was to be the case with the *licenciado* and doctoral studies included in the faculties of Sciences and of philosophy and humanities. Still, the new situation allowed for the continuation of the teaching of physics extension at the University of Havana, since that subject belonged to the period of preliminary studies (Leroy y Gálvez 1963).

[23] The quoted text continues thus: "When in 1888 Dr. Claudio André took charge of the Physics Chair [of the Institute for Secondary Education of Havana], he drew attention to the fact that most of [those] apparatus were in poor condition. They continued to be, without being either repaired or replaced, to the time of Dr. [Fernando] Reynoso, when a part of the academic fees was used to buy many of those it contains today, [which were then] classified, put into good shelves, though better suited for an exhibition than comfortable for daily use of the apparatus, and great care was taken for their preservation by naming for the purpose such an intelligent person as Mr. Plácido Biosca, who presently holds the physics professorship at the University of Havana, thanks to whose initiative many apparatus were acquired."

[24] By then, both the private schools and three of the island's four Secondary Education Institutes, considered to be "dangerous and too liberal," had been closed down and only the Institute of Havana was kept functioning though quite precariously, Buenavilla et al. (1995), 56–57. By way of an eloquent context, it is worth recalling that less than 7 weeks after Valmaseda's decree came into force, the most despicable and bloody act of colonial repression against Cuban university students took place, which was the execution by a firing squad, on November 27, 1871, of eight medical students who had been falsely accused of desecrating the grave of a journalist who had been an out–and–out champion of Spanish colonial rule.

When the first Cuban war for independence came to an end in 1878 after 10 years of armed struggle, the Spanish governor and captain general, Arsenio Martínez Campos, decided by a decree he signed on September 10 that year (sanctioned a year later by a royal order) to restore, as from the 1878–1879 academic course, the doctoral and other high level studies that had been suppressed by Valmaseda. He pointed out that the restoration was carried out

> … for the good of the studious youth and as a deference to the university, which from its very creation had the right to award doctoral degrees, until October 10, 1871, when it was deprived from this power that was one of its most esteemed merits. (Leroy y Gálvez 1963)

But restoration of doctoral studies did not apply to those studies done at the faculties of sciences and of philosophy and humanities: in these faculties, indeed, the highest degree awarded was that of bachelor (Universidad de la Habana 1873, 115). In fact, in 1877–1878 only one bachelor degree was awarded in the Faculty of Sciences. According to the Martínez Campos decree, this sufficed to "provide the country with adequate teachers for secondary studies." To attain the bachelor of sciences faculty degree it was required to pass the following study plan:

> First year: 1. complements of algebra, geometry and trigonometry (daily lessons); 2. physics (daily lessons); 3. geography (lessons on alternate days).
>
> Second year: 1. analytic geometry in two and three dimensions (lessons on alternate days); 2. general chemistry (daily lessons).
>
> Third year: 1. zoology and mineralogy (daily lessons); 2. botany with notions of geology (daily lessons).[25]

Students who wished to enter the Faculty of Sciences were required to have a bachelor of arts awarded by an institute of secondary education and, in addition, have some knowledge of linear and architectural drawing (Leroy y Gálvez 1963).

4.6.4 The 1880 Study Plan

It soon became obvious that the country's flawed educational system needed improving. With a view to improve the situation, a royal decree dated June 18, 1880, was issued that put higher and secondary studies under the same regime as the corresponding ones in Spain. In accordance with the decree, teaching in Cuba was reorganized so that, among other things, the creation of secondary education Institutes was authorized in the two provincial capitals where they did not exist yet, the post of auxiliary professor was created, the bachelor degree at university faculties was suppressed and the degree of *licenciado* degree was restored for studies done in all faculties, including those of sciences and of philosophy and humanities. The 1880 plan was the fourth and last one that was officially in force at the university during colonial times.

[25] Note that the subject called physics in the above was actually the one called experimental physics, as formerly taught by Antonio Caro.

> … the reform has taken the first step toward the assimilation of studies with the metropolis … This plan establishes in this university the faculties of philosophy and letters, and physical mathematical, physical chemical, and natural sciences, up to the degree of *licenciado* included. … The *bachiller* has been suppressed in the faculty (Universidad de La Habana 1881, 50).

On September 26, 1880, governor and captain general, Ramón Blanco, approved some provisional regulations required for the application of the royal decree that included new study plans for secondary and university education in the country. In particular, students doing the *licenciado* degree in physico-mathematical sciences had to take obligatorily, starting from the 1880–1881 academic course, a subject called Theory of Imponderable Fluids (caloric, luminic, electric and magnetic).[26] Its teaching was entrusted to Manuel Cañizares y Venegas (1833–?), *licenciado* of philosophy (section of physico-mathematical sciences) and doctor of medicine and surgery, born in Sancti Spíritus.[27] But this subject was never taught as such, since soon after it was eliminated from the study plan and replaced by another one called *Física Superior* (higher physics), following what had happened in Spain in August 1880.[28] In July 1881 Cañizares was officially appointed professor in charge of the subjects higher physics first and second courses, and practical exercises of higher physics. These new subjects simply amounted to an extension of the expanded experimental physics course, which in Spain was based on the French physics textbooks by Adolphe Ganot and Jules Célestine Jamin, to which another author Lozano was added later on, whose first edition was published in 1890 (Leroy y Gálvez (1963), 24 and Moreno González (1988), 392).

The above-mentioned provisional regulations approved by General Blanco for the application of the royal decree were generally supported by a royal order dated December 7, 1880, except that this stipulated that the three former sections of the Faculty of Sciences (exact, physical and natural sciences) would be reorganized into

---

[26] This subject clear testifies how at the end of the nineteenth century the teaching of physics was outdated with respect to most advanced theories in the most advanced countries: indeed, the "imponderable fluids" arrived in Cuba from Spain, although their ultimate origin was France. As mentioned, in the program of *Física Experimental* for the fifth course of secondary school in Spain in 1846–1847 "the role assigned in previous programs to the so–called imponderable fluids, incoercible or 'dinamídeos,' is reduced along with the more extensive denomination in our country deriving from French textbooks. This is a weak, but reliable symptom of modernity since, among other corrections to the physical theories, *molecular physics*, as it was first termed, began to spread as an explanation for electrical, magnetic, light and calorific phenomena starting from the motion of the molecules composing the substances. However, the contents of electricity and magnetism of the reproduced official program still necessarily retain the theories of electric and magnetic fluids … The program of experimental physics is similar, although lighter, to the index of the *Traité élementaire de Physique expérimentale et appliquée et de Météorologie*, by A. Ganot published in its 6th edition in Paris in 1856," Moreno González (1988), 299–300.

[27] In 1871 Cañizares had been appointed extraordinary professor in the Faculty of Sciences at the time of the Valmaseda reform. As such, from 1876 to 1879 he had taught three courses on expanded experimental physics in place of Antonio Caro, who had transferred to a chair in the Faculty of Medicine, as remarked in Note No. 23, Leroy y Gálvez (1979).

[28] This was in accordance with the stipulations of the Fermín Lasala Plan, Moreno González (1988), 391–392.

a section for physico-mathematical sciences and a section for physico-chemical and natural sciences.

From the 1881–1882 academic course, students of the *licenciatura* in physico-mathematical sciences, once they were awarded de bachelor of sciences degree, had to take the following courses:

> Differential and integral calculus, descriptive geometry, rational mechanics, geodesy, expanded practical work in physics, higher physics (1st and 2nd courses), and higher physics (1st and 2nd courses) practical work. (Universidad de la Habana 1882)

Another royal order, dated August 23, 1883, reestablished the doctor degree for all the faculties of the University of Havana and included a subject named mathematical physics in the study plan to be passed by students aiming at a doctoral degree in physico-mathematical sciences (like in the mother country in 1880). There is no indication that it was ever actually taught, though in 1892 it was put under the responsibility of Professor Cañizares' chair (Leroy y Gálvez 1979). Apparently, to obtain a doctoral degree the essential requirement was to approve a degree examination, which according to the 1880 Plan, entailed giving "a lesson composed by the graduate about some point either from the faculty's question paper or freely chosen" (Moreno González 1988, 391–398). A review of the doctoral theses during the period that are kept at the central library of the University of Havana clearly shows that there was no one dealing with physics, but even if there were one, it could be confidently presumed that its approach would have been merely qualitative and discursive.

The training in physical sciences of the students of physico-mathematical sciences surely experienced a considerable improvement after Juan Orús (1849–1911)—a Cuban-born graduate of the Professional School of Barcelona and a member of the previously mentioned Royal Academy of Medical, Physical and Natural Sciences in Havanna—was put in charge of the teaching of the curricular subject rational mechanics in 1883 (Universidad de la Habana 1902). His lessons finally were made to follow Ch. Delaunay's *Tratado de mecánica racional*, that was already used as a textbook at the universities of Madrid and Barcelona, and was translated from French into Spanish in 1866 (Moreno González 1988, 505, 508).

It is interesting to remark that these changes at the university were carried out in a context of substantial material development in Cuban society. The introduction of electric lighting in Cuba during the second half of the 1880s not only attracted strong public attention, mainly in Havana, where the first installations were made, but were also echoed in the teaching activities of professors Antonio Caro and Manuel Cañizares. Thus, for example, the latter dedicated to that theme his opening speech for the 1885–1886 academic course at the University of Havana (Cañizares 1885) while Caro became a sort of a journalist engaged in reporting and commenting on the most significant news concerning the public lighting system that was inaugurated in the cities of Havana and Cárdenas in 1889. The following can be taken as a sample of what he wrote on the subject, where he refers to a curious linkage between the technical applications he was describing and the physics lessons he taught at the university:

> [During the first test of the Thomson-Houton system at the Central Park of Havana] we have seen confirmed that its obvious superiority over all other systems is undeniable; and we

> have asserted this to our esteemed students at the university in their last course by drawing the most detailed parallels between all lighting systems [...]. (Altshuler and González 1997, 328)

Months before this note by Caro was published, the same Havana newspaper had published a scientific popularization article especially written by José Echegaray, the famous Spanish polymath, on the occasion of the inauguration of electric lightning in Cuba. This was aimed at explaining the main characteristics of the electric current and its use for lighting in a very readable form by referring to the classic hydraulic analogy. However, when referring to notions of electric current, resistance and potential, he warned the reader that

> It is not possible to explain in rigorous and scientific terms what this means in such an article as this, for what will it mean for the reader that we tell him, for example, that the potential is the integral of the electrical masses, divided by their respective distances to the point under consideration? (Altshuler and González 1997, 329)

It is not preposterous to imagine the possibility that such a reference to a fundamental concept in the theory of electricity could have awakened in some of the physics students an interest in deepening their knowledge of the concepts touched upon in the physics extension course. But of such a possible incidental effect we know nothing for sure, because from the available documentation it can only be gathered that toward the end of Spanish domination of the island and even later, physics subjects continued to be taught at the university without any appreciable increase in their scientific level.[29]

As a general conclusion, we could say that the changes we have just discussed, which were introduced in the last decades of the nineteenth century, actually had concrete connections with a growing need for the technical modernization of Cuban society. Still, the changes could not meet this need: the situation in fact reflected that of the motherland, which was relentlessly declining with respect to progress made in the most advanced countries, as the American military intervention in the Cuban-Spanish war was to demonstrate.

4.7 The University in the Early Neocolonial Period (1899–1922)

When Spanish colonial domination in Cuba officially ended on January 1, 1899, the United States proceeded to militarily occupy the country, with the implicit promise to later transfer its government to the nationals, who had been fighting for

[29] The number of *licenciados* and doctoral degrees awarded every year by the Faculty of Sciences (given in the *Memoria Anuario* of the University of Havana for the 1880s and 1890s) was quite low: the *licenciados* oscillated between 4 and 13, with an exceptional rise to 25 in 1889–90. The number of *licenciados* who received a doctoral degree was between 2 and 3, with a low of 1 in 1895–96, and an exceptional rise to 13 in 1899–1900 (in 1884–85 Plácido Biosca and Nicasio Silverio, whom we will discuss in the following, received the doctoral degree).

its independence for three decades. As a result of the heavy damage to agriculture (the main resource of the former Spanish colony) and the decimation of the population forced massive concentration of the rural population in cities, the situation of the only existing university became as precarious as that of the other national institutions.

The American military government named some Cuban officials to help with the handling of several areas. One of them, the distinguished lawyer José Antonio González Lanuza (1865–1917), who was appointed Secretary of Justice and Public Education, drew up a study plan for the University of Havana that came into force by Military Order No. 212 of November 4, 1899, and was complemented by successive regulations, although it was in place only for 8 months, for the academic year 1899–1900.

The new study plan did not introduce any really fundamental alteration relative to the former one, to the extent that it confirmed the generic professorship of all those who up to then had held such a position. This considerably increased the number of professorships, which had to be covered by increasing the teaching staff and redistributing the academic tasks among its members. As the subsequent secretary of education, Enrique José Varona (detailed in the following) wrote:

> It was a plan of proliferation. It increased the professorships and split them. But all that within the old lines and respecting the old teaching method. … It did not take into account that for such a luxury of professorships there would not be teachers, first, and then pupils (Delgado García 2008).

Thus, for example, Plácido Biosca, a professor of general chemistry since 1892, was transferred to the chair in charge of teaching the first physics course to the students in the preliminary period, and the two first physics courses for the *licenciado* degree students, while Manuel Cañizares was transferred to a chair dedicated to the teaching of trigonometry and astronomy, and Nicasio Silverio, who formerly taught chemistry, was appointed professor of the expanded physics chair, responsible for teaching the third physics course of the *licenciatura* and also for the teaching of meteorology, applied electricity and electrical measurements.

It very quickly became clear that the "Lanuza Plan" was unrealistic, because on the one hand it did not respond to current needs, and on the other, its implementation was exceedingly expensive for a country in such a weak economic situation. Accordingly, when on May 1, 1900, the distinguished Cuban polymath Enrique José Varona[30] was appointed Secretary of Public Education, he immediately took up the task of reorganizing higher education. His "Plan Varona" replaced the former

[30] Varona (1849–1933) was a distinguished Cuban writer, philosopher and pedagogue, an enlightened representative of the emerging Cuban middle class. He joined the ranks of those fighting for independence from Spain. For his ideas, he was sent into exile, and in New York collaborated with José Marti, and began in 1895 to head the newspaper *Patria*. During American occupation, he became Secretary of Education. With the establishment of the Republic in 1902, he dedicated himself to his work as a professor at the University of Havana, but then re-entered his political activities, founding the *Partido Conservador Nacional*, and becoming vice–president from 1913 to 1917 during the presidency of Mario García Menocal. Frustrated by the reality of the Cuban social situation, in 1921 they gave a lecture entitled "El imperialismo yanqui en Cuba." He became a

"Plan Lanuza" by Military Order No. 266, dated June 30, 1900.[31] In an article published shortly after, Varona stated:

> I think that our teaching should stop being verbal and rhetorical and be transformed into one that is objective and scientific. Cuba does not need more than two or three literary persons, but cannot do without a few hundred engineers. Here is the core of my reform. (Varona 1999, 207)

About the same time, Varona wrote to a colleague and friend to say that when drawing up his reform plan for Cuban education, he was guided by a spirit of "self-defense of the Cuban ethnic group, a defense such as it [was] possible and in the field where it [was then] possible." To this effect, he decided to reorganize national education "in the American way," assuming that Cubans had to "compete in the industrial field, which [amounted] to saying in the scientific field, with the Americans, if they [didn't] want to be dislodged from the field" (Varona 1999, 217).

No doubt, it was in view of the above consideration that, according to Varona's plan, a school of engineers, electricians and architects, and a school of agronomy were both included in a new faculty of sciences and humanities, together with the schools of humanities and philosophy, pedagogy and sciences. The other university schools belonged to two other faculties: medicine and pharmacy, and law. It seems clear that Varona was inspired by the goal to modernize the country, in some sense influenced by the example of the United States, in contrast to the backward Spanish system.

Varona was inspired by the need to make studies more effective and to speed up the preparation of a new generation of well-trained professionals. The "rationalization" and "speeding up" he sought, favoring professionalization, rather paradoxically turned out to be detrimental to the study of the sciences, since this was reduced to only 3 years and the three former specialized sections were eliminated, namely those of physical-mathematical, physical-chemical and natural sciences which had to be reintroduced in the academic year 1906–1907 (see below). Needless to say, this was a relevant deficiency and contradicted the Varona project: interested as he was in promoting the training of professionals with a practical background, he undervalued the relevance of the basic natural sciences (mathematics, physics, chemistry, biology). To be awarded a doctoral degree in the sciences, students had to pass a long-study plan of three academic years, two drawing courses done at the school of pedagogy, and some degree exercises. These consisted in giving one model lesson and presenting a thesis on a subject chosen by the student from a list produced yearly by an examination board.

Regarding the physics area, all the students of the school of sciences had to take the following subjects: mechanics (with related laboratory work) in their first year; physics 1st course (sound, heat and light, with related laboratory work) in their second year, and rational mechanics and physics 2nd course (electricity and

symbol of intellectual liberalism and in his later years was a source of inspiration for many young Cubans fighting against the dictatorship of General Machado.

[31] See, for instance, Delgado García (2008).

magnetism, with related laboratory work) in their third year. In addition, the new study plan required them to take, among other subjects, two semesters of mathematical analysis and courses in analytical and descriptive geometry, chemistry, astronomy, botany and zoology. This was a very heterogeneous study plan that in the end produced only four doctors of science.

Such was the context in which the teaching of physics and related sciences at the University of Havana was to proceed. As stipulated by the Platt Amendment, the United States's military occupation of the country was replaced on May 20, 1902 by a Republican self-government, but was forced to accept new interventions whenever the American government deemed it necessary.[32]

In the first week of May 1902, the university installations had been hurriedly moved from their original premises at the old convent of Santo Domingo or San Juan de Letrán to the buildings of the former military barracks, located on the hill presently occupied by the University of Havana (formerly known as Aróstegui Hill), in the city's Vedado quarter.[33] The physics laboratory was installed:

> in one of the old buildings beside the chair of the same name, equipped with eight departments of polarization and spectroscopy, six photographic dark rooms, water, gas and electricity services, and enough working tables to enable a hundred students to simultaneously do their practical training in physics. (Universidad de la Habana 1903)

Regarding the school of sciences, some of the defects of the original Varona Plan were soon evident, such as the elimination of the school's three traditional sections. These were re-instated during the second United States intervention (1906–1909), by Decree No. 737 dated June 29, 1907, so that thereafter the titles of doctor of physico-mathematical sciences, doctor of physico-chemical sciences, and doctor of natural sciences were awarded to students who had passed the corresponding 3-year study plans.

The approved study plans for the degree courses in physico-mathematical sciences and physico-chemical sciences still underwent some adjustments. In 1907–08 important changes took place, although the concrete results would be obvious only two decades later: the courses of *Física General* (one course, for students of medicine, pharmacy, veterinary medicine and natural sciences) and of *Física Superior*

[32] It should be noted that the Platt Amendment was a stipulation included in a law on army appropriations, approved by the United States government on March 19, 1901, which, among other things, "grants the United States the right to intervene militarily in Cuba when the American government considers that the citizens' life, property or rights are endangered; […] forces Cuba to concede parts of its territory to the United States for it to establish naval or coaling stations, and validate all the actions carried out by the military occupation government. This amendment had to be […] added as an appendix to the Constitution of the Republic of Cuba. […] Protest was very vigorous nationally […] But the United States sent an ultimatum [to the Cuban Constitutional Assembly] and the amendment was approved by a margin of four votes" Cantón Navarro (1996), 80.

[33] The laboratories, the chair, the classrooms and the rest of the assets dedicated to the teaching of physics were moved in 1916 to a new building, a twin of another one dedicated to the teaching of chemistry. In 1928, when the Alma Mater statue was placed on top of the present monumental entrance staircase, it stood symmetrically between the two buildings.

(two courses) were introduced.[34] However, the level of the teaching of physics did not improve substantially. The number of doctor titles issued by the *Escuela de Ciencias* remained low, between 0 and 4. The political situation, which was a caricature of a democratic republic, was not yet ripe: independence had disposed of the past, but had not provided the means to deal with the present and build the future.

More specifically, higher physics (1st and 2nd Courses) was taught not only to the physico-mathematical science students, but also to those doing physico-chemical sciences, civil engineering, electrical engineering, architecture, and agricultural and sugar engineering; and also to those working for their agricultural technician and chemical and sugar technician degree courses. A subject named general physics, of a lower level than that of the higher physics courses, had a place in the natural sciences, medicine, pharmacy and veterinary medicine study plans.

In July 1900, the courses of both the higher and general physics were ascribed to chair "D" of the school of sciences, Plácido Biosca (1862–1923), who was appointed by the American Military government. As different sciences in Cuba were taught in one academic discipline, it was possible for one single person to acquire doctoral degrees in different subjects. The Catalonian born Plácido Biosca, for example, had been successively awarded by the University of Havana a doctoral degrees in medicine, in physico-mathematical sciences and in pharmacy in the period 1883–1889, in addition to the titles of chemical technician and mechanical technician. Having received the general chemistry professorship in 1892 by passing a competitive examination, he was transferred in 1899 to the physics chair of the Faculty of Sciences when the ephemeral Lanuza Plan was implemented, as pointed out above (Leroy y Gálvez 1979, 19–20). In later years, his former student, Manuel Gran, made the following comment:

> Dr. Plácido Biosca was an extraordinarily cultured man, with a broad training, especially in chemistry. His classes were agreeable and full of intelligent remarks, but they were rather like popularizations. A different approach would have required a very serious mathematical knowledge and an iron will to impose a method with a hint of a university level in an environment not used to it, and without the accordance of other teachings. (Gran 1942)

From October 1903 on, Plácido Biosca deferred the teaching of the higher physics 1st course lessons to his assistant, head of the physics laboratory, Nicasio Silverio

[34] For the 1907–1908 academic course, the subjects of the physico–mathematical degree course were grouped as follows, Universidad de la Habana (1909):

First year: Daily lessons: (1) mathematical analysis (higher algebra); (2) geometry (higher and analytic). Alternate day lessons: (1) trigonometry (plane and spherical); (2) line drawing; (3) biology.

Second year: Daily lessons: (1) mathematical analysis (differential and integral calculus). Alternate day lessons: (1) descriptive geometry; (2) cosmology; (3) higher physics, 1st course (mechanics, heat, and sound); (4) general chemistry; (5) drawing from nature; (6) zoology.

Third year: Daily lessons: (1) mineralogy and crystallography. Alternate day lessons: (1) rational mechanics; (2) geodesy; (3) astronomy; (4) higher physics, 2nd course (light and electricity); (5) general botany.

This study plan underwent only minor changes in successive years. In particular, as from the 1914–1915 academic year, the 1st course in higher physics was placed in the first year of the degree course, and the 2nd course was placed in the second year (Universidad de la Habana 1916).

(1860–1926), who was a *licenciado* of medicine and doctor of physico-chemical sciences, who had taught chemistry at the University of Havana from 1885 to 1898 and was later responsible for some of the physics courses, as mentioned above (Leroy y Gálvez 1979, 21–23).

Since the subject known as rational mechanics can be considered as part of the principles of physics taught to the physico-mathematical students at the University of Havana, it is appropriate to recall that the corresponding lessons were given by Juan Orús[35] from 1883 until his passing in 1911. The baton was passed to doctor of physico-mathematical sciences and *licenciado* of medicine and surgery, Victorino Trelles (1870–1951), who was then promoted to the chair of rational mechanics, astronomy and cosmology (Parker 1919, 321–322). A comment is probably appropriate and meaningful on the above-mentioned person-dependent stories. Generally speaking, one of the worst ills of the old university was the fact that chairs were essentially autonomous and held for life by the professor in charge, whose knowledge and personal taste would determine the way his subject was taught. For decades, this was one of the main reasons for student protests.

While the study plan for the physico-mathematical sciences and physico-chemical sciences degree courses remained nearly unchanged from 1908 to 1933, some important changes took place during this period in the contents of certain subjects and in the qualifications of the teachers in charge, as will be seen in the following. Thus, for example, the teaching of mathematical analysis was greatly improved from 1913 on, when it was left in the charge of Pablo Miquel. As to the teaching of physics itself, it experienced a radical advance when it fell into the hands of Manuel Gran in 1923.

4.8 In Search of a Higher Level (1923–1958)

4.8.1 The 1923–1933 Period

The important qualitative change experienced in the teaching of physics at the University of Havana during the 1920s coincided in time and spirit with the first radical attempt to bring Cuban higher education up-to-date. Inspired to a large extent by the university reform movement that began in Argentina in 1918,[36] and

[35] Juan Orús y Presno was a Cuban-born industrial engineer, a graduate of the *Escuela de Ingenieros Industriales* of Barcelona. He was also a member of the *Academia de Ciencias Médicas, Físicas y Naturales de La Habana*.

[36] The university revolution (revolución universitaria), or Argentine university reform of 1918, was a general modernization of the universities, especially tending toward democratization brought about by student activism during the presidency of Hipólito Yrigoyen. The revolution set up the freedom for a university to define its own curriculum and manage its own budget without interference from the central government. This had a profound effect on academic life at the universities through the nationalization process that promotes academic freedom and independence

led by an outstanding student leader, Julio Antonio Mella (1903–1929), a similar movement began at the University of Havana in 1923, whose goal was to modernize higher education, eradicate the archaic teaching methods prevailing at the country's top educational center and to replace incompetent, absentee or corrupt professors. In this context, some of the members of the university teaching staff who were notorious for their lack of suitability for their posts were retired and replaced by more capable and responsible professionals, who were supported and at times even put forward by the students themselves.

Such was the case of Manuel F. Gran (1893–1962), who in March 1923 was put in charge of giving the higher physics lessons instead of the assistant professor, Nicasio Silverio, who in turn had replaced Plácido Biosca who was seriously ill at the time. Silverio's students boycotted his classes in protest of his incompetence. They also demanded that the corresponding lectures be provisionally entrusted to Manuel F. Gran, a brilliant graduate of architecture and civil engineering who had been awarded a doctoral degree in physico-mathematical sciences in 1922 with a thesis on the concept of central forces and its applications to celestial mechanics.

According to Gran, at that time the higher physics courses were taught in exactly the same way as they had been taught to him years earlier:

> in such an elementary fashion, that the level and extent of the teaching of many topics was substantially lower than those that the students had received at some secondary education institutes, e.g., that of Havana. The number of lessons was exiguous—12 or 14 in the first course [...]— and much less in the second course. Mathematics was rarely resorted to in the ratiocinations, and when used, was only of the most elementary kind and touched upon in a very timid way. (Gran 1942)

Particularly sensitive to the lack of rigor and elegance as he was, Manuel F. Gran introduced in his lectures on higher physics a radical spirit of renewal with respect to the preceding state of affairs, a consistent rigorous approach to the various subjects dealt with, a suitable use of mathematics and an effective connection with problem solution and laboratory work (Altshuler 2003, 155–182). However, this change in the approach to the subject required several years to fully develop.

> When we began our work, in the 1923–1924 academic course, we needed to undertake the solution to three problems: the creation of a higher physics course worthy of the name, the organization of the laboratory so that it might fulfill its purpose in the best possible way, and the organization of the corresponding course on laboratory work. The first task was the most difficult one and, consequently, the part of our effort we dedicated to its fulfillment we

---

throughout university life. The events began in Córdoba and spread to the rest of Argentina, and "the reform movement did not stop, but jumped across the Argentinian borders and had repercussions in Latin America which were significant in Peru, Chile, Colombia, Guatemala and Cuba. [...] Later on these extended to Uruguay, Paraguay, Bolivia, Puerto Rico, Costa Rica, Brazil, Panama, El Salvador and Venezuela," González Carbajal (1974). In what concerns Cuba, in the night of November 27, 1922, the university students of Havana became aware of the details of the Argentine university reform through the then rector of the University of Buenos Aires. His words "brought to blood red the already overheated atmosphere": however "his role was rather that of a fuse. Since a favorable complex of factors and a state of revolutionary spirit in the youths did not exist, his passing through the university simply originated a momentary combustion", Armas et al. (1984), vol. 1, 334.

> most appreciate [...] We were [...] in the process of explaining [...], no less than at the university and under the name of higher physics, two lessons a day for whose teaching we had no model to follow. Since this course could not be reduced to a reproduction of any foreign course, taking into account that previous student training was different from that taken for granted by professors at other universities, the difficulty was even greater, for in almost all cases it was necessary to explain at the same time a topic in higher mathematics followed by the physics subject that depended on it. (Gran 1942)

In the beginning, textbooks on general physics by Murani, Ganot, Ollivier and other authors were used as reference textbooks. But as time went by there was a growing influence of French textbooks, such as Jules Faivre-Dupaigre, Jean-Baptiste Lamirand and Léopold Brizard's *Cours de physique*, the excellent *Cours de physique générale* by Georges Bruhat, and Henri Bouasse's monumental work in more than 40 volumes, entitled *Bibliothèque scientifique de l'ingénieur et du physicien*,[37] characterized by its heterodoxy in the author's approaches and the ardor of his criticism of generally accepted viewpoints and ways of teaching.

By then, the novelties in physical sciences produced worldwide reached Cuba almost exclusively through popularization works and textbooks to whose study the better professors devoted themselves, since the absence of the most important specialized scientific journals in university libraries was compounded with the practical lack of scientific exchanges with foreign academic institutions and the rarity of visits to countries with international level physicists.

To illustrate the last point, it may not be out of place to recall that one of those rare visits was the very brief stay in Havana in 1927 of the Spanish physicist, Blas Cabrera, who had been invited by the *Institución Hispanocubana de Cultura*. He was known in Cuba mainly because of a popular scientific book on the theory of relativity he published in 1923, the year of Einstein's visit to Spain. One is free to assume that at least one of the two lectures he gave at the University of Havana, entitled "Magnetic properties of the (chemical) elements and atomic structure" (Álvarez Sandoval et al. 2002)[38] was quite informative and up-to-date, since Cabrera's own contributions to the study of magnetism were at an international level.

---

[37] These French textbooks, which have been through several editions over the course of time, are general treatises of physics, covering all the classical fields and characterized by very extensive and systematic treatments. The *Cours de physique* by Faivre–Dupaigre, Lamirand and Brizard was published since 1927 in several volumes and was re-edited up to the 1960s. Georges Bruhat (1887–1945) was a professor at the *École Normale Supérieure* and the *Faculté des Sciences de Paris*; he was arrested by the Nazi Gestapo and died in the Buchenwald concentration camp. His *Cours de physique générale* in four volumes (*Mécanique*, *Thermodynamique*, *Électricité*, *Optique*) had six editions (eight on electricity). Henri Bouasse (1866–1953) was a French physicist known mainly for his wide treatise *Bibliothèque scientifique de l'ingénieur et du physicien* in 45 volumes, first published between 1912 and 1934. In the preface he criticized scientific teaching in France and the "new physics" of the twentieth century (relativity and quantum theory).

[38] The title of the other lecture given by Blas Cabrera at the University of Havana was "Atom organization and the periodic classification." He gave a third lecture on the evolution of stars at the *Institución Hispanocubana de Cultura*. Accompanied by Fernando de los Ríos, Cabrera arrived in Havana on January 6, 1927 from Mexico where he had given 18 lectures and participated in various seminars.

In spite of the existing limitations, the level of the courses of the so-called higher physics courses continued to rise gradually, while the professor and his assistants went deeper into their theoretical and experimental aspects—by self-study, of course.[39] Toward 1929, both courses of higher physics (covered in four semesters, three 1-h lectures a week, complemented with experimental and problem solving sessions) had achieved, basically, the general characteristics they retained for the next three decades. Higher physics continued to be taught exclusively to students of physico-mathematical and physico-chemical sciences and those of civil, electrical and chemical-sugar engineering. According to Manuel F. Gran himself, their teaching:

> was still guilty of elementarism, as a specialist unaware of the prevailing conditions of the environment could think, but when we have traveled abroad, appointed by the university or not, we have seen that the great teachers only developed one chapter, at times along several courses, while ordinary teaching did not go beyond ours, in spite of the difference in training, resources and scientific tradition. (Gran 1942)

One factor that very much harmed the country's teaching, and in particular the development of physics teaching in the early 1930s, was the closure of the University of Havana decreed in December 1930 by the very repressive tyranny of President Gerardo Machado, by way of retaliation for the combative attitude against him by students, who were supported by the majority of the teaching staff. In fact, the death of the university student, Rafael Trejo, shot down by the police during a demonstration on September 30, 1930, was a turning point that generalized the struggle of the Cuban people against the tyranny in power. By way of an anecdote, it is worth recalling that such a state of affairs prevented what might have been a memorable lecture at the university by the world famous Albert Einstein during his brief stay in Havana on December 19–20, 1930 (Altshuler 2003, 143–154).

Since the closing of the top teaching center of the country decreed by the government was accompanied by the suspension of pay to the teaching staff, university teachers had to dedicate themselves to finding alternative jobs to make a living. In particular, Manuel F. Gran, accompanied by other colleagues, set up a school that did not succeed and became involved in the reform and expansion of another one, whose laboratory he managed to set up.

The situation and dynamics in the educational system in general were therefore still very unstable. They depended essentially on improvements that were promoted by some exceptional self-made fellow endowed with remarkable scientific knowledge who had no direct international contacts. The members of the middle class were still developing and were at the mercy of a very unstable political situation in which the resurgence of a dictatorial regime threatened to frustrate, at least momentarily, the goals of modernization.

---

[39] Among other tasks, Gran devoted himself to accurately measuring the value of gravity acceleration in Havana. His physico–chemical sciences degree thesis, presented in 1925, contains the description of the experimental procedure used for the purpose and the result obtained (978.8 cm/s^2), "which an American commission from the Smithsonian Institution repeated years later with higher precision instruments and better experimental conditions, and found the same value within the limits of experimental error [...]" Leroy y Gálvez (1954a), 150.

4.8.2 The 1934–1939 Period

The interruption of university teaching that began at the end of 1930 lasted just over 3 years, for even though the Machado dictatorship had been overthrown in August 1933, the turbulent period that followed prevented courses from beginning until January 1934. The reopening allowed for the completion of the course of the academic year 1933–1934, but not for the following one, because on March 1935 the University of Havana was militarily occupied by the new provisional government of the country in response to an open statement by the university council that contested it for being repressive and antidemocratic. The military occupation of the University of Havana took place in the context of the March 1935 revolutionary general strike against the dictatorship of the then colonel Fulgencio Batista, a situation that according to Gran, apart from paralyzing academic activities at the university and subjecting the teaching staff to prosecution, had made "real starving people of those who had dedicated all their energy to teaching instead of seeking out economic welfare for themselves by working as ordinary professionals" (Gran 1942).[40]

Classes did not start again until the end of March 1937, this time under the protection of an educational law dated January 8 of the same year, that granted autonomy to the university (formerly granted by a decree issued on October 6, 1933, by President Ramón Grau San Martín's provisional government). According to university statutes hurriedly cobbled together by a committee of professors, which later came into force by a presidential decree dated February 22, 1937, the University of Havana was restructured so that the three previously existing schools were replaced by 12 faculties, one of them being the Faculty of Sciences.

As far as university teaching was concerned, it was resumed in 1934 with the stipulation that the courses the students of physico-mathematical sciences had to take (now extended to *four* years) included for the first time theoretical physics 1st and 2nd courses, taught by Enrique Badell (1895–1947). A civil engineer and doctor of physico-mathematical and physico-chemical sciences, he had been awarded by way of competitive examination, the post of assistant professor of chair "D" (higher physics and general physics), headed by Manuel F. Gran.

The teaching of theoretical physics was introduced by taking advantage of the favorable atmosphere prevailing in the country after the downfall of the Machado tyranny, which allowed for the implementation of long yearned for changes, particularly at the University of Havana. An idea of how unstable the situation was at

[40]As seen by a veteran revolutionary militant, the above referred struggle: "was a heroic struggle, in which people went out to overthrow the military dictatorship, and when this proved impossible, a united and organized struggle continued against the militarism that had seized the educational centers, to take them out of its hands. Student decorum and dignity shined very highly. Students refused to undergo the ordeal of returning to classes at a university occupied by the army and at secondary education centers intervened by the police, and they also refused to leave in the lurch those teachers who had supported them in the struggle against Machado, both throughout the March general strike, and, especially, during the years that followed it", González Carbajal (1974), 496–497.

the time can be inferred from the fact that it was not until April 1938, when Badell was officially awarded the professorship of chair "J," which meant he was responsible for teaching two courses: theoretical physics and general physics.[41]

Gran plunged himself into the task of writing a textbook for the higher physics course "in spite of the enormous printing difficulties and the no lesser economical ones." In the end, only the first 256 pages were printed; the rest was not because the result did not please the author and he therefore decided not to make it public. "Now we have this course almost finished, partly published in mimeographed copies, waiting for formal publishing in the way we like," Gran remarked in the early 1940s (Gran 1942). Though his work was never formally published, among the mimeographed supplementary texts he managed to bring out as a substitute, there were several texts on complementary topics such as statics, harmonic motion, thermometry, geometrical optics, electricity and other subjects belonging to the courses in higher physics.[42]

Undoubtedly, a special virtue of these courses, such as they were taught, at least from the 1940s, was the careful and critical treatment of the subjects dealt with. But concerning its deficiencies, it must be noted that some topics were developed in so much detail that no time was left to duly tackle certain areas of greater conceptual significance, such as the foundations of thermodynamics or of physical optics.[43]

The concepts of theoretical mechanics dealt with in the higher physics 1st course were complemented with the contribution from the corresponding specialized (rational mechanics) course. After 1937, Rafael Fiterre (1900–1979) began to teach this subject with an approach that, from his viewpoint,[44] should serve "as a solid and *practical* basis for subjects in applied sciences," as it was of particular help to the engineering students who had to take the subject together with the physico-mathematical sciences students. Later on, these students had to take a subject called complements of rational

[41] This course on general physics belonged to chair "D," which was intended to take care only of the teaching of the two courses in higher physics and of a special physics course for students of optometry.

[42] Until about the mid-1950s, Manuel Gran's work *Elementos de física general y experimental* (elements of general and experimental physics), in two thick volumes, published for the first time in 1939–1940 for high-school students, was used as a complementary textbook for the subject of general physics, taken by the students of some degree courses at the University of Havana. When Alonso took over the teaching of the subject in the 1950s, he replaced Gran's textbook with his own four-volume high school textbook *Física, curso elemental* (physics, elementary course) published in 1946–1948 for high school students, Alonso (1946–1948).

[43] For more details, see the chapter by J. Altshuler in this volume.

[44] A civil engineering and physico–mathematical sciences graduate at the University of Havana, Fiterre did postgraduate studies at the Universities of Louvain at the Sorbonne, where he witnessed "the end of the struggle that for years had been fought in France between the 'vectorials' and their opponents." In 1937, by taking a competitive examination, he attained the professorship of chair "O," which included since the early 1940s, the subjects of rational mechanics, complements of rational mechanics, vectors, higher geometry, projective geometry and abstract groups, Fiterre (1942).

mechanics, where the same professor developed kinematics following the methods of Gaston Julia[45] at the Sorbonne (Fiterre 1942).[46]

About 20 years after joining chair "D," Gran commented on the teaching of higher physics. He stated that its associated laboratory had improved "in the proportion of a thousand to one," though it was still "far from having reached the [appropriate] level, because of the huge number of students [who had to take it], the staff shortage and its meager funding." He remarked that the chair had introduced as an added graduation requirement which was later made official (included for the first time in the final year of 1937–1938 study plan), that all students of the physico-mathematical and physico-chemical sciences must undertake some "practical training" in the laboratory. This amounted to carrying out certain experiments of a more refined and complex nature than those they had undertaken in the regular courses (Gran 1942).

Theoretical physics was taught to students in the two last years of the physico-mathematical sciences degree course following Leigh Page's standard textbook *Introduction to Theoretical Physics* (Page 1936).[47] This textbook was frequently complemented by alternative mathematical developments supplied by Badell, some of which were published later.[48]

---

[45] The French mathematician Julia gained attention for his work after the war when a 199–page article he wrote was featured in the *Journal de Mathématiques Pures et Appliquées*, a French mathematics journal. The article, which he published during 1918 at the age of 25 described the iteration of a rational function. The article gained immense popularity among mathematicians and the general population as a whole, and so resulted in Julia's later receiving of the Grand Prix de l'Académie des Sciences. Despite his fame, his works were mostly forgotten until the day Benoit Mandelbrot mentioned them in his works on fractals.

[46] The study plan for the doctor in physico–mathematical sciences degree course in force from 1937–1938 to the beginning of the following decade, gives a general idea of how rational mechanics and other subjects were coordinated with the teaching of physics, Universidad de la Habana (1939):

First year: (1) mathematical analysis, 1st course; (2) higher physics, 1st course; (3) analytic geometry; (4) higher geometry; (5) trigonometry; (6) line drawing; (7) biology.

Second year: (1) mathematical analysis, 2nd course; (2) higher physics, 2nd course; (3) descriptive geometry; (4) inorganic chemistry; (5) cosmography; (7) surveying.

Third year: (1) complements of mathematical analysis; (2) theoretical physics, 1st course; (3) rational mechanics; (4) mineralogy; (5) crystallography; (6) astronomy; (7) topography; (8) special didactics for secondary teaching (optional).

Fourth year: (1) complements of geometry; (2) theoretical physics, 2nd Course; (3) complements of rational mechanics; (4) geology; (5) complements of astronomy; (6) geodesy; (7) practical training and thesis preparation.

[47] Leigh Page (1884–1952), chairman of mathematical physics at the Sloane Physics Laboratory of Yale University, developed the theory of conformal invariance, deriving in 1912 a complete electromagnetic theory, including Maxwell's equations, from only Coulomb's law and the Lorentz transformation. He wrote some valued textbooks. As we will see in the following, Page's textbook of theoretical physics was translated into Spanish, with some additions by Marcelo Alonso in 1945. A reference work on theoretical physics much resorted to at the time for the course of theoretical physics was the treatise by W. Wilson (1931–1940), professor at London University, whose three volumes were published successively in 1931, 1933 and 1940.

[48] For example, Badell Portuondo (1955).

4.8.3 The 1940–1951 Period

Needless to say, the improvements introduced in the higher physics courses and the introduction of the theoretical physics courses to the study plan of the physico-mathematical sciences degree course considerably raised the scientific level of the new graduates, who generally worked as physics teachers at the country's secondary education institutes. No doubt, this contributed to the success of the reform of the high-school study plan introduced by the early 1940s, a reform whose most remarkable trait was that it was extended to 5 years, with the last of these dedicated to a specialization either in the sciences or the humanities, in place of the four non-specialized years required by the former Varona plan.

In order to enter the university degree courses in sciences, engineering, architecture, agronomy, medicine, odontology, veterinary studies and pharmacy, high-school graduates were required to have passed their fifth year in sciences, which meant taking, in addition to their previous studies, two semesters of five hours of lessons a week in each of the following subjects: biology, mathematics, chemistry and physics. With reference to physics, it should be noted that during the first 4 years of their high school training, *all* students were required to take four semesters (two in their third year and another two in their fourth year) of alternate day physics lessons, regardless of the specialty they wanted to follow in their fifth year. During the 1940s and 1950s, several valuable textbooks of physics were published by Manuel Gran (Gran 1939–1940), Alfonso Páez (1946), Marcelo Alonso (Alonso 1946–1948), and Virgilio Acosta (1950–1951), as well as an excellent mathematics textbook by Mario O. González (1946) written for the use of fifth year science students. These textbooks helped to define a relatively high level for the mathematics and physics training of Cuban high school graduates of science, especially in the later 1940s.[49]

The advances just mentioned in the Cuban educational realm took place in a period following a repressive militaristic regime. The whole of the country's public life and affairs were based on the new 1940 constitution of the republic, noted for its fairly democratic and progressive character, as far as the prevailing socioeconomic system would allow. However, the new situation did not prevent the campus of the University of Havana from going through a turbulent period of internal struggle, originating from the lack of institutionalization of justice to punish individuals responsible for bloody repressive action by previous governments (Armas et al. (1984), 503–518).

In line with the fully autonomic status of the university established by the new constitution, new statutes for the University of Havana came into effect in 1942. According to them, among other things, the 12 faculties established by the 1937 statutes were defined as "schools," with their total number increased by one since engineering and architecture became separate schools instead of belonging to a single faculty, as was previously the case.

[49] For more details, see the chapter by José Altshuler in this volume.

From the 1941–1942 academic course onwards, several adjustments and modifications of some importance were made to the study plans of science degree courses taught at the University of Havana, which were to be retained without any essential modification until 1959–1960.[50]

In the context of the new study plans, the teaching of physics was shared by chairs "D" and "J" (Gran and Badell, respectively), taking into account the actual possibilities of their staff. Thus, for example, Agustín Guitart, who officially belonged to chair "J" and taught general physics to the students of agronomical and sugar engineering, was also in charge of the problem solving sessions taken by the students of higher physics, while Miguel Ángel Maseda, who belonged to chair "D," had been teaching theoretical physics, 2nd Course since 1937 to fourth-year students of physico-mathematical sciences.

The year after Badell's demise in 1947, Maseda was awarded by a competitive examination, the professorship of chair "J" (theoretical and general physics) with the unusual compliment of the examining board "for his brilliant performance." Until his passing in December 1957, Maseda continued to collaborate with Gran's chair, while at the same time teaching both courses in theoretical physics and overseeing the experimental work the students of physico-mathematical sciences had to complete to obtain their degree.

Before dealing with further developments in the teaching of physics organically linked to the University of Havana, it is worth recalling, however cursorily, the emergence of a few other elements that in one way or the other contributed to an enhanced outlook for physics in the country in the period under consideration.

Thus, for example, in February 1942 the Cuban Society of Physical and Mathematical Sciences was officially created, with 34 founding members who elected Pablo Miquel (1887–1944), the respected professor of mathematical analysis at the University of Havana, as president of the society and director of its official journal—the *Revista de la Sociedad Cubana de Ciencias Físicas y Matemáticas*—whose chosen editor was Manuel Gran. In his presentation of the first issue of the journal, Miquel pointed out:

> As yet, the history of mathematics and physics in our country has not been recorded in written form, but those of us who have taken a quick look at archival documents and libraries have come to the conclusion that when such a history is written down it will be summed up

[50] The subjects for the physico–mathematical sciences were finally arranged in the study plan as follows:

First year: (1) mathematical analysis, 1st course; (2) higher physics, 1st course; (3) analytic geometry; (4) higher geometry; (5) trigonometry; (6) drawing.

Second year: (1) mathematical analysis, 2nd course; (2) higher physics, 2nd course; (3) complements of analytic geometry; (4) descriptive geometry; (5) vectors; (6) cosmography.

Third year: (1) mathematical analysis, 3rd course; (2) theoretical physics, 1st course; (3) rational mechanics; (4) group theory; (5) astronomy; (6) introduction to geodesy; (7) vector analysis and notions of tensor calculus.

Fourth year: (1) mathematical analysis, 4th course; (2) theoretical physics, 2nd course; (3) complements of rational mechanics; (4) projective geometry; (5) complements of astronomy; (6) geodesy; (7) practical training and thesis preparation; (8) didactics for high school (sciences) [optional].

> in a few pages where only details of the teaching of these sciences among us will appear, probably without the summary being very flattering. Those who know this will come to the conclusion, on seeing that this journal has appeared, that the change that has taken place is more a revolution than an evolution. Perhaps this is our typical form of behavior in all events of history; a form that corresponds to impatient and nervous temperaments and peoples. (Valdés Castro 2000)

While teaching had improved considerably in the period dealt with here, research played virtually no role in the activities of the professorship. This changed only after 1959.[51] The 1940s saw the publication of several excellent textbooks in both physics and mathematics, especially in the realm of high-school teaching as already pointed out. During the 1940s, important changes took place in the social composition of the university student population, thanks to the access to higher-level teaching of an ever increasing student minority with modest economic resources. This was made easier by a law on education approved by the government in 1937, which dictated that up to 20 % of the poorer student body could be awarded free tuition.[52]

The "upper class" of Cuba was worried by these changes in the composition of the national intelligentsia; the most reactionary sectors wanted to exclude, as far as possible, the most humble sectors of the population from higher education, reserving it for the most favored. With the aim to resist any progressive changes in the country that might develop in the future, and using as an argument, among others, the academic shortcomings of the University of Havana,[53] a sector of the more well-to-do social classes succeeded in obtaining governmental approval in August 1946 for the creation of the *Santo Tomás de Villanueva* Catholic University.[54]

In the end, a more far-reaching decision was the creation of the University of Oriente at Santiago de Cuba, which was made a public university on November 23, 1949. The university had actually been functioning unofficially since October 1947 as a private institution in Santiago de Cuba thanks to the initiative and participation

---

[51] See the chapters by Altshuler and by Baracca, Fajer and Rodríguez in this volume.

[52] This originated from a deal with the strong student movement of the 1930s, who first demanded a higher percentage.

[53] The April 29, 1951, the editorial of the Cuban newspaper *Diario de la Marina* (known for its extremely reactionary viewpoints) put it this way:

> "It is known [that the University of Havana] is home to a student population that exceeds three times its teaching capacity. Such a superabundance is produced, to a great extent, by the repeated practice of [awarding] free tuition to poor students [...]"

[54] This can be gathered from the wealthy and notoriously right-wing attitude of the patrons of the Villanueva University and the call by Cardinal Arteaga to Cuban Catholics in 1946 to send their children to the new Villanueva University (which aroused the open opposition of the Federation of University Students (FEU) and other organizations). "Notwithstanding, the project to set the University of Havana—laical and autonomous, with a long Cuban and revolutionary tradition—against a Catholic and class-based university, copied from the north–American model, did not materialize. Villanueva attained neither the prestige nor the rooting in Cuba that the bicentennial University of Havana did. In the year of its opening, Villanueva only had a matriculation of 34 pupils, and in its period of higher growth (1956–1957) a total 1,224 matriculated pupils", Armas et al. (1984), 491, 522–523.

of a group of intellectuals from the former Oriente province, who, well aware of the hindrance that the established "chair-for-life system" meant for the University of Havana, provided the new higher study center with a more modern and flexible structure, where teaching posts would be subject to formal contracting. This provision allowed the newly created university to set up in a relatively short time a high level teaching staff made up of Cuban and foreign born specialists (particularly, a number of refugees from the Spanish Civil War) for whom access to the University of Havana was practically blocked. Roberto Soto del Rey (1913–1995), who graduated as a doctor of physico-mathematical and physico-chemical sciences in 1939 at the University of Havana and was then a physics professor at the Institute for Secondary Education in Santiago de Cuba, was put in charge of teaching higher physics to the chemical and mechanical engineering students. A former student of Professor Manuel Gran, Soto del Rey was an outstanding lecturer who not only gave himself to the task of creating the physics laboratories for the institution and motivating the students, but also undertook the assembly of a pilot plant for the training of chemical engineering students (Méndez Pérez and Baracca 2001).

Around mid-1949, Marcelo Alonso (1921–2005), by taking a competitive examination, got the post of assistant professor of chair "J," whose professor in charge was Miguel A. Maseda. After graduating in physico-mathematical sciences at the University of Havana, Alonso had taken post-graduate physics courses at Yale University, and published in 1945 an updated translation into Spanish of Yale professor Leigh Page's already mentioned textbook on theoretical physics (Page 1945). He also taught courses on quantum mechanics, physical optics and statistical mechanics at the summer schools offered by the University of Havana in 1947 and 1948.

At that time, after the Manhattan project and the atomic bombs dropped on Hiroshima and Nagasaki, the field of nuclear physics and technology, although it began in the 1930s, became the new international frontier of physics. Apart from the bomb, radioisotopes were usually produced and used in many fields. The introduction of nuclear physics in Cuba actually complied with essentially political factors, corresponding to the substantial dependence of Cuba on the United States. The initiatives that followed were probably more important for the "Atoms for Peace" campaign of the United States than for the concrete situation in Cuba, but it seems important to recall them since they reproduced a situation that was quite common to all the countries in the Western world.

In February 1947, Eudaldo Muñoz (1901–?), who was in charge of chair "P" for physical chemistry at the University of Havana, delivered a series of lectures at the Cuban Society of Engineers entitled "Atomic structure and nuclear disintegration," followed by a paper on the radioactive process of the disintegration products of radium which was published in the journal of the above-mentioned Institute (Muñoz 1948).

In 1947, the decree 4054(2) of President Grau San Martín established the *Comisión Nacional de Aplicaciones de la Energía Atómica a Usos Civiles* (CNEA) (National Commission for the Civil Applications of Atomic Energy), appointed to the *Instituto Nacional de Hidrología* of the Ministry of Health and Social Care

(Castro Díaz-Balart 1990). Among the goals of the commission, the following were mentioned: to promote research in atomic energy and its applications for non-military use, fundamentally with medical goals and in other sectors of the economy; to investigate the existence in Cuba of "natural sources of atomic energy"; to distribute radioactive materials, substances and treatments in hospitals; to determine and prescribe the conditions of radiological protection; to control and verify the efficiency of the detection instruments and the preparation of radioactive substances in use in the country.

Until the mid-1950s, none of the goals of the commission had been achieved: its establishment had been bound in fact to political goals of the American government, which had sent its emissaries throughout the whole continent with the purpose of guaranteeing for its growing radioisotope production industry the assured market of Latin America. The *f* sub-paragraph of the decree established by the commission pointed out explicitly that investigations on the applications for peaceful use of nuclear energy in Cuba were to be performed only "according to plans and techniques elaborated jointly with the atomic energy center of the U.S."

Backed by his personal contacts within the American scientific milieus, Alonso became a particularly keen and influential promoter of the development of this new field of nuclear physics and technology in Cuba. In 1950, radioisotopes were actually used for the first time in Cuba for cancer treatment: these radioisotopes were obtained through an American pharmaceutical firm and manipulated in a rudimentary laboratory established at the Reina Mercedes Hospital. From that year on, Alonso made many trips abroad to participate in various activities related to the use of radioisotopes and other applications of nuclear energy, which had already found a nominal niche in the CNEA.

4.8.4 The 1952–1959 Period

At the secondary education institutes and at the universities as well, teaching enjoyed a period of relative stability until March 10, 1952, when the coup catapulted Fulgencio Batista, the former strongman and former constitutional president of the country, to power. From then on, the struggle against the repressive regime kept the country's academic activities in a permanent state of unease that grew with time. Still, on November 30, 1952, the Marta Abreu University was created as a public university in the province of Las Villas in the central part the country. Its teaching staff immediately joined the universities of Havana and Oriente in holding a position at odds with the regime. Given the situation of the public universities, the government, with the goal to weaken their public influence, created a whole series of new public or private higher education centers, whose political submissiveness was supposedly taken for granted, even when the government eventually closed the "rebel" universities.

Meanwhile, the struggle against the regime in power was heating up, especially after February 13, 1953, when Rubén Batista Rubio, a student at the University of

Havana, died a victim of police violence while participating in an anti-government demonstration. On the following July 26, after legal actions against the regime had proved hopeless, a group of courageous young people led by Fidel Castro attacked the Moncada barracks in Santiago de Cuba, the second most important military fortress in the country. While this attack did not fulfil its immediate plan, it later on contributed decisively to developing the revolutionary struggle against Batista tyranny, a struggle that included from 1955 the very important participation of the Revolutionary Directory, an underground organization created for that purpose by the University Students Federation (FEU).

In what concerns physics, at the end of 1952 the economic counsel of the University of Havana allocated to the school of sciences' chair "J" a modest budget for the creation of a radioactive isotope laboratory, proposed and organized by Alonso. It enabled a minimum of equipment to be purchased for the new laboratory, which included a gamma-ray spectrometer that cost around $20,000 (Universidad de La Habana 1949).

By mid-1955, the de facto government dissolved the previous National Commission for the Application of Atomic Energy and replaced it with a new entity, ascribed to the National Economic Counsel: the *Comisión Nacional de Energía Nuclear de Cuba* (CENC) (National Nuclear Energy Commission of Cuba), appointed to the *Consejo Nacional de Economia* (Castro Díaz-Balart 1990).

The already mentioned lecture series delivered by Eudaldo Muñoz in 1947, his published paper and the fact that he and his students of physico-chemical sciences had been working on topics related to nuclear chemistry, no doubt influenced the decision taken by the university in November 1948, to name Muñoz its representative in the National Commission for Atomic Energy (Universidad de la Habana 1937).

Despite the ambitious goals declared for the CENC, it lacked effective economic support and was in fact short-lived. Gustavo Gutiérrez, an economist, and Marcelo Alonso were respectively appointed president and secretary (later technical director) of the commission, whose avowed goal was to study the possible installation of nuclear reactors in Cuba, and to control and oversee everything related to the matter. The CENC was created in accordance with guideline NSC 5507/2 ("Peaceful Uses of Atomic Energy") issued by United States President, Dwight D. Eisenhower, with the goal of "using the exports of nuclear technology to promote the national and regional interests of the United States".[55] In June 1956, the United States signed with

[55] Lavoy (2003). The presidential directive NSC 5507/2 was based on the "Atoms for Peace" program announced by President Eisenhower in a speech delivered before the General Assembly of the United Nations in December 1953. The central idea of the initiative was to take away a fraction of the large quantity of fissile material produced for military purposes by the nuclear powers and establish a "uranium bank." Though the program maintained the superiority of the United States on nuclear weapons, it was presented as a viable alternative to the more radical Soviet proposal to ban the use and production of nuclear weapons and destroy all the existing ones. Despite the tremendous international promotion campaign which was immediately set in motion and which attracted general sympathy to the American proposal, the program did not make much progress until August 1954, when nuclear technology was allowed to be sold to American

the Cuban government an intergovernmental agreement for "bilateral cooperation," similar to those already signed with 36 other countries of the "free world." According to this agreement, Cuba was expected to buy from the United States a 1 MW experimental reactor, a Van de Graaff accelerator, a Cobalt-60 source and several auxiliary laboratories (Castro Díaz-Balart 1990, 334). In preparation for the setting up of an electronuclear plant in the country by the Cuban Electric Company (the American company that monopolized electricity supply in Cuba), an introductory course in nuclear energy for electrical engineers taught by Marcelo Alonso was organized by the middle of 1956. By the end of the same year, a so-called First National Forum on Nuclear Energy was organized by CENC with the participation of ten national and six foreign invited lecturers. One of these, who was presented as the director of nuclear engineering of Ebasco Service Inc. and consulting engineer on nuclear energy of the Cuban Electric Company, gave a general description of the a 10.5 MW electronuclear plant designed around a boiling water reactor that the company said would be put into operation by the end of 1959 (Reichle 1957).

In May 1957, Marcelo Alonso addressed to the rector of the University of Havana a proposal for the creation of an institute for nuclear studies,[56] accompanied by a laboratory that would include a particle accelerator and, possibly, a small nuclear reactor. But, given the expenditure required for implementing the project (according to Alonso around $100,000 for the laboratory), the university merely replied that the proposal would be taken into account when the economic situation of the center could support such studies (Universidad de la Habana 1949). In relation to Alonso, it is worth citing his modern and updated text on atomic physics, published in 1958 (Alonso 1958, volume 1, the subsequent one was not published). In 1957 Alonso sought support in foreign countries and undertook a trip to West Germany to meet scientists, among others Werner Heisenberg in Göttingen, and representatives of large companies such as Hoechst Pharmaceuticals and Leybold, specialized in the construction of physical-technical instruments.[57]

While this was taking place in Havana, confrontation between the revolutionary forces and the tyrannical government in power was escalating rapidly. On December 2, 1956, a group of 82 combatants led by Fidel Castro landed in the eastern part of the country, and some of them attempted to organize themselves in the Sierra

commercial companies, who saw the "Atoms for Peace" programs as "the umbrella under which a US dominated world nuclear market would be realized." See, for instance, Weiss (2003) and Baracca (2012).

[56] In this respect, it would be useful to remark on the political position of the professors of the University of Havana in that troubled period. Most surely the majority of them would not have been in favor of an initiative somehow related to a governmental institution, for this might have lead to criticism from the students, whose leaders were fighting to the death against the de facto government. Still, the argument put forward by the rectorate concerning lack of money was realistic enough.

[57] Meyer–Cording (Bundesministerium für Atomfragen), an das Auswärtige Amt, June 18, 1957. Politisches Archiv, Auswärtiges Amt, 215–82.SL/1–91.22. We would like to thank Albert Presas i Puig for providing us with this information. From the documents known to us about this trip the results of these meetings cannot be reconstructed.

Maestra mountains. The following Christmas day, 23 men of various political affiliations were murdered by repressive government forces in order to terrorize the population, and a few weeks later the Sierra Maestra guerrillas won their first combat against the military. On March 13 of the same year, members of the Revolutionary Directory attacked the presidential palace in Havana, an action that cost the lives of many revolutionary activists, including that of José Antonio Echeverría, a student of architecture and president of the University Students Federation (FEU).

In the wake of the occupation of the University of Havana by the police force from March 13, the FEU ordered the boycott of any eventual attempt to resume academic courses, which in fact remained interrupted for more than 3 years, until the fall of the de facto government. This situation was reflected at the universities of Las Villas and Oriente as well. By then, the Cuban Electric Company had long abandoned its much publicized plan to put into operation an electronuclear plant at the end of 1959, because the action was deemed too risky and financially unsound.

Instead, what actually took place in the early days of that year was the setting-up of a revolutionary government, which was led by the forces that had just vanquished the previous tyrannical regime and who were ready to introduce the basic socioeconomic changes that were much yearned for by the Cuban population. In such a context, it did not take too long to implement a radical reform of Cuban higher education, which, among many other things, created the necessary conditions for implementing a deep change in the teaching of physics in subsequent years. But this matter exceeds the limits of the present chapter and must be dealt with in other papers that follow.

4.9 Conclusion

In this reconstruction of the developments of physics in Cuba in the pre-revolutionary period, we have seen that the quality of the teaching of physics after almost a century of stagnation on an elementary and qualitative level began to change in the 1920s and 1930s under the personal initiative of a limited number of self-taught physicists. These physicists succeeded nevertheless in establishing new courses and programs and in stabilizing a rigorous approach to the discipline of physics. As has been emphasized, none of the activities surrounding physics in Cuba as yet included any kind of original research, in spite of a few rare attempts. Physics education was basically limited to the training of teachers for the secondary schools.

Nevertheless, probably more so than early attempts to establish new fields, the most important inheritance for Cubans was the advanced level reached by the teaching of physics and scientific matters in general in the secondary schools on the basis of high quality textbooks written in Cuba by the new generations whose commitment has been detailed above.

As will be seen in the following chapter by Baracca, Fajer and Rodríguez, this relatively broad environment of scholars with a solid background in basic physics turned out to be crucial in the early years after the Revolution, when they could provisionally make up (though with great difficulty) for the shortage of professors at the university, for the new challenging needs, for radical change and for the introduction of modern physical theories.

References

Alonso, Marcelo. 1946–1948. *Física, Curso Elemental*, vol. 1–4. Havana: Ed. Cultural, S.A.

Alonso, Marcelo. 1958. *Física Atómica*, vol. 1. Havana: Ed. Cultural, S.A.

Altshuler, José. 2003. *A propósito de Galileo*. Havana: Ed. Gente Nueva.

Altshuler, José, and Miguel González. 1997. *Una luz que llegó para quedarse: Comienzos del alumbrado eléctrico y su introducción en Cuba*. Havana: Ed. Científico-Técnica and Oficina del Historiador de la Ciudad.

Álvarez Sandoval, Orieta, José Manuel Sánchez Ron, and Roberto Diaz Martín. 2002. *Blas Cabrera en la Institución Hispanocubana de Cultura*. Havana: Sociedad Cubana de Historia de la Ciencia y la Tecnología.

Arboleda, Luis Carlos, and Diana Soto Arango. 1991. Las teorías de Copérnico y Newton en los estudios superiores del Virreinato de Nueva Granada y en la Audiencia de Caracas. Siglo xviii. *Quipu* 8(1): 5–34.

Armas, Ramón de, Eduardo Torres-Cuevas and Anna Cairo Ballester. 1984. *Historia de la Universidad de La Habana*, vol. 1: 1728–1929 and vol. 2:1930–1978. Havana: Ed. Ciencias Sociales.

Badell Portuondo, Enrique. 1955. Derivadas parciales de los vectores unitarios en coordenadas curvilíneas ortogonales. *Revista de la Sociedad Cubana de Ciencias Físicas y Matemáticas* 3(4): 103–108.

Baracca, Angelo. 2012. The global diffusion of nuclear technology. In *The globalization of knowledge in history*, Studies 1: Max Planck research library for the history and development of knowledge, ed. J. Renn, 669–771. Berlin: Edition Open Access.

Betancourt-Serna, Fernando. 2007. *La recepción del Derecho romano en Colombia (Saec. XVIII). Fuentes codológicas jurídicas I*. Seville: Universidad de Sevilla.

Buenavilla, Rolando, et al. 1995. *Historia de la pedagogía en Cuba*. Havana: Ed. Pueblo y Educación.

Caballero, José Augustin. 1999 [1801]. *Obras*. Havana: Imagen Contemporánea.

Cañizares, M. 1885. *El alumbrado eléctrico: Oración inaugural pronunciada en la apertura del curso académico del curso académico de 1885 a 1886 y memoria–anuario del de 1883 a 1884*. Havana: Imp. del Gobierno y Capitania General.

Cantón Navarro, José. 1996. *Cuba: el desafío del yugo y la estrella*. Havana: Editorial Si-Mar.

Castro Díaz-Balart, Fidel. 1990. *Energía nuclear y desarrollo: Realidades y desafíos en los umbrales del siglo xxi*. Havana: Ed. Ciencias Sociales.

Chávez Rodríguez, Justo A. 1992. *Del ideario pedagógico de José de la Luz y Caballero (1800–1862)*. Havana: Ed. Pueblo y Educación.

Damerow, Peter, and Wolfgang Lefèvre (eds.). 1985. *George Adams: Geometrische und graphische Versuche ... (Nach der deutschen Ausgabe von 1795)*. Darmstadt: Wissenschaftliche Buchgesellschaft.

Delgado García, Gregorio. 2008. El Plan Varona y el desarrollo de la enseñanza superior de la medicina en Cuba. *Cuadernos de Historia de la Salud Pública* 103, http://bvs.sld.cu/revistas/his/his_103/his15103.htm. Accessed 9 Oct 2013.

Díaz Molina, Libertad. 1991. La física en Cuba a finales del siglo xviii y principios del siglo xix. *Quipu* 8(1): 63–90.

Ely, Roland T. 2001. *Cuando reinaba Su Majestad el Azúcar*. Havana: Imagen Contemporánea.

Figueroa y Miranda, Miguel. 1975. *Religión y Política en la Cuba del Siglo XIX. El Obipo Espada visto a la luz de los archivos romanos 1802–1832*. Miami: Ediciones Universal.

Fiterre, Rafael. 1942. *Informe quinquenal*. Universidad de La Habana: Expediente administrativo nº 7818.

Garcia Pons, César. 1951. *El Obispo espada y su influencia en la cultura cubana*. Havana: Publicaciones del ministerio de Educación.

González Carbajal, Ladislao. 1974. *El ala izquierda estudiantil y su época*. Havana: Ed. Ciencias Sociales.

González del Valle, Francisco. 1942. Varela y la reforma de la enseñanza universitaria en Cuba. *Revista Bimestre Cubana* 49(2): 199–202.

Gran, Manuel Francisco. 1939–1940. *Elementos de Física General y Experimental*, vol. 1–2. Havana: Ed. Minerva.

Gran, Manuel Francisco. 1942. *Informe quinquenal*. Universidad de La Habana: Expediente administrativo nº 7838.

Gran, Manuel Francisco. 1945. Félix Varela y la ciencia. Vida y pensamiento de Félix Varela. *Cuadernos de historia habanera* 27: 7–28. Havana.

Guerra y Sánchez, Ramiro. 1964, *Sugar and society in the Caribbean. An economic history of Cuban agriculture*. New Haven/London: Yale University Press. (Translation from *Azúcar y población en las Antillas*. Havana: Imprenta Nacional de Cuba. 1961.)

Gruzinski, Serge. 2004. *Les quatre parties du monde. Histoire d'une mondialisation*. Paris: La Martinière.

Lavoy, Peter R. 2003. *The enduring effects of atoms for peace*. Arms Control Association: Arms Control Today. http://www.armscontrol.org/act/2003_12/Lavoy. Accessed 9 Oct 2013.

Leroy y Gálvez, Luis Felipe. 1954a. Las ciencias. In *Facetas de la vida de Cuba Republicana, 1902–1952*, ed. E. Roig de Leuchsenring, 141–164. Havana: Municipio de La Habana/Oficina del Historiador de la Ciudad.

Leroy y Gálvez, Luis Felipe. 1954b. *Breve reseña del origen y desarrollo de la química en Cuba*. Havana: Academia de Ciencias.

Leroy y Gálvez, Luis Felipe. 1963. *La Facultad de Ciencias de la Universidad de La Habana en el centenario de su creación*. Havana: Universidad de La Habana.

Leroy y Gálvez, Luis Felipe. 1976. Aristóteles en la Universidad de la Habana. *Revista de la Biblioteca Nacional "José Martí"* 18(1): 27–78.

Leroy y Gálvez, Luis Felipe. 1979. *Profesores de física de la Universidad de La Habana desde su secularización en 1842 hasta Manuel F. Gran*. Havana: Ed. Academia.

López Sánchez, José. 1973. Recepción de las ideas de Copérnico en Cuba. In *Nicolás Copérnico, 500 Aniversario, 1473–1973*, ed. Academia de Ciencias de Cuba, 3–16. Havana: Museo Histórico de las Ciencias "Carlos J. Finlay".

Luz y Caballero, José de la. 1832. Letter dated May 4, 1832, to Mr. D. Justo María Vélez, Headmaster of the San Carlos Seminar. *Revista Bimestre Cubana* 2(6): 94–132.

Marcet, François. 1850. *Curso de física experimental* (Translated from the third French edition by José Manuel Mestre). Havana: Del Tiempo Printing House.

Méndez Pérez, Luis, and Angelo Baracca. 2001. Cincuenta años de física en la Universidad de Oriente (Santiago de Cuba). *Rev. Cub. Fis.* 18(2): 146–154.

Moreno Fraginals, Manuel. 1978. *El ingenio, complejo económico social cubano del azúcar, 1*. Havana: Ed. Ciencias Sociales.

Moreno González, Antonio. 1988. *Una ciencia en cuarentena: Sobre la física en la Universidad y otras instituciones académicas desde la Ilustración hasta la crisis finisecular del XIX*. Madrid: Consejo Superior de Investigaciones Científicas.

Muñoz, Edualdo. 1948. Estudio del proceso radioactivo de los productos de desintegración del radio. *Archivo del Instituto Nacional de Hidrología y Climatología Médicas* 2(2): 162–170.

Niderburg, Nicolas. 1807. *Cultivo del galvanismo y uso de sus virtudes por la medicina*. Havana: Palmer's Printing Press.

Ortiz, Fernando. 2001. La Sociedad Económica de Amigos del País de La Habana en la formación de la conciencia nacional de Cuba. In *Universidad del Aire (conferencias y cursos)*, 119–141. Havana: Ed. Ciencias Sociales.

Page, Leigh. 1936. *Introduction to theoretical physics*. New York: D. Van Nostrand Co.

Page, Leigh. 1945. *Tratado de física teórica*. Havana: Cultural, S.A.

Parker, William Belmont. 1919. *Cubans of today*. New York/London: G. P. Putnam's Sons.
Plaza, Eugenio de la. 1804. Announcement. *Papel Periódico de la Havana*, Aug. 26: 274.
Pruna Goodgall, Pedro M. 1986. ¿Cómo se percibía la necesidad de una academia en la Cuba colonial?. *Conferencias y Estudios de Historia y Organización de la Ciencia* (50): 7–18 [Reproduced in *Estudios de Historia Social* (España) (44–47): 569–575, 1987].
Pruna Goodgall, Pedro M. 2010. La Real Academia de Ciencias de la Habana, 1861–1898. *Isis* 101(2): 441–442.
Pruna Goodgall, M. Pedro, and Rosa M. González. 1987. *Antonio Mestre en la cultura científica cubana del siglo XIX*. Havana: Editorial Academia.
Pruna Goodgall, M. Pedro, and Ovidio Ortega. 1985. La composición de la Real Academia de Ciencias de La Habana. *Revista Latinoamericana de Historia de las Ciencias y la Tecnología (Quipu)* 2(2): 251–262.
Puig-Samper, Miguel Angel, and Mercedes Valero. 2000. *Historia del Jardín Botánico de La Habana*. Aranjuez: Ed. Doce Calles.
Puig-Samper, Miguel Ángel, and J. Luis Maldonado. 2005. Ciencia y cultura en la época de Alejandro Ramírez en Guatemala, Cuba y Puerto Rico. In *Las antillas en la era de las luces y de la revolución*, ed. J.A. Piqueras, 159–183. Madrid: Siglo XXI.
Ramos Lara, Maria de la Paz. 1994. *Difusión e institucionalización de la mecánica newtoniana en México en el siglo xviii*. México: D.F., Sociedad Mexicana de Historia de la Ciencia y la Tecnología, A.C. and Universidad Autónoma de Puebla.
Reichle, L.F.C. 1957. Descripción de la central núcleo-eléctrica que se instalará en Cuba. *Ingeniería Eléctrica* 2(April–June): 25–28.
Rodríguez, José Ignacio. 1878. *Vida del presbítero D. Félix Varela*. New York: O Novo Mundo Printing House.
Saco, José Antonio. 1823. *Esplicación de algunos tratados de física*. Havana: Fraternal de los Díaz de Castro Printing House.
Saco, José Antonio. 1960 [1824]. Estado de las Ciencias Físicas en La Habana en los años de 1823 y 1824. *Colección de papeles científicos, históricos, políticos y de otros ramos sobre la Isla de Cuba ya publicados ya inéditos* 1: 21–44. Havana.
Sáenz, Tirso, and Emilio García Capote. 1989. *Ciencia y technología en Cuba*. Havana: Editorial de Ciencias Sociales.
Simpson, Renate. 1976. Francisco de Arango y Parreño. Sus esfuerzos en pro de la educación científica en Cuba. *Revista de la Biblioteca Nacional "José Martí"* 18(3): 13–51.
Simpson, Renate. 1984. *La educación superior en Cuba bajo el colonialismo español*. Havana: Ed. Ciencias Sociales.
Torres-Cuevas, Eduardo. 1995. *Félix Varela: los orígenes de la ciencia y con-ciencia cubanas*. Havana: Ed. Ciencias Sociales.
Universidad de la Habana. 1866. *Memoria del curso 1864–1865 y Anuario de 1865 á 1866*. Havana: Gobierno and Capitanía General Printing House.
Universidad de la Habana. 1870. *Memoria del curso 1869–1870 y Anuario de 1869 á 1870*. Havana: Gobierno y Capitanía General Printing House.
Universidad de la Habana. 1873. *Memoria del curso 1872–1873 y Anuario de 1872 á 1873*. Havana: Gobierno y Capitanía General Printing House.
Universidad de la Habana. 1881. *Memoria del curso 1880-1881 y Anuario de 1880 á 1881*. Havana: Gobierno y Capitanía General Printing House.
Universidad de la Habana. 1882. *Memoria Anuario 1881–1882*. Havana: Gobierno y Capitanía General Printing House.
Universidad de la Habana. 1902. *Expediente del Catedrático Titular de la Escuela de Ciencias, Sor. Juan Orus y Presno*. Expediente Administrativo nº 102.
Universidad de la Habana. 1903. *Memoria Anuario 1901–1902*. Havana: Avisador Commercial printing house.
Universidad de la Habana. 1909. *Memoria Anuario 1907–1908*. Havana: Cerqueda and Comp.
Universidad de la Habana. 1916. *Memoria Anuario 1914–1915*. Havana: El Siglo XX printing house.

Universidad de la Habana. 1937. *Expediente del profesor Eudaldo Muñoz Jústiz*. Expediente Administrativo nº 7,856.

Universidad de la Habana. 1939. *Catálogo General y Memoria 1937–1938*. Havana.

Universidad de la Habana. 1949. *Expediente del profesor Marcelino Alonso Roca*. Expediente Administrativo nº 9,056.

Valdés Castro, Concepcion. 2000. La primera publicación periódica cubana de ciencias físico-matemáticas (1942–1959)—Noticias y consideraciones. *Llull* 23(47): 451–468.

Varela, Félix. 1820. *Lecciones de filosofía, 4*. Havana: Palmer and Son's Printing Press.

Varona, Enrique J. 1999. Las reformas en la enseñanza superior, in Trabajos sobre educación y enseñanza. In *Varona—Política y Sociedad*, ed. Enrique José, 206–217. Havana: Ed. Ciencias Sociales.

Vélez, Justo. 1832. Letter to the Editor dated May 18, 1832. *Revista Bimestre Cubana* 2(6): 93–94.

Vitier, Medardo. 1938. *Las Ideas y la Filosofía en Cuba*. Havana: Instituto del Libro (reprinted1970).

Weiss, Leonard. 2003. Atoms for peace. *Bulletin of the Atomic Scientists* 59(6): 34–41, 44.

Wilson, W. 1931–1940. *Theoretical physics*, vol. 1–3. London: Methuen & Co.

Zambrana, Antonio. 1861. *Memoria del Sr. D. Antonio Zambrana, dando cuenta de los actos de su administración durante el período de su Rectorado*. Havana.

Zeuske, Michael, and Max Zeuske. 1998. *Kuba 1492–1902. Kolonialgeschichte, Unabhängigkeitskriege und erste Okkupation durch die USA*. Leipzig: Leipziger Universitätsverlag.

Chapter 5
Mathematics and Physics in Cuba Before 1959: A Personal Recollection

José Altshuler

On the occasion of the 50th anniversary of the formal inauguration of the Republic of Cuba (May 20, 1902), the then Historian of the City of Havana, Emilio Roig de Leuchsenring, decided to publish a memorial volume which would include various special studies on the progress made in the country during the preceding half century with regard to specific aspects of national collective life. The volume, entitled *Facetas de la vida de Cuba republicana, 1902–1952* (Facets of Life in Republican Cuba, 1902–1952) appeared in print in 1954. Its publication was delayed—as Roig explained in his prologue, somewhat cryptically—"due to causes beyond our control or of that of the Cuban writers to whom we entrusted the preparation of the said studies." Needless to say, the causes referred to were none other than those derived from the atmosphere of political unrest in the country after the military coup staged in March 1952 by former strongman, General Batista.

There is no question that the detailed study by professor Luis Felipe Le Roy included in the above-mentioned volume under the title *Las ciencias* (The Sciences) (Le Roy y Gálvez 1954) is required reading for anyone who wants to ponder the development of scientific activities in the country during republican times. He points out that it can be immediately inferred from his study that

> at the beginning [of the 20th century] the [country's scientific] production was nil or virtually nil, no doubt due to the state of political and social stagnation that though Cuba had just left behind when it ceased to be a colony of Spain, it unavoidably survived and would still influence for some time the cultural and scientific development of the country.

While I have no objection to this idea, I think another negative factor should be added to the scientific development of the country which, in my opinion, spread well beyond the early republican decades, at least until the onset of the 1959 Cuban Revolution. By this I mean the prevailing situation in that period which in a thousand ways promoted at the very heart of Cuban society the goal of achieving

J. Altshuler (✉)
Cuban Society for the History of Science and Technology, Havana, Cuba
e-mail: jea@infomed.sld.cu

A. Baracca et al. (eds.), *The History of Physics in Cuba*, Boston Studies in the Philosophy and History of Science 304, DOI 10.1007/978-94-017-8041-4_5,

personal "success," usually understood as the attainment of personal prosperity at the expense of higher and more useful values for the benefit of the nation.

Thus, for example, it is well-known that there were quite a few members of the higher education teaching staff who clung to their posts, not because of their vocation or their will to serve society, but simply to use these as a sort of a calling card to facilitate their participation in the business world or in national politics, also perceived as profitable business. Needless to say, such a mentality was hardly compatible with the full dedication typical of a real research worker. It must be said, however, that there were some exceptions to the rule—whom it is fair to remember with the respect they are entitled to—but these few certainly did not portray in the least the trend of the times.

In view of the above, and of what will be added hereafter, it is not odd for me to find rather debatable Le Roy's conclusion that "On the 50th anniversary of independent Cuba, the sciences in our country [were] at the same level of development as those in many of the nations of older civilization of the European continent..."

It is true that in various curricular aspects a valuable level had been reached in the country thanks, essentially, to the efforts of a group of highly regarded Cuban educators, some of whose textbooks achieved a high degree of excellence. To this I will come back later, and with no little satisfaction. But it is no less true that during the republican period, generally branded nowadays as Neocolonial or Conditioned, as far as physical, technical and mathematical sciences are concerned it was extremely rare to find any consistent engagement in scientific research work, which is the essential ingredient of any science deserving the name.

Surely enough, such a state of affairs as referred to above will be somewhat difficult to imagine for the present university students in Cuba and even for their younger teachers, currently subject to the requirement of having some novel results included in their so-called "diploma-work" which is needed to complete the degree course they have taken. It might be argued that before the University Reform of the 1960s, in order to obtain a university diploma, the student was required to defend an original thesis work before an examination board. In practice, however, that amounted to pure formality, for theses that contained original scientific results were truly exceptional. That was part of the established academic reality at the time.

As far as I recall, perhaps the closest thing then existing to independent scientific work done by students was the requirement established in the old Faculty of Sciences of the University of Havana to have the physico-mathematical undergraduates perform in their fourth and final year what was called a "degree practical work," intended for them to tackle by themselves certain experiments of some complexity. Of course, contact with practical work was better in the natural sciences and the physico-chemical degree courses, due to their very nature, closely linked to fieldwork in the former case and to the chemical laboratory in the latter.

By way of illustration, let me refer to a situation I know well, since it concerns something in which I was directly involved when I returned to Cuba by the end of 1955, after having spent a year of studies abroad thanks to a scholarship awarded by the University of Havana. I came back encouraged by the fact that one of my first scientific results, modest as it was, had been accepted for publication by a reputable journal in the field of electronics and telecommunications.

While there would be much to comment about the explosive political outlook in Cuba at the time of my arrival, what I want to stress here is that in the 1950s I did not find in my country a single scientific or technical journal where I could publish regularly even the modest results I set myself to obtain by swimming alone against the tide, if necessary. This was one of the reasons that prompted me to propose myself to the board of the Electrical Engineers Association to take the main responsibility for editing and publishing a journal, *Ingeniería Eléctrica* (Electrical Engineering), as the official organ of the Association.

Soon after the proposal was approved I devoted myself to systematically including in the above-mentioned publication whatever original scientific and technological papers related to electrical engineering I could get, aside from articles of general or professional interest to the membership. The result was that, while the journal was able to survive for more than 3 years, the job of getting manuscripts with original research or development results turned out to be much more difficult than I had anticipated. Faced with the virtual non-existence of such a type of material, it became necessary to approach a few authors from outside the profession to keep the level of the journal at a minimum of respectability. It is quite clear to me that something similar must have happened to another one of the then very rare Cuban scientific publications: the *Revista de la Sociedad Cubana de Ciencias Físicas y Matemáticas* (Journal of the Cuban Society for the Physical and Mathematical Sciences). This I gather from the explicit statement on the inside of its cover stating that the journal "is not periodical," added to the fact that the number of original results published was quite limited. I suppose that a motivation not too different from mine with *Ingeniería Eléctrica*, was what drove Dr. Mario O. González, an outstanding mathematics professor of the Faculty of Sciences of the University of Havana, to take on the burden of editing and publishing the *Revista de la Sociedad Cubana de Ciencias Físicas y Matemáticas* for many years. Still, it must be said that it did succeed in publishing the best of the scant research then done in Cuba on mathematics and physics subjects.

Typically, during the republican period before 1959, the meager scientific production of the country in the above-mentioned fields of study was generally due to the individual activities of a few lone workers. One example of this is the case of José Isaac Corral, a forest and mining engineer who worked as a civil servant at the Ministry of Agriculture, whose physical and mathematical papers appeared in local publications from the 1930s to the end of his life, in 1946. Another example is the case of Dr. Carlos Masó, who in spite of the fact that he had taken advanced courses on astronomy in France, had to count himself lucky when he was appointed professor at a provincial teacher training college where, nevertheless, in the 1940s he managed to create a mathematical seminar to expound to a handful of participants his solutions of some mathematical problems of his own choice.

The most important developments in the field of physics and mathematics which took place in the period dealt with here were actually implemented at the University of Havana, also centered around a few members of its teaching staff. An indication of this is the very fact that the teaching of the relevant subjects was carried for the first time to what can be considered, for the time, a genuine university level, thanks to the will, learning and talent of Drs. Pablo Miquel and Manuel Gran, two remarkable professors of the University of Havana who are generally remembered to this day as almost mythical figures.

In 1913, Dr. Miquel began to teach mathematical analysis in place of the regular professor, an engineer who had received his degree in the United States and took part in the Cuban war for independence, but who decided to leave the teaching profession to fully devote himself to republican politics. Miquel's textbook on higher algebra, published in 1914, became a landmark in the teaching of the subject in the country, and perhaps also in Latin America, especially after the much improved 1939 edition. Moreover, the two-volume textbook on differential and integral calculus he published in the early 1940s (Miquel y Merino 1941, 1942) was an excellent work which compared very favorably with the best of its kind at the international level. As to Dr. Gran, it is a significant fact that while he was a young graduate, like Dr. Miquel, in civil engineering, architecture and physico-mathematical sciences, he was promoted to a high level teaching post in 1923—the year of the first attempt at a general reform of higher education in the country—in response to the clamor of the students of physics subjects, who refused to continue taking classes from a glaringly incompetent teacher. From a humble background, and having achieved a self-acquired prestige and professional competence, Gran had to apply himself to the task of creating—as he would recount in later years—"a course of higher physics worthy of the name," and of organizing the laboratory "so that it would accomplish its purpose in the best possible way," and "preparing the corresponding laboratory guide" for the students. Noted for his interest in developing his students' critical awareness, he greatly influenced not only the teaching of physics at university level, but also at high school level, thanks to a carefully elaborated textbook on general and experimental physics in the two thick volumes he published in 1939–1940 (Gran 1939–1940), which brought to the explanation of the subject the scientific rigor and elegance lacking in previous works on the same subject. Professor Gran's textbook became an essential reference for other high school textbooks published later by Profs. Páez, Alonso, and Acosta, with alternative pedagogical approaches (Alonso and Virgilio 1950–1951; Páez 1946).

Let me stop here for a moment to take a glance at our secondary education in those days, for in spite of some of its shortcomings, I think it embodied quite a few commendable traits, especially after the early 1940s when its syllabus was overhauled and a final fifth year was added in two specialties—humanities and sciences—which could be freely chosen by the student. The new syllabus required that all students without exception should take two courses in physics and one in chemistry during the first 4 years of high school, while those who took the specialization in sciences had to take additional courses in mathematics, physics, chemistry and biology. Several textbooks published subsequently with a view to covering the needs of the students attained a remarkable level of excellence, as was the case of those by Rosell and González (mathematics), Ledón (chemistry), Velázquez (psychology), and Marrero (geography) among others. I believe that some high-school textbooks, published before the new syllabus was established such as those of Baldor on algebra and Galán on chemistry, were praiseworthy from a pedagogical point of view. On the other hand, it is fair to recall the early efforts by professor

Moleón, who in 1921 published the first high school textbook on general chemistry. To which I would like to add that in those times such competent professionals as Drs. Páez, Alonso, Acosta, Melgarejo and Soto del Rey in physics, and Drs. Souto, Labra, Paz, Álvarez Ponte, Fiterre, Rolando, Gutiérrez Novoa, Reguera, Davidson and others in mathematics worked as high school teachers. Later, some of them became respected university professors.

Returning to the development of mathematics and physics at the University of Havana after the above digression, it is fitting to point out in the first place that since the mid-1930s the physics curriculum had been enriched by the inclusion of the two-semester degree course physico-mathematical sciences in theoretical physics. The first full professor of the subject was Dr. Enrique Badell, from 1938 to his death in 1947. He was succeeded in the post by Dr. Miguel A. Maseda, an excellent teacher who was equally at ease in experimental work. Both were remarkable for their thorough command of the subjects they taught.

Throughout his tenure Professor Badell painstakingly dedicated himself to the step-by-step reconstruction of whatever mathematical developments were only outlined or incompletely specified in the available textbooks and to systematically solve all the exercises and problems he came across. As a result, after many years of tireless effort, he had accumulated an impressive collection of problems solved in great detail and frequently in two or more alternative ways. Something similar happened with Professor Luciano de Goicoechea, noted for his very thorough teaching of several courses on the theory of electricity to students of electrical engineering. Professor Gran accumulated a large collection of original physics problems, especially in the field of optics, while one of the most appreciated features of Professor Miquel's textbooks on higher mathematics was the wealth of exercises and proposed problems they incorporated, many of them created by the author himself. It is probably not an exaggeration to say that this laborious dedication to the development of educational problem development was the nearest thing to systematic scientific research attained by some of the best teachers of the older generation in the field of physical and mathematical sciences. I believe it is only fair that they should be remembered with utmost respect considering their hearty devotion to the task of raising the teaching level of their respective disciplines, in spite of the indifferent or even hostile cultural atmosphere that generally surrounded them at the time.

It was a time when the teaching of physics and mathematics was aimed at exclusively didactic goals; a time in which the most committed teachers spent a substantial part of their not very generous salaries to buy the latest book on their subject to appear in the catalogs of foreign publishers. On the other hand, it must be said that it was quite rare to find in the country's libraries any journals dedicated to the publication of original scientific work in the field of physics or mathematics.

With the inclusion of Drs. Mario O. González and Marcelo Alonso in the Faculty of the University of Havana, who graduated from the same university but had subsequently taken graduate courses abroad and published the results of their first scientific work in specialized journals, new subjects and styles were introduced into the higher education teaching of physics and mathematics. González, who succeeded Professor Miquel in 1947 as head of the department of Mathematical Analysis, was active in organizing various mathematical seminars, and published in 1952 an excellent textbook on the theory of complex variable functions (González 1952). Alonso, who had joined the department of theoretical physics around 1949, published in 1958 the first volume of a fine textbook on atomic physics (Alonso 1958) and laid the foundations for a modest atomic and nuclear physics laboratory. This, together with an introductory course on nuclear energy for electrical engineers, which he inaugurated in 1956, was closely related to the expectations created around the announcement by the American company that had the monopoly of Cuba's electricity services in those days. The company had intended to inaugurate its first nuclear plant in 1959, the implementation of which was never taken very seriously.

Needless to say, what actually did happen in 1959 was the inauguration of a period of deep revolutionary transformations in Cuban national life. As far as university education was concerned, those changes were expressed in detail for the first time in the Law for the Reform of Higher Education enacted in January 1962. Among many other regulations it stipulated that 5-year degree courses in physics and in mathematics should be established, to be taught at newly created schools within the Faculty of Sciences, with the explicit proviso that practical training, seminars and other activities should be organized with a clear view to promoting scientific research. As a result, in subsequent years there was a remarkable increase in the number and qualification of students who completed their physics or mathematics degree courses at national universities and pedagogical institutes.

Contrary to common practice in the country before 1959, starting in the 1960s it became a sine qua non requirement for the completion of any university degree course that a proper academic board should approve a "diploma work" containing original results obtained by the student. A national commission was created in 1977 to regulate and oversee the awarding of doctoral degrees by individual institutions, to which it may be added that scientific research, fundamental as well as applied to the solution of problems of national interest, was then considered more than a purely academic endeavor. This was indicated by the fact that many newly created institutions were dedicated to research and development work, while the post of scientific researcher was given professional standing, in other words put on the same level with that of the university teaching staff. It was taken for granted that specialists might be paid not only to do full-time research on applied subjects, but also on subjects from whose eventual results one could hardly expect to have a practical impact in the short run. A comparison of this situation with the former situation is quite revealing, among other things, because it causes one to appreciate the value of the work done by the very few who dedicated themselves to scientific work at that time.

To conclude, let me recall that in "Cuba's Great Leap," a four-page "commentary" published by *Nature* in 1983 (Ubell 1983), its author reported on the considerable scientific and technological advances achieved by the country in a brief period, especially in the fields of biomedical and agricultural research, though they were not what they are today. Still, a comment on the progress of the physical and mathematical sciences already achieved in Cuba in the same period was conspicuous by its absence. In view of the above, I like to think that it would not have been out of place to briefly mention therein the progress in those disciplines already achieved in the country at the time.

References

Alonso, Marcelo. 1958. *Física atómica*, vol. 1. Havana: Cultural, S.A.

Alonso, Marcelo, and Acosta Virgilio. 1950–1951. *Introducción a la física*, vol. 1–2. Havana: Cultural, S.A.

González, Mario O. 1952. *Fundamentos de la teoría de funciones de variable compleja*. Havana: Cultural, S.A.

Gran, Manuel F. 1939–1940. *Elementos de física general y experimental*, vol. 1–2. Havana: Minerva.

Le Roy y Gálvez, Luis Felipe. 1954. Las ciencias. In *Facetas de la vida republicana, 1902–1952*, ed. Emilio Roig de Leuchsenring, 141–164. Havana: Town Council of Havana, Office of the City Historian.

Miquel y Merino, Pablo. 1941. *Curso de cálculo diferencial*. Havana: Cultural, S.A.

Miquel y Merino, Pablo. 1942. *Curso de cálculo integral*. Havana: Cultural, S.A.

Páez, Alfonso. 1946. *Física elemental, 2do curso*. Havana: Cultural, S.A.

Ubell, Robert. 1983. Cuba's great leap. *Nature* 302: 745–748.

Chapter 6
A Comprehensive Study of the Development of Physics in Cuba from 1959

Angelo Baracca, Víctor Luis Fajer Avila, and Carlos Rodríguez Castellanos

6.1 General Introduction

6.1.1 The Context

A foreign scientist visiting Cuba will be able to appreciate the generally excellent teaching and research levels of its higher education institutions, as well as the large quantity and high quality of the existing scientific research centers. On the whole, these advances are the fruit of five decades of a clear and continued policy based on the principle of universal and free education for all, and the backing of scientific research as a fundamental tool for social and economical development. Because a good example of this development of science and higher education (along with its

---

We dedicate this chapter to the memory of

Andrea Levialdi, an extraordinary figure consistently engaged in the struggles for social justice of his time, a militant antifascist and an outstanding scientist, who, terminally ill as he was, accepted an invitation to give a graduate course in Cuba. This, however, he could not finish because he passed away on December 8, 1968 in Havana, where his remains lie.

Fernando Crespo Sigler, whom we had the pleasure and the honor to collaborate with before his demise on June 12, 1997. He was a person and a scientist intensely committed to the best development of his country. We think the result of this historic research would have been much better had he lived long enough to follow it to its completion.

A. Baracca
Department of Physics, University of Florence, Florence, Italy
e-mail: baracca@fi.infn.it

V.L. Fajer Avila (✉)
Department of the Physics, Center for Technological Applications and Nuclear Development (CEADEN), Havana, Cuba
e-mail: vfajer@infomed.sld.cu

C. Rodríguez Castellanos
Physics Faculty, University of Havana, Havana, Cuba
e-mail: crc@fisica.uh.cu

A. Baracca et al. (eds.), *The History of Physics in Cuba*, Boston Studies in the Philosophy and History of Science 304, DOI 10.1007/978-94-017-8041-4_6,

peculiarities) can be seen in the field of physics, it is interesting to examine the details of the historic development of physics in Cuba, its general and specific elements, and the periods in which it happened.

In order to appreciate this achievement, one may recall that prior to the Revolution, Cuba had four universities—three public and one private—with a total of about 15,000 students; by 1985 the country had 46 such institutions with a total of 200,000 students. According to official data the amount of state resources allocated to scientific activities has been considerable: in 1997 more than 60,000 specialists worked in 422 academic centers of diverse disciplines. With 590 physicians per 100,000 people, Cuba is the leading country worldwide in the quantitative provision of medical personnel (CEDISAC 1998).

Another aspect that has characterized contemporary Cuba is the extension of the education system and the very high percentage of the population with a degree. In 1961, a general campaign of widespread mobilization eradicated illiteracy in the country. In the following decades, quite an advanced research and higher education system was built up, starting from relatively precarious levels. In addition, Cuba has an exceptionally large number of university graduates with a solid level of training, which is at present over 8 % of the national population.

Physics is a quite peculiar aspect of this whole picture, but very interesting and significant. The physicists are a small part of the Cuban scientific community, but they played an important role in the development of science and its applications in Cuba. Because physics is a basic tool in all branches of the natural sciences and technical disciplines, physicists are usually employed in many of them. In order to graduate in any of these disciplines, the students have to follow one or more courses of physics, as almost all the experimental and measurement devices (apart from those that have a purely chemical basis) are based on physical principles, and the presence of physicists is needed for their setting up and control. It is impossible to conceive of the development of scientific disciplines in a country that heavily lacks physicists.

After the Revolution in 1959, there were no existing activities in the country that deserved the name of scientific research in the area of physics; Cuban universities did not train physicists in the modern sense,[1] and the number of physicists in the country ranked in the order of dozens. In view of these initial conditions, it is quite remarkable that – in a country that at the time had around 6.2 million inhabitants, barely 2.2 % of the world population – in the relatively short period of only 15–20 years, a higher education system with an international standard was developed, simultaneously with some branches of scientific research that could compete with those of much larger Latin American countries with a stronger scientific tradition. In the following years, the Cuban scientific system was expanded and strengthened,

[1] See the contribution to this volume by Baracca and Altshuler. A report by the ad hoc Truslow Commission of the International Bank for Reconstruction and Development, which had travelled to Cuba to study the provision of loans, stated unequivocally in 1950 that "in the field of applied research and labs, there was no development at all in Cuba" (Sáenz and García-Capote 1989; Clark Arxer 2010, 123).

and assumed a modern and efficient structure.[2] The number of graduated physicists reached 429 in 1976 (see in the following and Appendix A.1), and the accumulated number of physicists graduated in Cuba and abroad today ranks on the order of 2,500 (more than 10 % of them with a higher degree equivalent to a PhD), plus a similar number of graduates as professors of secondary school teaching. One can conclude that since 1959, more than one million professionals in all specialties have graduated from Cuban universities, and the total number of physicists graduated in Cuba and abroad, including those graduated in physics and teaching, is less than 0.5 % of the total number of professionals.

An international report on physics in Latin America (Morán-López 2000) places Cuba (a country with some 11 million inhabitants), together with Chile, Venezuela, and Colombia, among the countries with an intermediate number of PhDs in physics (between 100 and 500), after Brazil (3,000), Mexico (2,200), and Argentina (2,000). Moreover, this report places the University of Havana among the outstanding regional institutions in the advancement of physics, and recognizes the contribution of Cuban physicists to such areas as condensed matter and materials physics, solar energy, optoelectronics and medical physics.

These aspects assume a more general sense and relevance when considered in relation to what frequently happens in underdeveloped countries. Scientific and technological development is linked to an understanding of progress according to the model presented by the richer countries. However, it is not always clear whether the advancement of science and technology is a necessary condition or a result of that type of development. In this sense, the Cuban case also gives rise to interesting considerations. The Latin American experience, in particular, was generally different from the Cuban case (Gálvez Taupier 1986). Latin American countries often saw the creation of advanced research centers dedicated to research barely related to national needs, while other institutions of higher learning with educational functions did not reach the same level.[3]

---

[2] For evidence, see the remainder of this paper and Part III of this volume.

[3] Argentina, for instance, has a long history in physics starting at La Plata University in 1909 when Emil Bose arrived. Up until the 1950s, this university was the foremost center for the study of physics in Argentina. In the fifties, the University of Buenos Aires and the University of Cordoba created physics departments. At about the same time, the Atomic Energy commission laboratories at Bariloche and in Buenos Aires were established. The physics department of the Universidad de Buenos Aires as a whole has a very broad spectrum of activities which essentially cover all areas of physics. Its biggest deficiency is the relative lack of experimental programs. This is a result of the historical lack of funds available for the purchase of equipment.

In a recent study, Hurtado De Mendoza and Vara (2007) underline some features that may help in appreciating differences in the Cuban case (apart from the most developed countries): "… some developing countries decided to invest a significant part of their rather modest science budgets in building many-million-dollar facilities. A comparative approach to the study of the first stages of the Argentine TANDAR heavy ion accelerator and the Brazilian National Laboratory Synchrotron Light (LNLS) projects may help to understand specificities in patterns of organization of big science in peripheral contexts. Oversimplification of the decision-making processes linked to authoritarian political contexts—which allowed to overcome the lack of consensus within the physics community as well as financial uncertainties—seem to have been a necessary condition for TANDAR and LNLS, which differentiated them from big science in developed countries."

In comparison, the Cuban case presents quite different and original features that we will discuss in detail for the field of physics. The effort has constantly been made by both the Cuban government and the national scientific community to link science to the most urgent needs of the country's further economic and social development, although several different phases must be set apart, corresponding to growing needs and changes of the decision structures and social actors. Soon after the Revolution, intense discussions were held on the need to promote education and scientific research, and on the branches that could be most useful for the promotion of general development. As we shall see below, this is evident from the decisions that were made from the beginning in the case of physics, which was considered (although probably sometimes implicitly and for the interest and participation of prominent foreign specialists in this field) a strategic discipline for supporting both technological development and other scientific branches, such as biology and medicine. Such discussions were not limited to a narrow scientific and academic environment, but involved, on the one side, the student component, which had played an important role in the political and social events that preceded the Revolution and actively helped in overcoming the deep deficiencies of the university, and on the other side, the leadership and decision levels of the country, whose intellectual protagonists—in particular Ernesto Guevara and Fidel Castro—had already acquired a clear awareness of the importance of the development of science.

In the 1960s, the challenge of developing a modern scientific and high-level education system, and the choices of which research fields had to be promoted, was tackled through a wide debate which involved all the components of Cuban society—political leaders, intellectuals, professors and students, as well as foreign advisers—and paid special attention to the bonds of the branches of physics with the national needs of the country. In the following section, we will sometimes refer to this period as the "romantic" phase, because of the enthusiasm for the debate and action by the Cuban actors and by the international left wing milieus (in which, as already mentioned, physicists showed a remarkable interest). While a special effort was made to reach a critical mass of well-trained physicists, new research centers were created with the same aim of fulfilling basic national needs, and they were integrated with other institutions of higher learning and research. In this respect, advantage was taken of both the support of "western" visiting professors and of the opportunities offered by sending students to graduate in Soviet universities).

Ultimately, the Cuban government and Cuban scientists tried to build a balanced system of sciences, avoiding over-specialization in one or the other field, as had happened in many developing countries. An example of the correspondence between science and public needs is the close collaboration between meteorological institutes and civil protection, which makes helps avoid the worst consequences of natural catastrophes, such as hurricanes. Another example is the collaboration between biotechnological research and the public health system. In spite of this emphasis on public needs and applied science, the fundamental fields of physics and other sciences have not been neglected.

In a general sense, the highly participatory processes in the post-revolutionary climate were not limited to the educational sector, and this made Cuba one of the most egalitarian societies in the world, and resulted in acquiring a rather high ranking in the Human Development Index elaborated by the United Nations Development Program (UNDP 2003).

When, in the mid-1970s, a basic modern education and scientific system was shaped, a new phase began: some prominent activities had reached a competitive level in Latin America, in particular microelectronics, and the university curriculum and teaching body were stabilized. Aside from the intrinsic need for re-organizing and coordinating branches, institutions and initiatives that had emerged mainly with their own specific tasks, some improvisation, loose coordination and more general factors intervened during this time.

The initial plans of a rapid industrialization, conceived to overcome economic dependency and backwardness, had not been achieved. Even the subsequent re-orientation of the economy toward the sugar sector at the end of the 1960s was met with the failure of the *Zafra de 1970.*[4] New economic strategies had to be implemented. Meanwhile, the state structures had reached a considerable level of development. The links with the Soviet Union had become much stronger, the economic planning system was introduced, and from 1972, Cuba was integrated into the Council of Mutual Economic Assistance (COMECON). In this new context, a "Process of Institutionalization" initiated in the 1970s, soon involved the educational and the scientific systems as well. Organizational directions and forms of rigid control of teaching activities came from above. Government institutions were created for the supervision, control and organization of teaching and scientific activities and the first general assessment of the state of development of research in physics was arranged (Pérez Rojas et al. 1976). Although this process of centralization led to contradictions, it must be acknowledged that "for the first time in Cuba, contrary to the situation of other Latin American countries, science received decisive support and occupied a first order place" (Gálvez Taupier 1986).

These achievements allowed for further extension and differentiation of the Cuban scientific system in the 1980s. The development of new scientific and technical fields was deemed necessary for the future needs of the country, such as nuclear energy (the decision to build a nuclear power plant in Cuba) and electronics. Other fields, both old and new, were also promoted as will be discussed. The Cuban government made strategic decisions for the development of these new areas. In the case of electronics, there was strong participation of the budding scientific community in this field and specialists from industry in the elaboration of proposals, which were later assimilated into governmental policies. The early development of the nuclear sector in the previous decade, which was promoted by collaboration with the Sovict Union and the active and decisive work of then recently graduated Cuban physicists, benefitted an autonomous organization the 1980s under the general direction of the physicist Fidel Castro Díaz-Balart.

[4] "Zafra" is the Latin American term for sugar harvest. The term "Zafra of 1970" refers to the target of reaching the record sugar production of 10 million metric tons during the 1969–1970 Zafra. This target was not reached: the Cuban economy was severely affected and this led to important changes in economic policies, with a progressive approach toward the Soviet model.

Around the mid-1980s, the Cuban government promoted a *Proceso de Rectificación* (Process of Correction), which re-centralized the economy and returned ideologically to a strong emphasis on the original character of the Cuban Revolution. This in turn began a process of "Correction of Errors and Negative Trends", which implied a certain dose of self-criticism regarding the rigid adaptation since the 1970s of Soviet models. Nevertheless, the reliance on scientific and technical development was reinforced as a result of this revision process and led to the creation of new centers and programs, particularly in the area of physics.

It is difficult, if not impossible, to provide an appraisal of the effectiveness of this revision process, in particular, of its possible effects on scientific fields due to its occurrence at a time when Gorbachev's *perestroika* policies put a particular strain on political and economic relations between the two countries. Between 1986 and 1989, the Soviet Union gradually reduced its economic support for Cuba.

When the Soviet Union collapsed, many people thought that Cuba's social organization, sciences and higher education system would not survive. But this was not the case. Without doubt, the new situation had a negative impact on the development and organization of scientific research and teaching, but science, in particular physics, did not collapse and even registered some progress. In the decade between 1991 and 2000, the existence of a "core" of scientists and technologists, together with a consolidated system of science and higher education, enabled Cuban science and technology to withstand and overcome obstacles that appeared insurmountable. From 1960 to 1990, about 1,300 Cuban physicists graduated (836 of them from Cuban universities) and a similar number of secondary school physics teachers graduated from higher institutions of pedagogy. The number of PhDs in physics registered in 1990 by the National Commission of Scientific Degrees was 129, about 10 % of the total number of physicists, and the number was growing at a rate of ten PhDs per year. In the 1990s, 518 new physicists and 58 PhDs received their degrees in Cuba; this is not dramatically less than the 103 physicists of the 1980s. In addition, MSc programs in physics, which had an ephemeral existence in the 1970s, were launched again in 1994 and produced 108 specialists in various branches of pure and applied physics. In the last two decades a "brain drain" occurred in the form of scientists who decided to leave the country: this problem, however, did not exist before the 1990s.

Concluding these preliminary remarks on the peculiarity of the educational and scientific development of Cuba with respect to the other Latin American countries, it seems pertinent to add that the choices made in Cuba seem rather original in comparison to other developing countries facing problems concerning the so-called "underdevelopment", the technical and scientific gap, and generally the relationships and reciprocal influences between scientific, economic, cultural and social development. An appraisal of these choices, their effects and effectiveness will be attempted in the course of this analysis of the development of physics. In this respect, it must be specified from the beginning, however, that, as mentioned, the disintegration of the Eastern Bloc occurred too early and too suddenly, at a time when the most important developments were still underway. This prevented their full enforcement. An undoubtedly positive effect of the Cuban choices and efforts, however, was the fact that, in spite of the tremendous shock, its scientific system did not collapse and even showed signs not only of resistance, but also of renewal and revival.

6.1.2 Methodological Note and Structure of This Chapter

The present chapter aims to reconstruct the main stages of the above-mentioned process from the Cuban Revolution (1959) to the present time.[5]

A general premise is necessary on the documentary sources used and the method adopted in this research. The written documents available on the development of physics in Cuba related to the early two decades are quite scarce, so that the reconstruction presented here is based mainly on oral history research. Most of the information presented in this study is sourced from interviews with Cuban colleagues who played leading roles in the events described here. This approach obviously implies drawbacks since the information obtained relies on the personal memories and views of the interviewees, and in some cases uncertain or controversial aspects arise. Despite this drawback, every effort has been made to verify the information.

A second premise is that the level of the results presented in this analysis is not homogeneous, since a verification of the information could only be applied in a systematic way to the first 20 years of the whole period under consideration, i.e., the 1960s and 1970s. This period has some peculiar aspects because it corresponds to the launching of the process, which was undertaken under extremely difficult conditions for the country as a whole, and for universities in particular. In this sense, the direct witnesses we approached were irreplaceable for transmitting the liveliness, fantasy, and creativity that characterized the early steps and consolidation of the results and structures. This could hardly be reproduced in official documents and resolutions. Improvisation was essential but always guided by the goal to achieve rigorous standards, which, however, were not always easy to adhere to. The vitality of the personal reminiscences, however, was generally compatible enough with a reasonable level of detachment in order to render a good historical perspective.

Concerning the later decades, from the 1980s to the present, the level of this investigation is less supported and complete. This is due, in the first place, to a lack of time to proceed systematically with the above-discussed method (although personal records were verified), and also because of other factors that complicated

[5] Partial results of this research have appeared in several journals: Baracca (1999), Baracca et al. (2001, 2004), Baracca and Méndez (2001), Baracca (2005), and Baracca et al. (2006).

the picture and the work: the higher complexity, articulation and amplitude of the mature phase of the Cuban scientific system, combined with a smaller temporal distance, all of which makes the necessary historical perspective more problematic. Perhaps the greatest difficulty that characterizes the most recent period, and makes it difficult for an historical evaluation of the previous developments was the drastic turn between the 1980s and the 1990s sparked by the collapse of the Eastern Bloc. After three decades of efforts to build a modern and efficient system from scratch, it now seemed necessary to begin anew. The challenge for a historical reconstruction of these dramatic events is given by the close time proximity and the risk of falling into ideological or apologetic interpretations. Despite this, we consider the following analysis, based on the soundness of the developments achieved in the preceding decades, to be reliable enough to explain the resistance of the Cuban scientific system to the effects of the crisis induced by Soviet collapse.

6.1.3 The Phases of Development in Physics in Cuba After 1959

Since the Revolution, the development of physics in Cuba has gone through several phases of change, growth and settling: these were characterized, as anticipated, by somewhat different processes, motivations, mechanisms and actors. At times, the very spirit with which problems and difficulties were tackled changed from one phase of development to another. This took place, in particular, during the initial stages when the difficulties of developing an advanced education and research system from practically nothing but with a collective and cooperative enthusiasm which is labeled as the romantic phase.

The distinction of a sequence of phases of development exposes historians to the dangers of arbitrariness, abstraction, schematics or approximation, since changes do not happen abruptly but are the result of a merging process. Different approaches and proposals coexist in every phase, often as many as the actors involved. Nevertheless the approximation proposed here is the result of efforts to identify the root processes, their evolution and incorporation in the Cuban scientific system and methods. Using this approach, we will discuss the developments comparatively before reconstructing the development of each separate branch over half a century and the level reached by almost all branches of physics at each stage.

It is useful to state beforehand that the scientific system in Cuba, despite its limits in comparison to larger, more developed countries, has reached a considerable level of complexity and interconnectedness. More so, because the promotion or creation of new organs—as well institutional changes that sometimes appear rather arbitrary—tell much about scientific choices. Every effort will be made to separate conceptually the unavoidable analysis of the entanglement of institutions and organizational initiatives. This will be undertaken by evaluating their contribution to specific scientific achievements, although this may lead to repetition.

6.1.3.1 The 1960s: Early Efforts and Improvisation

When the revolutionary forces took over, the three public universities existing in Cuba had been closed since the end of November 1956 when the University Council suspended academic activities because of the brutality of police repression of students. At that time, the universities had just over 15,000 students, most of them enrolled in humanities degree courses. The task of the universities had been to educate the neocolonial elite and to teach scientific courses to secondary school teachers.[6] As a result, a good level of secondary teaching of physics had been attained in the decades preceding the Revolution. This provided a solid basis for the new university, though the number of both professors and graduates was rather limited and thus delayed the development of a new generation of well-trained physicists.

In the period between the academic years 1959–1960 and 1970–1971, enrollment rose by only 10,000 students. This increase took place mostly because of the greater opportunities offered to the relatively few students who had completed higher secondary education. Most important was that this modest increase was accompanied by a substantial change in the enrollment structure, which favored scientific and technological degree courses. The real quantitative step forward took place in the following decade, when the wave of educational growth that had begun with the 1961 literacy campaign reached the universities, thus increasing enrollment to 155,000 students (MES 1997).

During the 1960s, advances took place at the University of Havana, particularly at its Physics School (*Escuela de Fìsica*), founded in 1962. Great efforts were made to implement an updated syllabus for the *licenciatura* in physics, to stabilize the teaching staff, to create the first new laboratories and workshops, to set up research activities and to graduate, at the turn of the decade, the first hundred physics *licenciados*, who would be capable of boosting higher-level education and scientific research in various branches.

The situation of the Physics School remained very unstable during the first half of the 1960s, mainly because of a lack of adequate teaching staff and aggravated by the emigration of some former staff members. The situation was such that practically all students from the second year and above had to teach what they could to the students in lower years. This was despite the fact that since 1962 several foreign physicists, both Soviet and western, were involved in relevant academic activities for relatively long periods, as will be detailed later. In 1961, a process began which was to play a fundamental role in later years: a substantial number of Cuban students took scholarships offered by the government to study physics or physics-related degree courses in higher education centers in the Soviet Union.

After extended debates, the enactment of the Higher Education Reform Law in 1962 contributed to the stabilization of the ongoing educational process at a time of violent external aggression (Comisión de Investigaciones 1963). To grasp the

[6] This is analyzed in detail in the chapter by Josè Altshuler and Baracca in this volume.

irreproducible atmosphere and enthusiasm of those years, one must consider the intellectual and international origin (Guevara was an Argentine physician) of the leaders of the revolution, whose trips abroad brought them into contact with international effervescent milieus and the most advanced revolutionary debates and projects. Moreover, one must also consider the central role that the larger part of Cuban students, intellectuals and also workers' leaders had played in the revolution. As a consequence, among the leading actors, the need to build a modern university and to develop modern and advanced scientific research, which was almost completely absent in Cuban universities before 1959, was widely acknowledged.

Keeping this in mind, one may understand how the goal of developing research in the university was at the core of the lively discussions that prepared the 1962 reform. The reform was elaborated through intense debates involving all of the above-mentioned components of Cuban society. This made scientific research pre-eminent as an expression of the desired link between theory and practice and as a way for the academic community to participate in the social and economic development of the country. In turn, this gave rise to a group of revolutionary transformations that radically modified the university in less than a decade. In Cuba, there was a clear perception of the relevance of physics as a basic tool for every further scientific and technical development, and efforts concentrated in this discipline together with other basic sciences. Given the post-war international climate, however, this choice should not be too surprising. For better or worse, physics brought striking advances during the war and in the rhetoric of the leading countries, including the socialist ones, it promised miracles, both in itself and as a basic resource for fantastic progress in every field.

The first steps in developing scientific research activities advanced quickly, in spite of difficulties and deficiencies of every kind. In the Physics School of the University of Havana, this was linked especially to experimental work. In spite of its substantially voluntaristic drive and the modest results obtained, this early introduction of the new spirit played a very important role and was to leave an enduring tradition. The establishment of several new laboratories and workshops provided the effective and necessary base for the future development of research work.

Two other foundational acts that were significant for the further general development of physics and science were again linked to the University of Havana. The first was the inauguration at the end of 1964 of the José Antonio Echeverría[7] University City (CUJAE), the new site of the Faculty of Technology. The second was the merging of the then recently-founded National Center for Scientific Research (CNIC) with the University of Havana. The center would develop fully-fledged courses in basic sciences and play a central role in training research workers in the biological, medical, veterinary, chemical and other scientific branches.

---

[7] José Antonio Echeverría Bianchi (1932–1937), a student of architecture, president of the University Students Federation (FEU) and of the *Directorio Revolucionario*, was a leader of the student fight against the tyranny of Fulgencio Batista. He led the actions of March 13, 1957, which included the attack on the presidential palace in Havana. He died the same day in a clash with the police near the walls of the University of Havana. Cf. Sect. 6.6.3 of the chapter by Altshuler and Baracca in this volume.

The clear idea of the need to develop an articulate system of scientific research spread beyond the borders of the universities. Thus, the new Cuban Academy of Sciences (ACC) was founded in 1962, which aimed at promoting research in various fields.[8] In 1962–1963, the Ministry of Industries, then headed by Ernesto Guevara, set up a group of institutes dedicated to doing research related to sugar production, sugar cane-derived products, metallurgy, standardization, industrial chemistry and mining. When the Latin American Center for Physics (CLAF) was created in 1962 as an intergovernmental organization for the promotion of physics in Latin America, Cuba became one of its founding members. All of these were seeds that started to evolve into a broad base and a scientific culture which later were to yield important results.

In order to grasp some of the basic features of this analysis, it is worth emphasizing from the beginning the basic factors that contributed to promoting and shaping Cuban physics. Obviously, the enormous efforts made by Cuban physicists, and early on by the students, were fundamental. From the very beginning, they relied on their own knowledge and abilities to advance the development of autonomous work and basic tools. These efforts left a kind of enduring imprint in the attitude of Cuban physicists from this era. It is interesting to note that this attitude of making the best of a situation corresponds to a peculiar and common feature in every aspect of Cuban life, exemplified, for instance, in the continuous use of old-time American cars, despite the lack for more than half a century of any spare parts. One of the first occurrences of this kind of do-it-yourself method in physics was the use of the most advanced American textbooks of physics and mathematics of the time—regardless of copyright—in some cases piecing together single chapters into a mosaic textbook, sometimes known as "Frankensteins." Suitable for modernizing the teaching of physics, these textbooks were printed and distributed freely to students.

As a second basic factor, the support given by the Soviet Union and the East European countries can be hardly be overstated. However, this became increasingly relevant only in the subsequent years when the bonds between Cuba and those countries became closer, and the participation of Soviet and East European specialists increased dramatically. Cuban graduates in those countries, especially in the Soviet Union, but also in East Germany, Czechoslovakia, Hungary and Poland, began to return and their numbers increased. Cuban scientists were also hosted by high-ranking Soviet and eastern academies and institutions.

In the early years, and until the early 1970s, there was another extremely relevant but lesser known group of actors who supported and promoted the development of higher-level teaching and scientific research activities in Cuba, and who played a concrete role in molding the mentality and attitude of Cuban physicists and was key for certain branches and initiatives. As early as 1961, a remarkable number of professors and experts from many western countries, such as the United States,

[8] The first academy was established in Cuba in 1861 as a scientific society under the name *Real Academia de Ciencias Médicas, Físicas y Naturales de La Habana* (cf. the chapter by Altshuler and Baracca in this volume). With the independence of the country in 1902, the academy lost the appellative *Real*. In 1962, the Academy of Science was re-established with a national character and with the task of developing research centers.

France, Britain, Italy, Argentina and Mexico, visited the Physics School of the University of Havana for more or less extended time periods, providing expert advice, giving advanced courses, organizing student laboratories and workshops and promoting the first scientific research activities. Moreover, between 1968 and 1973, French physicists organized summer schools and, together with colleagues from other countries, gave courses and supplied much-needed apparatus and materials.[9] In particular, we will refer to the role played by French physicists by the end of the 1960s on the development of electronics and microelectronics, before the cooperation was consolidated with such institutions as the Moscow "Lomonosov" State University, the Leningrad "Ioffe" Physico-technical Institute, and other teaching and research centers in the Soviet Union and the East European countries. From 1969, several Cuban physicists were trained at teaching centers and laboratories in Italy (MASPEC University and laboratory at Parma), Sweden (Uppsala University) and France (Paris Sud and Montpellier Universities). In the 1970s, a UNESCO program made it possible for Cuban physicists to collaborate with their Canadian colleagues, mainly from the University of Toronto. While some of these links weakened by the mid-1970s, contact was maintained, though on a smaller scale, with scientists, academics, and professionals from both eastern and western countries.

Anticipating some of the conclusions of this analysis, the influence of the latter two factors and actors differed from field to field. In the early phase, the development of semiconductor physics, solid-state electronic devices and microelectronics was supported at the Physics School by western specialists: its most advanced achievements were made by the French physicists in the summer schools. Only in a subsequent phase did the relationship with the Soviet "Ioffe" Institute take a leading role. But, as the contribution by Fabrizio Leccabue to this volume emphasizes, as late as 1981 the aim to develop advanced research on solid-state physics and items of technology (calcopyrites and related alloys for photovoltaics, magnetic materials, sensors, magnetically coupled mechanical systems, piezoelectrics, ferroelectrics) met with the difficulty that "scientific exchanges with the USSR were useful, but could not cover the whole set of conditions for the development of this field of research," and wider international contacts and collaborations had to be promoted, also to "help in providing and shipping scientific equipment" (Italy, Spain, France, Austria, Portugal, México, Venezuela, Brazil). On the other hand, the fields of nuclear physics and technology were thus developed and organized with the overwhelming support of the Soviet Union.[10]

In the second half of the 1960s, the first graduates to have obtained their degrees in the Soviet Union returned to Cuba and the first *licenciados* trained in physics were incorporated. The Physics School was thus able to stabilize its teaching staff and rigorously update its syllabus. The graduates trained in the Soviet Union introduced a much more systematic and rigorous method, which led to some conflicts. The student component, which played an important role in the early activities and

[9] See the chapter by Jacqueline Cernogora in this volume.

[10] See the chapter by Leccabue in this volume.

shared considerable influence in its operation, had a more relaxed attitude toward learning and teaching methods. A clash was unavoidable and the influence of the students on teaching methods subsequently declined.

After the development phase and the stabilization of the curriculum in the second half of the 1960s, the number of students grew, to the extent that in 1970 the first "mass" graduation of 67 *licenciados* occurred. From these, a "critical mass" of physicists with good basic training was selected either to increase the teaching staff of the Physics School itself or to staff other recently created institutions. In those years, new research centers or groups related to physics were established within the Academy of Sciences and at the Physics Schools of the universities of Oriente (Universidad de Oriente n.d.) and Las Villas. The first graduation of the Physics School of the University of Oriente took place in 1970. At the same time, dozens of Cuban students were studying degree courses in physics in the Soviet Union, East Germany and other European socialist countries.

With the support of visiting professors and autonomous resources, the first research activities began to materialize in the Physics School and the first results were attained. It was in this period that the debate took place concerning the most appropriate direction to steer in physics so that it would meet the basic needs of the country. Apart from the political leaders, this debate not only involved the teaching staff, but spread beyond the university to involve some of the physicists who had participated in the Cultural Congress of Havana in January 1968, which was attended by 483 renowned figures.[11] These included the Italians Roberto Fieschi, Bruno Vitale and Daniele Amati, the Frenchman Jean-Pierre Vigier and the Soviet Boris Pavlovich Konstantinov.[12] To put this into the cultural and political context of the time, one has to keep in mind the movements against the aggression of United States in Vietnam, and also the French riots of May 1968. In view of the existing limitations of available resources and the accepted policy of achieving effective collaboration between the universities and the research institutions, there was a consensus on the need to concentrate efforts in a certain directions. For instance, it was decided to discard the construction of charged particle accelerators, and the pursuit of sub-nuclear physics.[13] While complying with its duty to train a wide range of *licenciados* in physics, the Physics School of the University of Havana, the country's leading institution in physics, still pursued research and postgraduate studies in (mainly) experimental and theoretical solid-state physics, specializing in semiconductor materials and devices, metals and magnetism. As will be shown later, this choice had important consequences.

It was in these years that the long-term directions of scientific research were defined for the majority of physics-related institutions in Cuba. Other institutions were created, expanded or transformed in order to develop other fields of scientific activity that were considered important for national needs. These institutions,

[11] For more on the choice of the research field, see Sect. 6.3.2.1.

[12] The General Declaration of the Congress was published in various languages, among these in the Journal "Vida Universitaria": Vida Universitaria (1968).

[13] Other socialist countries chose to pursue all fields, thereby overstretching their capacities.

moreover, came under the increasing influence of Soviet and other socialist schools, as these trained a large number of the Cuban physicists. At the CUJAE, the focus was placed first on research and development of microelectronics, and then later on optics. Geophysical, astronomical oceanological and meteorological work began at the Academy of Sciences, which was extended to nuclear physics in 1968 and to quantum optics, acoustics and field theory from 1970. Nuclear physics and its applications were also dealt with at the University of Oriente (as well as X-ray and metal physics, optics and spectroscopy), at the CNIC and at Havana's Oncology Hospital. A Center for Metallurgical Research and an Institute for Metrology grew at the Ministry of Industries, as well as a Central Laboratory for Telecommunications at the Ministry of Communications. Thus, a network of more or less specialized technoscientific institutions came into being, within which groups of physicists were created to tackle research and development programs which, when put together, covered a wide spectrum of physical sciences.

Toward the end of the 1960s, scientific exchange with Soviet universities and institutes of the Soviet Academy of Sciences began to be institutionalized. Rather than undergraduate training, it aimed instead at the postgraduate training of young Cuban *licenciados* in physics and at the strengthening of the country's incipient research groups. Collaboration with the University of Parma and the MASPEC Institute of the same city began during the same period.

6.1.3.2 The 1970s: Shaping the Scientific System in Cuba

The 1970s witnessed the decisive takeoff of research in physics and other areas as well as the achievement of the first important results. The advances achieved in microelectronics at the Physics School of the University of Havana and at the CUJAE were remarkable: there was already a project for a pilot plant to produce semiconductor devices and integrated circuits. We have already stressed the role of French physicists in the summer schools in introducing the most advanced materials and techniques. In the first half of the decade, the activities in this field reached an acceptable level by international standards, with the ability to completely manufacture medium integration circuits. When the 4th Latin American Symposium on Solid State Physics was held in Havana in 1975, it could be noted that much progress had been made in this field in a very short time. At the same time, remarkable advances were made in other branches such as nuclear physics, optics, meteorology, geophysics and medical physics. In the early 1970s, postgraduate courses yielded the first master degrees (in Cuba and abroad) and doctorate degrees (abroad). By the end of the decade, the Cuban Society of Physics had been founded.

In under 20 years of growth, the Cuban scientific system had achieved its first institutionalization as part of the Cuban state organization. In 1975, the Communist Party of Cuba held his first congress, and in 1976, the Constitution of the Republic of Cuba was enacted. As we have mentioned, the "process of institutionalization" and the adoption of economic planning led to stricter controls by the government, with the creation, or reform, of institutions responsible for drawing up national scientific policy, supervising its execution, and generally guiding the country's scientific efforts (in 1974, the National Council for Science and Technology, CNCT; 1976, Ministry of Higher Education, MES). While to a certain extent this institutionalization numbed the romantic, improvising and enthusiastic spirit that had fueled the initial stages, it introduced the organization and rationality required to enable the full development of the national scientific system in the following decade.

Collaboration with leading institutions in the Soviet Union and in many other socialist countries was substantially strengthened, as were the final choice of economic and political bonds. Nevertheless, as mentioned, Cuban physicists never cut contact, collaboration or exchanges with physics institutions and centers in capitalist countries; in several cases these increased. As will be discussed in more detail later, despite the tremendous prestige and influence of Soviet science, its methods of scientific organization and approach were never fully adopted in Cuba. Instead, they were adapted to the characteristic Cuban do-it-yourself approach.

6.1.3.3 The 1980s: Progress, Criticism, Reorganization and Crisis

Thanks to the foundations laid during the two preceding decades, the 1980s saw the full development of the Cuban scientific system, and of physics in particular. A new organization arose in which science as such was embedded in the national economic planning, especially when required to support the processes of industrial technology transfer from the Soviet Union and Eastern Europe. Thus, for example, the strategic governmental decision to build a Soviet electronuclear power plant, together with a wide range of nuclear technologies, led to an accelerated development of the techno-scientific nuclear sector, which was radically reorganized in an autonomous and top down manner, raising some contradictions in the Cuban community of physicists. But the efforts were not limited to the nuclear sector and involved other strategic fields, such as investment for the development of the electronics industry and its related research. This increase in scale posed new challenges and raised new contradictions.

Around the mid-1980s, Fidel Castro and the Cuban Communist Party promoted the *Proceso de Rectificación de errores y tendencias negativas* (process of correction of errors and negative trends), mentioned above, and the reorientation of the Cuban economic system in order to correct the errors derived from the uncritical implementation of the Soviet model. This process was contemporaneous to Soviet *Perestroika*, but was aimed to strengthen socialism and not to introduce capitalist reforms. It included practically all spheres of society, but one of its main aims was undoubtedly to 'close the cycle of scientific research.' In spite of the remarkable results achieved thus far, the low level of application of the results obtained was in fact criticized, together with the dispersion and lack of integration of efforts made. Still, the dissatisfaction with the scant economic and social contribution of science did not cause a lack of confidence, but rather strengthened its position. During this process, important changes were introduced to Cuba's scientific policy, and relevant investments were made in the creation of centers of research "of a new kind," which aimed to ensure conditions for the introduction of scientific results into social practice. Scientific poles and national, territorial and branch technoscientific programs were created, along with an innovation movement through the organization of science and technology forums (Rodríguez Castellanos 1997). The main positive results were the creation of the scientific pole of biotechnology and of the medical- pharmaceutical industry.

Many of the changes introduced during the rectification process will be evaluated in the following, although this poses difficulties since they overlap with the loosening of economic bonds with and diminished support from the Soviet Union of Third World countries. This in turn hampered the full development of projects and the fulfillment of scientific objectives in Cuba: the construction of a nuclear power plant, for example, was at first put on hold and then finally cancelled. In general, the transformations that were already underway helped the scientific institutions to increase the applications and direct economic contribution of their results, especially in the *Período Especial* that was to follow. However, in many cases basic research and the training of doctors were weakened.

6.1.3.4 Cuban Physics During and After the "Periodo Especial"

There is at least one final historical judgment that can be drawn. The entire Cuban research and education system experienced tremendous cutbacks in the financial and material resources at its disposal, and important international scientific collaboration links were suddenly cut off. Many physics programs were reduced or redirected, since priority was given to sectors less related to physics, such as tourism, food, biotechnology, medicine and pharmacy (Rodríguez Castellanos 1997). But, in spite of this, the solidity achieved by the national scientific system during the three previous decades, added to the attained critical mass, professional level and social engagement of its scientists, enabled it to basically resist this tremendous blow. Needless to say, the related changes and consequences have been quite marked, the experimental activity has diminished considerably, a significant number of physicists have emigrated, the enrollment, graduation and graduates working for their PhD has decreased, but the core of the community has been able to resist, sometimes by redirecting its activities while still working at a high professional level. New relations and exchanges have been established with quite a few countries (in particular, Spain and other countries in Latin America). In some branches, new activities have started up. All in all, physics in Cuba remains vital and internationally applicable—in spite of the temporal proximity or the contemporaneity of these events—as will be outlined in the final part of this paper.[14]

[14] This is also discussed in most of the contributions by Cuban physicists in the second part of this volume.

6.1.3.5 The Actors in the Development of the Scientific System

The understanding of the peculiarities of the emergence and development of the Cuban scientific system, outlined above for the case of physics, can be improved by adding some remarks about the actors involved in this process. We have stressed the fact that both the revolutionary movement and the new generations were consciousness of the need to develop a modern scientific research system, in particular, of the importance of physics as a fundamental basis, although each retained specific views and were inspired by specific types of logic. For instance, the views of the students who taught in the Physics School differed from those who graduated in the Soviet system. However, both constituted the first generation of Cuban physicists in the new university. The early teachers at the school came in part from the elite component of professors at secondary schools and had been educated at the pre-revolutionary university, whose influence by the American scientific system and organization had been overwhelming since the neo-colonial period.[15]

The leaders of the Revolution had different cultural and educational backgrounds. Ernesto Guevara trained in medicine in Argentina, which was already a fairly advanced country in the first half of the twentieth century. He later traveled the continent and lived in Mexico, where together with Fidel Castro, he came into contact with left-wing international milieus. Castro in contrast was taught at private Jesuit boarding schools and later at the law school of the University of Havana. Guevara was the main promoter of sending Cuban students to graduate in the Soviet Union.

The main scientific Cuban institutions had at least a double origin: many of them arose from the government and advised by the scientific community, which probably had more direct influence in the early phase, before the process of institutionalization. Apart from prioritizing the development of solid-state electronic devices, Guevara promoted courses in advanced mathematics at the Ministry of Industry when he headed it, given by professors from the University of Havana. In speeches he held at the country's three universities from 1959 to1960, he insisted that university programs and enrollments had to correspond with the needs for the development of the country, and that this applied to the establishment of research centers within the Ministry of Industry. But some of the most influential Cuban institutions arose from the intuition and direct initiatives of Fidel Castro himself.

6.1.4 Summary of the Original Traits of Cuban Physics

We have repeatedly referred to the original traits that can be identified in Cuban physics, its method and approach. Considerations of this order are always an abstraction: strictly speaking, each scientist has his own personal approach and style of research, and physics (and science in general) is becoming increasingly

[15] The pre-revolution university has been analyzed in the chapter by Altshuler and Baracca in this volume.

standardized on a global level. Nevertheless, in our opinion some common features can be pinpointed. We have already remarked that what has happened with physics shows some similarity with what happened with practically all the cultural expressions of Cuban culture and mentality, where different tendencies and influences have amalgamated to produce an original result.

Cuban physicists were protagonists in the revolutionary environment, which strongly promoted the development of science. They had free or inexpensive access to international physics textbooks. In decisions made about the choices and directions of development, there was a wide collective participation, and subsequently the government insisted on and promoted the application of physics in activities concerning both the economy and society at large.

With regard to the methodological approach, the Cuban physicists have kept, and fused together different traditions (in particular, Soviet and "Western"), with a touch of typically Cuban fantasy and "do-it-yourself". The lively and combative student component directly engaged with the renovation of the Physics School of the University of Havana constituted the first generation of the new physics graduates in Cuba. They came abreast to the first graduates of the Soviet Union, who were formed with a very different spirit: nevertheless it is certain that both transmitted their own attitude in their subsequent teaching and research activity.

Furthermore, the numerous western physicists active in Cuba up to the early 1970s certainly transmitted their own attitude (besides western techniques). We will see in more detail how the first invited professors implemented do-it-yourself activities, starting practically from nothing. This was a characteristic that integrated well with the innate Cuban spirit and constantly recurs in many of the activities and initiatives of Cuban physicists.

We will now give a more detailed and chronological description of the innovations and developments in physics, of the growing complexity of initiatives, and finally of their implications.

6.2 Up to the Mid-1960s: Coping with the Difficulties

6.2.1 Early Initiatives for Reform in Physics

The University of Havana reopened in January 1959, and on May 11, the leader of the Revolution, Fidel Castro, inaugurated the new academic courses. As mentioned already, the leaders of the Revolution and active participants from the intellectual milieus understood that general instruction and higher education were necessary conditions for building a modern country and they had a clear vision of how to do this. Around the mid-twentieth century belief in science as an instrument for liberation and development belonged to the tradition of the left wing and of progressive thought. The events of war and post-war reconstruction highlighted even more the crucial role of science and scientific research in the modern world. At that time, the prestige of Soviet science was immense.

The revolutionary government announced that from then on education in Cuba would be free of charge at all levels and accessible to all people, whatever their social economic status; it was destined to play a very relevant role in the country. Alongside the ongoing political struggle, an intense debate was held on university reorganization and the need for modern higher education and a scientific research system linked to the development of the country (De Armas et al. 1984). This debate involved different levels of Cuban society. Beside that of teachers and various professional sectors, student participation carried great weight since they had been particularly active in the revolutionary struggle. The implementation of a very comprehensive scholarship plan and other forms of economic support served to open doors at the university to many talented young people who in former times had no access to higher education studies, thus changing the social composition of the student body.[16] Several revolutionary leaders, especially Fidel Castro and Ernesto Guevara, insisted in their public speeches on the importance of transforming the universities and developing scientific research. On January 15, 1960, Castro delivered a speech at the site of the old Academy of Sciences in which he asserted that "The future of this country must be a future of men of science, men of thought," which later on inspired the declaration of January 15 as the "Day of Cuban Science." Between the end of the 1950s and the early 1960s, "Ché" delivered speeches at the three public universities, putting forward his ideas on the role to be played by the universities and the importance of increasing the enrollment in scientific and technological degree courses. Later on, as minister of industries, he proposed several areas to be tackled by the universities to support the technological development of the country, which included cutting-edge branches such as solid-state electronics.

Although the Cuban Higher Education Reform Law, (Consejo Superior de Universades 1926), drawn up by a commission of professors and students, was not enacted until January 10, 1962, many of the ideas it contained were based on a consensus that had been already reached among the actors involved, and these had already been implemented by 1960–1961. Quite a few initiatives that were institutionalized later on had already been put into practice during this period. The reform was the result of a process that was quickly and increasingly radicalized as time went by, in step with the transformations taking place in the country while facing political, economic and military aggression from abroad.

6.2.1.1 1960: The Early Reform Movement in the School of Electrical Engineering

The recurrent emergencies of the country naturally promoted the early efforts. Being a country isolated from the rest of the world, under critical circumstances that could escalate at any time, in that moment it was deemed quite urgent to develop short wave radio communications; such a scenario urgently required in the first place the special training of highly qualified personnel, such as engineers. It turned

[16]The interview with Melquíades de Dios Leyva in this volume is particularly eloquent on the importance of the unprecedented opportunities provided to the lower classes.

out that this triggered the first substantial reform of a degree course syllabus, which included basic courses and physics teaching. This was first taught in 1960 at the engineering faculty of the University of Havana (Altshuler 2006, chap. 7) Most of the established teaching staff were too conservative and unable or unwilling to implement the required changes, and consequently the students responded by taking the initiative to provisionally hire several competent professionals for the purpose. In the first half of the 1960s, students had a large influence on the direction of the university, sometimes more than that of the academic authorities. This role can be understood against the background of their leading participation in the fight against the Batista dictatorship, directed by the *Federación de Estudiantes Universitarios* (FEU). The students had direct relationships with Fidel Castro and other leaders of the Revolution, and they had economic resources and a broad autonomy. This situation began to change in the second half of the 1960s, with the establishment of the Communist Party at the University of Havana and the strengthening of new academic authorities: nevertheless the participation of the student organizations, FEU and *Unión de Jóvenes Comunistas* (UJC), remains at every level a peculiar feature of the Cuban higher education system.

The first radical changes were introduced in the electrical engineering degree course, which included a sharp increase in the scope and quality of the physical and mathematical training of the students, with the time dedicated to physics almost twice what it was before. Due to the scarcity of able personnel, the toughest problem was the teaching of the newly designed four semester course in physics. To quickly supply the provisional textbooks needed, individual chapters of renown textbooks (mostly Sears, and Sears & Zemansky's[17]) were put together to match the general conception of the new syllabus, and printed in book form at a printing workshop organized by a group of students. The next big challenge was to create a staff capable of satisfactorily teaching the matter for all engineering students: headed by Ramón Ventoso, an electrical engineer with a remarkably mature approach to general physics through self study, it included some of the most brilliant senior engineering students and several high school physics teachers. At the beginning, quite a few encountered great difficulties in mastering the subjects they were supposed to cover, especially solving the textbook problems (Altshuler 2006, chap. 2). It took a much longer time for the experimental work in the student laboratories to reach a satisfactory level.

6.2.1.2 The Reform Movement in the Faculty of Science

The reform movement reached the Faculty of Sciences later than the Faculty of Engineering (Altshuler 2006, chap. 2). With a view to reforming its teaching system, models applied at several western universities were examined, but the shortage

[17] Francis Weston Sears and Mark W. Zemansky wrote a widely disseminated textbook on physics, which had innumerable editions and translations: the Spanish translation was reprinted in Cuba in the 1960s.

of material resources made it impossible to follow them. As analyzed in Chap. 4, Sect. 4.8.3, Marcelo Alonso had been exploring the possibility of importing surplus equipment from American laboratories that had been working on nuclear research during the war, but to no avail. However, he did manage to introduce physico-mathematical students to some experimental and thesis work on atomic physics before the students struggle against the dictatorship in power led to the interruption of academic activities. He had made several trips abroad before 1959 to attend all kinds of meetings on nuclear subjects, which made him well acquainted with American and other colleagues working in this area. At the end of that year, he applied for a leave of absence to honor a contract he had been offered as scientific advisor to the Organization of American States in Washington (Boletín 1959, 1960). This was granted and he traveled to the United States where he finally settled.[18] Very few members of the original staff remained in the country during the early years of the Revolution. In 1960, Nilo Blanco set up for the Energy department of the Ministry of Public Works the first laboratory to study non-conventional energies (solar and eolic) at the old physics building of the University of Havana (A. Cerdeira, personal communication 1997). This work, which counted on the collaboration of a group of electrical engineering students, came to a halt when Blanco left the country.

The reform movement attached a very special role to scientific research, as expressed in the Reform Law: "In today's Cuban society the university is the link through which modern science and technology, in its highest expressions, should be put to the service of the Cuban people" (Consejo Superior de Universidades 1962). In the section dedicated to the aims of the university, it can be read: "(c) to carry out scientific research, develop a positive attitude toward research among university teaching staff and students, and collaborate with scientific institutions and technical organizations beyond the university system" (p. 1); and further on: "The function of the scientific research institutions must necessarily be linked to the institutions of higher education" (p. 10). "Research work should be closely linked to the training of scientists." To take this into consideration, as far as teaching and research is concerned, the concept of "department" and not of "chair" should be taken as the foundation of the university structure.

6.2.2 Creation and First Steps of the Physics School of the University of Havana (1961–1965)

Despite enormous initial difficulties, the most significant and enduring reforms and innovations in physics were introduced at the University of Havana. In December 1961, the Physics School was created in the Faculty of Sciences of the University of

[18] Marcelo Alonso, who authored or coauthored several widely used physics textbooks (Alonso 1958), was one of the Cuban physicists best known in the West. In his later years, he visited Cuba twice. His impressions on Cuban physics can be found in an article he wrote for *APS News*, which is reproduced in this volume.

Havana.[19] Apart from material resources, the new school lacked an adequate scientific tradition and professional staff to deal with the new teaching and research tasks. The retirement (and later passing) of Manuel F. Gran[20] (after having been appointed Cuban Ambassador to France), and the emigration of Marcelo Alonso and quite a few other qualified personnel decimated the staff.[21] In the words of the first director of the Physics School, Rubén Martí:

> … the whole of the new Physics School would fit into its director's wallet, *initial conditions* being the worst. There was no laboratory equipment, nor professors, nor well-trained students, nor appropriate installations and housing (Martí and Auchet 1962).

As a result of lack of professional teachers, senior students were appointed as assistant students and undergraduate instructors to work as laboratory assistants and to teach students in the lower years of their degree courses. At the time, not even an established school library existed because it was extremely difficult to get much needed modern textbooks in English as a result of the heavy economic restrictions imposed on Cuba by the US government. A bold decision was made by the revolutionary government to prevent the reform process at the universities from stalling: it was not too long before university students could have at their disposal many American and other textbooks photocopied by "*Ediciones Revolucionarias*" which, given the national emergency conditions, disregarded copyright and distributed books free of charge to those who needed them, despite their high prices in the world market. By the end of this period, the library of the Physics School had accumulated some 3,000 volumes (Pérez Rojas et al. 1976). After the promulgation of the University Reform Law, the first restructuring of the physics degree course was made, but continuous difficulties prevented the stabilization of the curriculum, and additional remedial changes had to be introduced in the years that followed.

> The first course of the *licenciatura* in physics began in 1962 with an enrollment of almost 70 students. The diversity of its composition, added to the difficulties faced by the newly formed school, resulted in the promotion of only *two* students to the second course. At the same time there were three students in the third course and four in the fourth, who were taking specially designed transition plans that took into account what had already been learnt by the students of the former physico-mathematical degree course. The number of students who graduated in this three-year period was only about 6 or 7. (Pérez Rojas et al. 1976)

It must be taken into account that students were returning to the classrooms after having lost contact with formal studies for several years while the universities were closed. Many of them had been involved in the revolutionary struggle and when they returned, the political and military events, the intensive work demanded by the

---

[19] See the interview with Hugo Pérez Rojas in this volume.

[20] On the figure and the role of Manuel Gran before the Revolution, see the chapter by José Altshuler and Angelo Baracca in this volume.

[21] For the history of the Physics School at this early stage, here and in the next paragraphs, see also the interview with Hugo Pérez Rojas in this volume.

reorganization of the country and the debate about its future development absorbed most of the capabilities of quite a few of the best. Various contradictions also cropped up between a teaching staff that stuck to an essentially academic attitude and the enthusiasm of the students who most identified with the revolutionary process.

With the scarce resources available, a great effort was made to direct the physics degree course "toward the modern conceptions of physics, placing research above (ordinary) teaching" (Martí and Auchet 1962). While the effort made to give the school a stable structure did not yield the expected result until the end of the 1960s, in the meantime some important preliminary results were obtained. Two factors were very relevant. Firstly, in 1960, the first Cuban students were awarded scholarships in the Soviet Union: some of them studied physics at the "Patrice Lumumba" University for the Friendship of Peoples and at the "M.V. Lomonosov" Moscow State University.[22] Six students expressly sent by the then Minister of Industry, Ernesto Guevara, to study electronic engineering, managed to follow a physics degree course instead. In the following period, students were sent each year first by the hundreds, and later on by the thousands to socialist countries, most of them to the USSR. A second factor had an immediate positive impact of considerable importance: the presence of quite a few foreign teachers and specialists, partly Soviet, but also from several "western" countries.[23] Thus, for instance,

---

[22] The Peoples' Friendship University of Russia is an educational and research institution founded in 1960 and located in Moscow, which is ranked by the Ministry of Education of Russia as the country's third-best university after Moscow State University (see below) and Saint Petersburg State University. Its stated objective at the time was to help nations of the Third World, mainly in Asia, Africa and South America, by providing higher education and professional training; many students from developed countries also attended. The Lomonosov Moscow State University is the oldest and largest university in Russia. Founded in 1755, it was renamed in 1940 in honor of its founder, Mikhail Lomonosov (1711–1765), a polymath and writer of Imperial Russia.

[23] Some general political background information here will situate and interpret foreign initiatives and collaborations with Cuba, as well as Cuban development. The Cuban Revolution obviously attracted worldwide attention and a deep interest in the leftwing milieus (although with contrasting evaluations, due to the unorthodox origin and development of the Revolution). In France, there was quite an authoritative group of leftwing scientists in the French Communist Party, who organized strong coordinated actions, in particular in support of the efforts of the Cuban scientists, especially physicists, to promote the technical scientific development of the country. Probably the most influential among the French physicists was Jean-Pierre Vigier (1920–2004). An active supporter of communism throughout his life, Vigier was a proponent of the stochastic interpretation of quantum mechanics, after the ideas of de Broglie and David Bohm. Vigier was invited to be Einstein's assistant, but because of political controversy concerning the Vietnam War, the US State Department refused to issue him with a visa). Vigier was probably the driving force behind the organization of French scientists in the summer schools in Cuba (see Cernogora's recollections in this volume). In Italy, too, the interest of scientists toward the Cuban events was strong: the first case was that of Andrea Levialdi, described in the present chapter. In the summer schools, Italian physicists organized courses of teaching: among the leading lights were Bruno Preziosi and Bruno Vitale. Throughout the 1960s and at the beginning of the 1970s, international interest was revived by the war in Vietnam.

Amanda Blanco, who had been a Spanish civil war refugee in the USSR since childhood, arrived at the University of Havana in 1962 to lecture on acoustics and electronics, and to help organize the respective laboratories. Others, like F.D. Kochanov (methodology and philosophical foundations of teaching) and I.V. Ponomarenko (student laboratory work), acted as experienced advisers on various teaching activities. Vladimir Grishin, a highly qualified theoretician and expert in particle accelerators, visited the country a couple of times, staying in Cuba for 4–5 years altogether. He lectured on quantum mechanics, nuclear physics and atomic physics, and helped organize the undergraduate laboratory for atomic and nuclear physics.

Beginning in 1961–1962, several subjects were dealt with in one way or the other by visiting professors and specialists from various western countries. Some stayed in Cuba for a relatively brief period (among others, the British professor Trevor W. Marshall,[24] the American theoretician Richard Bourret, the Israeli nuclear physicist Eleazar Baruch, the French professor Michel Degallier and the Mexican electronic engineer, Angel Zapata), while others stayed for quite a few years, as was the case of Claude Monet-Descombey, a French electronic engineer who very thoroughly contributed to developing a good electronics laboratory (Monet-Descombey 1963). The American electronic engineer Theodore Veltfort, who had fought against Fascism in Spain as a member of the Lincoln Brigade, arrived in Cuba in 1962 willing to help develop the country in the field of technology. After contacting the central planning board and having talks with the then Minister of Industry, Guevara, and later with the vice-rector of the University of Havana, José Altshuler, an agreement was reached that he would settle in the Physics School. Together with the Argentinian physicist Dina Waisman, he lectured until 1968 on solid-state physics and semiconductor devices, while contributing to the organization of a special electronics workshop, and developing the first research activities aimed at the manufacture of semiconductor components.[25]

The Physics School of the University of Havana was initially comprised of three "official" teaching and research departments: solid-state physics, general physics and theoretical physics. In addition, an "unofficial" nuclear physics department was created to take care of the instruments and equipment. In fact, activities in nuclear physics were at the time among the most important at the Physics School. Among the few documents available from this period are reports that detail the main focus areas in scientific research among the academic authorities (Comisión de Investigaciones 1963; Montes 1963).

The professors that were invited from developed countries brought a broader and more comprehensive vision of the importance of developing and organizing modern research activities in the field of physics. In 1963, French expert Claude Monet-Descombey delivered a plan for the electronics laboratory to the director of

[24] Marshall would later author the so-called Stochastic Electrodynamic, and would make relevant contributions also to the problems of the foundations of quantum theory. Some of his early contributions are Marshall (1963, 1965) and Marshall and Claverie (1980).

[25] See Veltfort (1998), reprinted in this volume.

the Physics School. The director then passed this plan along to the rector. This plan exemplifies a general, articulated and long-term vision:

> [the laboratory] should rather be put in charge of the maintenance and design of experimental equipment for physics work under the immediate direction of the interested physicists, so that this personnel could be the nucleus from which a staff dedicated to taking care of the electronic circuitry required for any type of scientific research, be it in solid-state physics [transistors] or above all, in nuclear physics if a particle accelerator or something similar are bought. (Monet-Descombey 1963)

In addition, it was suggested that, in collaboration with the electrical engineering school of the Faculty of Technology, a technological research laboratory should be created which

> would rather play the role of the nucleus of a technological research group aimed at solving problems in electronics arising in the country, and as well as at taking care of the design of equipment intended for industrial manufacture [e.g., radio communication] (Monet-Descombey 1963).

The proposal also discusses a further development scheme for the initial joint work of both laboratories. In order to be able to design and build new equipment, the proposal states:

> It would be good for us to have a team in charge of bringing to fruition—however difficult it may be, for it would require time, lots of bibliographic research, and many preliminary trials—something that was the result of all concerns, so that it may become a collective work par excellence, so that it would prompt spontaneous consultation among the staff, favoring teamwork and eliminating any possible discrepancies, and in the end, obtain equipment, not only useful but also with a "history." For this, we think than in a period of no less than two years, an analog computer with 8 or 12 amplifiers coupled to a register could be procured to begin with, subject to progressive improvements.
>
> This project may seem too difficult to implement under the present conditions at this university, but we believe that if the problem is properly considered, it is possible to see quite clearly that this will have to be done some day if we want to develop in Cuba a minimum of fundamental or applied research. (Monet-Descombey 1963)

The above project was not implemented as a result of a need to tackle other urgent priorities. However, it is still important to note the effort made by the Physics School of the University of Havana to help other institutions and to generally disseminate science and its use, despite the meager resources available at the time. In the above-mentioned "Memorandum" by the director, several activities dealing with equipment checks, the set-up of laboratories and the determination of physical properties of samples were carried out by some of the laboratories of the Physics School. This was done in cooperation with the Ministry of Public Works, the civil airport at Ciudad Libertad, the Ministry of Industry and the General Calixto García hospital. While Minister of Industry, Guevara, asked for help to give Saturday morning courses on scientific matters to his board of directors.[26] A similar sort of cooperation also existed with other ministries and with the Central Planning Board (Jiménez Pozo and Sánchez Fernández 1993).

---

[26] Cf. the interview with Hugo Pérez Rojas in this volume.

The first students obtained their physics degree by the mid-1960s. Some of them, such as José Antonio Tabío, Adriana Fornés, Alfredo de la Campa and Blanca Reyes, were to have long and dedicated careers as professors in the Physics School. Only at the end of the 1960s did a large enough mass of physics graduates become available to cover the more pressing needs of the universities and of the country (Sect. 6.2.1).

6.2.3 Initial Efforts to Promote Scientific Research in the Physics School

The efforts to launch proper research activities at the Physics School began despite enormous difficulties and lack of resources. The first results were not achieved until the following decade.[27] There was an initial attempt to develop a system of scientific journals to stimulate people to write and publish scientific papers. This attempt was made because until that point the only existing journal with a similar purpose was the *Revista de la Sociedad Cubana de Ciencias Físicas y Matemáticas*. This journal was first published in 1942 by professors Pablo Miquel and Manuel Gran, and edited by Mario O. González from 1944 to 1945.[28] Promoted by the Scientific Research Commission of the University of Havana, the first issue of the scientific journal, *Memorias de la Facultad de Ciencias*, was published in 1963. Its purpose was to publish the results of scientific research produced by the staff of the various schools belonging to the Faculty of Sciences of the University of Havana. In the few years that it existed, the journal included special issues dedicated to mathematics, geology and biology, but none to physics due to lack of suitable material. José Altshuler, who was at the time vice-rector and chair of the University of Havana's commission for research, made remarkable progress in promoting the generalization of scientific and technological research activities at the university. The department heads were asked to send their reports on the research work already begun or planned for the near future. The goal was to coordinate and evaluate the totality of the then available research projects, however incipient or simple (for example, the design and realization of certain electronic equipment proposed by the Physics School and the electrical engineering school). Student research groups were promoted to stimulate among their participants, a preference for research work and to become familiar with its peculiarities. There was also a call for the presentation of research projects, in spite of the limited resources available for backing any new ones. To promote familiarity with scientific research among the students, and even a substantial part of the university staff, the commission published in October 1963 a widely distributed booklet entitled *Scientific Research – an Outlook*, containing a collection of brief and

[27] See also Vigil's article in this volume.

[28] Cf. Altshuler and Baracca's chapter in this volume.

easy-to-read essays by prestigious authors on the nature of scientific research (Comisión de Investigaciones 1963).

As mentioned previously, at the Physics School the work of two foreign professionals, Dina Waisman and Theodore Veltfort, played an important role in helping to develop a scientific mentality and to take the first steps forward in semiconductor research. In order to put Waisman's experience to good use, namely the application of a certain technique for growing germanium monocrystals, a radiofrequency oven was purchased. When Boris Pavlovich Konstantinov, then vice-president of the USSR Academy of Sciences, visited Havana, he invited both scholars to visit his institute in Leningrad in order to establish strong collaboration within the field. Facilitated by the Cuban Academy of Sciences, who had already officially established a fluid collaborative relationship with the Soviet Academy, Waisman and Veltfort arrived in the USSR in 1964 for a 2-month visit that included not only the Ioffe Institute,[29] but also other important laboratories in Moscow and Tallinn. In addition to sending basic materials and equipment required for setting up the semiconductor laboratory at the University of Havana, advanced information on semiconductor research, scientific literature and didactic material was obtained. Regrettably, despite these positive results, full collaboration with the Ioffe Institute was slow to develop due to organizational problems between the University of Havana and the Cuban Academy of Sciences.[30] As a result, the first activities related to semiconductor research completed in the Physics School had to continue without support for some time.

These autonomous efforts—both in teaching and in research, as well as in the relationship between the two activities—left the Cuban physicists with a lasting impression whose diverse characteristics was to re-emerge in subsequent moments.

6.2.4 *Other Initiatives and New Scientific and Institutions of Higher Education*

In the unrest of developing initiatives, new scientific and higher education institutions were founded. In addition, new activities and fields were promoted.

---

[29] The Ioffe Physical-Technical Institute of the Russian Academy of Sciences (Ioffe Institute) is one of Russia's largest research centers specialized in physics and technology. The institute was established in 1918 in Petrograd (St. Petersburg) and was run for several decades by Abraham Fedorovich Ioffe (1880–1960) who was a prominent Russian/Soviet physicist, expert in electromagnetism, radiology, crystals, high-impact physics, thermoelectricity and photoelectricity.

[30] The collaboration between the Cuban institutions and the Soviet Academy of Science was coordinated at a governmental level by the academies. In the beginning, it was difficult for the Cuban universities to participate in the collaboration with institutes of the Soviet academy, since the ACC prioritized the participation from the centers pertaining to the ACC itself.

6.2.4.1 The Cuban Academy of Science

In 1962 the new Cuban Academy of Sciences (ACC) was created with the purpose of supporting several scientific branches and scientific research in the country. In 1963 the mathematics department was established, which included a cybernetics group. Other areas related to physics research dealt with at the academy or other Cuban institutions are summarized below. It seems important to remark that, unlike the promotion of activities in the school of physics of the University of Havana, these initiatives in the academy (though they came somewhat later) were promoted through Cuban initiatives, and relied heavily on collaboration with the Soviet Union and other East European countries.

Geophysics

There was strong motivation for the development of geophysics: mine prospecting, international communications and, to a lesser extent, seismic risk. The Soviets too were interested in observations made from the geographic position of Cuba. One has also to recall that the first president of the ACC, Antonio Núñez Jiménez, was a geographer. This explains at least in part why, since the early years of the academy, the development of areas of physics were linked directly with geography. Examples of this occurred in the departments of geophysics, astronomy, metereology and atmospheric physics, whose relevance the president certainly understood better.

The department of geophysics of the Cuban Academy of Sciences was inaugurated in 1964, with a personnel of only 11 non-graduate members and 9 technicians. This inauguration was preceded by a newly created 5-year degree course in geophysical engineering within the Faculty of Technology of the University of Havana (Universidad de La Habana 1963), which benefited from the very effective cooperation of the Czechoslovakian J. Hladík. An International Geophysical Year National Committee was established in collaboration with the Academy of Sciences of the USSR. With the support of these groups, it was possible to set up geomagnetic, ionospheric and seismologic stations in a relatively short period of time, to which an ionosphere vertical sounding station was added later (Historial de Geofísica n.d.). With the help of Soviet specialists, a seismological station was installed in 1965 in the eastern city of Santiago de Cuba. Thanks to cooperation with Soviet firms and the application of physics techniques in mining and oil prospecting with nucleonic instruments, mineral deposits were discovered at Cristales, Majagua, and Cuenca Central between 1963 and 1965.

Astronomy and Meteorology

In 1964, while astronomical calculations services were increased at the National Observatory, at that time part of the Cuban Revolutionary Navy, a group dedicated to astronomy and meteorology was created at the Cuban Academy of Sciences

(Historial de Geofísica n.d.; González 1985; Ortiz 1987). Data was published for moonrise and sunrise and moonset and sunset times, and for tidal times in the *Suplemento al Almanaque Náutico.* In collaboration with the Soviet Academy of Sciences, Cuban technicians were trained to operate the first Cuban station for visual satellite tracking. The first regular observations began in April 1964 and transferred to a specialized center in Moscow for further analysis. As a member of *Interkosmos* (Intergovernmental Program for the Exploration and Peaceful Use of Outer Space) Cuba had participated since 1967 in its permanent working group on space meteorology (Ortiz 1987). In 1965, the astronomy section of the National Observatory was transferred to the Cuban Academy of Sciences to become its Astronomy Group (lodged at the National Capitol, which turned into the central site of the Academy), and the Meteorology Institute (in Casablanca, across the bay of Havana), so that they became two independent entities. Later in the same year, the group was upgraded to an astronomy department (Pérez Doval 1991). At the same time, a very extensive network of meteorological stations was gradually spread across the country. This was a genuine necessity since, due to lack of government funding, the collection of nationwide meteorological data had previously rested almost exclusively on the cooperation of amateur meteorologists and a few institutions who voluntarily sent them to the National Observatory with variable regularity and questionable reliability. This data was added to the local meteorological information gathered by the National Observatory in Havana (founded in 1902). Curiously enough, while all the said data continued to accumulate at the observatory for years, it came as a surprise to discover in the early 1960s that they had remained essentially unprocessed, not even subjected to the most elementary form of statistical analysis (which was undertaken for the first time by a few students of electrical engineering). In the same year, 1965, Laszlo Destre, director of the Budapest Konkoly Observatory in Hungary, visited Cuba with the purpose of establishing a collaboration agreement with the academy to study variable stars.

Space Communications

As stated above, after the Revolution, the need arose to urgently provide the country with an autonomous communication system. This situation encouraged the reorientation and updating of the electrical engineering syllabus, which in turn soon stimulated the reshaping of other engineering syllabuses in 1960 and later on (Sect. 6.2.1.1). The October 1962 missile crisis made dramatically clear the need for long distance communication able to operate reliably under "exceptional conditions." In November 1965, some 6 months after the Soviet Union had put their first communications satellite, the Molniya-1, into orbit, representatives of the Cuban government participated in a multinational gathering held in Moscow. They discussed a Soviet proposal to use the assets their country had developed "for the peaceful study and use of outer space" (Altshuler 1997). It was not too long before an agreement was reached among the participant socialist countries to implement what was called the Interkosmos Program (Sect. 6.2.4.1), which included a working group dedicated to research on space communications. Cuba became involved between 1966 and 1967, initially through the Ministry of Communications, and with the strong backing of Minister Jesús Montané, and Vice-minister, Pedro Luis Torres, an electrical engineer. This was the basis for the establishment of a satellite communication system, which could be put to good use not only for day-to-day international communications, but also under "exceptional conditions." However, it was not implemented until the early 1970s, after the creation of the *Intersputnik* organization, of which Cuba was a member.

Medical Physics

From the early 1960s, a small group of physicists began to work at Havana's National Institute of Oncology, mainly on matters relevant to the planning and dosimetry of radiotherapy treatment.

Metrology

The metric system of units had, since colonial times, been far from consistently used in Cuba, but its unification and rigorous application became a strict requirement for the modern industrial development of the country.[31] However, the systematic use in the country of SI units, consistent itself with metric units, was enforced by Law 915, enacted on December 1960, which allowed a 3-year period (which in 1964 had to be extended for another 2 years) for its full application. In 1962, the Standards and Metrology Directorship was created in the Ministry of Industry, followed by a Metrology School in 1963, as the first Cuban went to Romania on a scholarship to study metrology. In February 1965, the first metrology laboratory was inaugurated. The first measurements were duly verified against standards received from East Germany and the Soviet Union in 1963. Conspicuously, these related to the verification of the size of the country's main fuel tanks (especially those at the oil refineries) and to the accuracy of the industrial balances used in sugar factories. In this context, the pioneering work of a physicist and

[31] Before 1959, a truly chaotic situation existed in Cuba in what concerns units of measurement. There was an arbitrary mix of units of Spanish, English and Cuban origin. It was usual to use these in trade: for units of weight pounds, *arrobas* and *quintals*; for units of length in the textile trade *varas* (Spanish or Cuban) and yards; for areas and ground measurements *caballerías* and *cordeles*, while the architects used meters and distances were expressed in kilometers. This situation caused many problems in exports, with the differentiation of markets. For instance, in sugar production/export, one had to specify if one was dealing with "a ton" (2,000 lb) or "a metric ton" (1,000 kg). The lack of a suitable means for the country to verify its own instruments of measurement also led to commercial losses. On the other hand, the Soviet Union and the other socialist countries had adopted the international system of units, and the rapid development of trade with them created pressure to do the same in Cuba. It is probable that the creation of the Ministry of Industry under Ernesto Guevara was key in this matter.

electrical engineer, Ángel Álvarez Ponte, must be mentioned, as he was an extremely knowledgeable and persistent promoter of scientific metrology in Cuba. A national register of measuring equipment began in 1968. Metrological standardization made gradual progress in Cuba and eventually reached a remarkable level of development.

Starting from a primitive situation, the spectrum of initiatives to develop a modern and articulated scientific system was really broad from the very beginning. Besides keeping up with the more advanced countries, they were able to cope with the multiple needs that the development of a modern, industrialized country would require, not only the foreseeable ones (as in the case of meteorology in a tropical country), but also unexpected ones.

6.2.4.2 The Faculty of Technology of the University of Havana

The need to prepare engineers and to modernize their careers was also a strict requirement, as we have already seen from the early attempts at the School of Electrical Engineering (Sect. 6.2.1.1). Due to lack of space for the expansion of University of Havana campus, which was located at the very heart of the Cuban capital, the construction of a new campus began in 1960, originally intended to become the new site of the entire university. In 1964 Fidel Castro opened the Ciudad Universitaria "José Antonio Echeverría" (CUJAE). This was to be the new site of the Faculty of Technology of the University of Havana, which at the time had some 5,000 students. A few years later it would become an independent polytechnic university—almost as large and important as the two-centuries old University of Havana—where new generations of engineers and architects would be trained.

6.2.4.3 The University of Oriente

The spirit of renovation and modernization also reached the University of Oriente in Santiago de Cuba.[32] This university, founded in 1948, did not yet have an independent physics degree course, and the development of a general physics department was slower than that at the Physics School of the University of Havana. In 1962, the dean of the Faculty of Sciences visited several higher education centers in East Germany and Italy, and signed a collaboration agreement with the Dresden Technical University (Tratado 1962).[33] In a report in 1926, José Fernández Bertrán mentioned quite a few Italian physicists he had met, who had declared their willingness to visit the University of Oriente and cooperate with its staff, but these offers did not bear

[32] More information is given in the following, though this is dealt with in detail by Cabal and Méndez Pérez in this volume.

[33] The Technische Universität Dresden is the largest institute of higher education in Dresden, one of the oldest colleges of technology and oldest universities in Germany. In 1961, it was given its present name, the Dresden University of Technology.

fruit (although some of them did visit the University of Havana later on) (Fernández Bertrán 1962). Still, as a result of these contacts, some Italian specialists actively collaborated with the University of Oriente, for example, the young Italian nuclear physicist Piero Basso and the Italian electrical engineer Mario Chirco, who both had a solid physical and mathematical background and worked at the university for several years. Additional collaboration arrangements were implemented as a result of visits paid at the time to Poland, the Soviet Union and Bulgaria by Roberto Soto del Rey, head of the physics department of the University of Oriente (Soto 2000).

6.3 The Second Half of the 1960s: Stabilization and Take-Off

Early efforts to introduce modern physical content and research activities were to reach the first stable level of success in the second half of the decade, providing the basis for further progress.

6.3.1 Stabilization of the 'Licenciatura' at the Physics School of the University of Havana

Around the mid-1960s, the Physics School of the University of Havana endured a critical period when several of its teachers left the country.[34] Despite this critical period, the will to achieve a stable structure was very strong. From 1966 onwards, a staff reinforcement process began with the inclusion of several new graduates (Fernando Crespo, Eddy Jiménez, Medel Pérez, Arturo D'Acosta, José Roig, Mercedes Alonso, Fernando Comas). A decisive factor was the return to the country, in 1966–1967, of the first Cuban physicists to have graduated from Soviet universities (Daniel Stolik, Armando Pérez, Juan Fuentes, Elías Entralgo, Antonio Cerdeira, Magaly Estrada, José Marín). These graduates immediately joined the Physics School staff thanks to steps taken by its director, Hugo Pérez Rojas, who was competing with other institutions that were in great need of capable professionals. During the next 5 years, other Cuban physicists obtained their degrees in the USSR (Pedro Díaz, Andrés Martell) and in the US (Elena Vigil Santos), and a part of the new *licenciados* came from the Physics School itself (Melquíades de Dios, Miguel Ramos, Luisa Noa, Joaquín Torres, Luis Hernández, Luis Falcón, José Matutes, Mario Brizuela). In 1968, the dean of the Faculty of Sciences, Ruth Daisy Henriques, became director of the Physics School. Soon after, Heliodoro Medina, an engineer who had graduated in Czechoslovakia, became the new director, accompanied by one deputy director for education (Elías Entralgo) and another for research (Daniel Stolik). These young men and women introduced a qualitative change, acting as pioneers for today's Physics School.

[34] Cf. the interview with Hugo Pérez Rojas, the second director of the Physics School, in this volume.

We have already commented on the contradictions concerning the curriculum and education system that arose between the graduates from the Soviet Union and the students who had played a key role in the early activities at the Physics School and who became its first graduates. Despite these differences, both the organization and the subject matter of physics studies acquired a strong Soviet influence, inspired by the basic cycle of the Lomonosov Moscow University,[35] and its rigorous rules that were subsequently introduced. Of the ten students to complete the 1966–1967 academic course, four defended their dissertation before a formal board. This amounted to a first step toward the inclusion of research into the undergraduate teaching process, which unfortunately could not be consolidated until much later (Pérez Rojas et al. 1976). The reason was that the incipient research activities developed at the time were not strong enough to cope with the rapidly increasing number of physics students, which shot up from about 100 in 1965 to more than 400 in 1970. To this it must be added that many of these students combined personal study with some form of teaching work, either at the university or some other education center.

Gradually, the Cuban graduates took on most of the weight of the teaching activities, while the students helped by taking care of complementary activities (problem solving classes, laboratory practice, and so on). Finally, in 1970, the first "mass" graduation of 67 *licenciados* in physics took place. Twenty of them were assigned teaching posts at the Physics School, some of whom have continued to work as highly qualified members of the staff for more than 40 years (Manuel Hernández Calviño, Félix Martínez, Esperanza Purón, Carlos Rodríguez, Carlos Trallero, Julio Vidal).

As a result, the Physics School staff attained a definitive stability. For the first time all the fundamental teaching activities could be performed by national graduates (the assistant student movement continued but now with the purpose of basic education), making it possible to dedicate an important part of their working time to research and to the postgraduate training of the young staff. This group also took on the responsibility of teaching the physics courses included in the course plans of other degree courses of the Faculty of Sciences. They also supervised the physics laboratory practice undertaken by the students of agricultural sciences. In addition, ten *licenciados* in physics were incorporated into the staff of other university faculties, research centers and institutions, which were able to receive for the first time a substantial addition of well-trained young physicists.

In 1967, the Physics School moved to the old building of the School of Engineers and Architects, which had been left by the Technology Faculty when it moved to

[35] The schedule of classes was very intense with more than 30 h per week of classes and compulsory attendance. These schedules were meant for students who did not work and dedicated themselves exclusively to study (although as part of their education, the students did participate in teaching activities and other community activities). Several texts and programs of the Soviet universities were used, mainly from the University of Moscow, in addition to some western textbooks. The *Editora Revolucionaria* printed English, Spanish and French translations of Soviet textbooks or foreign editions in Cuba. The general concept was to give the student a strong physical–mathematical basis—both theoretical and experimental—before they began any specialization.

the CUJAE. This provided ample space for improving the school's workshops and laboratories. An important investment was made under the management of Magaly Estrada, as a result of which ten student laboratories had been built and equipped by the end of 1970. In addition, electronics, mechanics and glass workshops were developed that substantially increased the local capacity for making the equipment needed to support teaching and research activities. The first research laboratories were also set up. This development was accompanied by a significant increase in the number of qualified workers and technicians attached to the new laboratories and workshops.

In a general sense, it does seem to be an exaggeration to assert that by 1970 undergraduate studies at the University of Havana's Physics School had succeeded in achieving relative stability, comparable to international standards, and in particular with respect to the more developed Latin American countries. Although it refers to a somewhat later period,[36] Leccabue emphasizes that "the preparation of Cuban graduates or PhD students and young researchers was of a very high level with high degrees of autonomy and critical skills being developed for designing and carrying out research projects."[37]

Although there was still much to be done, the school could count on an enrollment of some 400 students, a syllabus compatible with the most advanced international standards, a material base of excellent textbooks and adequate undergraduate laboratories. Added to these features was a staff consisting of more than 50 graduates, who, although still very young and inexperienced, were well trained for undergraduate teaching purposes, and enthusiastically prepared themselves to tackle the new challenges represented by serious research work and postgraduate studies.

6.3.2 *The Take-Off of Solid State Physics Research at the Physics School of the University of Havana*

6.3.2.1 The Wide Debate on the Choice of the Research Field

Let us now tackle a very relevant and very peculiar process that involved and animated the whole of the Cuban scientific, political and intellectual circles—not to mention foreign advisers—about the future of the development of physics. Throughout the 1960s, an extended debate in which professors and students participated took place on the choice of the most suitable research lines to follow. Widely differing ideas were brought into the discussion. Some preferred to leave the choice to individual researchers. Still, the development of the country called for solutions to a variety of specific problems. Thus, it was reasonable to concentrate the country's limited resources on strongly related research areas in order to help provide

[36] Cuban students had been integrated into high-level studies in physics at the University of Parma since the very end of the 1960s (Sect. 6.3.3).

[37] See the contribution by Leccabue to this volume.

these solutions. Although at the time there were no precise governmental guidelines on the subject, the university authorities and the country's leaders promoted, demanded and in fact personally participated in the search. As we have already remarked, from his position at the Ministry of Industry, Ernesto Guevara had made it clear that solid-state electronics was destined to play an increasingly important role in technology and industry, and that in such an area Cuba could make original contributions within the framework of the socialist camp. On the other hand, the long-term solution of the country's energy problems was seen at the time to be linked with the peaceful uses of nuclear energy. Accordingly, at the Physics School, serious research activities began in two departments that had been created at the beginning of the decade: nuclear physics and solid-state physics. As will be seen later, other institutions undertook the development of other branches of physics.

The choice referred to above did not happen quickly; it took some time and came to fruition only at the end of the decade. There were exchanges between the Physics School staff and specialists and executives from national industries, as well as with foreign visitors and institutions. Two events had a specific influence on the final decision. The first was the First Cultural Congress of Havana, which took place at the beginning of 1968 with the participation of several professors from the Physics School together with prestigious scientists from abroad, among them the Italian physicists Roberto Fieschi, Daniele Amati and Bruno Vitale, the French physicist Pierre Vigier, and Boris Pavlovich Konstantinov, vice president of the USSR Academy of Sciences. In the context of the gathering, a document was drawn up that stressed the importance of developing physics in Cuba; it recommended working on the development of solid-state physics and expressed the willingness of the undersigned to do what they could to back the development. The second influential event was the inauguration, in January 1969, with the presence of Fidel Castro and other leaders of the Revolution, of an Institute for Nuclear Physics (IFN) at the Cuban Academy of Sciences. The institute's inauguration was perceived at the Physics School as a strong signal that the resources required to develop that new field would be allocated to other institutions.

As a result, it was decided to concentrate the Physics School's research and postgraduate activities on solid-state physics, with the goal of supporting the future national development of the electronic, metallurgical and chemical industries. Some members of staff already working at the department of nuclear physics changed their former research subjects to techniques used for solid-state physics, such as Mössbauer spectroscopy and nuclear magnetic resonance, among others.

6.3.2.2 Redirection and the Promotion of Semiconductor Electronics in the Physics School: The French Contribution

As mentioned earlier, initial research activities on solid-state physics at the Physics School was autonomous; the visit by Veltfort and Waisman to the Ioffe Institute did not produce immediate results. The do-it-yourself organization of research at the school began with the mentioned purchase of the radio frequency oven from England

and a Zeiss vacuum chamber apparatus for measuring semiconductor characteristics from the GDR, which made it possible to grow semiconductor crystals. Other equipment and materials were acquired with great difficulty, or were brought into the country by foreign specialists or Cubans who had traveled abroad. Some equipment was built by a development group created at the Physics School (Alamino Ortega 2005) in cooperation with the electronics, glasswork and mechanics workshops. Subsequently, a semiconductor laboratory was created at the Physics School where first research activities started in that area. The first germanium monocrystals were obtained and characterized, followed by silicon monocrystals a few years later. An East German Zeiss vacuum chamber was added as well as a 20 kW air conditioner.[38] After overcoming other obstacles, it was finally possible

> … to activate the crystal growth oven, grow an electrically pure germanium crystal, and make a germanium diode from it (Crespo et al. 1968): this piece, humble and primitive as it was, was the first semiconductor device never before obtained in other Latin American country from local abilities and facilities, i.e., without importing a ready made production plant. (Veltfort 1998)

When Veltfort and Waisman returned to their respective countries in 1968, young Cuban *licenciados* continued semiconductor research. They began by following the same germanium technique that had been used before. A semiconductor group was created which sought to obtain germanium monocrystals, make semiconductor devices (diodes and transistors), and obtain thin semiconductor films for various applications. Marcos Lage, the vice rector for research at the University of Havana, promoted the manufacture of alloy diodes and transistors, inspired by a model set up at the University of Peking, whose physics faculty ran its own factory for semiconductor devices. The Cuban Ministry of Communications supported this activity. Practically all the equipment needed for the manufacture and testing of the germanium alloy devices was made at the Physics School itself. In 1969, germanium monocrystals were grown as well as alloy diodes and transistors with suitable electrical characteristics.[39]

The year 1969 was of crucial importance. As a result of exchanges with the French specialists participating in the Cuban summer schools (Sect. 6.3.4), a change in the direction of research was decided by the group, which would substitute silicon for germanium as the substratum of planar technology. In 1970, in order to move forward in this new direction, work began by setting up a Planar Technology Laboratory close to the Physics School. To give a boost to the new activities, it was decided to set up two research groups, one for semiconductor devices, and another for semiconductor materials. There was also a group that worked on thin films and another dedicated to measurements. The idea was to integrate a full cycle from the material used for the device. It should be noted that while the influence of French

[38] Cf. Veltfort's report and the reconstruction by Elena Vigil Santos, both in this volume.

[39] As a curiosity, it may be pointed out that for the manufacture of the first masks, a postage stamp copying camera was used. Donated by the Ministry of Communications, the same camera had been used years earlier for reproducing a celebrated Cuban postage stamp which featured the text of the First Declaration of Havana in very small print (Arias 1997).

physicists was determinant regarding the devices, cooperation with institutions from the USSR was of great help, especially when two Soviet specialists, Valentina Ostroborodova and Vladimir Dik, visited the Physics School in 1969–1970, and helped to optimize the crystal growing system.

The equipping of the Planar Technology Laboratory was largely set up at the Physics School itself, where the high temperature ovens were made, including their electronic controls and a very primitive aligning device. Cuban specialists visited France to obtain technical information (including the control system for a high quality commercial oven), while monocrystal growth was done entirely in Havana.

6.3.2.3 Reorganization of the Teaching and Research Activities in the Physics School

The specialization of teaching and research in the Physics School of the University of Havana required reorganization. With the purpose of finding the organizational form best suited to local conditions, the structure of several foreign laboratories in France, Italy and the USSR were studied. As a result, in 1969 an important change was introduced to the Physics School. This change involved suppressing the former departmental structure and introducing a double organization that differed from all others in Cuban higher education institutions. The change consisted in creating research groups and teaching section, side by side, each one headed by its own deputy director. This made it possible to provide the boost required for research work, without neglecting teaching activities.

Research groups were formed for semiconductor materials, semiconductor devices, metals, crystals, magnetism and microwaves (this one did not last long). The creation of a theoretical physics group took some time, because it was deemed convenient to give priority to experimental and applied research, linked to the needs of national economic development, as envisaged at the time, and avoid a disproportionate growth in the number of theoretical physicists, typical of some countries in Latin America.

All of these efforts laid the foundations for the establishment, achieved by the mid-1970s, of a fairly strong teaching and research center with a marked applied profile, but also with a good theoretical basis. At the time, it had become a prominent center of its kind in Latin America and was following an increasingly successful path.

6.3.3 International Collaboration

By the end of the 1960s, all undergraduate teaching was already taken care of by Cuban professors. Foreign collaborators dedicated themselves to postgraduate teaching and research and stayed in the country for shorter periods of time. As has been said above, C. Monet, T. Veltfort and D. Waisman returned to their respective countries, as well as V. Grishin, who had headed the department of nuclear physics from 1965 to 1968.

In 1967 a delegation of the Physics School's staff visited Moscow University:

> This step gave rise to an increment of the collaboration between the physics faculties of the Universities of Moscow and Havana. First rank professors of the former began to visit the Physics School regularly, and the school's professors began to travel regularly to Moscow to take postgraduate courses and begin PhD studies (Pérez Rojas et al. 1976).
>
> In that very year, a delegation of professors from Moscow's Lomonosov University and other Soviet higher education centers (Kochanov, Kolesnikov, Ostrovorodova, Ponomarenkov, Smirnov, Solutsev, Sontsev, G. and M. Strukov, Timushev, Valdanov, Zhipopitsev) visited the Physics School with a view to strengthen its teaching activities. (Pérez Rojas et al. 1976)

Starting in 1969, longer stays (up to 18 months) at the physics faculty in Moscow became systematic for the specialization and training of the younger staff members of the Physics School. As part of that exchange, professors from the Moscow university visited the Physics School to give postgraduate courses and help with research work. Most of them came for short periods, although in some cases, such as that of V. Dik and N. Ostroborodova, they stayed for 1 year.

Furthermore, several physicists from western countries continued to visit the Physics School. Among them, the visits from the Italian physicist Andrea Levialdi, professor at the University of Parma, deserve special mention.[40] Though sick with lung cancer, he arrived in Havana on November 5, 1968, with the purpose of establishing a stable cooperation. He died in the Cuban capital city on December 8, while giving a postgraduate course on solid-state physics.[41] He conceived the ambitious project of establishing a long-term collaboration between the Universities of Havana and of Parma,[42] but this could not be accomplished because of his passing. During the month in which he was able to develop these activities, he sent notes to his colleagues in Parma asking them to do their best to send Cuba badly needed materials and equipment. After his demise, his colleagues in Parma managed to establish a "Levialdi Scholarship," through which around 20 Cuban physicists have been able to get specialized training at the University of Parma and at MASPEC.[43] A fruitful collaboration continues to this day.[44] It is worth recalling that by 1968 the Cuban physicists who had visited Parma received such good training there that it allowed

---

[40] Cf. Waisman and Fieschi, both in this volume. Ms. V. Kleiber, Andrea Levialdi's widow, was kind enough to supply documents on his life and activities.

[41] See Levialdi 1968; cf. Fieschi's memories in this volume.

[42] Hand-written letter by A. Levialdi (November 14, 1968), provided by Ms. Kleiber.

[43] "*Materiali Speciali per Elettronica e Magnetismo*" (special materials for electronics and magnetism, today IMEM. The Italian physicist Andrea Levialdi (whose collaboration with Cuba will be discussed in the following) had begun in the years 1965–67 to develop at the University of Bari the production of monocrystal semiconductors for applications in photoconduction. Called subsequently to the University of Parma, he transferred this activity there, and its institute for physics became one of the most qualified centers in this technology. On this basis, the Italian research council (CNR) founded in those years the MASPEC Institute.

[44] Cf. Leccabue, in this volume.

Table 6.1 Summer school results, 1968–1972 (Vida Universitaria 1970)

Year	Courses	Lecturers	Enrollment	Cuban institutions
1968	15	12	138	*No data available
1969	24	129	471	*No data available
1970	57	172	1,182	38
1971	52	152	1,357	40
1972	45	88	1,118	54

them to fruitfully engage in research soon after their return to Cuba. In 1969 another collaboration agreement was also signed with Paris Sud University at Orsay.

In 1970 Cubans participated for the first time in a Latin American gathering on physics: the 2nd Latin American Solid State Symposium, which took place in San Carlos de Bariloche, Argentina. On this occasion, links were revitalized with CLAF (Latin American Center for Physics), an organization of which Cuba had been a member since it foundation in 1982. The Cuban professor Daniel Stolik (along with the Chilean Miguel Kiwi, the Brazilian Sergio Mascareñas, the Argentinian Antonio Misetich, and the Mexicans Edmundo de Alba, initially, and Feliciano Sánchez, later) became a member of its Commission for the Development of Solid State Physics in Latin America. This was the beginning of a fruitful relationship that resisted all attempts to isolate Cuba from other Latin American countries. For four decades, CLAF has provided quite effective links between Cuban physicists and their Latin American colleagues. In 2012, a Cuban physicist, Carlos Trallero Giner, was elected director of CLAF, a position previously held only by physicists from Brazil, Argentina and Mexico.

6.3.4 The Summer Schools (1968–1972)

At the above-mentioned Havana Cultural Congress, participants also agreed to organize *summer schools* in Cuba on physics, chemistry and other areas with the participation of foreign professors. These schools took place between 1968 and 1972 and their establishment became a massive phenomenon that played a key role in the training and updating of Cuban scientific personnel, as well as in the development of a positive approach to research. The courses given covered nearly all the university branches. Hundreds of lecturers from many countries (172 in 1970) and thousands of Cuban students participated (Table 6.1).

In the 1972 summer courses

> 88 lecturers from 15 countries gave 45 courses, each 4–6 weeks long, to 865 students from 54 production, research and development entities. [...] 910 of the 1118 initially enrolled are less than 40 years old, and from these 603 are less than 30 years old [...] the courses are satisfying the real needs of our economy and development [...] more than 41 % of the enrolled correspond to professional cadres of our university centers, which means the increase of the knowledge level imparted in our classrooms and, consequently, the further

> reproduction of the benefits they represent [...] Moreover, after 1970 fuller and more detailed information has been supplied on the theoretical contents of the summer courses to the various institutions that may be interested in them. [In] 1970, from the 1182 initially enrolled, only 401 (31 %) represented institutions other our university centers, while in 1972 the number of those initially enrolled coming from other institutions is 655 (81 %), and, more significantly, more than 10 % belong to the sugar industry. (Universidad 1972)

As far as the physics summer school courses are concerned, the French professors lectured the advanced courses (Vigier had in mind the idea of creating a center for advanced research in mathematics and theoretical physics), while the Italians (Bruno Vitale and others) in 1970 and 1971 were in favor of promoting basic teaching. The courses led by the French physicists (C. Weisbuch, G. Lempel, J.M. Debever, B. Cocquelin, J. Cernogora, J.F. Jaquinot, J. Pollard, J.P. Pinceaux, J.P. Cervan, and D. Bois, among others) covered mainly electronics, semiconductors and magnetism. The summer schools were extremely profitable for the development of solid-state physics at the Physics School, due to the theoretical courses imparted, and to the laboratories set up with materials and equipment brought by the foreign professors and donated by them as well.

6.3.5 The Take-Off of Physics at the University of Oriente

In spite of even greater difficulties, the efforts made at the University of Oriente, in Santiago de Cuba (Sect. 6.2.4.3), brought the first consistent results within a short period of time, although the development of physics was slower than at the University of Havana. The school of basic science of the technology faculty, created in 1966, had serious difficulties in teaching physics to a growing number of students who were taking their degree courses in engineering, agricultural sciences and medicine, and at the Pedagogical Institute as well.[45] Through contacts with the University of Havana, a Soviet nuclear physicist (Valeri Smirnov) and two Cuban *licenciados* (Jorge González and Homero Fuentes) arrived at the University of Oriente to help, but the real problem was that a degree course for physicists was lacking. Accordingly, in 1967, 20 "physicist-engineers" were trained at the university's technology faculty with an applied profile in three areas: optics and spectroscopy, nuclear physics, X-ray physics and metals (Pérez Rojas et al. 1976). Valeri Smirnov organized this plan, with the collaboration of Cuban colleagues and physicists from the Dresden Technical University. As candidates for this degree course, students were chosen from among the best of those taking degree courses in engineering and chemical sciences. The goal was to train them to become physicists in a relatively short period of time. Thanks to the dedication of teachers and students, for example, who waived vacations and recess periods, the first graduations were accomplished in 1 year less than normal. In November 1970, the University of Oriente had its first (eight) graduates.

[45] This is discussed in more detail by Cabal and Méndez Pérez in this volume.

That was the first step in the gestation of the Physics School of the University of Oriente, which was first conceived in 1968 and implemented in 1970, with the departments for nuclear physics, optics and spectroscopy, X-ray and metals, general and theoretical physics and electronic physics. The structuring of the Physics School and of its first and only graduation of physicist engineers may be taken as the turning point between the end of one historical stage and the beginning of the following one in which the full development of physics took place at the University of Oriente.

6.3.6 Development of Other Activities and Institutions

6.3.6.1 Further Initiatives at the Cuban Academy of Sciences

The Cuban Academy of Sciences was the institution within which other scientific initiatives and activities were promoted (Sect. 6.2.4.1). In part, they were often promoted by the Cuban presidency or government in view of the future needs of the country. But they were also promoted under the stimulus of the Soviet institutions and collaborators, with which toward the end of the 1960s there was an increasing integration, and which were engaged in diffusing Soviet scientific and technical advances to the socialist countries. Among these, further and new activities such as nuclear physics, communications and space research, and electronics were prominent.

Nuclear Physics: The Establishment of the Instituto de Física Nuclear (IFN)

In 1966, a nuclear energy group was created at the Cuban Academy of Sciences (Pérez Rojas et al. 1976), having obtained from the USSR the donation of various pieces of equipment suitable for teaching and research work. In 1968, President Fidel Castro put forward the need for the development of nuclear energy in Cuba, since the country was too dependent on imported oil. In order to implement this development, the country needed the technoscientific foundation required to support the envisaged nuclear activities.[46] This vital foundation was created on January 8, 1969, and was called the Institute for Nuclear Physics (IFN) of the Cuban Academy of Sciences. It was located on the premises of an old military school in the town of Managua, not too far from Havana. An indication of the high priority given

[46] Nuclear technology's most impressive developments were actually military focused, with the production and use of nuclear weapons at the end of the Second World War. However, nuclear technology has a wide spectrum of applications, some of them starting before the war. These ranged from civil energy applications (launched by Eisenhower's 1953 "Atoms for Peace" campaign, and linked to military applications due to the intrinsic 'dual-use' of this technology), medical and industrial uses of radioisotopes, and other techniques such as nuclear spectroscopy, nuclear resonance and the Mössbauer effect.

to this event was that beside Fidel and Raul Castro, several high-level Cuban officials and Alexander Novikov, vice president of the USSR cabinet, were present at its inauguration. The institute was equipped with laboratory materials supplied by the Soviet Union, including a subcritical nuclear reactor. Basically, the IFN dedicated itself to training and teaching, with a view to offer specialization in reactor physics, nuclear spectroscopy and radioisotopes to graduates from the physics and chemistry schools. Initially, the institute's staff numbered 28, composed of university graduates, technicians and administrative personnel, but by September 1968, 11 Cuban physicists, who had graduated and specialized at the Moscow State University and mentored by Soviet specialists, were added to the staff (Pérez Rojas et al. 1976).

Space Research

After the early developments discussed in Sect. 6.2.4.1, in 1966 the Cuban National Commission for the Exploration and Peaceful Use of Space was created under the Ministry of Communications to participate as a founding member of the Interkosmos Program, an international scientific collaboration program that incorporated four working groups: 'space physics,' 'space communications,' 'space meteorology,' and 'space biology and medicine,' to which a 'remote sensing' working group was added later on. There was substantial progress made in solar radioastronomy thanks to a strong collaboration established between the Geophysics Institute of the Cuban Academy of Sciences and the Pulkovo Observatory[47] in Leningrad (now St. Petersburg). In 1967, a satellite tracking station was inaugurated at Cacahual, in Santiago de las Vegas, near Havana. In 1969, Cuba became a member of the International Astronomical Union, and the following year, via collaboration with the USSR, received the first radio telescopes and used them to make observations of a solar eclipse. In the same year, the Institute of Astronomy was created at the Academy of Sciences, and a Cuban representative participated in the 14th General Assembly of the International Astronomical Union, which took place in London.

While the country had been participating since the mid-1960s in the Interkosmos working group on space meteorology, the first meteorological image taken from space was received in March 1969, soon after the first meteorological satellite tracking station had arrived from the USSR. This was installed on the premises of the Institute of Meteorology, which was created at the Academy of Sciences in 1964 in place of the old National Observatory, which was formerly attached to the Navy. The institute made great efforts to set up an extensive network of meteorological

[47] The Central Astronomical Observatory of the Russian Academy of Sciences at Pulkovo, near St. Petersburg, is the principal astronomical observatory of the Russian Academy of Sciences.

stations throughout the country (Ortiz 1987). Some of the first graduates of the Physics School of the University of Havana (Rosendo Álvarez, Mario Álvarez Guerra, Miguel Angel Portela) became pioneers of the new meteorology studies and contributed to the massive training of meteorologists at that time.

Electronics

In 1967, a small electronics working group was created at the Cuban Academy of Sciences, led by José Altshuler, with the collaboration of Roberto Díaz, Fausto Rodríguez and Miguel González, who set up the basic laboratory facilities required while initially carrying out original research in electric network theory. The results were published from 1967 to 1982 in *Comunicaciones*, a journal created in 1965 by the Ministry of Communications for the publication of research and development work in the area. It was not long before a basis for fruitful collaboration was established, among others, with the Institute of Radio Engineering and Electronics of the Czechoslovak Academy of Sciences and the Institute for Fundamental Technical Research of the Polish Academy of Sciences, which widened the scope of the research and included work on physics.

6.3.6.2 The Instituto Técnico Militar (ITM)

The year 1967 also witnessed the founding of the Technical Military Institute (ITM), the first higher education center of the Cuban Armed Revolutionary Forces, whose purpose was to train young officers in several much needed engineering branches. This training was possible under the guidance of a staff that included several young physicists graduated from the Physics School at the University of Havana between 1967 and 1970 (Carlos Álvarez, Ramón Buergo, José Matos, Domingo Jacob, Mercedes Carnero, Juan Monzón, Edwin Pedrero, Gudelia Ortega, Miguel Ángel García) who became regular members of the teaching staff and developed the center's physics department. Various Soviet collaborators visited ITM to teach formal courses and to organize seminars. Serious discussions also took place over what research lines to choose, as a result of which it was decided to concentrate on quantum radiophysics and laser development. Later on, work began in the areas of holography and optical communications. In the 1970s the staff of the physics department at ITM included more than 20 physicists. Some worked on laser applications (whose first significant result was the construction of the first CO_2 laser designed and built in Cuba in July 1974), while others did significant work on the methodology of physics teaching and made a remarkable contribution to the development of physics departments in other Cuban technical training centers.

6.3.7 The Establishment of the Centro Nacional de Investigaciones Científicas (CNIC) to Promote the Development of Science

The National Center for Scientific Research (*Centro Nacional de Investigaciones Científicas*, CNIC) deserves special mention since it has become a fundamental institution for the development of science in Cuba. It was created in 1965 by presidential resolution to accomplish the following aims and functions (Memoria 1966–67):

- To carry out scientific research in the field of natural, biomedical, technological and agricultural sciences that may benefit national interests;
- To organize and develop postgraduate training in the different branches in which it will develop its activities;
- To deal with any scientific activity of a national or international nature that may be deemed necessary for the accomplishment of its goals.

After its creation, the CNIC was incorporated into the University of Havana, yet it still retained its own legal and economic status, autonomy and the direct support of the government. The link with higher education helped to realize its goals, especially the joint work with different university areas dealing with research, undergraduate and postgraduate training as well. To accomplish the basic aim of training high level specialists, a considerable initial investment was made in equipment (the first electronic microscope in the country was acquired by the CNIC in 1965), and specialists were brought from abroad, mainly chemists, from East Germany (scientific personalities from West Germany were also contacted), the USSR, Spain and the United States (in the neurosciences sector). CNIC established numerous collaboration arrangements and international agreements, among others with the Polish Ministry of Science, Higher Education and Technology, the Chemical-Military Complex of East Germany, the Physics Institute of the Soviet Academy of Sciences, Lomonosov Moscow State University and the National Center for Scientific Research (CNRS) of France (Memoria 1976–77).

In the beginning, the CNIC's research activities concentrated on the fields of biology, medicine, chemistry and on some agricultural problems, with the participation of physicists, mathematicians, physicians and mechanical engineers, among others. Throughout its development process, the CNIC became the generator and developer of other important institutions, such as the Center for Neurosciences, the National Center for Animal Health, the Center of Genetic Engineering and Biotechnology, and the Center for Immunological Tests. Starting with a small group of workers in 1965, the CNIC's personnel grew to a staff of 1285, among them, 320 professionals, 304 technicians and 167 university students, including personnel from other institutions (Memoria 1976–77).

Physics was mainly an element of support at the CNIC. In the beginning three departments dealt with physical techniques: radioisotopes, X-rays and corrosion.

There were physicists working on X-ray diffraction, fluorescence, electron microscopy, and Mössbauer spectroscopy, while others worked in the department of electronics research. Techniques for nuclear analysis were developed at the Industrial Radioisotope Laboratory (LRI), created at the CNIC in October 1966 (Díaz and Meitin 1975). From 1965 to 1968, it profited from the expert advice of two Soviet physicists: Boris Bierjovski and Lev Medsel. The further developments that took place at the CNIC and its relevant contribution to Cuban science up to the present day will be discussed in Sect. 6.6.6.

6.4 The Consolidation of the Cuba Scientific System in the 1970s

The 1970s was a decisive decade for the development of physics in Cuba, and for the national scientific research system in general. The efforts made during the previous decade were effective in achieving a stable structure for the training of *licenciados* in the main branches of science, the creation of a critical mass of university graduates with good basic training and a strong feeling for research, a group of scientific centers and institutions—some of them dedicated to fundamental research and others dedicated to specific branches—as well as an extended network of international cooperation. Though some of these centers were still in their infancy, they had great potential. Taken together, these activities were characterized by what could be called a pioneer, or romantic, spirit marked by enthusiasm and initiative. Students compensated for the lack of adequate resources and sufficient training with their imagination and confidence in the usefulness and need for modern scientific development that would improve the country's future.

In the 1970s, mainly in the second half of the decade, the benefits of previous work could be reaped with a reorganization of the Cuban scientific system, which was integrated into the planning of the main branches of social and economic development. This period also included some elements aimed at promoting excellence, such as the creation of a national system of scientific degrees and the growth in the number of scientific societies, publications and journals. By the end of the decade, the Cuban Society of Physics had been created.

As far as international cooperation was concerned, the collaboration with the Soviet Union and other socialist countries of Europe was stabilized and became predominant in the second half of the decade. Relations with the main Soviet state universities and the institutes of the USSR Academy of Sciences were institutionalized, especially between the Lomonosov Moscow State University and the Zhdanov Leningrad University as well as the Leningrad Ioffe Institute and the Moscow Lebedev Institute of the Soviet Academy of Sciences and the Joint Institute for Nuclear Research in Dubna, USSR. These links laid the foundation for Cuban scientists to have access to elite Soviet institutions of science. Also boosted was the collaboration between the academies of sciences and universities of various

European socialist countries, particularly the Humboldt University of Berlin,[48] the Dresden Technical University and the Rossendorf Center for Nuclear Research in East Germany.[49] Many Cuban physicists who received their PhDs at these institutions gave a decisive boost to research in various branches of physics.

In spite of these prevalent collaborations and links, it is very important to take into account, as has been emphasized, that Cuba never severed its scientific exchanges with capitalist countries. In fact, during the 1970s, relations in the field of physics flourished with Parma's MASPEC, the Swedish Uppsala University, and others. There was increasing participation from Cuban specialists in Latin American scientific gatherings, especially in solid-state physics symposia (SLAFES) and CLAF meetings. Links were strengthened with several Latin American universities, especially UNAM and IPN in Mexico after the 4th SLAFES was held in Havana.

We will now examine in detail the achievements of the academic institutions and the research centers discussed above, as well as the promotion of further initiatives and organizations throughout the 1970s. The following sections will deal with the scientific fields, and particularly with specific, old or newly established institutions where the fields of activity were concentrated.

6.4.1 Consolidation and Linkage of the System of Higher Education

6.4.1.1 The Physics School of the University of Havana

After 1970, the staff of the Physics School of the University of Havana continued to improve, as did the syllabus and teaching laboratories. In addition, all subjects were taught by Cuban graduates. The above-mentioned reorganization of the school into research groups and teaching sections (Sect. 6.3.2.3) proved fruitful.

The first research seminar of the sciences faculty and the first meeting of Cuban physicists were organized in 1972 and 1974, respectively. The Physics School had already become a well-established physics teaching and research center. It was

[48] The Humboldt University of Berlin (*Humboldt-Universität zu Berlin*) is one of Berlin's oldest universities, founded in 1810 as the University of Berlin (*Universität zu Berlin*). After the war the Soviet military administration in Germany ordered the opening of the university in January 1946, redesigning the Berlin university according to the Soviet model, but for political reasons insisting, however, on the phrasing "newly opened" and not "re-opened."

[49] Since 1956 the Forschungszentrum Rossendorf near Dresden has been a research center geared toward nuclear research. It was founded as a result of the International Conference for Peaceful Uses of Atomic Energy held in 1955 in Geneva. The Soviet Union supplied the center with necessary equipment such as a particle accelerator and reactor. For the Dresden Technical University, see fn. 33.

perhaps one of the most vigorous in Latin America at that time, with a strong experimental and applied profile as well as a focus on theoretical foundations. Cubans continued to travel to foreign countries, with the majority going to the USSR and European socialist countries. Some also traveled to Italy (Parma) and France (Université de Paris Sud, Orsay). It should be noted that, as a rule, all those who had graduated in Cuba had received solid basic training, which enabled them to be successful in their future careers. MSc dissertations were successfully defended in Cuba: the first one, in 1972, was by Melquíades de Dios Leyva.[50] The first physicists to defend their doctoral dissertations in the USSR began to return to Cuba in 1974, where they subsequently supported and stimulated the progress of research activities.

In the first half of the 1970s, some 50 MSc degrees were awarded for physics, and the first Cuban physicists traveled to the Soviet Union to work for their PhDs. Throughout the decade, 30 graduates were awarded with a doctoral degree.

The general attitude that prevailed among Cuban scientists in those years was one of full dedication to the solution of problems to promote the development of the country. Their main concern was not the publication of research results in scientific journals, in particular international journals. More importance was attached to the development of "know how," research infrastructure and applications than to the publication of original results. The first scientific publications in international journals date back to 1975 (Castaño et al. 1975). Throughout the decade, some papers were published in *Physica Status Solidi*, and in Soviet and Cuban journals (Table A.2).

By 1976, the Physics School had 76 teaching staff of a high scientific level for Cuban standards at the time; 42 of them had specialist degrees (MSc), and 4 had candidate of sciences degrees (doctors or PhDs, according to western terminology). A substantial part of this group was already working on their doctoral degrees in Cuba or abroad, and 337 students had already graduated as *licenciados* in physics (Table 6.2). The country's most complete physics library contained 14,000 volumes, which included a considerable bibliography on semiconductors, metals, plasmas and nuclear physics. The scientific journal collection at the faculty of sciences of the University of Havana had a fairly good assortment of the most important titles. However, as a result of inadequate funds between 1961 and 1966, many collections were incomplete, especially those of journals from the West. In spite of the above results, it was pointed out that in 1976:

> … the collaboration plan signed with the Ioffe Institute has not yet been duly exploited, in spite of the interest shown by both sides [and] in the Physics School, though the existing equipment made it possible to reach an initial development in research and guarantee the minimum undergraduate teaching. At this very moment a series of basic equipment and instruments is lacking, which constitutes an obstacle for the development of research and postgraduate studies (including the PhD graduation plans), as well as the teaching of certain undergraduate study subjects at an adequate level. (Pérez Rojas et al. 1976)

---

[50] Cf. the interview with Melquíades de Dios Leyva in this volume.

Table 6.2 Some stable research groups in physics and related fields ca. 1990

N°	Institution	Field of research
1	UH. IMRE Faculty of Physics	Semiconductor lasers
2		LEDs
3		PV cells
4		Crystal growth
5		Solid state lasers
6		Magnetism and ferroelectricity
7		Superconductivity
8		Structural analysis (XRD, EM)
9		Ceramics
10	UH. Faculty of Physics	Complex systems
		Condensed matter theory
11	ICIMAF	Ultrasonics
12		Quantum field theory
13		Piezoelectric ceramics
14		Optical communications and robotics
15	CEADEN	Solid state techniques (DRX, NMR, MBE)
16		Experimental nuclear physics
17		Nuclear electronics
18		Nuclear theory
19	UCLV	Metal physics and welding
		Theoretical cosmology
20	CNIC	Physical chemistry and physical methods of analysis
21		Neurophysics
22	ISCTN	Nuclear methods of analysis
23		Physics and technology of nuclear reactors
24		Nuclear theory
25	UO	NMR and Medical physics
26		Solid state physics
27		Theoretical physics
28	CEDEIC	Optics
29		Electronics
30	Institute of Geophysics	Seismology
31	and Astronomy	Geophysics of magnetosphere and ionosphere
32		Solar activity
33	Meteorology	Physics of the atmosphere
34	ISPJAE	Microelectronics
33		Optics
34		Physics teaching
35	ISPEJV	Dielectrics
36		Theoretical physics
37		Didactics of physics
38	Institute of Oceanology	Ocean physics
39	UC	Thermal energy
40	CIME	Metallurgy
41	ICID	Electronics and microelectronics
42	Institute of Metrology	Metrology. Standards of physical quantities

6.4.1.2 Other Centers of Higher Education

University of Oriente

The Physics School of the University of Oriente was created in 1970 (Sect. 6.2.4.3), the same year as when the first and only graduation of their physics engineers took place. The school's staff grew rapidly with the inclusion of the first 19 graduates and the definition of a work profile related to physics methods of analysis.[51] By 1976 the total number of physics graduates amounted to 40 while the enrollment was some 50 students distributed among all the years of the degree course. All the graduates had defended their diploma work or dissertation, many of which dealt with nuclear physics.

The Physics School pioneered work in optics and spectroscopy in the country and trained the first graduate specialists in the area. A certain development was also reached in the areas of nuclear physics and of X-rays and metals. Scientific research began in collaboration with several professors from Leningrad University (V. Smirnov, A. Petrov, V. Niementz, N. Panichev, M. Braun, L. Lavzovsky, among others), who periodically visited the university while staff members visited the universities of Leningrad, Dresden, Stockholm and Italy. Priority was given to experimental and applied research in optics and spectroscopy, in connection with the nickel industry,[52] though work on nuclear physics also developed. The introduction of the subject called medical physics gave birth to an interesting development in biophysics that would become very important in the years to come. The first PhDs were completed abroad between 1976 and 1977 and specialist exchange increased, primarily with the Soviet Union and East Germany.

The following 10 years marked the highest level of development in physics at the University of Oriente. Starting in 1977, the then department of nuclear physics developed a project named "Introduction of Nuclear Techniques in the National Economy," that was approved by the National Commission for the Peaceful Use of Atomic Energy in 1979, and financed by United Nations Program for Development (UNPD), (UNDP 1979).

La Villas Central University

The Physics School of this university was created in 1970 with the enrollment of 31 students (19 of them in the first year and 12 in the second year), who came from the university's engineering degree courses or were students from the Las Villas province who had already begun their physics studies at the University of Havana

[51] For more detail, see the chapter by Cabal and Méndez in this volume.

[52] Cuba has one of world's best deposits of ore reserves rich in nickel and cobalt, localized to the north of its eastern region. At the time of the Revolution, two processing plants existed on North American property, which were nationalized. At the end of the 1980s, a third plant was built using Soviet technology. No refining plant exists. At present nickel is Cuba's primary export.

(Pérez Rojas et al. 1976). The staff was made up of Las Villas-born physics professor (José Villar) and others from the University of Havana (Eudaldo Tarajano, Irma González). Several physicists from Havana came to give courses here. By 1976, 12 *licenciados* had been awarded their degree, eight of which became members of staff. A research line on metal physics was begun and some physicists went to the USSR to gain PhDs in the subject. In 1976, the school was closed because of low enrollment, with only the department for general physics left to teach this subject to students taking other degree courses. Later on, when an interdisciplinary center for research on welding was created, those physicists who had specialized in metals played an important role.

6.4.2 Research on Microelectronics: Relevant Results and Fluctuations

The case of the research effort made in the area of microelectronics deserves special analysis, not only because of the remarkable results achieved in Cuba, but also because of the impact this branch had on the scientific programs of many developing countries, in particular on the search for ways to reduce the technological gap. In fact, the activity in this branch developed in Havana in the first half of the 1970s, was one of the country's first research projects in a highly technological area. The goal was to develop production at an industrial level and achieve an outstanding position in the area, as far as Latin America and the market served by the Council of Mutual Economic Help of the Socialist Countries (COMECON) were concerned.

The initiatives and activities in the summer schools, discussed in Sect. 6.3.4, became a determining factor, thanks to the exchanges between the physicists from the Physics School of the University of Havana and the French specialists; the exchanges, information, materials and equipment that the French physicists brought by the late 1960s led to the shift from germanium to silicon technology and the introduction of planar technology. In 1970, the Planar Technology Laboratory (LTP) was created within the Semiconductor Devices Group (GDS) of the Physics School. There was some financing from the university authorities and materials donated by the French. A large part of the equipment for the LTP was made in the workshops of the Physics School.

A particular fact marked the development and work style of the activities on microelectronics from the very beginning. In 1969, almost at the same time when the first steps were taken in this branch at the Physics School, the electrical engineering school of the technology faculty created the Center for Research on Microelectronics (CIME), made up of engineers, physicists and students belonging to the school and to the ministry of communications. The aims of CIME were practically identical to those of GDS (Arias 1997). An evaluation of this decision and its effects is not easy. The research workers of the Physics School were not happy with the creation of such a center at the time, because they thought it meant a division of resources. Subsequently, a great rivalry developed between the two institutions. The competition that resulted likely became an incentive to work harder. Prime Minister Fidel Castro visited CIME in the early 1970s and allocated funds for acquiring a complete system for the manufacture of integrated circuits. In the first half of the 1970s, both CIME and GDS had laboratories suitable for device manufacturing and characterization. The facilities for making masks and encapsulate devices were concentrated at CIME, but were used by both groups.

These efforts fell within a wider framework, which was the so-called electronics branch. Under the direct care of the rector José Miyar and the vice rector for research, Marcos Lage, it coordinated and tried to integrate the efforts made by various university areas with the common purpose of developing electronics at the national level. In this way, a close integration was achieved, probably unthinkable in other countries, with the participation of:

— the Chemistry School, which tried to obtain electronic grade silicon from Cuban sand but made very little progress. This was taken up again in the late 1980s, in the framework of a United Nations Development Program project (UNDP), which was also unsuccessful;
— the Physics School, particularly the Semiconductor Devices Group (GDS) worked on the development and manufacturing of integrated circuits, while other groups dealt with obtaining materials for electronics and studying them;
— CIME of the Ciudad Universitaria "José Antonio Echeverría," CUJAE, with similar aims as GDS;
— the Mathematics School, to take care of the development and introduction of appropriate computer software;
— the Systems Development Center of the CUJAE, to engage in the development of application systems for computers;
— the Digital Research Center (CID), later Central Institute for Digital Research (ICID), that had already developed the first Cuban computer and had a pilot plant for the development and manufacture of minicomputers;
— the Department of Electronic Research of CNIC, whose job was to develop and manufacture advanced electronic equipment for various applications. (Cerdeira, personal communication 1997)

As far as microelectronics is concerned, results followed quickly and were relevant both as physical achievements, and on a technical level in relation to future applications and possible developments (Cerdeira, personal communication 1997; Arias 1997). After the inauguration of the LPT in 1970, devices with a growing integration level and commercial characteristics were made, including MOS channel P transistors and MOS integrated circuits in 1970, and the first MOS channel N and N-P-N bipolar transistors on normal substrates and on epitaxial layers grown at the same laboratory in 1973. A MOS-P integrated circuit, a silicon solar cell (1975), and channel P and N MOS transistors with a polysilicon gate—this last one deposited in equipment developed at the LPT—were manufactured in 1974. In 1976, I^2L integrated circuits and bipolar transistors were made on epitaxial layers obtained at the LTP. In 1975, the first silicon solar cells and in 1977 the first programmable logical array (PLA) were obtained, designed and made entirely in Cuba, containing 1,100 components per chip in a 2.5×2.5 mm^2 wafer.

Important results were also achieved at CIME, where the first pilot plant for semiconductor devices and integrated circuits was installed (Arias 1997).

A bilateral cooperation plan (Plan CUSO) was started in 1974 with Canada,[53] and an MSc course was offered by prestigious Canadian professors with the participation of several specialists from the Physics School's GDS. That same year, CIME started a cooperation arrangement with Belgium to make silicon gate MOS-P and MOS-N transistors and integrated circuits.

The results appeared mainly in monographs published by the science faculty of University of Havana, in the proceedings of national scientific meetings, in journals published by the Cuban Academy of Sciences, and later on, in the journal of the Cuban Society of Physics. These were the first scientific publications on these subjects (Estrada et al. 1972). While at that time there was no special interest in publishing in international journals, the results obtained (Cerdeira, personal communication) and the basic studies made while working on the various subjects resulted in several Master and Doctoral dissertations (the first defended in the USSR and the later ones in Cuba).

In a very brief time, Cuban physicists had independently achieved developments in microelectronics at a relatively high level with respect to Latin America and even to European socialist countries, counting mostly on material help from the French, long before the connections with the USSR in this branch had really taken off:

> However, collaboration with the Ioffe Institute did not yield its first results until 1973, when a mission of theirs visited Cuba, later followed by a specialist; still, in 1976, after a collaboration agreement was undersigned through the Academy of Sciences, collaboration was not yet satisfactory. (Pérez Rojas et al. 1976)

As mentioned in the introduction, there was an international meeting in January 1975 which brought Cuban physics to the physics elite of Latin American: the 4th Latin American Symposium of Solid State Physics, the first international high level scientific gathering in physics, which was organized by the Physics School of the University of Havana and took place in Cuba. The Physics School staff presented 20 of the 26 papers presented by Cuban authors (Pérez Rojas et al. 1976). This meeting served to attest that Cuba had reached an outstanding level in solid-state physics as far as Latin America was concerned, only surpassed by countries of incomparable dimensions and scientific traditions, such as Argentina, Brazil and Mexico.

In the subsequent years, several institutional changes occurred. Up to 1976, the Center for Research on Microelectronics (CIME) and the Semiconductor Devices Group (GDS) were laboratories localized in different faculties of the University of Havana, which completed substantial work in their fields despite little cooperation and rivaling for the same resources. In 1976, GDS became a "science and technology unit" under the name of *Laboratorio de Investigaciones en Electrónica del Estado Sólido* (LIEES, Laboratory for Solid State Electronics Research). Shortly afterward,

[53] In the time of the Prime Minister Pierre Trudeau relations between Cuba and Canada were excellent.

the institutional changes which led to the creation of different universities independent from the University of Havana, included decisions which sought more rationality, avoiding the duplication of efforts and trying to optimize the use of available human and material resources. In this context, it was decided to unite research in microelectronics at CIME. The development of silicon-based devices was unified at CIME, while the development of optoelectronics based on III–IV semiconductor compounds continued at LIEES. However, personal differences hindered a real unification of the two groups. In practice, the unification at CIME was essentially formal, since only a few specialists from LIEES transferred to CIME, and then only for a short time. Some of them continued to be linked to microelectronics work in other research centers or entered the projects on industrial applications that are discussed below, while others decided to engage in alternative activities, also related to electronics.

The university physicists and engineers, who were engaged in microelectronics research, actively participated in the conception of a national semiconductor-based industry. A commission visited some West European countries to get acquainted with state-of-the-art technology for producing semiconductor devices, and a suitable industrial plant was finally bought from a Spanish company. This company, however, was bought some time later by an American company, which canceled the agreements previously reached with Cuba; as a result, the factory was built but could never operate according to the original project. This marked the onset of the crisis that developed in the microelectronics sector. After 1979, research in microelectronics lost the impulse that in its first decade of development had been based to a large extent on western cooperation. As a result, Cuban microelectronics began to fall behind. A second impulse in the 1980s was associated with Cuba's entry into the Council of Mutual Economic Aid (COMECON) market. In this context, a technological reconversion effort was made in the 1980s, while at the same time another plant dedicated to the manufacture of hybrid circuits was erected. For years there was no adequate national support for it until the creation in the 1980s of the so-called "electronics front." (Sect. 6.6.3). However, in the end, the downfall of the Soviet Union and the East European socialist system ground these later projects to a halt. While research work continued until the early 1990s at CIME and ICID, they never fully recovered the positions they had achieved in the 1970s.

In the end, all of the above efforts and successes in the development of microelectronics met with an unforeseen obstacle and ended in complete failure, not only in Cuba, but also in the majority of countries that sought a way toward economical progress in that branch. That obstacle was the fact that the high-speed advances in high and very high integration microelectronics demanded huge investments that were well beyond the economic possibilities of Cuba, which was subject to tight restrictions imposed by the United States. Finding a general balance seems quite problematic. In principle, the attempt to develop and compete internationally in microelectronics appeared at that time to be a good choice for underdeveloped countries. No one could have foreseen the extraordinary technological drive that took place parallel to the spectacular development of high and very high level integration.

Nevertheless, from the viewpoint of the advancement of physics in Cuba, the projects that had been previously tackled for the development of microelectronics left a very positive mark on the field, since they stimulated the search for excellence, contributed to the training of several generations of Cuban high-level physicists, and allowed for the accumulation of unprecedented experiences in the history of a country with no tradition of scientific research. Some areas of solid-state physics that grew under the influence of the effort to develop microelectronics continued to be vital and are now finding other spheres of application, for example, in the field of photovoltaic energy. Fortunately for the country, national efforts in other high-technology sectors were successful, as exemplified by the case of biotechnology, which is nowadays a growing and highly productive sector of Cuban industry.

6.4.3 Nuclear Physics During the 1970s

In comparison to microelectronics, the early developments of nuclear physics had quite different motivations and actors, which would lay the basis for its main development in the subsequent decade when more ambitious programs arose. At the time, however, it appears that the original developmental aims of this field of activity were more general and basic. We have already mentioned in Sect. 6.3.6 the creation in 1969 of the Institute for Nuclear Physics (IFN) of the Cuban Academy of Sciences. Its staff increased from the initial 28 members (university graduates, technicians and administrative personnel) to 82 in 1971 and 113 in 1972 (Pérez Rojas et al. 1976) distributed among six departments: nuclear reactors; radiochemistry; dosimetry and radiological protection; engineering; nuclear research techniques; and management.

Between 1971 and 1973, a research project on the analysis of neutron activation in metallurgy and other fields (supported by the International Atomic Energy Agency) as well as the construction of a subcritical uranium-graphite reactor were initiated. In this period, from 41 technical-scientific tasks (i.e. activities planned as parts of a larger project, such as the assembly of a technique, the development of a program, or the fabrication of a piece of equipment) ten were finished and 11 MSc dissertations were successfully defended.

Early in 1974, the center was renamed Institute for Nuclear Research (ININ) and its first scientific conference was held, in which 106 scientific presentations were given by specialists from the center and from other national institutions dealing with areas related to the nuclear branch. Ten MSc dissertations were defended and a theoretical nuclear physics group was created within the ININ.

Despite the existence of collaboration agreements with the USSR State Committee for Atomic Energy and with the Unified Institute for Nuclear Research (IUIN[54]) in 1976 "serious difficulties that the ININ has faced up to now due to lack

[54] The Joint Institute for Nuclear Research (JINR) in Dubna, Russia was established on the basis of the convention signed by the plenipotentiaries of the governments of the Member States (Albania, Bulgaria, China, Czechoslovakia, East Germany, Hungary, Mongolia, North Korea, Poland, Romania,

of material resources" were regretted, and it was expected that "with the supply plans and support of the UNPD the institute may complete to a large extent the basic equipment required for its development during the following five year period" (Pérez Rojas et al. 1976).

The ININ had a specialized library (in 1976, 3,000 monographs and symposia, a collection of scientific and technical reports of more than 40,000 titles, several thousand copies of periodicals), as well as basic installations and instruments: a subcritical reactor, neutron and gamma radiation detectors, radiochemical installations (these were responsible for the implementation of the first work with radioisotopes), gamma radiation sources and analyzers for spectrometric measurements. All the instruments, detectors and radioactive sources came from the Soviet Union. Because of its contribution to the later development of the country's own scientific instrument manufacturing, it is worth mentioning the nuclear electronics laboratory, headed by José L. Díaz Morera, who was later the founding director of the Special Bureau for Instrument Making with Attached Production (BECICPA), (Sect. 6.4.8).

The fundamental areas were: nuclear reactor calculation, VVER reactor design (in collaboration with an international group in Hungary which designed reactors to be produced in Eastern Europe), radiochemistry, nuclear electronics, Mössbauer spectroscopy, neutron physics, analysis of minerals, nuclear agrophysics, fast neutron spectrometers, absolute measurement methods of neutron sources, and neutron spectrometry. For the latter, monocrystal neutron spectrometer, a flight time spectrometer and a Monte Carlo program for the determination of the efficiency of scintillating organics were developed. The Monte Carlo program—requested for use at the Oak Ridge Laboratory—was developed by Víctor Fajer and Lilliam Álvarez. The nuclear electronics group produced a group of instruments known as the nuclear chain, which included preamplifiers, counters and low and high voltage sources. At the same time, work on dosimetry resulted in the development of film dosimetry for individual Geiger-Müller counters, dosimeters and standards for establishing radiological protection controls in all of the institute's installations, as well as for the release of radioactive waste.

Work on nuclear physics was not limited to IFN-ININ. At the *Centro Nacional de Investigaciones Científicas*, CNIC (Sect. 6.3.7) nuclear analysis techniques were developed, and counters, discriminators, high and low voltage nuclear modules, and instruments for oil prospecting were produced. At the Industrial Radioisotope Laboratory in Cuba (LRI, Sect. 6.3.7), autochthonous developments and applications of nucleonic instruments, based on the characteristic properties of ionizing radiation that allow contactless measurements, were made for the first time (Maggio and Desdín 2000).

The Oncology Institute installed high-dose gamma radiation sources for radiotherapy (Sect. 6.2.4.1), film dosimetry and radiologic protection systems. Also,

USSR and Vietnam) of the JINR in March 1956 in Moscow. The JINR was created in order to unify the intellectual and material potential of the Member States in order to study the fundamental properties of matter. This initiative was also a response to the creation in 1954 of the European Organization for Nuclear Research (CERN), near Geneva, to unite the efforts of Western European countries in this field of research.

the Center for Energy Research of the *Instituto Superior Politécnico "José Antonio Echevarría"* (ISPJAE, see Sect. 6.4.5) carried out research on nuclear energy and trained specialists in the area. Work on applied nuclear physics was also done at the University of Oriente.[55]

The National Commission for the Peaceful Use of Atomic Energy had been created in 1974, and in 1976 an important intergovernmental agreement was signed with the USSR that included the construction of the first electronuclear power plant in the country. At the same time, Cuba entered the Joint Institute for Nuclear Research (JINR, Dubna) as a member country, and several Cuban physicists, chemists and engineers started to work for long periods of time at that center. A remarkable transformation and growth in the nuclear sphere took place in Cuba in the 1980s as a result of the development of the ambitious national nuclear program.

6.4.4 Geophysical, Astronomical and Meteorological Activities at the Cuban Academy of Sciences

We have seen in Sect. 6.2.4.1 how the first activities in these branches were created and institutionalized within the Cuban Academy of Sciences (ACC). From 1972, the academy emphasized a more rigorous approach to both ongoing research and personnel training. A more advanced scientific council was created and researchers were encouraged to work to obtain their PhDs at academic institutions in European socialist countries. Among them, were physicists working at the institutes of nuclear physics, meteorology, atmospheric physics (later merged with meteorology), geophysics, astronomy, and electronics (Memoria 1973–74). In 1972 the department of geophysics became an institute whose main research areas were seismology, sun-earth relationships, short radio wave propagation, and other areas of geophysics.

In 1972, the Institute of Astronomy began work on optical astronomy by taking photographs of solar spots. It established a collaboration agreement with the Crimean Astrophysical Observatory, which led to the installation of a solar radio telescope at the institute by a Soviet team. In 1973, a second team of 15 Soviet radio astronomers made observations of a partial solar eclipse from Santiago de Cuba (Historial de Geofísica n.d.).

The Institute of Geophysics and the Institute of Astronomy merged in 1974 to form the present Institute of Geophysics and Astronomy (IGA), which had 130 staff members. Its fundamental research lines were the study of the sun, the magnetosphere and the ionosphere, the endogenous physical processes and the deep structure of the Cuban archipelago (Historial de Geofísica n.d.). Such research lines were related to the country's economy, aimed at improving radio communication links, determining seismic risks and drawing magnetic and gravimetric maps. Until 1990, the IGA

[55] See the contribution by Cabal and Méndez to this volume.

maintained close relations with similar institutions of the socialist countries, which led to the installation of stations for continuous Faraday effect recording, the reception of telemetric data from satellites, the inclined sounding of the ionosphere, satellite tracking, a short wave transmitter and a spectral analyzer, among other experimental facilities, all located at the institute.

In 1977, specialists from the Pulkovo Astronomical Observatory installed at El Cacahual near Havana a horizontal solar telescope with a spectrograph coupled to it (Pérez Doval 1991). Meanwhile, a first generation "Krypton" laser radar was installed at El Salado satellite tracking station, a result of the joint collaboration established between the USSR Astronomical Council and the Interkosmos Program.

A relevant event boosted the development of the country's meteorology, when in 1970 the World Meteorological Organization approved a request for the implementation of a project to extend and improve the meteorological service in Cuba through the UNPD (Ortiz 1987). In the Institute of Meteorology and its national network of stations, long-range radar stations, synoptic stations, and climatology and agro meteorology stations were set up with high quality equipment. Thanks to tracking equipment donated by East Germany, images in the visible and infrared spectral bands provided by the NOAA meteorological satellites were used from 1974 to obtain an overall view of atmospheric phenomena, including clouded areas (Ortiz 1987).

6.4.5 Activities for Engineers in the Departments of Physics

In the 1970s, physics teaching in the department of basic sciences in the technology faculty at the CUJAE (Sect. 6.2.4.2) evolved considerably. There were some 10,000 engineering students in Cuba, most of them enrolled at the CUJAE. It was not easy to consolidate a stable and updated curriculum where theoretical and practical aspects would work hand in hand, and to properly organize the new student laboratories. Yet, all of this was gradually accomplished. Quantum physics was introduced into the basic course taught to engineering students, and in 1975 a department of nuclear engineering was created that started the regular training of nuclear engineers.

The department of basic sciences also began to get involved in research activities. Groups were formed to deal with lasers, holography and microelectronics—this last one in close collaboration with the Center for Research on Microelectronics (CIME, Sect. 6.4.2), also belonging to the CUJAE. By the mid-1970s, an MSc program in microelectronics run by Canadian specialists was started. Several members of the Cuban staff defended their MSc dissertations at the department. Other staff members went to the USSR to work for their doctoral degree. Work on optics was very much strengthened in the 1980s, as will be seen later on.

In 1976, the technology faculty of the University of Havana, already fully established at the CUJAE, became the "José Antonio Echeverría" Higher Polytechnic Institute (ISPJAE), an independent higher education center attached to the Ministry of Higher Education (Sect. 6.5), which had just been created. A subsidiary of ISPJAE was created in the city of Matanzas, which later became an independent university.

6.4.5.1 Work on Optics

Work on the optical characterization of semiconductors, spectral analysis, photoluminescence and interferometry had been going on at the Physics School of the University of Havana since 1968. But the development of optoelectronics properly began in 1970. MSc dissertations on light emitting diodes were defended, followed by research on the optical properties of III-IV semiconductors. Photodiodes, photovoltaic cells, phototransistors, semiconductor lasers, and optical guides were made in close collaboration with the Ioffe Institute, especially with the Laboratory of Contact Phenomena headed by the Soviet academician Z.I. Alferov, who would win the Nobel Prize for physics in 2001. The physics chair of the Military Technical Institute (ITM) engaged in work on holography and optical communications. Work was done on Q-switching modulators using lithium niobate electro-optical devices, and magneto-optical ones with lead molybdate. Research was carried out on multispectral recognition. It must be said that work by the ITM staff made a significant contribution to the development of optics in Cuba.

6.4.6 Medical Physics

As a result of the pioneer work accomplished by the first group of physicists and other specialists at the National Institute of Oncology and Radiobiology in the 1960s (Sects. 6.3.6 and 5.4.3), radiotherapy services spread throughout the country and nuclear medicine was introduced. From the beginning, the Ministry of Public Health included specialist courses on medical physics in its basic modules for radiotherapy, nuclear medicine and radioprotection, which substantially increased the number of physicists doing practical work and engaged in research in clinical environments. By the end of the 1970s, more than 30 medical physicists were already working in Cuban hospitals, especially in the area of nuclear medicine—a substantial figure in comparison to the rest of Latin America.

6.4.7 The Institute for Fundamental Technical Research (ININTEF)

The Institute for Fundamental Technical Research (ININTEF) was founded in 1975 at the Cuban Academy of Sciences as a multifaceted outgrowth of the electronics department (itself a more mature version of the original electronics group), where several neglected branches of physics were incorporated in other scientific institutions. As indicated by its name, the ININTEF's main, but not single, purpose was to tackle the so-called "oriented" or "strategic fundamental research" needed to support applied research of technological interest (ININTEF 1982).

Among the research and development areas that involved physics at the ININTEF, the following stand out: quantum electronics and holography, ultrasonics, solar energy, remote sensing and quantum field theory. Physicists also collaborated with other groups dealing with engineering problems, such as electronic circuits and systems, and time and frequency standards. The solar energy group was the basis for the Institute for Solar Energy created in Santiago de Cuba later on.

A few graduates dedicated to theoretical physics were taken on right from the beginning, initially to deal with stochastic processes. Later on, they moved to 'field theory' mainly through close collaboration with the Lebedev Institute of the Soviet Academy of Sciences. The group ended up as one of the country's top groups dealing with theoretical physics. The ININTEF was amalgamated in 1983 with the Institute of Mathematics and Computing, another center of the Cuban Academy of Sciences, where other physicists were working on modeling problems.

In spite of a chronic scarcity of material resources, the ININTEF managed to produce some practical results related to the exploitation of solar energy, remote sensing, industrial applications of ultrasonics, high precision time and frequency measurement equipment and industrial controllers. The ININTEF participated very actively in the design and preparation of five original experiments intended to be carried out in orbit during the USSR-Cuba Interkosmos joint space flight that took place in September 1980 (Sect. 6.4.10) (ININTEF 1982; Altshuler and Serafimov 1991). This research allowed the training of specialists and the development of advanced technologies, some of which were transferred years later to industry and the health system.

6.4.8 The Special Bureau for Instrument Making with Attached Production (BECICPA)

It is worth mentioning the creation in 1978, out of previously existing entities, of the Center for the Construction of Scientific Instruments. The center had two aims: (1) to realize a complete cycle in the design and production of optical and scientific equipment (electronic devices, lasers for physiotherapy and analytical instruments, ordinarily produced in developed countries); and (2) to reinforce the impact of physics and technology in society. This institution experienced its main development in the 1980s and 1990s, as will be explained later on.

6.4.9 Activities at the National Center for Scientific Research (CNIC)

Some 20 physicists worked at the CNIC in different areas of applied physics. Apart from its activities in nuclear physics, mentioned earlier, it included groups dealing with physical metallurgy and physical chemistry of surfaces. The CNIC had at its disposal laboratories that made it possible to make use of a wide range of physics techniques (X-ray diffraction, electron microscopy, nuclear magnetic resonance, mass spectrometry and Mössbauer spectrometry) with which it was able to provide services and implement applied research work for industry, especially mining. Moreover, it attracted many physicists who worked in other institutions, but carried out part of their experimental work using equipment at CNIC, which also offered a wide range of postgraduate courses. In 1971, with the goal of creating the conditions for the design and manufacturing of electronic and electromagnetic instruments suitable for different kinds of analyses, and equipment required for scientific work, a Department of Electronic Instrumentation (DIE) was created at the CNIC, a product of the development of the technical group associated with the center's electronics and mechanics workshops (Memoria 1976–77). Nuclear analysis techniques were developed, and counters, discriminators, high- and low-voltage nuclear modules, instruments for oil prospecting (among many others) were manufactured.

6.4.10 Space Communications and Experiments in Space

Cuban physicists participated in the country's space communications activities which thrived in the 1970s in close collaboration with the USSR. We have already discussed (Sects. 6.2.4.1 and 6.3.6.1) the need the country had to set up a communications system via satellite. Previous collaboration among participants (Sect. 6.2.4.1 and "Space Research" in Sect. 6.3.6.1) in the program facilitated the creation in 1971 of Intersputnik (Sect. 6.2.4.1), an intergovernmental organization created to provide and manage satellite, long distance telephone, telegraph, TV and data communication channels on a commercial basis. In the framework of a bilateral agreement between the communication ministries of Cuba and the Soviet Union, 1972 saw the beginning of the construction of the "Caribe" station in Cuba, with a 12-m diameter parabolic antenna, capable of simultaneously transmitting one TV channel and 60 telephone channels in the 4–6 GHz frequency range. It worked initially with Molniya 2 type satellites, which were later replaced by Soviet-made geostationary satellites. In November 1973, the first color television programs were received from Moscow. This system was effective in establishing a reliable wide band communication link between Cuba and the USSR and East European countries, and through them, also between Cuba and France, Italy and Spain (Altshuler 1997).

However, the limitations of the Intersputnik system for wide band communications with Latin American and Western European countries became critical in 1979, when Havana was chosen as the site for the celebration of the Non-aligned Countries' 6th Summit. It was not until the 1980s that the much higher capacity geostationary communication satellites replaced the Molnya satellites. Accordingly, an Intelsat satellite communication system was added to the Caribe station.

In 1974, the institutional affiliation of the Cuban Interkosmos Commission was transferred from the Ministry of Communications to the Cuban Academy of Sciences. Two years later, an intergovernmental agreement was signed by each of the countries involved in the Interkosmos collaboration program pledging to send a cosmonaut into space to personally participate in carrying out, in orbit, suitable experiments put forward by his country's scientists. Cuba managed to set up a 200-member special team, which in collaboration with space scientists from other socialist countries succeeded in preparing some 20 original experiments. Except for experiment "Hologram," which flew a few months later, the other experiments were successfully carried out during the Soviet-Cuban space flight that took place in September 1980, with the participation of the Cuban cosmonaut-researcher Arnaldo Tamayo (Altshuler and Serafimov 1991). Four important techno-physical experiments designed by Cuban physicists were prepared for the flight.[56]

6.5 Settling of the Cuban Scientific System

6.5.1 1976: The "Process of Institutionalization"

During the 1970s, the organization of science underwent important institutional transformations at a state level. Actually, scientific activity had reached a fairly advanced level of development and achieved considerable expansion. The relatively spontaneous initiatives that characterized the earlier period of development and the dramatic difficulties to be solved were replaced by a general organization and more systematic process, directed in a much stronger way by governmental institutions. This change happened in the context of a general reorganization of the state structure and the economy at all levels, in the context of the so-called "Process of Institutionalization" which took place in Cuba in 1976. Given the considerable growth reached by the country at all levels, the process was deemed necessary. Besides this growth, the entry of Cuba into the Council for Mutual Economic Aid (COMECON) in 1976 strengthened the country's political and economical integration with the USSR and East Europe. An economic management and planning system similar to the one established in those countries was adopted, which included long-term (a few decades), middle-term (5 year), and short-term (yearly) development plans.

6.5.2 Changes in the Organization of the Scientific System

The state structure created in Cuba in 1976 established—in a growing hierarchic order—national institutes, ministries and state committees similar to those in the Soviet Union and other socialist countries. The state committees set and controlled aspects of the global politics of the government, such as science, finance

[56] These experiments are described in the chapter by Juan Fuentes et al. in this volume.

and labor policy, in every institution of the country. The ministries and the institutes corresponded to the different sectors or branches of the economy or other social activities.

Within this structure, the scientific sector of the Academy of Sciences was granted the level of a national institute, with centers of research associated with it, as shown above. A *Comité Estatal de Ciencia y Técnica* (CECT, State Committee for Science and Technology) was created, which governed the Academy. This choice shows the high priority assigned to science and research, but overinflates its bureaucratic influence in a small country where science was only just beginning to develop. For this reason, the CECT disappeared in 1980, and the ACC, now with a ministerial rank, assumed its functions. Further changes occurred in the subsequent decade, as will be discussed in the next paragraph.

As a part of the 1976–1980 national five-year plan, a plan for science and technology was drawn up which encompassed several research and development objectives related to physics. These included: the production of electronic components; the application of chemical and physical analytic methods for the nickel extraction industry; the development of experimental physical techniques; technological calculations and studies for nuclear reactors; radiological protection and nuclear safety; nuclear techniques applicable to the country's economy; sun, magnetosphere and ionosphere studies; studies on the exploitation of solar energy; quantum electronics and laser uses; and environmental pollution.

6.5.3 1976: A General Study of the Results Obtained by the Cuban Scientific System

Under the new structure, the CECT arranged in 1976 for a general study charged with the task of providing an assessment of the results obtained by the Cuban scientific system. This study gives an idea of the great advances achieved in a period of only 15 years (Pérez Rojas et al. 1976). This study mentions the main institutions that dealt with physics research in Cuba: The Physics School and the Laboratory for Solid State Electronics Research (LIEES) of the University of Havana; the Physics Schools of the Universities of Oriente and Las Villas; the Department of Basic Sciences and the Center for Research in Microelectronics (CIME) of ISPJAE; the Institute for Nuclear Research (ININ), the Institute of Meteorology, the Institute of Geophysics and Astronomy (IGA); the Institute for Fundamental Technical Research (ININTEF); some departments of the National Center for Scientific Research (CNIC); the "Enrique José Varona" Higher Pedagogical Institute (ISPEJV); the physics department of the Ministry of Education (MINED); and the Institute of Oncology and Radiobiology of the Ministry of Public Health (MINSAP). The report most likely missed the presence of physics in some laboratories and multidisciplinary centers attached to industry, among them the Center for

Metallurgical Research, the Metrology Laboratory, the Central Telecommunications Laboratory, the Research Center for the Mining and Metallurgical Industry, the Cuban Institute for Sugar Research, and several hospitals where physicists were employed.

The study then discusses some of the main reasons for the importance attached to the development of physical sciences. These reasons show that the historical actors held a strong belief in the pivotal role of physics as a catalyst for both scientific and technological progress; they also emphasized the successful applications already made. In addition, they underlined the role of scientific achievements in physics for national prestige and its potential for supporting the ideas of dialectical materialism. Remarkably, while the prestige argument is still used today, the connection of physics to dialectical materialism has been replaced by the cultural impact of physics:

- from an *educational* point of view, because of the growing importance of physics in all the branches of science and technology;
- from the economic and techno-scientific point of view, because of its contributions of a fundamental, oriented and applied character. The following points were made: (a) "nuclear physics, whose importance for the economy is great because of the prospects of installing nuclear power plants in our country in the immediate future, (not to mention) the utilization of radioisotopes in agriculture, medicine, biology, etc.," and (b) "solid-state physics, of utmost importance for the development of electronic and mineral-metallurgic industries, and in which we have been able to achieve a remarkable position in Latin America; this branch may have a wide range of applications in very diverse fields in Cuba, such as the chemical industry, the building industry, etc.";
- from a political standpoint, "especially if one considers that in 10 years we have reached a position close to third place in the Latin American context, only behind Brazil, Argentina and México (countries with a much larger industrial potential than ours) as far as solid-state physics is concerned";
- from an ideological standpoint, because of the importance of having a sound scientific model to help acquire a deeper understanding of dialectical materialist philosophy (Pérez Rojas et al. 1976).

Quantitative Achievements and Growth of Higher Education and Research

The study also presents some quantitative data on the number and growth of physics graduates in the various universities (a total of 429 up to 1976), the scientific publications by the various institutions, and the growth of PhDs in physics (a total of 26). This data is effective in showing the considerable achievements of Cuba in only 15 years: it is presented in Appendix A.1 at the end of this chapter.

6.5.4 The Creation of the Ministry of Higher Education and Changes in the Universities

The fast growth of enrollment and the growing linkages between universities and other state institutions led not only to the subdivision in 1976 of the three largest universities (Havana, Oriente, Las Villas) from which an extended network of higher education centers emerged, but also to the creation of the Ministerio de Educación Superior (MES, Ministry of Higher Education). The faculties of pedagogy became pedagogic institutes, attached to the Ministry of Education, and those of medical sciences were transformed into the Higher Institute of Medical Sciences, attached to the Ministry of Public Health. The faculties of technology and agricultural sciences of the University of Havana and the University of Oriente were transformed into the Instituto Superior Politécnico "José Antonio Echeverría" (ISPJAE), the Instituto Superior de Ciencias Agropecuarias de La Habana (ISCAH), the Instituto Superior Politécnico "Julio Antonio Mella" (ISPJAM, in Santiago de Cuba), and the Instituto Superior de Ciencias Agropecuarias de Bayamo (ISCAB). Other universities were created in several provinces from the previously existing university campuses, which implied an increase in the number of physics departments. The MES introduced unifying standards for curricula, organization of teaching, and research activities, a unification which at first gave rise to contradictions between previously existing approaches, development levels and organizational forms. The former school system was replaced by a structure similar to that of soviet universities, divided into faculties and departments (chairs). These last ones were the basic units that included both the teaching and the research activities belonging to a particular discipline or degree course.

The Physics School of the University of Havana was replaced by three departments and a so-called science and technology unit (LIEES); all of these were absorbed by the faculty of exact sciences and geography, which unified the old schools of physics, chemistry, mathematics and geography (Memoria 1976–77). Consequently, it became necessary to discard the previous system, based on teaching sections and research groups, which had fairly successfully allowed specialized attention to be given to each of these activities. As a result of this choice, the organizational unity that had brought together all the physicists was eliminated (Memoria 1976–77). This structural change did not last long and one after the other the faculties of chemistry and of geography became independent until the faculty of physics was created in 1984, now directly under the authority of the Rector. Despite these difficulties, this stage of development was characterized by an increase in the number of PhDs, a boost in postgraduate studies, the inclusion of a diploma thesis as a condition to obtain the *licenciado*, and the expansion of collaborative relations with Soviet institutions. In August 1979, "the first doctoral thesis (on microelectronic devices) entirely undertaken and discussed in Cuba" was defended (Boletín 1979, 32–33).

On the other hand, the innovations mentioned above opened up a process of positive growth at the University of Oriente. There was a progressive increase in the scientific-academic level and rigor, which was made easier by the increased exchanges with East Germany and the USSR.[57] ISPJAE became the reference center for engineering training in Cuba. The number of degree courses grew, and courses in basic physics were adapted for individual engineering branches. In the wake of this trend, the first physics textbooks for engineers were written by Cubans.

Creation of the Cuban Physical Society

Shortly after the Revolution, the old Cuban Society of Physical and Mathematical Sciences practically disappeared due to the fact that an important part of its membership had left the country. Toward the end of the 1970s, several scientific societies were created in Cuba. On June 24, 1978, almost 200 physicists met to create the Cuban Physical Society (though the final resolution that created it was officially registered only a year later), followed by the creation of the Cuban Mathematics Society 5 days later (Jiménez Pozo and Sánchez Fernández 1993). Daniel Stolik was elected as its first president.

6.5.5 A First Balance of the Period

Although the achievements of the Cuban scientific system have been discussed in stages, specifically for the field of physics, some general considerations may be useful before looking at the subsequent phase of development. The commitment of the Cuban community of physicists (including the students, mainly in the early phase) combined profitably with the clear consciousness of the leading group, and succeeded in helping the huge effort of building, in a short period of 15–20 years, the basis of a modern and articulated scientific system, not to mention remarkable results in scientific research.

The importance of support given by foreign specialists, both from Western and socialist countries, can hardly be overestimated, but the Cuban physicists integrated their (different) contribution into a scientific organization and method that was characterized by originality. In the difficult economic and geographic conditions, material support was obviously of utmost importance. However, we have discussed how support from the Western visiting professors, and from the Soviet Union and the socialist countries (the second obviously much greater and with more continuity) had different relevance and consequences for physics: the first one was decisive from a technological point of view in the field of microelectronics; the second was overwhelming in structuring other fields, for example, nuclear physics (with major consequences in the 1980s). The scientific contacts and collaborations continued with all of these countries, although they became increasingly important with regard to the socialist countries, where an increasing number of Cuban physicists were

[57] Cf. the chapter by Méndez Pérez in this volume.

trained, and where Cuban scientists were welcome and integrated in the most elite institutions and academies.

The achievements of Cuban physicists consisted of, on the one hand, remarkable scientific results at a relatively higher level compared to larger and more advanced Latin American countries; and, on the other hand, the construction of a modern and very articulated system of higher education and scientific research, starting practically from nothing. The "process of institutionalization" was necessary due to the maturity reached by the Cuban institutions, and in this sense it posed the basis for further developments of the subsequent decade. It also led to the end of the early "romantic" period, and introduced to Cuban science a certain level of bureaucracy, characteristic of hierarchical organizations.

6.6 The Growth of Cuban Physics Until 1990: The Maturity of the Scientific System

6.6.1 The General Context

Compared to the level attained in the preceding decade, the 1980s saw a substantial increase in research and postgraduate studies in the sphere of physics. New research centers were set up and branches that had already reached a certain level of development received additional support. Changes, in the 1980s, to the organization of the Cuban economy and the adoption of a planning system similar to that of the Soviet Union and the socialist countries in the previous decade (Sect. 6.5) had severe consequences for the organization and development of science.

Science was included within the centralized planning mechanism so that it would accompany the country's programs for economic and social development. International collaboration agreements multiplied at all levels, which fostered the access to leading scientific and higher education institutions abroad, and increased its participation in joint programs. Unlike what happened in the previous period (leading up to 1975), when some individual institutions played the main role in conceiving and developing scientific policy, the general planning and main decisions were now taken over by the Science and Technology State Committee (CECT), and later on by the Cuban Academy of Sciences (Sect. 6.5.2).

At least in its initial stage, this system only allowed scientific research at universities to function on a limited basis, which meant the training of specialists and engaging in basic research. This seemed to follow the corresponding model in European socialist countries at the time, in which three scientific sectors with supposedly different functions coexisted: higher education institutions, institutes belonging to the academies of sciences, and branch research centers linked to industrial, service or military sectors. At times, this model implied duplication or even trebling of human and material resources in certain areas of knowledge, something that made dubious its applicability in a small country like Cuba.

While collaboration with the USSR and COMECON extended to practically all sectors and institutions in the country up to 1990, some of the investment programs for industrial development within such a framework greatly influenced the progress of physical sciences in Cuba. The investment programs had a great impact on the advance of the country's nuclear and electronics programs, and to a lesser degree, on those branches of physics related to the mining-metallurgical and the iron-mechanical sectors. These programs counted on supposedly guaranteed long-term credit facilities, technical assistance, supplies and markets. It will be shown later that the development of the health, biotechnology, and medical-pharmaceutical sectors, where physicists played a role, proceeded in quite a different way.[58]

Each program had its specificities, but in a general sense, the principal task assigned to Cuban science was to support the investment process and technology transfer sectors. This meant the training of high-level specialists, the creation of laboratories to support industry, receiving a considerable amount of foreign technical assistance and doing a lot of research work. Applied research activities were prioritized, many of them related to the learning, adaptation and integration of new technologies to local conditions, though a certain space was granted to basic research aimed at providing a solid base for the training of specialists and the future development of Cuban technologies. In some cases, the technologies transferred were not among the most advanced at the global level, which expressed itself in relatively high-energy consumption, a low automatization level or excessive aggressiveness toward the environment. Still, for Cuba they seemed to represent a considerable leap and a great challenge for its young scientific and technological community.

Research was not organized by scientific branches, but by spheres of application, and research centers became mainly multidisciplinary. In this context, physicists were scattered, with a more or less even distribution, among a wide variety of institutions whose numbers and activities were not necessarily recognizable as related to physics. Their contributions were published in journals with a wide variety of profiles. In many cases, research was valued mainly for its newly created capacities, the group's "know-how" and its contribution to generating products or services, rather than for its publications.

[58]The program in biotechnology and the medical-pharmaceutical industry was initiated in the 1980s by the Cuban government to meet the needs of the national health system, and to create an export sector based on scientific achievement. The government strongly supported the creative groups in this field, which were producing new drugs, vaccines and immunoassay kits for detecting congenital defects in pregnancy. The investments in this sector were made exclusively by the Cuban state. In the second half of the 1980s, the integration of Cuban biotechnology products in programs of commercial interchange with the COMECON countries was pursued with the aim to provide the high technology market with Cuban products. This process ended with the construction of the *Centro de Inmunoensayos*, el *Centro de Ingenieria Genetica y Biotectnología* and others, with huge investments being made in the worst years of the special period. The research performed in these and other related and newly created scientific centres resulted in new advanced products, instruments, assay kits, vaccines, drugs and technologies, which at present are commercialized around the world.

Another general trait of this stage was the boost given to doctoral degrees. According to the Cuban law for scientific degrees, the recipients were called doctoral candidates in (specific) sciences, while the title of doctor of (specific) sciences was reserved for those who had earned a second (higher) doctoral level, following the Soviet nomenclature. The number of students in Cuban higher education institutions grew considerably, and though the massive training of *licenciados* and engineers in European universities continued, the weight of educational collaboration, at least in the field of physics, shifted to postgraduate studies. According to data furnished by the National Commission for Scientific Degrees (CNGC), between 1980 and 1990 the number of doctoral graduates in the physical sciences increased from 26 to 129, a sizable number of whom were awarded in high-level Soviet institutions and socialist countries. The true figures were probably somewhat higher because not all of those who were awarded with degrees abroad bothered to validate their titles at the CNGC. In addition, the physicists who obtained degrees in the technical sciences or in other specializations were not included in these figures.[59]

Despite the unquestionable advances achieved, toward the mid-1980s a critical attitude toward scientific research developed throughout the country, especially in view of the low level attained in the application of its results, its thematic dispersion, and the overall lack of an integrated view of many of the efforts made. This criticism became part of a more general trend, which led to the process of so-called "rectification of errors and negative tendencies," comparable with "Perestroika," where many of the original traits of the Cuban socialist experience (with respect to the East European model) were reflected. In Cuba, dissatisfaction with the social contribution of science led not to a loss of confidence, but rather to its strengthening.

As a result, from 1985 on, important changes occurred in the country's scientific policy. These changes, which aimed at closing the research-application cycle, led to the creation of research centers of a new type and of the so-called 'scientific poles,' the increased utilization of the universities' scientific potential and finally to the growth of the innovation movement linked to the so-called science and technology forum. Most visible among these transformations was the emergence in Cuba of a high-tech export production sector associated with biotechnology and the medical-pharmaceutical industry which today has an important and growing influence on the country's economy. During these years, however, the defense of doctoral theses and the publication of scientific papers were discouraged in favor of the rapid practical application of results whose quality was not always assessed by independent experts (Rodríguez Castellanos 1997).

New research centers, generally formed from preexisting groups, emerged in the above-mentioned context in order to strengthen the original collectives by providing them with greater capabilities for applying their scientific results. The creation of IMRE (Institute for Materials and Reagents for Electronics, 1985), CEADEN (Center for Studies Applied to Nuclear Development, 1985), and CBFM (Center for Biophysics and Medical Physics, 1993), which will be analyzed in detail in the following sections, were among the most prominent institutes with a close relation to physics.

[59] Cf. the contribution by de Melo Pereira and Sánchez Colina to this volume.

6.6.2 Experiments in Space

Perhaps the most direct expression of the magnitude of collaboration efforts between Cuba, the USSR and other socialist countries in physics is represented by the scientific experiments designed by Cuban and Soviet physicists and carried out on board the Salyut-6 orbital station. During the 1980 joint space flight in which the Cuban cosmonaut, Colonel Arnaldo Tamayo Méndez, participated from December 18 to 26, three out of around 20 experiments carried out had been proposed and designed by Cuban physicists and engineers who had been working on them with Soviet scientists for almost 5 years. The purpose of one of the experiments was to obtain new semiconductor materials under microgravity conditions. The other two were designed to study the effect of microgravity on the growth of sucrose crystals, including the molecular kinetics and the micro topography of the crystal obtained in the process. These two were relevant to industrial sugar processing. By the end of the year, two experimental tests on holographic information transmission between the space station and Earth were made, which had been prepared in collaboration with the Ioffe Physico-Technical Institute of Leningrad. In March 1981, a time-series of holograms was taken on board the station of the dissolution of a salt crystal in a liquid, and delivered to Earth so that specialists could study the dynamics of the process.[60]

6.6.3 Physics and Physicists in the Electronics Program: The Development of the Institute for Materials and Reagents for Electronics (IMRE) and the Development Center for Scientific Instruments and Equipment (CEDEIC)

The independent efforts that Cuba had previously made in the fields of electronics, microelectronics and computing were linked in the early 1980s with the corresponding COMECON programs. A plant for manufacturing silicon microelectronics components, under construction in the province of Pinar del Río (whose equipment was bought from a Spanish firm that was later absorbed by the US-controlled firm Motorola) had to be restructured technologically before starting up so that it would be able to operate with raw materials and technologies from the COMECON countries. At the same time, investment plans were made for the manufacture of electronic equipment, computers, hybrid integrated circuits and printed circuits. Industrial automation was boosted and a software industry began to develop. All of this was incorporated into a National Program for the Development of Electronics, under the then vice president of the Council of Ministers, Pedro Miret, who led the so-called "electronics front," in whose meetings and commissions all the

[60] For more information on this subject, see the chapter by Juan Fuentes et.al. in this volume.

organizations participated: the directing institutions for science and technology, industry, institutions of higher education, research centers, technological schools, import firms, software development centers, etc. Physicists played a substantial role in this program.

Nearly 30 young physicists who worked mainly in the technology, quality control and development areas were included in the personnel of the Pinar del Río electronics plant, officially inaugurated in 1987 and named "Comandante Ernesto Che Guevara" Electronic Component Complex (CCE). Production activities in the assembly and encapsulation areas had actually started in 1984. In 1988, Cuban personnel fully developed linear bipolar integrated circuits, which went into production in the GDR prior to its dissolution. In addition, for the first time the 741 operational amplifier was entirely manufactured in the country.

The pioneers of microelectronics research were concentrated at the Center for Microelectronics Research (CIME) of the "José Antonio Echeverría" Higher Polytechnic Institute (ISPJAE) and at the Microelectronics Department of the Cuban Institute for Digital Research (ICID). They worked with the purpose of transferring their developments to the country's industry or to the industries of other COMECON countries. The first high-level integration microchip for specific use, the LOCICHIP, was developed at the ICID. It was designed in Cuba and manufactured later on in Bulgaria. A group of ten newly graduated physicists were assigned to a future hybrid integrated circuit manufacturing plant. Planned for eventual inclusion into the "Copextel" industrial complex, it was never brought to fruition.

During the 1980s, scores of physicists engaged in the development of electronic devices and equipment, telecommunications, software and robotics at the Central Laboratory of Telecommunications (LACETEL), at institutes such as the Academy of Sciences as the Development Center for Scientific Instruments and Equipment (CEDEIC) and the Institute for Fundamental Technical Research (ININTEF, Sect. 6.4.7), at the National Center for Scientific Research (CNIC, Sect. 6.3.7).

6.6.3.1 The Institute for Materials and Reagents for Electronics (IMRE)

At the University of Havana, the solid-state physical research laboratories existed in the physics faculty together with other groups from the chemistry faculty; these were incorporated into the Institute for Materials and Reagents for Electronics (IMRE). This institute was created in 1985 with the purpose of concentrating resources, attracting state investment and assuming important engagements with the National Program for the Development of Electronics. The IMRE was created after a number of different institutions and initiatives were amalgamated to form a more coordinated and effective organization.

Halfway through the 1980s, the Laboratory for Solid State Electronics Research (LIEES) dedicated itself to obtaining and characterizing semiconductor materials and optoelectronic devices (solar cells, electroluminescent diodes, lasers, photodiodes, phototransistors, etc.), first led by Pedro Díaz Arencibia, and subsequently by Juan Fuentes Betancourt. LIEES was staffed by full-time research personnel, and participated in the collaboration established with the academies of sciences of the socialist countries, with the Interkosmos Program, and so on. In particular, the collaboration with the Ioffe Institute, especially with the Laboratory for Contact Phenomena led by Zhores Ivanovich Alferov, was wide-ranging and effective. The contribution of LIEES to the design of experiment "Caribe" represented a significant addition to the scientific prestige of the laboratory.[61]

Another unit that flourished in the period was the magnetism laboratory, led by Oscar Arés Muzio, where magnets and ferrite nuclei were manufactured from imported raw materials. As a result, a good command of laboratory-level technology for obtaining ferrites suitable for various applications was obtained. At the same time, the Laterite Laboratory of the chemistry faculty, led by Leonel Pérez Marín, managed to obtain good quality iron oxide from nickel mining waste. In view of all this, both laboratories decided to jointly explore the practical possibility of manufacturing ferrites from Cuban iron oxide. In view of the positive results obtained, Pérez Marín, on behalf of the University of Havana, presented this achievement at a meeting of the Electronics Front held on February 17, 1984, where raw and other materials required for the development of the electronics industry in the country were discussed. It was concluded that a research institute dealing with materials for electronics should be created within the main campus of the University of Havana, based on the existing facilities. It was also agreed that industry should support the new university institute and regard it as its own research center in the area of materials. The corresponding proposal was submitted to the vice president of the Counsel of Ministers, Pedro Miret, and was finally approved at the April 9, 1984 meeting of the Electronics Front. It was also requested of the University of Havana that it participate in research work aimed at obtaining quartz suitable for the manufacture of optical fibers and resonators from Cuban raw materials (a task finally undertaken by LIEES).

The deans of the physics faculty (Carlos Rodríguez Castellanos) and the chemistry faculty (Jacques Rieumont Briones) were quick to back the proposal for the creation of IMRE, which was then endorsed by the rector of the University of Havana, Fernando Rojas Ávalos, and sanctioned by the minister for higher education, Fernando Vecino Alegret. A commission was created which included both deans and Leonel Pérez Marín, and was chaired by Vice Rector Armando Pérez Perdomo. It undertook the design of the structure and working rules of such a *sui generis* center, subordinated to both the University of Havana and the Electronics Front, where professors and students from the physics and chemistry faculties were to collaborate with research workers and technicians belonging to the institute's own staff.

All the relevant workshops and experimental research groups that were previously subordinate to the physics faculty, as well as research laboratories for chemical analysis, polymers, inorganic chemistry and others from the chemistry faculty were

---

[61] This experiment investigated the growth of semiconductor crystals under microgravity and was successfully carried out during the 1980 Soviet-Cuban space flight. See the contribution by Juan Fuentes et al. in this volume.

gradually incorporated into IMRE, which had close relations with the university faculties. The most important decisions were made by the collegial leadership, which consisted of the director of the institute and the two deans. Both faculties supplied professors who took care of IMRE's main responsibilities. They also supported the institute's staff, students, delivery of postgraduate courses and international relations. Elena Vigil Santos from the faculty of physics became deputy director of IMRE while Juan Fuentes Betancourt and Oscar Arés Muzio became the heads of two of its departments.

From 1986 to 1991, IMRE grew considerably and received investment such that it was able to increase its staff and infrastructure and purchase new research equipment and transport. Two visits by President Fidel Castro, one in 1987 and another in 1989, brought additional support to the center.

During its first working years, IMRE kept very close working relations not only with the units belonging to the Electronics Front, but also with several institutions of the public health system and other research or production centers. To these it offered a wide range of services of substantial economic and social value. The analytic capacity that was created turned out to be of strategic importance for the country.

Jobs of great practical value were done at IMRE, for example, the characterization of raw materials bought by industry in hard currency, with a view to eventually replacing them with materials produced in Cuba or other COMECON countries. Numerous raw materials of national origin, such as sand containing quartz, zeolites and residue of metallurgical plants were evaluated. Procedures were worked out for their transformation into materials for use in electronics. A pilot plant was built and began to produce polymer materials used in industry. Technology for the industrial production of electroluminescent diodes, solar cells and photodiodes were transferred to the CCE, and the first lasers for medical and industrial applications were manufactured.

In the same period, there were two events of importance that were related to one another: the obtaining of the first superconducting ceramics and the setting-up of a plant for the liquefaction of helium. We will briefly refer to these in the following. Prior to this time, no experimental research was carried out on superconductivity in Cuba since liquid helium was essential for the job and it was not available at any local institution. At the end of February 1987, Paul Chu announced in Houston, Texas that he had obtained a copper oxide, barium and yttrium superconducting ceramic whose critical temperature was above 90 K. Chu's discovery was easy to replicate as it involved cooling the sample with liquid nitrogen, which in turn made the field of superconductivity more accessible. Work with superconductivity had, up to that point been reserved for those laboratories which had liquid helium, a much more expensive refrigerant, at their disposal. Laboratories worldwide were able to reproduce the above finding based on the available information about the new materials. As a result of the experience accumulated at the magnetism laboratory in dealing with ceramic materials, Oscar Arés Muzio was asked to try to create a high temperature superconductor. Under his leadership, a team managed to obtain and produce (barely 2 months after Chu's announcement) the first YbaCuO superconducting ceramics in Cuba. This was given widespread coverage in the national media in early May, and generated numerous talks and seminars referring to the new scientific development. It was now clear that IMRE was now capable of engaging not only in solving specific practical problems, but also in obtaining world standard scientific results. Not unexpectedly, a new group was created which from the very beginning attracted many brilliant young researchers, and was endowed with resources for setting up a reasonably good laboratory that enabled some ambitious projects. A few months later, YbaCuO superconducting ceramics were obtained at CEADEN, where research on superconductivity had also started.

About the same time, but unrelated to the above, steps were taken to purchase a nuclear magnetic resonance tomograph for the "Hermanos Ameijeiras" hospital (HHA) whose superconducting electromagnet required liquid helium cooling. In light of this, the directors of IMRE and HHA requested that the top government authorities acquire a helium liquefying plant to be installed at IMRE, which would be responsible for the supply of the required liquid helium, including supplies for IMRE own research work. The final decision for the purchase of the RMN tomography and auxiliary equipment depended on President Fidel Castro, who wanted details of the proposal. To this end, he paid a visit to IMRE on July 10, 1987. Other participants in the meeting were Fernando Rojas, rector of the University of Havana, and other vice rectors. Leonel Pérez expounded at length on the subject, answered questions from the President and was not reluctant to offer guarantees of achieving the proposed objective. Once the plant set up, and details were agreed upon, Castro asked to be shown the superconducting piece referred to in the press. The group moved to the Magnetism Laboratory located in the physics building, where Sergio García carried out an experimental demonstration of the Meissner effect and responded to Castro's countless queries on the nature of the phenomena themselves and their possible applications. When the visit was about to end, President Castro asked one of his aides to take care of IMRE, identify its main needs and support the continuity of its research on superconductivity. This finally led to the building of a new superconductivity laboratory and to the acquisition of equipment valued at more than half a million US dollars.[62]

On February 9, 1989 President Castro discussed a plan for the further development of IMRE, including the buildings that now house the institute. The Council of State subsequently supported the construction of these buildings.

In short, the creation of IMRE was the result of a long chain of efforts aimed at two great objectives. Firstly, to engage the scientific potential of university staff members and students, not only in their traditional academic role, but also in tasks more directly related to the economic development of the country. And secondly, to provide higher education institutions with the infrastructure, logistics and management capability required to achieve results of the highest level, and to supply quick, efficient and integral answers to the queries raised by production, which in turn should have a positive influence on the quality of the training received by the new generations of professionals and scientists.

[62] Cf. the contribution by O. Arés and E. Altshuler in this volume.

6.6.3.2 The Development Center for Scientific Instruments and Equipment (CEDEIC), Originally BECICPA

The Special Bureau for the Construction of Scientific Instruments with Adjunct Production (BECICPA, *Buró Especial para la Construcción de Instrumentos Científicos con Producción Adjunta*, Sect. 6.4.8) had been created in 1978 with the objective indicated by its long name, which was taken from a report that put forward the idea. This objective was in line with the COMECON policy of generating interfaces between research institutes and production units to speed up the introduction of scientific and technological developments to society, a process in which the socialist camp lagged behind western developed countries. Since the new unit started from zero as far as infrastructure was concerned, during its early years it dedicated itself mainly to producing equipment, materials and machine tools needed to set up its own laboratories and production workshops. Wilfredo Torres, then president of the ACC, offered strong support for the development of the infrastructure of the center, assigned its initial place, and favored collaboration with the USSR.

Still, from the very beginning, short series were designed and produced for equipment required for more general application, such as environmental ionizers and equipment for the electrification of cattle fences. The idea behind this was to start training BECICPA's personnel in the series production of instruments and equipment, and also to show some results with a social impact. This was done under the leadership of the director, radio physicist José L. Díaz Morera, with the valuable contribution of fellow radio physicists Justo Ravelo Triana and Oscar Hernández Jiménez, all graduates from higher education institutions in the USSR.

In 1986, the BECICPA was given the new name CEDEIC, *Centro Especial de Desarrollo de Equipos e Instrumentos Científicos* (Center for Development of Scientific Equipment and Instruments), and four specific objectives were defined for it:

- to engage in research work related to lasers and their application in medicine, the pharmaceutical, food, electronic and sugar industries;
- to develop scientific equipment based on the results of the above research work and promote its introduction into practice;
- to contribute to the guidance of national policies related to laser use, optics and their applications;
- to engage in research work related to high-technology mechanics and electronics.

In accordance with these objectives, a complete cycle was implemented which encompassed research, design, production and marketing of instruments and equipment based on lasers, optics, modern electronics and precision mechanics. Four optics laboratories, a machine tool workshop and an equipment assembly workshop were set up at the premises of an old building in Old Havana that had been the site of the Belen Catholic School a long time ago.

By the end of the 1980s, the center had already designed, made and introduced into social practice the following optical and electronic instruments in addition to those previously referred to: the "Lasermed 1," a device used for laser physiotherapy and acupuncture that replaced the classical acupuncture needles; the "Laserpol 1," a device employed for measuring the glucose content of urine, which worked with practically no chemical reagents thus being the most economic control alternative when analyses are made on a large scale (it was used in thousands of urine analyses at the Havana Anti Diabetes Center); the "Laserpol 101 M" polarimeter, introduced in the Cuban pharmaceutical industry; the Cell Counter employed in several hospitals and polyclinics for counting up to a 100 different components, and—among other devices—the "Paraltex-1," for measuring the quality of the fibers produced by the textile industry. Physicists José L. Díaz Morera and Andrés Combarro Romero played a very active role in the development of laser physiotherapy equipment, while the physicist Víctor Fajer Avila and the engineer Niclolás Duarte Marrero developed the design of the automatic laser polarimeters applied in hospitals, pharmaceutical and sugar factories. These instruments of high precision are designed for analytical control measurements. All of the instruments mentioned so far were important substitutes for imported equipment; this practice has still been maintained up to this day and has even allowed for the exportation of this equipment, although in limited quantities.

In the CEDEIC, under the direction of the physicists Rolando Díaz and Justo Ravelo and a multi disciplinary group, highly complex work was performed for the development of an automatic laser ellipsometer, for the measurement of thin films and for the production of integrated circuits in the electronic industry, all of which had strong support from CITMA and thus established an important collaboration with the Ministry for Electronics of the Soviet Union and the Ioffe Institute in Leningrad. On the Cuban side, headed by the physicists Rolando Díaz and Justo Ravelo, a multi disciplinary group participated. Once finished, it was planned to reproduce it for the industries of the COMECON countries; although it reached a good degree of advancement, its conclusion and development was stopped by the dissolution of the USSR.

The "Lasermed 1," which used a He-Ne laser as a light source, was widely applied in physiotherapy and passed thorough tests at the Center for Clinical and Surgical Research (CIMEQ). The "Lasermed 401 M" used an infrared laser diode as a source and was used to control most of their national production.

6.6.4 *The Nuclear Sector Faces New Ambitious Programs: Enlargement and Reorganization*

The programs in the nuclear field were particularly ambitious, and required a reorganization of the whole sector, which was given a special and favored place and structure. One may remark that something similar always happens for the nuclear sector, which presents peculiar aspects of complexity and danger. Furthermore, the choices in Cuba in this period must be evaluated, taking into account that the model was the particularly centralized one of the Soviet Union, where most Cuban specialists had been trained. In any case, this process, and the deep changes associated with it, also raised, as we will see, some discontent and contradictions inside the Cuban scientific community.

The ambitious Cuban nuclear program, as agreed with the USSR in 1976, was aimed at the construction of a nuclear power station in Juraguá (province of Cienfuegos), the introduction of nuclear technologies in the country's economy and the creation of a system of nuclear radiological protection and safety. The goal was to reduce dependence on imported oil and the project met with little opposition in Cuba.[63] Moreover, there was enormous trust in Soviet nuclear technology.

These programs involved enormous efforts to develop nuclear sciences and technologies. The need to create a solid infrastructure for the assimilation of nuclear energy and the introduction of nuclear sciences and technologies into the country's economy led to important political and organizational decisions since such a task required investment operations, training of cadres, research, collaboration and other matters.

As a part of this reorganization, a technical evaluation was made of the personnel belonging to the Institute for Nuclear Physics (IFN, Sect. 6.3.2.1), which led to the decision to relocate a group of its workers who were not considered suitable for the job. This decision was very much disputed because it led to the disposal of the services of some people that had already acquired experience, most of whom performed excellently in their new workplaces.

The previous nuclear activity was restructured in the early 1980s starting with the creation of a new body: the Cuban Commission for Atomic Energy (CEAC). The CEAC was responsible for the coordination and control of national efforts of the main institutions involved in nuclear activities, and for advising the Government on the policy to be followed in this sphere. This inter-institutional commission of the Central State Administration was headed by a vice president of the Council of Ministers and included the ministers of Basic Industry and of Higher Education, the president of the Cuban Academy of Sciences and the Executive Secretary of the Commission, who also headed the Executive Secretariat for Nuclear Affairs (SEAN), an institution in charge of "applying in a professional and systematic way the approved policy" (Castro Díaz-Balart 1990, 353–354).

As mentioned earlier, a complex mandate was given to the nuclear sector, corresponding to its different and very specialized aspects, from the training of scientific and technical personnel, to overall safety and protection. Several institutions were created for the scientific and technological support of the program: the National Center for Nuclear Safety (CNSN), the Center for Radiation Protection and Hygiene (CPHR), the Isotope Center (CENTIS), the Higher Institute of Nuclear Sciences and Technologies (ISCTN), the Center for Studies Applied to Nuclear Development (CEADEN), and the Information Center for the Nuclear Sphere (CIEN). These developments are analyzed in detail in the following.

The nuclear program relied on the participation of hundreds of Cuban physicists, including SEAN's executive secretary, Fidel Castro Díaz-Balart, the eldest son of

[63] The considerable oil deposits in the Gulf of Mexico, recently discovered though still not exploited, were unknown at this time and renewable sources did not appear as a viable alternative.

President Fidel Castro, who had graduated in physics at the University of Moscow, and who received a PhD in physics at the Kurchatov Institute of the Soviet Academy of Sciences. Physicists were present in almost all the aforementioned institutions, especially the ISCTN and CEADEN, and tackled tasks of a very different nature in the planning process, the training of cadres, research, the creation of specialized services and in establishing international collaboration. Some 20 physicists defended their doctoral degrees in various nuclear specializations during the decade.

The nuclear program was developed in close collaboration with the USSR, the European socialist countries and the International Atomic Energy Agency (IAEA). Cuban students and researchers gained access to excellent universities and research centers, and to IAEA projects. Participation in the Joint Institute for Nuclear Research (JINR) at Dubna, that began in 1970 (Sect. 6.4.3), increased to the point that a Cuban physicist, Elías Entralgo Herrera, was appointed as one of its vice directors.

During the 1980s, hundreds of the best Cuban students who had finished high school education all over the country were awarded scholarships for studying specializations in the nuclear sciences and technologies, including nuclear physics, in the USSR and other socialist countries. While studying abroad, they were looked after by the personnel of their Cuban institutions of origin, who checked their educational performance and made sure that they were duly trained in specializations related to the Cuban program. This evaluation was performed through diplomats or other officials who resided in annexes of the Cuban embassy, in some cases these persons were professors or researchers who simultaneously developed other scientific activities in Soviet institutions. To guarantee an adequate preparatory training for students, the "Martyrs of Humboldt 7" Exact Sciences Senior High School[64] was created with an excellent teaching staff and very good material conditions. This so-called "Humboldt" experience was later reproduced in 14 other Exact Sciences Vocational Senior High Schools (IPVCE) in the country (one per province), which to this day have been the main source of Cuban participants in national and international scientific Olympics, and of students for science degree courses in national universities as well.

Significant changes were also introduced at the university level: in 1981 a Faculty of Nuclear Sciences and Technologies (FCTN) was inaugurated at the University of Havana whose purpose was to train specialists in nuclear physics, radiochemistry and nuclear engineering. The initial nucleus of the faculty was made up of staff from the nuclear physics department of the university, from ISPJAE and from other centers. The dean appointed was José Roig Núñez, a physicist who had taken his degree course at the University of Havana and obtained his PhD in the USSR. The FCTN offices and first teaching spaces were installed on the campus of the University of Havana at the Quinta de los Molinos, separate from the faculty of physics. The building of additional laboratories and facilities began

[64] "Humboldt 7" is the address of the building in the Havana district of Vedado where four university students and members of the Revolutionary Directorate were hidden, survivors of the failed assault in March 13, 1957 on the Presidential Palace in Havana. After a delay, the building was invaded by the police and the students were killed.

immediately. The first students graduated from the FCTN in the mid-1980s. Continuity was given to the creation in the FCTN of experimental and theoretical research groups in nuclear physics, reactor technology and nuclear methods of analysis. In 1987 the FCTN detached itself from the University of Havana and became the Higher Institute of Nuclear Sciences and Technologies (ISCTN), directly attached to SEAN. Evelio Bello, a physicist who had obtained his PhD in the USSR, was appointed rector of the ISCTN. The ISCTN was to receive many of the best students from the secondary schools, and in particular, from the expressly created IPVCEs.

Another initiative, as mentioned, was the creation of the Center for Studies Applied to Nuclear Development (CEADEN), heir to the IFN and the ININ, which began its activities in 1985, though the official inauguration took place on October 28, 1987 at new premises in the presence of President Fidel Castro and the Director General of the International Atomic Energy Agency (IAEA), Hans Blix. Its goal was to engage in applied research, development work, technology assimilation, technical and scientific services and other areas supporting the national nuclear program. It incorporated well-equipped laboratories for solid-state physics, nuclear physics, radiochemistry and nuclear electronics. Its appointed director was Daniel Codorniú Pujals, a physicist who had graduated from the Central University of Las Villas and obtained his PhD in the USSR.

One of CEADEN's achievements is particularly noteworthy: the assimilation and development of preparatory techniques for marked substances widely used in medicine and in biomedical research, which allowed for a replacement of imports and an increased use of radioisotopes in this realm. At the end of 1988, even before all the necessary equipment was at hand, production of the gamma variant of ATP marked with P-32 began, whose quality was similar to that produced by recognized firms. It was produced for the Center of Genetic Engineering and Biotechnology (CIGB). The systematic production of insulin marked with I-125 to satisfy the demand of the "Hermanos Ameijeiras" hospital (HHA) was implemented, and marking techniques were developed for another group of hormones.

Irradiation techniques showed very promising results in the sterilization of bio preparations, in the obtention of new varieties of sugar cane, pasture and rice, and in the preservation of film archives of high historic value. The development and production of electronic equipment for promoting the use of nuclear techniques in the country grew considerably from 1988 onwards, well beyond the collapse of the socialist bloc, up to the present day (Sect. 6.7.2.1). Ten single-channel radiometers, a prototype of a multichannel analyzer, dosimeters, a low background radiation dosimeter, detectors of superficial barrier ionizing radiations and other equipment were produced. Despite this abundance of new initiatives, it was decided to terminate the research in applied nuclear physics at the University of Oriente: this resulted in the exodus of specialists and probably in the dispersal of precious accumulated professional skills.[65]

[65] For details, see the chapter by Cabal Mirabal in this volume.

To sum up, similar to what happened in many countries that had developed nuclear programs, even if solely for peaceful use, the need to concentrate enormous resources and to guarantee the safety of all the processes involved led to the organization of an articulated system of institutions and regulations specifically dedicated to the nuclear area, which somewhat separated "nuclear" from "non-nuclear" sciences. The main decisions in this field were likely made at the top (as is always the case in plans for nuclear power); discussions within the scientific community were deficient, and criticisms and discontent were not sufficiently heard. This generated tension, since some thought that it would have been more rewarding for the new institutions to profit from the experience accumulated in some of the preexisting ones, and the latter in turn would benefit from the resources allocated to the nuclear program. In fact, some of the newly created training and research centers essentially constituted an unwanted duplication of efforts and programs. One of the most visible examples of this was firstly the creation of the FCTN next to the main campus of the University of Havana, and later of the ISCTN, for the purpose of teaching degree courses in nuclear physics and radiochemistry. Thus, an elite group of physicists, chemists and engineers with a different and more specific profile were being trained.

A critical evaluation of these choices, as in the previously discussed case of microelectronics, is not easy. While in the latter case, the advent of high-level integration frustrated the (however intelligent) attempts made by several developing countries, the developments of the nuclear programs in Cuba (mainly the huge and extremely complex project of building a nuclear power plant) were soon cut off by the dissolution of the Soviet Union and the socialist bloc, as will be analyzed in the next sections.

6.6.5 *Physics at the Universities*

We will now look what occurred at Cuban universities during the 1980s, especially in relation to the development of the field of physics. So far we have discussed the Physics School of the University of Havana in the 1970s (Sect. 6.4.1.1) and the Eastern University (Sect. 6.4.1.2), summarizing the results they had obtained (Sect. 6.5.1) before the establishment of the Ministry of Higher Education in 1976 (MES, Sect. 6.5.2). In terms of research, we have presented the creation in 1985 of the Institute for Materials and Reagents for Electronics at the University of Havana (IMRE, Sect. 6.6.3). We will now analyze the main changes that occurred in the structure of these universities during the 1980s.

6.6.5.1 The Creation of the Physics Faculty of the University of Havana

The "period of institutionalization," with the creation of the Ministry of Higher Education in 1976 (Sect. 6.5), started—for better or worse—a phase of organizational changes in the university. During a period of eight years, the physics departments of the University of Havana were mixed with those of other specializations, first with the faculty of exact sciences, and then with the faculty of physics, mathematics and computing. Finally, in 1984, the physics faculty was created as a continuation of the Physics School founded in 1962, under the leadership of Carlos Rodríguez Castellanos (dean), Carlos Trallero Giner and Luis Hernández García (deputy deans).

During the 1980s, there were important advances in the teaching of degree and postgraduate courses both in the physics faculty and theoretical physics research (Rodríguez Castellanos 1984–5). The program of electronics and the creation of IMRE led to the enlargement and equipping of the research laboratories of the University of Havana, thus improving the conditions under which teachers and students performed their research work. Moreover, this multiplied the job offers for physicists both in the academy and in industry. In the second half of the 1980s, important investments were made in repairing the faculty buildings, and in equipping the teaching laboratories for physics and computation. Plans and programs for studies were elaborated by a commission with a wide presence of physicists who worked in the industry, the health system and in external research centers; this reflects the degree of development reached with respect to relations between the faculty of physics and other institutions of Cuban society. A so-called "C" syllabus was worked out which stayed operative for 15 years. Moreover, the faculty began to play an important role in the postgraduate training of physicists who worked in other institutions, especially in the pedagogical institutes, thus helping to improve the quality of teacher training and high school physics teaching.

From 1981 to 1990, 204 *licenciados* who had passed a 5-year syllabus and successfully defended their diploma thesis and around 20 PhDs, received their degrees from the physics faculty. By the end of the decade, the faculty had a teaching staff of about 70, 30 of whom had completed their doctoral degree studies.

In 1989, the theoretical physics department had its twentieth anniversary. It had a staff of 15. Of these, nine had doctoral degrees (five graduated in the USSR, one in East Germany and three in Cuba). It had been led initially by Melquíades de Dios Leyva, and later by Rolando Pérez Álvarez. In that year, 28 articles were published in international specialized journals. Besides teaching high-level degree and postgraduate courses in theoretical physics and mathematics, the staff supervised around ten collaborators from other institutions (essentially staff members of pedagogical institutes) who worked toward their doctoral degrees whilst being tutored by professors of the department.

By this time, the research lines of the department began to diversify beyond solid-state theory, and the first papers were produced on non-linear dynamics and complex systems. At the international level, links were developed with many prominent figures and institutions in Europe and Latin America. Outstanding among these were the Ioffe Physico-Technical Institute in Leningrad, the Joint Institute for Nuclear Research in Dubna, the Humboldt University in Berlin, the International Center for Theoretical Physics[66] (ICTP) in Trieste, the Paris-Sud University, the

[66]The ICTP was founded in 1964 by the late Nobel Laureate Abdus Salam, see http://www.ictp.it/homepage.aspx. Accessed October 17, 2013.

Max Planck Solid State Physics Institute in Stuttgart, the CSIC Institute for Materials in Madrid and the Campinas University in Brazil. Two professors belonging to the department were elected associate members of ICTP.

As a whole, in 1985–1990 the physics faculty managed to expand its relations with European and Latin American institutions. It also supported the activities of the Cuban Physical Society, whose first three presidents were members of the faculty. Several high-level national and international scientific gatherings were organized, among them the 10th Latin American Solid State Physics Symposium in 1987, which was attended by more than 200 participants.

By the end of the decade, several hundred square meters of new physics laboratories and workshops had been built for IMRE in previously unused areas of the physics faculty premises. The building itself underwent overall structural repairs and the students laboratories were newly equipped.

6.6.5.2 Physics at the University of Oriente

With the incorporation of new graduates and specialists who had obtained their doctoral degrees in the USSR and East Germany, the staff of the physics departments of the University of Oriente gradually grew and advanced to the extent that in 1985 their staff had already grown to 35, of whom 8 had PhDs. A laboratory of physical methods of analysis was created in 1983, which was able to apply atomic spectroscopy, X-ray diffraction, neutron activation analysis, radiometry, nuclear magnetic resonance and other techniques to the study of soils, minerals, and so forth. Theoretical research on atomic and molecular physics was also undertaken. From 1981 to 1990, the staff published more than a 100 articles in national and international specialized journals. The 3rd Symposium of the Cuban Physical Society, the only one held outside of Havana, took place in Santiago de Cuba in June 1985, which is indicative of the favorable situation at the time for physics at the University of Oriente.

In the mid 1980s, the decision was taken to set up an independent physics department at the newly created "Julio Antonio Mella"[67] Higher Polytechnic Institute (ISPJAM), to which the technology faculty of the University of Oriente had been transferred. On the other hand, the national reorganization of research activities in the nuclear sphere that took place at the time made the work on applied nuclear physics come to an end. All this had the effect of weakening physics work at the University, especially since 17 members of its physics staff had to leave for the ISPJAM while some others had to change their scientific area of research.

[67] Julio Antonio Mella (1903–1929) was a Cuban student leader and revolutionary, founder of the *Federación Estudiantil Universitaria* (FEU, University Student Federation) and of the first Cuban Communist Party. Due to his political activities, he was expelled from the University of Havana in 1926 and persecuted by the Gerardo Machado dictatorship. He moved to Mexico, where he continued his activities until he was murdered in 1929.

Still, by the second half of the decade, there were some relevant scientific results obtained by the groups working on medical physics and nuclear magnetic resonance, which had developed strong ties with health organizations in the province of Oriente and in Havana. In 1989 these groups joined forces and formed the core of what would later be the Center for Medical Biophysics, officially inaugurated in 1993. Work on materials science started in 1987, soon after the arrival of an electron microscope suitable for the task.

6.6.5.3 Physics at the "José Antonio Echeverría" Higher Polytechnic Institute (ISPJAE)

The physics department of the "José Antonio Echeverría" Higher Polytechnic Institute (ISPJAE) grew considerably and its teaching staff finally topped 100 members. Being the national reference center for the teaching of physics to engineers, it gave a major boost to research on the didactics of physics. Some of its members were linked to the field of microelectronics at CIME, a center that played an important role in the electronics program. However, the main scientific results were attained in the realm of optics.

In 1975, a small team made up by Ángel Augier, Jorge Alum, Beatriz Moreno, Luis Martí and others was formed with an initial goal of producing a dye laser. The group grew, trained its first PhDs in the USSR, and, after bringing about its first achievements, was transformed into a Science and Technology Unit for Coherent Optics attached to ISPJAE's electronic engineering faculty. Besides engaging in postgraduate training, it dedicated itself to the design, construction and application of lasers, speckle interferometry, holography, and optical communications (Martí 2009, private communication). The group, headed by Luis Martí, actually achieved outstanding scientific and technical results, such as laser scalpels that were successfully used in standard surgery at the "Calixto García" hospital, high power lasers and other laser and holography techniques.[68] Further work on holography was developed, with the support of He-Ne lasers and Agfa-Gevaert photographic plates donated by Canada. Under the direction of Beatriz Moreno, holography was put to practical use in the preservation of the optical image of high value objects.[69] A holography laboratory was set up at the National Museum of Fine Arts.

Non-destructive optical testing of mechanical systems, and digital simulation of speckle pictures were also done. Under the leadership of Jorge Presmanes, an

[68] Among the other main achievements of the group, let us mention a 400 W laser technological installation, Nd-YAG lasers (neodymium-doped yttrium aluminium garnet, a crystal used as a lasing medium for solid-state lasers) in free generation regime, and ruby modes synchronism with free generation of 0.5 J pulses regime, as well as a colour center switch for use in holography.

[69] Among them, the Denisyuk holograms of relics of the fighter for Cuban independence, General Antonio Maceo, personal objects that belonged to the national hero, José Martí, and Ernest Hemingway's Nobel Prize medal.

experimental channel for optical communication between the headquarters of the Cuban Academy of Sciences at the Capitol building, and other centers that belonged to the Academy was created.

6.6.5.4 Other Universities

Another peculiar process testifies to the increased interest in physics, its study and applications, resulting in the generation of a complex network of institutions and initiatives that intertwined in flexible and complex ways. The reorganization and expansion of the Cuban system of higher education at the end of the 1970s led to the emergence of a high number of new higher education centers specialized in the training of engineers, physicians, agronomists and high school teachers.[70] As a result, there was an increase in the number of physics departments dealing with the teaching of the subject, which were chiefly composed of young physicists graduated from the universities or pedagogical institutes, but also engineers. With the creation of a variety of military centers for higher education, several professors from the physics department of the "José Martí" Technical Military Institute (ITM), some of them already PhDs, transferred to other military centers or became civilians.

Various institutions and mechanisms contributed to the promotion of links and exchanges among the above physics departments: the National Commission for Physics attached to the Ministry of Higher Education (MES), various syllabus-unifying commissions, frequent inspections among educational centers where teachers could take on the role of inspecting other universities. Teachers from all over the country came to the faculty of teacher improvement attached to the University of Havana to take courses designed to improve their work, and to take part in national student contests and forums and the activities of the Cuban Physical Society (SCF). All of these mechanisms created a strong network for the development of physics.

Many of the young physics teachers began to engage in research related to the profile of their institutions (technological, agricultural and livestock, medical, sports, pedagogical, and so on); others were connected with physics research groups at the universities, the CNIC or the institutes of the ACC. Various agreements established with Soviet institutions were used to train the first PhDs.

By the end of the decade, a higher academic level had been reached among the various staff members, and several physics departments had set up their own research groups. The existing tradition of metals physics was kept at the Central University of Las Villas (UCLV), but other groups appeared such as the one studying thermal physics at UCLV and at the University of Camagüey, the groups on

[70] In 1976, the political-administrative division of the country was modified, expanding from 6 to 14 provinces. In this context, several centers, which worked as branches or offices of the existing universities, were changed to independent universities and received investments and more human resources to assist with their development.

dielectrics and theoretical physics at the "Enrique José Varona" Pedagogical Institute, the group researching soil physics at the Higher Agricultural Sciences Institute of Havana (ISCAH), and finally the group looking at sugar crystallization physics at the University of Matanzas.

One has to mention also the increase during the 1980s of civil, teaching and academic collaborations of Cuba with foreign countries. During the 1980s, Cuba had a thriving civil collaboration[71] with Angola, Ethiopia and Nicaragua involving several dozens of Cuban physicists who accomplished international missions as teachers at the universities of these countries under generally difficult conditions.[72]

6.6.6 *Physics at the National Center for Scientific Research (CNIC)*

With the restructuring of higher education in 1976, the National Center for Scientific Research (CNIC), which since its creation had been attached to the University of Havana (Sects. 6.3.7 and 6.4.9), separated from this institution and was attached directly to the Ministry of Higher Education (MES); however it continued to play a very important role in the postgraduate training of teachers at the higher education level. Though scattered in several areas and initially conceived as a complement to chemical and biomedical research, in the 1980s physics developed its own profile and reached a high level at the CNIC. Some 20 physicists worked at the center, in addition to several physicists from other institutions who were allowed to use CNIC instruments and equipment for part of their experimental work, as well as those working toward their doctoral degree.

At the CNIC research on applied physics and technology were mainly developed: applications for the exploitation and utilization of Cuban natural resources (Marrero et al. 1969; Meitín and Roig 1971; Cardero et al. 1973; Marrero and Meitín 1976), which included the introduction of X-ray fluorescence and methods based on Mössbauer spectroscopy into the nickel industry; studies on the sulfurous polymetalic minerals of western Cuba as a potential source of such elements as indium (In), gallium (Ga), germanium (Ge) and other metals for the emerging electronics industry; research on the utilization of Cuban zeolites for environmental

[71] The military and civil collaboration of Cuba with the African anti-colonialist movements began in 1961. During the Angolan war (1975–1989), there was also a wide presence of Cuban doctors, teachers and builders in civil missions. After the signature of peace and the withdrawal of the Cuban and South African troops, the collaboration between Cuba and Angola continued to grow. In many other African countries, there is now a wide presence of Cuban doctors and other professional performing missions according to intergovernmental cooperation agreements, Gleijeses 2002.

[72] On March 25, 1984, Héctor Alfredo Pineda Zaldívar, a young Cuban physicist, fell during the defense of the city of Sumbe, in Kwanza South, Republic of Angola, where he was working as a teacher. After his death, the Cuban Higher Pedagogical Institute for Professional Technical Teaching was named after him.

repair, animal feeding, soil treatment, and controlled liberation of fertilizers and medicaments. Physicists working at the center also dealt with certain subjects in metallurgy (surface treatment, effects of welding on the properties of steels, metal shaping with explosives) and with studies on the lifetime of materials under the corrosive conditions of Cuba's tropical, damp and coastal climate. Among the strictly biomedical activities in which the CNIC took part, it must be recalled that during this decade the neurosciences group grew stronger and dedicated itself to modeling and studying the electrical activity of the brain with a view to developing diagnostic equipment and methods for various pathologies.

Until the early 1990s, the CNIC had at its disposal a sizable infrastructure in equipment for electronic microscopy, X-ray diffraction, mass spectrometry, nuclear magnetic resonance, infrared spectroscopy, X-ray fluorescence and Mössbauer spectroscopy, all supported by the presence of physicists, from which many of the country's institutions benefited. In 1990 the group doing Mössbauer spectroscopy, in collaboration with researchers from the University of Havana in this area, organized the second Latin American Conference on the Applications of the Mössbauer Effect (LACAME'90), which took place in July at the CNIC with participation of research workers not only from the region, but also from the rest of the world. A sign of the level reached by physics at the institution was the fact that the CNIC journal—perhaps the most significant in the country with the longest continuity—was published in three series: biology, chemistry and physics. This last one disappeared in the late 1990s due to the inclusion into the chemistry series of the articles dealing with physics.

6.6.7 Physics at the Cuban Academy of Sciences (ACC)

6.6.7.1 Creation of the Cybernetics, Mathematics and Physics Institute (ICIMAF)

The Cybernetics, Mathematics and Physics Institute (ICIMAF) was created in 1986 from a merger with the Mathematics, Cybernetics, and Computing Institute (IMACC) and the Fundamental Technical Research Institute (ININTEF). The latter inherited the larger part of its physicists from those who had belonged to the departments of theoretical physics, ultrasonics, holography and remote sensing. Some 20 physicists either continued working at the first two departments, which remained at the ICIMAF, or participated in its newly created automatic control and electronics multidisciplinary groups.

The theoretical physics group, led by Hugo Pérez Rojas, engaged in research on quantum field theory and its applications to the physics of condensed matter, nuclear physics, particle physics and astrophysics. At the time, it was included in the ICIMAF. It had already reached a good international scientific level, continued to collaborate closely with the Lebedev Institute of the USSR Academy of Sciences, and had published more than 30 papers in specialized international journals. Toward the end of the decade, the group included five members with a doctorate degree and several newly graduated physicists, including some from other institutions. The group supervised degree students every year, but since their numbers were small, formal courses were rarely organized: instead, each student received a study program and a list of books and papers useful for self training under the supervision and counsel of a tutor, so that they could start their doctoral thesis work. In this way, successive generations of Cuban physicists received specialized training in field theory and related matters.

The ultrasonics group was devoted to the development of ultrasonic equipment and sensors for industrial, medical and scientific applications. In relation to this, a group emerged for the obtention and study of piezoelectric ceramics.

6.6.7.2 The Geophysics and Astronomy Institute

The Institutes of Geophysics and Astronomy belonging to the ACC merged in 1974 into the Geophysics and Astronomy Institute (IGA), which started with 130 workers. The subjects and tasks it dealt with fell in one or the other of the following so-called principal problems:

- Study of the sun, the ionosphere, the magnetosphere and their mutual relations.
- Study of the endogenous physical processes and the deep structure of the Cuban archipelago.

Research of economic importance was done in both these directions, linked to the development of radio communications in the first case, and in the second, to the determination of seismic risks and the production of magnetic, gravimetric and other maps.

The IGA collaborated closely with homologous institutions in the USSR and other socialist countries. As a result, a large amount of scientific equipment and facilities were installed and systematically used, such as a Faraday effect recording polarimeter, a telemetric station for working with Interkosmos and other satellites, a station for inclined ionospheric sounding, a satellite tracking station, a special spectrum analyzer and others.

6.6.7.3 The Solar Energy Research Center in Santiago de Cuba

At the beginning of the 1980s, the Cuban Academy of Sciences created a research center in Santiago de Cuba dealing with solar energy. This developed in 1984 into the Solar Energy Research Center (CIES), a unit aimed at promoting energy saving and the use of solar energy in the country, contributing to the related scientific and technical development in its eastern provinces, on the basis of the intellectual potential already created at the institutions for higher education.

From the beginning, the CIES was conceived as a closed cycle research-production center. Its first building was erected in 1982, as was the first stage of a small experimental polygon. Later, workshops for experimental production and prototypes were created, the definitive laboratories were set up and the experimental polygon was completed.

6.6.8 Overview of Cuban Physics Toward 1990

By the early 1990s physics had reached a very promising level of development in Cuba. Besides their important contribution to education at various levels, physicists made a significant contribution to electronics and nuclear programs, with a recognized presence in interdisciplinary groups related to biomedicine, meteorology, computing, and the country's program for mining and metallurgy. As a result of the preceding three decades of work and of the international collaboration described above, important goals were achieved, some of which will be pointed out in the following.

6.6.8.1 Training and Upgrading of Professionals and Scientists

- From 1959, some 1,200 *licenciados* in physics graduated at Cuban and foreign universities (generally after passing a 5 year degree course and defending a dissertation), more than 10 % of whom already possessed a PhD or DrSc in physics. A considerable number of these had already gained a doctoral degree in other sciences, technical, pedagogical, or natural.
- A similar number of *licenciados* in education, with a specialization in physics, had graduated from the pedagogical institutes. Several dozen of these had already obtained their doctoral degrees in pedagogical sciences and specialized in physics teaching.
- At the universities of Havana and Oriente from the 1960s, and at the Nuclear Sciences and Technologies Higher Institute in the 1980s, the programs for training physicists grew stronger and as a whole some 60 *licenciados* graduated every year.
- The above-mentioned courses were strongly influenced by the Soviet school, which was characterized by strong basic training, both theoretical and experimental, and included research work done by the student. Generally speaking, they compared favorably with the international standard.
- All physics graduates were guaranteed a paid job according to their professional profile, mainly in higher education and research centers, and to a lesser extent in secondary education centers. An increasing number of physicists working in industry and health.
- The first master's degree courses were organized in the 1970s, and doctoral dissertations started to be prepared and defended in the country in the 1980s.
- At the pedagogical institutes, the scientific level of staff and curricula had risen too, which expressed itself in better-trained high school physics teachers.

- The established Vocational Pre-university High Schools in Exact Sciences (IPVCE) together with student competitions and the scientific Olympics became an excellent source of new students for physics degree courses. The fact that in 1991, the 22nd Physics International Olympics took place in Cuba was a reflection of the country's advances in the field and implies international recognition.

6.6.8.2 Research

- According to a report by the Cuban Physical Society, in 1990 no less than 40 groups existed in the country, each of which included five or more physicists who systematically did basic or applied research. Some 400 physicists, more than 120 of them holding a doctoral degree, participated in these groups, which existed mainly in universities and research centers.
- The most active areas were solid-state physics, nuclear physics, optics, earth and space sciences, mathematical physics, field and particle physics, and medical physics.
- There was an increasingly important presence of physicists in multidisciplinary research groups dealing with electronics, computing, biotechnology, neuroscience, meteorology, geophysics, astronomy, materials science, various medical specializations and other areas.
- Experimental research aimed at concrete applications prevailed. The number of publications in international journals was modest.
- While sophisticated experimental facilities did not exist, enough equipment and supplies could be counted upon for regularly doing experimental research. This was complemented with work visits to laboratories abroad.
- Some of the internationally recognized scientific results achieved by the country in the 1980s may be taken to illustrate the advances attained in the area of physics at the time. These may be exemplified by the "Caribe" and the "Hologram" experiments successfully carried out in space, and the manufacture of high critical temperature superconductors soon after their discovery.
- The Cuban Physical Society, created in 1978, had some 500 members, published the *Revista Cubanade Física* every 4 months from 1981, and organized its congresses and symposia biannually. Juan Fuentes Betancourt was its president at the end of the 1980s.
- International scientific gatherings were organized and took place in the country on a regular basis.

6.6.8.3 International Relations

- Strong cooperation links were kept with high-level institutions, especially those of the USSR, but also of East Germany, Italy, Spain, Mexico and Brazil. In this framework, Cuban physicists were able to receive training and promotions, participate in joint research work, and obtain scientific and technical information, while the material basis for teaching and research laboratories could be developed. The brain drain phenomenon was practically nonexistent.

- There was participation in the main regional and world organizations related to physics (CLAF, FEIASOFI, ICTP, IUPAP, IAEA, ICO, IPhO),[73] as well as organizations from socialist countries (JINR, Interkosmos Program, etc.).
- With respect to Latin America, only Brazil, Argentina and Mexico had larger human resources than Cuba as far as physics was concerned. Still, because of the quality of the basic training received by its specialists, the encouragement given to experimental and applied research, as well as the links established between the physicists and production and services activities, Cuban development was not behind that of the countries in the region with a stronger tradition in the field (Morán-López 2000).

6.6.9 A Structure Resistant to the Collapse of the Soviet Union

It is evident from our reconstruction that in these 30 years of efforts to build a strong scientific structure and to develop advanced scientific research and technological applications (excepting changes or internal contradictions) were inspired by big programs, ambitions and hopes. On the one hand, the enduring post-war trust in the progressive role of science had not yet been shaken. On the other hand, up to the mid-1980s, the Soviet Union and the socialist bloc appeared as a very solid strategic alternative to the capitalist model: Cuba cherished projects covering some sectors of highly specialized production, such as electronics, microelectronics and biotechnology. The Soviet Union had actively supported, both economically and through its high level scientific institutions, technical and scientific development in the socialist countries, partly to emulate and contrast the politics of the United States, and partly to feed the Eastern market. It is difficult to overemphasize the relevance of such an important choice such as the construction of a nuclear power plant for a small country like Cuba, with the extremely complex and articulated system of technical, intellectual, health and social support it requires. One must take into account that before the 1986 Chernobyl nuclear accident (and in part even after it), nuclear power appeared as the most advanced technology and the choice for the future; opposition against it was almost completely absent in the socialist countries and existed only among a minority in the Western world.

An historical evaluation of the possible outcomes of these choices is made extremely problematic by the unexpected turn of world events at the end of the 1980s. One must take into account that no one expected the collapse of the Soviet Union and the disintegration of the socialist bloc. This obviously had dramatic effects on every aspect of the Cuban economy and social life, and an analysis of these effects reaches beyond the scope of this work. However, the Cuban scientific system had reached such a level of development, extension, vitality and robustness

[73] Acronyms respectively for: Centro Latino Americano de Física, Federación Latino Americana de Sociedades de Física, International Center for Theoretical Physics (Trieste), International Union of Pure and Applied Physics, International Atomic Energy Agency (Vienna), International Commission for Optics, and the organization which holds the international Olympiads in physics.

that—apart from the unavoidable drastic consequences, cuts, reductions, reorganization and reorientation imposed by the sudden crisis—it succeeded in resisting the shock and even displayed signs of recovery and progress. This appears as one of the most remarkable outcomes of the 30-year effort to develop the Cuban scientific system.

6.7 Physics in the "Special Period"

6.7.1 The General Context

The collapse of the USSR and the downfall of the Eastern bloc in the 1990s had a disastrous effect on the Cuban economy and the living conditions of its people. This long period of economic crisis and deprivation, which began in 1991, was officially called the "Special Period." The country had to cope with the sudden disappearance of important markets, supply and credit sources, followed by the intensification of the US embargo/blockade via the Torricelli (1994) and the Helms-Burton (1996) laws. During the first 4 years, the national GDP dropped more than 36 % and the standard of living diminished considerably. The Cuban economy began to recover from 1995 on, but at such a slow pace that it was only in 2007 that the GDP reached the former 1990 value (Banco Central de Cuba 2000).

The crisis gave rise to a complete reorganization of the Cuban economy. While some sectors were virtually paralyzed, others such as tourism, oil extraction, the nickel industry, biotechnology and food production were given a fresh impetus. Foreign investment grew and the country found new markets and sources of finance.

One strategy employed by the Cuban government to cope with the crisis stands out above the rest. This was the boost given to scientific research and technological innovations that could contribute either to the development of prioritized sectors or to solving the problems associated with economic and social issues. During the worst years of the crisis, the budget allocated to science was untouched and investments in this area grew, especially in the biotechnological and medical-pharmaceutical sectors. The universities were part of this process and new full-cycle research groups emerged or flourished at the time.

Needless to say, financing was concentrated in a small number of sectors, namely in those projects that promised the fastest results in the shortest possible time. Other scientific activities were seriously affected by the decline of working and living conditions and by the loss of collaborative relations with scientific institutions in the USSR and other socialist countries. Many research groups were either weakened or simply disappeared. The exodus of scientists to better paid sectors grew considerably. International scientific collaboration was reoriented, in particular toward Spain as well as Latin America, Western Europe and Canada. The processes of brain drain and talent theft began to increase substantially, phenomena that had not been of great significance in Cuba until then. Still, it can be said that most of the Cuban scientific community made great efforts to preserve the achievements of previous decades and made a considerable contribution to the recovery of the country, despite of the difficult conditions it had to endure at the time.

Financial systems were gradually introduced that allowed institutions to generate resources for their own support on the basis of scientific and technical services commercialization, specialized production and consultancies in foreign or national markets. Aside from their contribution to the Cuban economy and society in general, these activities had a positive impact on the working conditions and income of research workers. They led to an increase in some aspects relevant to technological innovation, such as the protection of intellectual property, the preparation of technical documentation, the implementation of quality systems, the registration of products and equipment by regulatory offices, the participation in commercial fairs and exhibitions, and so forth. But in some cases, there was a negative side to this that expressed itself in a reduction of both scientific publications and the training of new PhDs, which in turn led to a weakening of both research activities and the professional competence of part of the available scientific personnel.

In line with the priority given to science and technological innovation, the Ministry of Science, Technology and Environment (CITMA) was created when the central state administration was restructured in 1994. The Cuban Academy of Sciences remained as a consulting institution consisting of scientists elected as academics for a 4-year period, thus carrying on the tradition of the first academy of sciences established in Cuba in 1861. A wide variety of functions were given to CITMA, including those of SEAN, of the Cuban Office for Industrial Property, of the National Commission for the Environment, and others. It also implemented a system for the evaluation, selection, financing and control of scientific research and technological innovation projects attached to national, branch or territorial programs that expressed the state priorities for science.

The real priorities, however, were concentrated in the Western Scientific Pole, a network of dozens of research and production centers in the sphere of biotechnology and the medico-pharmaceutical industries, which were directly taken care of by the State Council. Those centers became important exporters of vaccines, medication and other products, services and technologies, which had increasing significance for the Cuban economy.

During the 1990s, the Science and Technology Forum (FCT) became another factor that boosted research and innovation activities in the country. It was a mass movement directed by the Party. Created in the 1960s as the Spare Parts Forum, its goal was to promote the securing and generalization of technical solutions for problems in various production and service areas. Almost all of the state institutions and organizations that acted in Cuban society participated in one form or another in the FCT. Its initiatives covered a wide spectrum of activities related to innovation, some of them at a high scientific level, others with a more local relevance; ultimately its goal was to promote the application and generalization of the best initiatives. To this end, the FCT organized yearly gatherings at basic levels (firms, schools, universities, research centers, and so on), as well as at municipal, provincial and national levels, where more than a million people participated, including workers, technicians, students of all levels, professors and researchers. Prizes were awarded by juries that evaluated the economic or social impact of the presented results. At the national level, the FCT was closely observed by the government and authors of prize-winning research received important moral and material incentives. Most of the scientific institutions paid great attention to the results presented at the FCT. The number of prizes awarded was taken as a measure of performance, and was frequently considered to be a better indication of success than the number of scientific publications or other more traditionally used means. The activities of the FCT continued to develop and still exist today, although at present they have no influence at a national level, nor do they maintain the level of influence they had in the 1990s.

According to data furnished by the National Office for Statistics,[74] at the end of the 1990s Cuba had more than 60,000 workers (1.15 % of the economically active population) involved in scientific and technological activities, of whom some 31,000 had university degrees and more than 50 % were women. The running expenses in these activities fluctuated around 1 % of the GDP. Half of these were dedicated to R+D, and financed from three sources: state (60 %), state firms (30 %) and external sources (less than 5 %).

6.7.2 *Cuban Physics and the New Priorities*

As explained above, toward 1990 the electronics program and the nuclear program were the main driving forces behind the development of physics in Cuba. Since both programs had been conceived in the framework of the collaboration with the USSR and COMECON, for obvious reasons their scope and orientation had to be radically changed from the beginning of the decade. As a result of these changes and of the new development directions that had to be taken by the country's economy in the context of the Special Period, physics was not among the high priorities of Cuban science. Still, the fresh impetus given to the development of biotechnology and the medico-pharmaceutical industry turned out to be an important stimulus for the further development of physics during the 1990s.

6.7.2.1 The Nuclear Program

The main objective of the nuclear program was to set up and operate the Juraguá electronuclear plant. However, when its construction was interrupted in the early 1990s and definitively put to rest in 1998, the priorities of the program had to be radically changed. They were shifted to the application of nuclear techniques in other production and service sectors, especially public health. During those years collaboration was strengthened with the International Atomic Energy Agency (IAEA), which contributed to the financing of numerous projects. As part of the then ongoing restructuring of the state central administration, SEAN was dissolved.

[74] See http://www.one.cu

The responsibility for the activities and centers under its control was given to the Agency for Nuclear Energy and Advanced Technology (AENTA), dependent of CITMA, which brought together an important group of institutions related to nuclear science and other branches of physics (Hardy et al. 2006).

CEADEN, the center in the nuclear sphere most closely related to physics, continued doing basic research in nuclear and solid-state physics, although it concentrated on equipment manufacture (semiconductor detectors, lasers, instruments for measuring residual tension in welds, instruments for measuring energy and power in lasers, densitometers, and so on) and also on the provision of scientific and technical analysis and measurement services. In later years, it merged with CEDEIC (though keeping its name), and its manufacturing line was extended to high quality equipment with polarimeters and lasers for acupuncture or physiotherapy, which were introduced in the country's industry and health services and also exported to several countries in the region.

The ISCTN continued to train *licenciados* in nuclear physics and radiochemistry, physicist engineers, nuclear energeticists (until 1993) and nuclear engineers (from 1993 to 2003). Gradually, it extended the profile of its training and scientific activities beyond the field of nuclear specialties.

The Isotope Center was inaugurated in 1994. It is a scientific-production institution that presently produces more than 80 % of the radioisotopes required by the country, which amounts to a substantial import substitution for the expensive health care services employing these reagents. Jointly with other specialists, physicists have played an important role in the development and establishment of radioisotope measuring systems.

6.7.2.2 The Electronics Program

The collapse of the USSR found the Cuban electronics program in the process of culminating the main investments that it had begun in the previous decade. The production of semiconductor devices and integrated circuits (cycle 1) was already in the start-up phase at the "Ernesto Che Guevara" Electronic Components Complex (CCE) at Pinar del Río and the hybrid circuit factory of the Copextel Corporation in Havana, whose staff included close to 40 physicists. Further work had to be stopped due to the sudden interruption of supplies and technical assistance, in addition to the disappearance of prospective markets. There were several attempts to reactivate these installations by modifying them to manufacture products for which reliable supplies and markets could be found, but these efforts did not succeed. For instance, by the mid-1990s, specialists from IMRE, CCE and Copextel managed to set up a successful production line for crystalline silicon solar cells with serigraphic contacts, but this never moved beyond the experimental stage. As time went by, the installations became obsolete and had to be dismantled. Little by little, the staff members took up alternative functions. Presently, the CCE is dedicated to various areas of production, such as the assembly of photovoltaic panels for national consumption and export, but activities in the microelectronics area have ceased.

Investments dealing with the production of electronic equipment, printed circuits, computers and software were more fortunate since they succeeded in reorienting toward new markets, especially toward the production of medical equipment and other objectives related to the development of biotechnology, the medico-pharmaceutical industry and tourism.

It was not long before the research activities in the microelectronics area disappeared altogether at the ICID. The institute acquired the status of a company and was dedicated to the development and production of medical equipment for the national health system and for export. As to ISPJAE's CIME, its research work was greatly affected too, but it continued to work on the design of microelectronic sensors for use in microbiology, medicine and other areas. It also increased its training activities by offering degree and graduate courses, especially through its MSc and electronics specialization programs.

At the University of Havana, the activities of IMRE and the physics and chemistry faculties linked to electronics were considerably affected. Initially, great effort was made to transfer to the productive sector whatever technological development could contribute to the substitution of imports, the generation of exports or to solving some social need. However, due to the lack of supplies and spare parts the experimental possibilities of IMRE rapidly deteriorated and in the end most of the efforts had little effect. Gradually, the previously established links with the semi paralyzed or reoriented entities belonging to the Electronics Front were reduced, while those with other sectors were reinforced, especially with the health sector, to which IMRE and the physics and chemistry faculties contributed important services. The development and manufacture of optoelectronic equipment for industry and medicine was stimulated. Outstanding at the time were the manufacture of PLC lasers for the engraving and cleaning of works of art, the "Fototer" phototherapy equipment and the "Saturomatic" and "Colormatic" intelligent photo colorimeters for the sugar industry. Graduate training activities were extended through the organization of international summer schools and other gatherings, and scientific publications grew to about 150 a year by the end of the decade. The IMRE gradually shifted its profile to materials science while diminishing its involvement in the production of electronic devices, apart from photovoltaic cells and some types of sensors.

6.7.2.3 Physics and Physicists in the Development of Cuba's Biotechnology and Medico-pharmaceutical Industry

From the late 1980s, the rapid growth of biotechnology and the medico-pharmaceutical industry attracted dozens of newly graduated physicists to the new centers situated within Havana's Western Scientific Pole. Most of these young people continued their training in molecular biology and became specialists in this field, some in a very distinguished way. One example is Rolando Pérez Rodríguez, who became one of the scientific leaders of the Center for Molecular Immunology. A few of these graduates remained closer to their original training profile in research areas related to analytic techniques, modeling and simulation of biologic systems, as well as in software development and the production of equipment for diagnostic and therapeutic purposes. Together with other physicists and engineers working at the Center for Immunoassay, Miguel Ángel García Álvarez stands out for the high economic and social impact of his participation in the production and successive improvements of the SUMA (Ultramicroanalytic system) equipment. Originally developed at the CNIC, this has seen widespread use in massive disease investigation programs in the country as well as in many other countries profiting from Cuban healthcare collaboration.

Many other institutions related to physics directed their efforts totally or partially toward biomedical applications. Their most frequent activities were in the realm of equipment manufacture and software design. Outstanding at the height of the Special Period was the creation in 1993 of the Center for Biophysics and Medical Physics of the University of Oriente, led by the physicist Carlos Cabal and dedicated to R+D in the area of magnetic resonance and its biomedical applications. Its most relevant contribution was the production of three NMR tomographs at low fields (0.005 and 0.1 T) for use in Cuban health institutions.[75] High level research work in biophysics and neurophysiology was done at the Center for Neurosciences, originally attached to the CNIC, where a small optoelectronics group, led by Luis Martí, developed and produced laser pain meters, scalpels and vibrometers. At the same time, they engaged in other important studies whose results were published in numerous scientific papers, and also made a substantial contribution to postgraduate training (Martí 2009, private communication). The "LASERPOL" polarimeters, which to this day are the analytic support of glucose measurements in the country, were improved at CEADEN. Laser physiotherapy and acupuncture equipment were installed in numerous hospitals and polyclinics. These and other new instruments, conceived, designed and produced with the participation of the country's physicists, continue to be introduced nationwide and in other countries where Cuban medical collaboration is present.

6.7.3 An Overview of Cuban Physics in the 1990s

In the late 1990s, CITMA, MES and various scientific societies completed several reports on the state of basic sciences in the country. As far as physics is concerned, the general characterization reflected by those reports will be summed up in the following.

6.7.3.1 Physics Teaching and the Training of Physicists

In spite of the difficult conditions prevailing at the time, the country made great efforts to preserve education at all levels. Still, the teaching of physics at the intermediate level suffered serious drawbacks related to the exodus of teachers to other sectors, the

[75] See the chapter by Cabal Mirabal in this volume.

deterioration of laboratories and teaching media, the reduction of physics class hours from 600 to 470 between grades 7 and 12, and the elimination of physics from the group of subjects to be examined for entry into other university degree courses. However, the IPVCE students continued to train and participate with good results in national, Ibero-American and international physics Olympiads, though in some of the related activities the number of participants were frequently reduced for financial reasons. On the other hand, in 1991 Havana hosted the XXII International Physics Olympics, the first one held on the American continent. It was successfully organized with the support and participation of many Cuban physicists called together for this purpose by the Ministry of Education.

Training programs in physics continued to develop. During the 1990s, 550 physicists graduated from the universities of Havana (243) and Oriente (165), and from the ISCTN (110 *licenciados* and 32 physics engineers). However, the number of new graduates continued to decrease following the general reduction in the number of students enrolled in higher education that took place at the time. Only 30 physicists graduated in the year 2000. While the general quality of the training received by these professionals was maintained and skill levels for computing techniques were increased, experimental training was affected by the aging and deterioration of laboratory instruments and equipment. The relative number of purely theoretical dissertations began to grow. On top of this, a significant reduction of content was made in the training programs central to the higher pedagogical institutes for secondary level physics teachers, which included the elimination of theoretical physics to make room for pedagogical matters and pre-professional practice.

The number of physicists hired by some important industrial branches, which had expanded greatly in the previous decade, diminished rapidly as a consequence of the paralysis or reorientation of the related industrial activities.[76] Though there was a slight growth in the number of physicists working in hospitals and research centers belonging to the health sector, the roles of staff froze at most of the universities and research centers. Still, in spite of the depression in the job market, the Cuban state continued to guarantee a job to every new graduate thanks to funds allocated for the purpose by a specially created "scientific reserve" program. This allowed for the provisional incorporation of new graduates in physics to university faculties and research centers where they could continue their professional training without burdening the normal staff budget of the host institutions.

The creation of a national board before which doctoral dissertations in physical sciences had to be defended, promoted the uniformity and increasing quality of doctoral degrees. While only 58 graduates were awarded doctoral degrees in the decade, compared to 103 in the 1980s, the overall number of dissertations defended

---

[76]The presence of physicists in industry is greatly benefitted by the cooperation with academic physicists. When many physicists left industry, the academics lost their main interlocutors. Due to Cuba's size, the absolute numbers are small: the number of physicists in industry were on the order of about one hundred at the beginning of the 1990s, which could have reduced to 20–30 by the end of the decade.

in Cuba grew. MSc programs reappeared in the 1990s, producing about a hundred graduates, many of whom later took up doctoral studies in the country or abroad.

The number of graduated physicists from 1962 numbered almost 2000 *licenciados* (866 from the University of Havana, 377 from the University of Oriente, 12 from the University of Las Villas and 230 from the Higher Institute of Nuclear Sciences and Technologies, plus those who graduated overseas). A similar number must be added to the *licenciados* in education who had specialized in physics at the higher pedagogical institutes. The number of those who held doctoral degrees in different branches of the physical sciences and were registered at the National Commission for Scientific Degrees was 187. Their branch distribution was: solid-state physics and materials, 91; nuclear physics, 26; mathematical physics and field theory, 14; optics, 12; geophysics and astronomy, 12; meteorology and atmospheric physics, 10; molecular and atomic physics, 7; other branches, 15. Their average age was already over 50. Some 200 physicists left the country in the 1990s, including no less than 30 with a doctoral degree.

6.7.3.2 Scientific Research

In a general sense, it can be said that scientific research in the field of physics was not brought to a complete stop but managed to continue, though it was subject to very strong limitations. The deterioration of research laboratories and workshop equipment added to difficulties in the acquisition of other supplies; long and frequent blackout periods and other factors considerably affected the capability for carrying out experimental work. About one third of the experimental groups existing in 1990 had disappeared by the year 2000.

Institutions like the IGA of the Cuban Academy of Sciences, which flourished in the context of the collaboration formerly developed with Soviet research institutions, managed to survive thanks to their ability to engage in the provision of scientific and technical services (for example, participation in local geoecological consulting and involvement in electrical grounding and lightning protection installations) without abandoning its former research work, which in the case of the IGA dealt with solar activity, ionospheric and geomagnetic field studies, and the seismic potential of the Cuban archipelago. In 1992, the seismology area separated from the IGA to become the Center for Seismology Research, permanently situated in Santiago de Cuba. At the same time, some laboratories dedicated to the development and construction of equipment flourished, especially in the field of optics, and the already mentioned Center for Biophysics and Medical Physics of the Universidad of Oriente was created.

The growth of public awareness of ecological problems, especially after the 1992 Rio de Janeiro Summit encouraged related research by physicists in several institutions, especially on the development or use of physical methods of analysis of environmental samples. At Havana's Agricultural University a soil physics group was created. Also, R+D work on renewable energy sources was also promoted, especially thermal and photovoltaic, the use of hydrogen, biomass, solid-state batteries and eolic energy, among others. However, this failed to get the substantial material support required to produce a significant leap in the area.[77]

Progress in computational means combined with the depression experienced in work on experimental physics contributed to the progress of research in theoretical physics, now extended to other areas such as complex systems (UH), gravitation and cosmology (UCLV) and molecular physics (ISCTN) beside those dealt with in the existing groups for condensed matter (UH and UO), field theory (ICIMAF), and nuclear physics (ISCTN and CEADEN). The new groups developed important international collaboration links and generally stood out because of the valuable scientific papers they produced and their proficiency in doctoral degree training. The support of the International Center for Theoretical Physics (ICTP) in Trieste, which began in the mid 1980s, played an important role in their development.

Paradoxically, the number of scientific publications grew to about 200 papers appearing each year in specialized international journals. This increase was linked to the broadening of collaboration with colleagues from western countries, which led Cuban physicists to pay more attention to publication, something that had been frequently neglected under pressure to generate practical results. Meanwhile, as has been emphasized throughout this chapter, physicists make up only a very small part of the Cuban scientific community. Their publications, however, represent a relatively large fraction of the national total. A similar situation prevails concerning the prizes awarded by the Cuban Academy of Sciences, the Science and Technology Forum and other national or international awards for scientific work.

National and international scientific gatherings were frequently organized during the 1990s, though strictly subject to the condition that they should finance themselves and abstain from generating additional expenses for the state. At the University of Havana, beside the summer schools on materials science organized by IMRE (Sect. 6.3.4), other Latin American scientific gatherings were held, dedicated to surface physics, magnetic materials, optics physics teaching, and so forth (OPTILAS 1995; TECNOLASER 2000). The Cuban Physical Society was active in organizing symposia and congresses, publishing the *Revista Cubana de Física* and supporting

[77] There were many discussions with respect to this, however, the point of view prevailed that renewable energy sources—with the exception of cane biomass traditionally used in Cuba for electric generation—would not have a significant impact in the short term, and that the urgent energy needs of the country, in particular in the area of electric generation, had to be solved through the efficient use of fossil fuels (both imported and from national production). However, there were no investments made that guaranteed the future development of renewable energy sources. Electric generation continued on the basis of large thermoelectric plants integrated into a national distribution system and the contribution of cane biomass in the sugar plants. The recourse to solutions based on other energy sources was favored only for isolated communities with no access to the national network. These politics changed radically in 2005 when, during a electricity generation crisis, the so-called "Energy Revolution" began. The national electric energy system was decentralized, and hundreds of small electricity generators using diesel or fuel oil were installed to complement the production of the large-scale plants. Domestic electric equipment and devices were replaced on a massive scale with more efficient ones. The introduction of renewable energy sources was accelerated and their growth planned.

many activities related to physics. It was chaired by Carlos Rodríguez Castellanos, María Elena Montero Cabrera and Víctor Fajer Ávila throughout the decade.

As mentioned, the collaborative relations that had developed over three decades with very high-level scientific institutions in the USSR and other East European countries were almost totally interrupted in the 1990s. But Cuba never abandoned its links with Latin American centers (especially with Mexico, Brazil and Argentina) and with West Europe (Spain, Germany, Italy and the ICTP, among others). Some of these, while initiated in the 1960s, expanded under the new circumstances and attained a higher relevance by playing an important role in the survival of Cuban physics. Contacts with the American Physical Society (APS) were re-established in 2000. With the active support of Nobel Prize winner for physics Leon Lederman and of the director for international relations at APS, Irvin Lerch, exchange agreements were signed between the APS and the SCF.

Some external impressions on the situation of Cuban physics at the end of the twentieth century can be found in two articles published abroad, one by Morán López (2000) and another by Cetto and Vessuri (1998), as well as in the comments published soon after their visits to Cuba by Irving Lerch and by Marcelo Alonso in the August/September 2002 *APS News* issue, both reproduced in the present volume.

6.8 The Twenty-First Century Begins

The first decade of the new century shows a contradictory balance and some encouraging perspectives with respect to the development of physics in Cuba. On one hand, some of the ill effects accumulated in the Special Period were felt even more strongly. On the other hand, some recovery signs and a stronger ability of the state to allocate resources for development appeared toward the middle of the decade when the Cuban economy managed to grow by 12.6 % (2006). However, the effects of the hurricanes that devastated the country in 2008 and the negative impact of the ongoing worldwide economic crisis endanger the continuity of these efforts.

6.8.1 General Overview of Cuba's Economic Efforts in Scientific Research

In 2009, Cuba spent 0.93 % of its GDP (61 billion US dollars) on scientific and technological activities, 0.64 % of which on R+D (Estrada 2007), see Figs. 6.1 and 6.2 below. In 2010, Cuba dedicated 18.4 % of its budget to education, 14 % to public health and 1.4 % to science (78 % of the total funding for scientific and technological activities (ONEI, 2010). [78]

[78] See www.one.cu. (Oficina Cubana de Estadisticas e Informacíón).

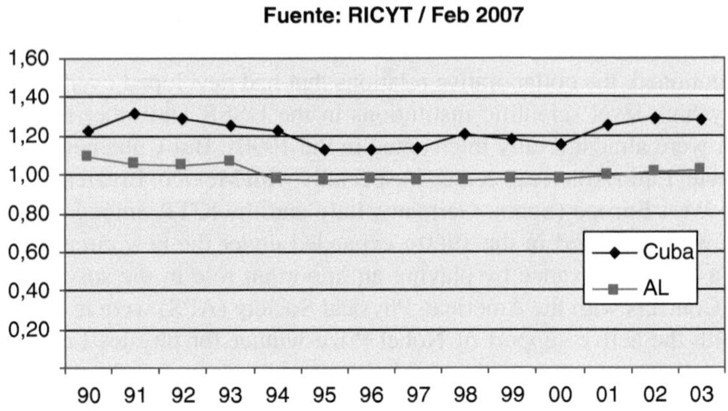

Fig. 6.1 Researchers per 1,000 members of PEA (economically active population) in Cuba and Latin America (*AL*) (Estrada 2007)

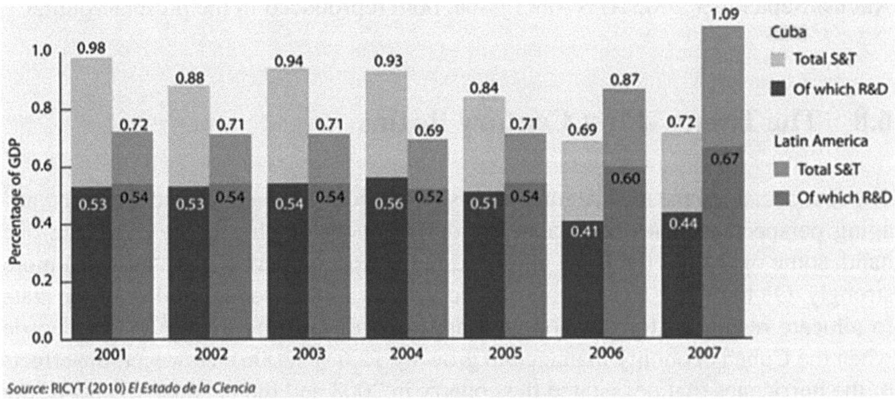

Fig. 6.2 Financial investment in R&D (Clark Arxer 2010, 125)

6.8.2 *The Teaching of Physics and the Training of Physicists*

In secondary education, the dominant tendency was the deterioration of the quality of physics teaching and of science teaching in general. Many experienced professors left the classrooms. The massive introduction of TV classes and other modern techniques were not as effective as expected. Specialized training in physics virtually disappeared. Physics laboratories in senior high schools ceased to exist. On the whole, the quality of high school physics teaching was negatively affected by the absence of the subject in the university entrance examinations.

During the academic courses between 2008 and 2010, there was a severe reduction in the number of hours dedicated to physics teaching at mid-level (from 710 to 438 class hours: see Table A.4, Appendix A.2). Moreover, the physics content that was taught in the 8th and 9th grades was diluted into a subject called natural sciences. Even though the total numbers still compare positively with the international average, these changes amounted to a backward step with respect to what was achieved in previous decades.

The data indicate that by the end of the decade a nationwide effort was made to recover the lost levels. The main problem now was not the number of class hours, but the lack of teachers with the required training. It must be remembered that in Cuba, general education is free and compulsory until the 9th grade, and that the great majority of Cuban youth goes to school for 12 years, either at senior high school or through programs for mid-level training of technicians, teachers and art instructors. New study programs have been created for the higher pedagogical institutes to train better quality *licenciados* in education who will be dedicated to mathematics and physics teaching. The recovery and re-equipping of laboratory classrooms for the teaching of physics at senior high schools has also been initiated. Those students with a special interest in taking science degree courses are now completing their 12th grade at the universities, a brand new situation that has a positive effect on their education. The participation of Cuban students in national contests and Ibero-American and international Olympiads has continued to be encouraged and this has generally resulted in the achievement of good results (see Table A.5, Appendix A.2).

As far as higher education is concerned, the average number of graduate physicists was a little more than 25 per year (127 at the UH, 65 at the UO and 124 at INSTEC), much less than in the preceding decade (see details in Table A.6, Appendix A.2).

Since 2005, special entrance requirements have been introduced for the degree course (*licenciatura*) in physics that stimulate better training and choice of new students. Physical engineering was inaugurated at the Universities of Havana (UH) and Oriente (UO) in 2007, as was the *licenciatura* in physics at the "Marta Abreu" Las Villas University (UCLV). While these measures have shown a modest (quantitative and qualitative) effect on enrollments, as yet they have not had an impact on graduations. Moreover, the decision to relate the university enrollments to the explicit social demand of professionals may result in the closing down of degree courses that do not have a long tradition, such as physical engineering. In recent years, a great quantity of equipment has been made in China to renew the physics teaching laboratories in all universities. However, the condition of the research laboratories has weakened the experimental training of physics students and has continued to result in the increase of the proportion of purely theoretical dissertations.

International standard MSc and PhD programs have continued to develop. Generally, even if the defense of the corresponding dissertation is to be done in Cuba, a stay abroad for 1 year or longer is included to access the necessary information and complete the experimental work at an adequate level. According to data provided by the National Commission for Scientific Degrees (CNGC), 61 graduates were awarded doctoral degrees in physics from 2001 to 2010, averaging 6 per year, a somewhat higher number with respect to the preceding decade. More than half of the theses (34) dealt with condensed matter topics. If the exodus of no less than 50 physics doctors during the last two decades and the approximation to retirement age of a sizable part of the graduates who qualified in the 1970s and the 1980s is taken into account, the present graduation rate does not guarantee a reproduction of the high level of human resources that have been trained to date in the field of physics.

As pointed out above, the training of physicists is presently carried out at UH, UO, UCLV and INSTEC. The physics faculty of the University of Havana continues to be the largest and most representative faculty in the field of physics. In 2002 it commemorated the 40th anniversary of the inauguration in the country of the *licenciatura* in physics. It now has a teaching staff of some 60 people, half of them holding doctorate degrees. It is in charge of the physics *licenciatura,* the physical engineering degree course, and the MSc and PhD training in physical sciences, as well as the physics teaching for students taking other degree courses in natural sciences at the University of Havana. It has diversified its research lines, which are now taken care of by working groups dealing with condensed matter, atomic physics, complex systems, biophysics, statistical physics, electronics and physics teaching. These groups have good scientific production and good participation in international collaboration, having distinguished themselves for their publications, graduate training and national and international prizes awarded to them. The physics faculty is also engaged in cultural extension activities through the Félix Varela Scientific Culture Chair, the Physics and Music Chair, and the Solar Energy Chair.

The IMRE, another research center closely linked to the University of Havana and located on its campus, was renamed Institute for Materials Science and Technology in 2005. Presently, it has a staff of 50 researchers (mainly physicists, chemists and engineers) and about the same quantity of associate professors. There are 30 physics PhDs on staff who work in research laboratories dedicated to laser technology, structural analysis, solar energy, zeolites and molecular physics, as well as to traditional areas of solid-state physics (semiconductors, magnetism, Ferro electricity and superconductivity). It is also active in the development of MSc and PhD programs dealing with materials science. While local experimental activities remain low, its personnel manages to publish about a hundred scientific papers each year thanks to the strong collaboration links established between IMRE and foreign research institutions. The institute also takes care of graduate students working for their MSc and PhD degrees, develops innovative projects for various industries and other institutions, and engages in widespread scientific popularization through the media and otherwise. Recently, the Ministry for Higher Education allocated sizable funds for the creation at IMRE of a laboratory for analytic services required by universities (project LUCES), which should be of great help to physicists, chemists and other specialists working in the materials area.

An excellent and ambitious project for the restoration and remodeling of the physics building of the University of Havana (which used to house a large part of IMRE's laboratories) is being implemented. To allow for this, a provisional spread of classrooms, laboratories and workshops all over the campus was unavoidable, which makes working conditions difficult, especially since the architectural work proceeds at a very slow pace due to financial difficulties.

The physics department of the University of Oriente's natural sciences faculty is presently staffed by 25 professors, 10 of whom hold doctoral degrees. The department teaches physics for specializations in the natural sciences, including the *licenciatura* in physics. It also includes research groups on materials science, superconductivity, digital holography and physics teaching. Another group exists at the University of Oriente, which teaches physics to the engineering degree courses. There is also an important presence of physicists in two research units belonging to the same university: the Center for Medical Biophysics and the Center for Applied Electromagnetism.

The Las Villas Central University's physics department is presently staffed by 19 professors, 9 of whom hold doctoral degrees. This department teaches physics for degree courses in the natural, technical and agricultural sciences, and for a small group of 25 students working on their *licenciatura* in physics. Research activities concentrate in three areas: gravitation and cosmology (including astrobiology), physics of materials and physics teaching. Physicists are also present at the University's Center for Welding Research.

The ISCTN (Higher Institute of Nuclear Sciences and Technologies), which in 2003 was renamed INSTEC (Higher Institute for Applied Sciences and Technologies), is nowadays a university by itself, attached to CITMA (Ministry of Science, Technology and Environment), and since 2011 attached to the MES (Ministry of Higher Education). Besides the *licenciadura* and the *maestría* in nuclear physics, it implements several pre- and postgraduate programs that are not necessarily related to the nuclear area. A degree course in meteorology was started there in 2005. Forty-six physicists work at INSTEC, 11 of whom hold doctoral degrees. It has continued to train *licenciados* in nuclear physics, though in practice these are physicists with a wide profile who, once graduated, are able to work in any area of physics. Research work at the institution has also diversified its own profile, which now includes theoretical and experimental nuclear physics, atomic and molecular physics, and the development and application of nuclear techniques for health, industry and protection of the environment.

6.8.3 Research

As pointed out in Sect. 6.7.3, at the end of the 1990s CITMA, MES and various scientific societies completed several reports on the state of basic sciences in the country. Following the recommendations derived from them, CITMA organized some science and technological innovation programs which supported research in physics, chemistry and other basic sciences, such as the "new and advanced materials "program, the "basic research in physics, chemistry, mathematics and computer sciences" program, the "nuclear branch program" (also backed by the IAEA), and the "optics and laser" program. This led to the financing of some basic scientific activities, promoted cooperation between institutions working on related subject matters, and bestowed a greater recognition on physics research among the officially defined priorities. The financial backing of these kinds of programs, however, is still quite modest with only a small fraction of it in hard currency, which is not very helpful for the improvement of the research capability of the groups involved. Since the contribution of hard currency has been very small, it has been necessary to look for funding from other sources.

An area of opportunity with a large presence of physicists, which has received some stimulation in recent years, deals with renewable energy sources, especially eolic, solar-thermal and photovoltaic energy. Physics projects have also found a space in national programs such as "climate change" and "renewable energies." The acquisition and installation of modern equipment has improved the research possibilities of physicists working in the field of meteorology who have made important contributions to the establishment and development of meteorological services. Cuban specialists participated in an International Panel on Climatic Change that was awarded the Nobel Peace Prize in 2007.

Generally speaking, those research groups with a greater tradition at the Universities of Havana and Oriente, the ICIMAF, CEADEN and others, have remained active. Some younger research groups, such as INSTEC's molecular and atomic physics group, the Las Villas University's cosmology and gravitation group, the "Henri Poincaré" complex systems chair, and the laser technology laboratory (the latter two belonging to the University of Havana) show increased activity. Developments in the areas of medical physics and nanotechnology will be dealt with in subsequent sections.

The number of scientific papers published in international journals, which grew substantially in the 1990s, has remained at a constant of about 200 a year on average. A world ranking by countries or territories of publications and citations, published in 2007 by The Thompson Corporation, puts Cuba in place 66, with 796 publications and 2663 citations in indexed physics journals during the preceding 10 years, behind Brazil, Mexico, Argentina, Chile, Colombia and Venezuela. It is believed that Cuban physicists actually publish a similar or perhaps larger quantity of papers in journals specialized in other disciplines or interdisciplinary areas such as materials science. Still, the presence of physics in the yearly research prizes of the Cuban Academy of Sciences and other institutions continues to be relatively high, maybe because other scientific branches face similar problems. Recent research shows that in 1997–2007, Cuba published more physics articles in ISI journals than biology, biochemistry and neuroscience articles put together, though less than articles in the medical sciences. Still, the number of citations per article in other areas is always larger.

In most physics areas, experimental research is lacking for the time being, and in many cases international collaboration is essential for its survival. This is due not only to lack of costly equipment, but also to poor availability of materials, industrial gases, reagents, electronic devices and other supplies needed for experimental work, as well as due to difficulties in accessing technical services, restricted electricity consumption, and so on. All of this, added to the fact that in the last two decades scientific equipment has experienced an important qualitative jump worldwide, has deepened the present technological backwardness of the country's experimental physics. On the other hand, although access to information found in electronic journals and specialized databases has been expanded in various ways, its use is limited by low-speed connections to the internet. This is expected to change in the near future when wideband international communications will be generally available to Cuban research centers.

Experimental physics laboratories dedicated to equipment development and production are the ones that enjoy better working conditions, since the commercialization of their products enables them to obtain the resources needed for their activity. For example, IMRE's Laser Technology Laboratory has succeeded in developing systems based on solid-state lasers for engraving, surface cleaning, filleting, environmental sample analysis, optical tweezers and other applications, for which it has won national and international recognition. These achievements and their transfer or commercialization have enabled the laboratory to obtain the hard currency financing required for its infrastructure and equipment development. CEADEN and the Immunoassay Center laboratories have also profited from the commercialization of their production, as a result of which their production areas are provided with adequate mechanical and optical technology, electronic assembly facilities and reference measurement equipment.

Theoretical physics has continued to develop. At the University of Havana, the theoretical physics department has renewed itself and extended its research and postgraduate work to statistical physics and complex systems, in addition to its traditional line on condensed matter. In 2005, it organized a Latin American conference on interdisciplinary applications of statistical mechanics. It now has a 15 member staff, nine of whom are doctors, while the rest are young people working toward their PhDs. Furthermore, several students and collaborators from other institutions are working in the department for their *licenciatura*, MSc or PhD dissertations. At INSTEC, the molecular and atomic physics group has increased its publications, its doctoral training work and its organization work for international scientific gatherings on photo dynamics. The gravitation and cosmology group, which emerged in 1999 at the Las Villas "Martha Abreu" Central University, has experienced rapid development too. It has trained three doctors, regularly publishes papers in specialized international journals, publishes a newsletter entitle *Correo de Gravitación y Cosmología*, actively participates in international collaboration and has organized two international conferences in this field. The theoretical physics group of the ICIMAF has continued its research work on particles and fields and also on condensed matter. It has developed extensive collaboration activities, national and international, and contributes to postgraduate training in theoretical physics of young physicists from other national institutions. Thanks to its initiative, Cuba was admitted as the 29th member of the CERN's project ALICE for high energy physics, with ICMAF and CEADEN participating in related experimental and theoretical work. Supported by the ICTP, it has organized four Caribbean workshops on quantum mechanics, and a school on string theory. It has proposed its own transformation into a theoretical physics chair of regional character. All of these groups have been awarded several prizes by the Cuban Academy of Sciences for their work during the decade.

Another institution with a long tradition, the Institute of Geophysics and Astronomy (IGA), is staffed by some 120 people, 76 of whom are technicians, divided into four departments: astronomy, space geophysics (which includes ionospheric physics), physical environment (for soil, water, contamination and other studies) and regional geophysics (dedicated to the study of the Earth's gravitational and magnetic fields, its electrical conductivity, and so on.). Nine astronomers and 11 engineers and technicians staff the astronomy department. Its main work areas continue to be radio astronomy, optical spectroscopy of the sun, satellite observation and various problems related to computational astronomy. A report on the visit to Cuba of two representatives of the International Astronomical Union (UAI), which took place in 2005, recognized the quality of the existing basic infrastructure and of the personnel working in this field, especially at the IGA. On the other hand it pointed out that the failure to renew the existing equipment and personnel had led to stagnation in the field. To overcome this, the report recommended extending the presence of astronomy and related disciplines in Cuban universities, and activating the links with the UAI and other forms of scientific exchange with the international astronomical community (Hearnshaw and Fierro 2005). During the recent International Year of Astronomy, numerous scientific and popularizing activities took place in Cuba, including the inauguration of a modern planetarium located in the Old Square of Havana's historical center.

6.8.3.1 Medical Physics

Medical physics and biomedical engineering experienced a significant growth during the decade and contributed to the development of medical services of great social impact. Among the main achievements of Cuban development and innovation activities in this field, the following results stand out: the construction of a third MRI tomograph and the implementation of a biomedical image transmission system (CBM-UO), the development and production of new microfluorimeters and spectrometers of the SUMA series (Immunotests Center), and the production of new laser medical equipment for analysis, physiotherapy and acupuncture (CEADEN), and for contactless blood extraction (IMRE).

By 2009, radio diagnostic and radiotherapy services had extended throughout the country, supported by the installation of 64 computerized tomography units, 20 gamma cameras, 11 radiotherapy units (including four linear accelerators) and 13 MRI units. In 2008 some 80 medical physicists were working for these services (López 2009; Pérez et al. 2011). About the same number of physicists, engineers and biologists linked to this area worked at the universities, research centers, the isotope production center and other workplaces involved in the control, regulation and security of these activities.

Important services and development work have been contributed to this field, such as a system for the calculation of patient personalized doses; the cards and software packages for gamma camera image acquisition, manipulation and processing;

the production, design and shielding of nuclear medicine installations in hospitals; the development and validation of a method for the optimization of the activity; the study and evaluation of radio isotope-marked monoclonal antibodies; the design and validation of relative methods for the quantification of blood circulation in the brain; the production of detectors for digital radiographic images; the national production, evaluation and validation of radio medications; and the implementation of national protocols for the auditing and quality control of nuclear medicine and radiotherapy services.

MSc programs on biomedical image technology for engineers were opened in several universities. A diploma course on physical aspects of nuclear medicine, another one on radiotherapy physics, and an MSc program on medical physics were implemented for physics graduates working in the biomedical field. In 2004, ISPJAE inaugurated a degree course in biomedical engineering.

The Cuban physicists made an important, though indirect, contribution to the so-called "Operation Miracle." Thanks to this Cuban-Venezuelan program of free surgery for vision ailments for low-income patients in third world countries, from 2004 to October 2008, some 1,314,000 people from 33 Latin American, Asian and African countries had surgical treatment for cataracts, glaucoma, strabismus and other eye ailments, mainly through the use of refractive laser surgery techniques. The program aims at having about six million patients surgically treated by 2014. Since 2005, under the leadership of Germán Muñiz Planas, a group of staff members of the ISPJAE physics department, supported by other specialists from CEADEN, AENTA and INSTEC, organized and taught a diploma course on optoelectronics and lasers. Between 2005 and 2007, 175 engineers who took this course have been in charge, under the guidance of experienced physicists, of the assembly, start-up, maintenance and repair of the equipment installed in ophthalmology centers created in Cuba (10) and abroad (49). Apart from this, an evaluation of the systems for their use under tropical conditions was completed, a monograph on ophthalmological equipment was prepared, and a postgraduate course was designed and delivered to 1,200 ophthalmologists. Manufacturing firms have certified the ability of many of these specialists to install, maintain and repair surgical excimer lasers, derived from their excellent technical and professional training.

6.8.3.2 Nanotechnologies

In Cuba, as in other countries, nanotechnologies sprang from the convergent development of research on materials physics and chemistry, microelectronics, supramolecular chemistry, microbiology and molecular biology. However, the physicists were those who led the spreading, promotion and national integration of this emerging field of science and technology.

Theoretical and experimental work on semiconductor nanostructures physics, initiated in the 1980s, grew at the University of Havana in the 1990s. Led by

Cuban physicists, the CYTED network (dedicated to the study, manufacture and characterization of semiconductor nanostructures for micro- and optoelectronics) was organized and remained operative from 1998 to 2003, with the participation of eight Ibero-American countries. It organized several courses and scientific gatherings, edited a book on the subject, and supported scientific collaboration among participant institutions.

Starting in 2001 there was an increased promotion of nanotechnologies in the country. Summer courses on nanoscience and nanotechnology were organized at the University of Havana's IMRE, and an international conference on nanoelectronics was organized by ISPJAE's CIME. In the same year, on the initiative of the CNIC and IMRE, a national workshop on "Nanotechnologies in Biotechnology and the Medico-Pharmaceutical Industry" was held with the participation of several centers belonging to Havana's Scientific Pole.

In 2002, the Ministry for Higher Education created a network for nanotechnologies, with 12 participant institutions coordinated by IMRE, whose objective it was "to promote national and international cooperation in the nanosciences and nanotechnologies." This network has organized several national meetings and coordinated the participation of Cuban specialists in the Cuba-Mexico (2003, 2009), Cuba-United Kingdom (2004), Cuba-Brazil (2007) meetings, and in the 2006 meeting of the Latin American Macrouniversities Network.

Also in 2002, CITMA's Science and Technology Observatory organized a team consisting of many of the physicists who prepared the document "Initial Elements for the Analysis of Nanotechnology in Cuba," addressed to the government and other decision-makers. A second prospective project on "Nanomaterials" was carried out in 2005–2006 within the framework of the "New Materials and Advanced Materials" Science and Technology National Project.

Throughout these years, research and doctoral training in nanophysics extended to magnetism, polymers, porous materials, photovoltaic cells, crystallography, the use of nanoparticles in medical applications and other areas.

The spreading and popularization of concepts and achievements related to nanosciences and nanotechnologies among various audiences has extended throughout the decade and has been noted by conferences on the topic in different institutions, newspaper and journal news, TV classes, and so on. Promoted by IMRE, the University of Havana created a nanotechnology association, which brings together specialists from the natural sciences and the social sciences. It has participated in the "Nanotechnology and Society Latin American Network" and the CYTED Network for the Spreading of and Training in Nanotechnologies. A book is being prepared for Cuban university students on a wide range of specializations.

In 2005, a national project began to develop the construction of a national multidisciplinary center for nanotechnology, where physics and its methods will certainly have an important space, especially in the areas of interest related to nanobiotechnology and nanomedicine. Within the framework of this project, an intensive training program of young Cuban physicists at European universities is underway, and some international scientific meetings have been organized.

6.8.3.3 Other Activities

During the last few years, national and international scientific meetings organized by Cuban universities and research centers or by the Cuban Physical Society (SCF), have continued to take place at a rate of about two per year. Besides those mentioned above, the following ones stand out: the SCF triennial symposia, the International Summer School Symposia on Materials Science, the Medical Physics International Workshop (2002), the Ibero American Workshops on University Physics Teaching (2003, 2007, 2009), Technolaser (2003, 2005, 2007, 2009), the Inter American Conference on Education in Physics (2003), the Nanomagnetism International Workshop (2004), the 17th Latin American Solid State Symposium (2004), the 10th International Nuclear Physics Workshop (2005), Photosciences (2005), the Optics meetings and the Life and Patrimony meetings (2004 and 2009), the "Materials Science in the Nano Era" conference (2009), the Nuclear Physics Workshops and the Symposia on WONP-NURT Nuclear and Related Techniques (2007, 2009). Cuban physicists played host to many distinguished foreign visitors, such as physics Nobel Prize winners Leon Lederman (2003), Zhores Ivanovich Alferov (2007 and 2010) and David Gross (2010); chemistry Nobel Prize winner Robert Curl (2009), and Christopher Geber, inventor of the AFM, and to recognized authors of widely used physics textbooks, like Marcelo Alonso and Robert Resnick.

Links with international networks and organizations related to physics (CLAF. IUPAP, FEIASOFI, ICO, IUCr) are still active. The *Revista Cubana de Física* has experienced considerable improvement and its electronic version is easily accessible.[79] Originally presided over by Osvaldo de Melo Pereira and presently by Augusto González, the Cuban Physical Society established the "Manuel F. Gran" annual Physics Prize, which was awarded for the first time on the occasion of the society's congress held in March 2011, to four distinguished Cuban physicists: Melquíades de Dios Leyva, Hugo Pérez Rojas, Carlos Trallero Giner and Carlos Cabal Mirabal.

After the agreements reached in 2000 between the Cuban Physical Society and the American Physical Society, two meetings were jointly organized in Havana: The Medical Physics Workshop (2002), in which 30 American physicists participated, and the Inter American Conference on Education in Physics (2003), in which 34 American professors took part. However, the restrictions imposed by the US Government's blockade on Cuba have prevented a stable collaboration between Cuban and American physicists. Soon after the terrorist attack on New York's twin towers, the Institute of Electrical and Electronic Engineering (IEEE) cancelled its Cuban chapter, which had been established in the preceding decade.

A national committee was created in Cuba for the celebration of the World Year of Physics (2005), within whose framework many related activities were programmed. These included seven national or international scientific meetings, TV spots, press meetings, interviews, the printing of two telephone cards and of two postage stamps commemorating the 75th anniversary of Einstein's visit to Cuba in 1930. Related documentary films were shown in the media and an itinerant

[79] See http://www.fisica.uh.cu/biblioteca/revcubfis. Accessed October 17, 2013.

exhibition on twentieth-century physics was set up. Cuban physicists were involved in the project named "A Light Ray through the World," three of them participated in the Paris "Physics for Tomorrow" conference that inaugurated the year, and another four participated in the Durban "Physics and Sustainable Development" international conference which closed it. An article entitled "A Look at Physics in Cuba" was published in the September 2006 issue of *Physics Today* (Baracca et al. 2006).

Physicists are quite an active minority among the Cuban scientific community. Their presence is well known in the media and other activities related to the popularization and cultural impact of science. They have contributed in particular to the general awareness and promotion of the emerging field of nanotechnology, to confronting pseudoscience, informing about global climate changes, employing renewable energies and other subjects of great relevance.

6.9 Final Remarks and Conclusion

We have presented in this paper a reconstruction of the birth and growth of a modern scientific system in Cuba, specifically in the field of physics, starting from 1959, the year of the Revolution, through its subsequent phases, until the dissolution of the Soviet Union and the socialist bloc and the deep difficulties of the following period, and finally up to the present. Since the reconstruction of the structure of this system and its subsequent phases is complex, we have included partial evaluations and assessments over the course of this analysis to help the reader conceptually and to avoid the dangers of following a pure chronology or a list of institutions or initiatives. However, at the risk of repetition, it seems necessary to draw some general conclusions. It is important to recall what we have stressed from the beginning, which is that the scarcity of documentary material and the temporal proximity of the events analyzed make it more difficult to draw general assessments on this topic.

6.9.1 Pushing for the Development of a Strong Scientific System

A first evaluation concerns the choices and efforts made in Cuba, viewed as a whole, in order to promote the development of science and to build a strong, modern and articulated scientific system. We have repeatedly insisted on the constant commitment to orienting the difficult scientific choices (in the extremely problematic Cuban situation after 1959) toward the serious and urgent social and economic needs of the country. Such a concern was shared by the political leaders, the intellectual and scientific milieus, and other social protagonists like the student movement, although the relative weight, influence and interplay of these subjects in the decision-making processes have changed in the successive social and political phases.

In this context, we will comment in particular on the widespread consciousness among the Cuban leaders and intellectuals of the relevance of physics, and the consequent choice of favoring the development of pure and applied physics. Although the community of physicists is only a relatively small part of the whole Cuban scientific community, it constitutes a solid frame and a flexible (thanks to the malleable and manifold approach discussed in this paper) reservoir for feeding and supporting the whole system. There is no doubt that, among the natural sciences, physics is the one that has developed the most rigorous approach, both at the basic and the applied levels. Consequently, one may consider that the insistence on physics was extremely important and influential—even before its applications or foundation for every scientific field—in developing a rigorous mentality, approach and methods in a country that had previously only prepared teachers (albeit very competent ones) for the secondary schools. We have stressed that even for the measurement systems, there was deep confusion and arbitrariness in the period before 1959.

6.9.2 Difficulties of Assessments, Influence of Unexpected Historical Changes

The whole picture is made even more complex by the strong polarizations and tensions in international relations, and the dramatic (often unforeseeable) political and economic changes that the post-war age underwent. All of these factors make it more difficult to draw final conclusions on the assessment of the choices that were made in Cuba after 1959. On an historical level, the results, or the consequences of these political choices, cannot be assessed in an abstract sense: their outcome depends on the historical events. Unexpected or unpredictable changes of the general situation can bring to an end important efforts, or radically change the actors or the conditions, causing some basic factors which were the basis for marking these choices to expire or to change.

We have concretely seen and discussed at least a couple of revealing examples of this process in the context of this analysis. Several developing countries, in particular in Latin America, made the basic choice in the 1970s to favor their engagement in microelectronics, which appeared as a new, technically and industrially affordable, low-cost and low energy consumption field, in which they could compete with the more advanced countries. How can one evaluate the soundness of such a choice? Within a few years, many of these countries (including Cuba, which had started in 1959 practically from nothing) in fact reached scientific and technical achievements comparable to those obtained in the most advanced countries. But the unforeseeable advent of high- and very high-level integration completely changed the situation, requiring a leap in the technical level and in costs, which were affordable only for the most powerful countries (even the Soviet Union—but also Europe as a whole—fell behind in such high precision fields, while it could compete in more complex techniques). Another matter to evaluate is the legacy that this effort left behind in these countries. At least in the case of Cuba, the skills and professionalisms that were created (taking into account moreover the starting point of the absence of both a basis for quantum physics and for scientific research) set up a strong basis for further developments in specialized sectors such as electronics, optoelectronics, photovoltaics, laser technology and other fields and applications.

The second case, of nuclear physics and technology, differs even more. The general evaluation of nuclear technology, in particular of nuclear power, is strongly debated internationally at a public level: such an evaluation exceeds the aims of the present paper. In any case, it is evident that nuclear power constitutes a strong basis for the most developed countries, for better or for worse, that is, apart from total costs, dual-use and nuclear armaments, waste disposal, and so on. This was all the more true in the 1970s and 1980s, before the major nuclear catastrophes, when nuclear power was still perceived, with a few exceptions (although oppositions existed), as the most advanced, progressive and promising technology. This was even more true in the case of the Soviet Union, which saw this field both as an area of competition with the U.S. and other Western countries (mainly for military applications), and as a factor of influence on the socialist countries. This was mainly the case for civil applications of nuclear power, but also for their ideological impact, since the US had used the "Atoms for Peace" program as a kind of "Trojan horse" to extend its influence in the Third World and in non-aligned countries. With this in mind, the Cuban choice seems hardly criticizable. But in this case, it was the unforeseeable collapse of the Soviet Union that brought to an end the ambitious Cuban projects, such as the subsequent unsuccessful efforts to find other partners to complete the Juraguá nuclear power plant. In the case of Cuba, it is impossible to say what could have been the outcome of this choice if the foreseen conditions had continued. Nevertheless, also in this case one must acknowledge that the scientific expertise and skills developed in this field, non-existent before the 1960s, laid a solid basis for other fields and applications. Take into account, moreover, that (civil) nuclear technology is not only nuclear power, but has a wide range of applications, from industry to science and health. In a large measure, the development of medical physics in the last two decades was an inheritance from the Cuban nuclear program.

6.9.2.1 Tentative Assessment of the Cuban Scientific Choices, Especially for Physics

Even more difficult would be the question of whether different scientific and technical choices could have been more profitable for Cuba in the long run, a question, moreover, which cannot be posed by leaving aside the concrete historical situation. The Cuban projects in this field (but probably not only this field) were quite ambitious. One can imagine—and some declarations of its leaders that are reported in this paper seem to confirm this hypothesis—that the projects were entertaining ideas of developing products and services exportable to the COMECON market based on the intensive use of a highly qualified labor force, which compensated for the Cuban weakness in traditional industries. One could say that this effort was successful in itself, that is, considering the results obtained from a scientific perspective. Much more difficult would be to assess this effort at a general level, or to assess whether different choices could have led to better and more substantial results. But which choices would these be? As we have stressed, no one could foresee the dissolution of the socialist bloc. The general remarks that one could make might concern on the one hand the extraordinary health standards reached by the country, which rate among the highest in the world. These are marked by features such as low death rate at birth, high life expectancy in spite of the unavoidable deterioration of social conditions in the past two decades. This provides a further indirect confirmation of the soundness of the health structures, and also of the scientific bases underpinning them. On the other hand, one can consider that the strong choice made in Cuba for developing basic and applied science has produced important and lasting results of international relevance, which are also sources of revenue in these difficult times (much more than the weakened sugar industry). Above all, the most successful are the sectors of biotechnology, pharmaceuticals and medical equipment. Last but not least, one must acknowledge that the robustness achieved by the Cuban scientific and higher education sector has allowed it to withstand the shock produced by the dissolution of the Soviet Union, despite the unavoidable drawbacks and reorganizations that occurred during this period. We have seen this concretely in the case of physics.

6.9.2.2 Other Peculiarities of the Cuban System and Its Scientific Approach to Physics

In respect to Cuba's scientific approach to physics (and probably not only to physics), we would like to point out more factors that are related to its peculiarity. In the first place, we refer to the multiplicity of contacts, collaborations and support that Cuban physicists have kept alive. This multiplicity was more pronounced in the first decade after the Revolution, but was never abandoned, even in the subsequent period when integration in the socialist bloc became stronger. We have stressed this variety of contacts in the successive steps of our analysis: for example, after the initial presence of Western visiting professors (Sect. 6.2.2); the contributions from French physicists (among others) in the summer schools (Sect. 6.3.4); the collaborations with Parma (Sect. 6.3.3) and with Canada in microelectronics (Sect. 6.4.2); of IMRE with institutions in Berlin, Trieste, Paris, Stuttgart, Madrid, Campinas in Brazil, Argentina, Mexico (Sect. 6.6.5.1); or Cuba's participation in the main regional and world organizations related to physics (CLAF, FELASOFI, FEIASOFI,[80] ICTP, IUPAP, IAEA, ICO).[81]

Concerning the soundness of the Cuban system to resist and react to the downfall of the Soviet Union, at least in the case of physics, these contacts turned out to be extremely important when it was no longer possible to rely on the support of the socialist countries, and it was much easier to reinforce collaborations and exchanges

[80] Respectively, the Federation of the Latin American and of the Ibero-American Societies of Physics.

[81] See as well the contributions to the third part of this book, in particular, the chapter by Leccabue.

with Spain, or with other Latin American countries. In the latter case, the soundness of the preparation of Cuban physicists turned out to be a true ace in the hole. Cuban physicists are in fact in high demand in larger countries as Mexico and Brazil, which have without doubt a much bigger scientific and education system, but suffer in terms of a shortage of scientific personnel, and a lack of homogeneity in terms of its level of training. As we have pointed out in the introduction to this paper, Cuba managed to avoid such problems.

We have already commented, but it is worth revisiting, some peculiar features of the training and general outlook of Cuban physicists, in spite of the substantial homologation of physics all around the world. The initial influence of the numerous Western professors who both worked in Cuba and taught in the summer schools (in addition, the approximately 20 students who studied and worked in Parma), as well the influence of the Soviet system, in form of both the presence of Soviet advisers in Cuba and the massive training of Cuban students in Soviet institutions, left some permanent traces.

We have insisted, for instance, on the quite different derivation of two of the main fields we have analyzed, that is, microelectronics (and in general solid-state physics) and nuclear physics. In the former case, the role of the advice of Western specialists was crucial at the beginning, and also subsequently when the French specialists introduced the new and decisive technologies and materials in the summer schools. But as late as 1981, when links with the "Ioffe" Institute had already become dominant, and as Leccabue testifies in his contribution to this volume, the "scientific exchanges with the USSR were useful, but could not cover the whole set of conditions for the development of this field of research," and wider international contacts and collaborations had to be promoted, also for "providing and shipping scientific equipment." In the case of nuclear physics and technology, on the contrary, the role of the Soviet Union was crucial.

The conditions prevailing in a small and developing country like Cuba, subject for more than half a century to the relentless blockade of a superpower, impose a pragmatic view of its economic and social development. In such a context, the future of Cuban physics will depend to a large extent on the country's possibilities to implement long-term strategic programs in fields that may offer opportunities for physics, as well as on the ability of Cuban physicists to identify and profit from the opportunities offered by the country's priorities, and at the same time to actively participate in international scientific collaboration. An essential condition for this is to continue to raise the high standards already achieved in the preceding decades through the teaching of physics at all educational levels and the graduate training of physicists.

Appendix

Data on the development of physics in Cuba are given in this appendix to support and document the analysis carried out.

A.1 Data on the Quantitative Achievements and Growth of Higher Education and Research up to 1976, from the Study Examined in Sect. 6.5.3

Tables A.1 and A.2 show the total number of graduates in physics and of publications on physics subjects. Table A.3 shows that starting in 1978, there was a remarkable increase in the number of physics PhDs.

Table A.1 Number of physics graduates (Pérez Rojas et al. 1976)

Number of physics graduates (1976)	Origin
337	University of Havana
35	University of Oriente
12	Las Villas Central University
35	Abroad
10	Physico-Mathematical
429 (Total)	

Table A.2 Scientific publications (Pérez Rojas et al. 1976)

Publications (1976)		
Institutions	National	Foreign
Physics School, University of Havana	100	20
Inst. of Geophysics and Astronomy, ACC	17	32
Inst. of Nuclear Research, ACC	14	
Inst. for Fundamental Technical Research	10	7
Total	141	59

Table A.3 Physics PhDs registered at the National Commission for Scientific Degrees (Pérez Rojas et al. 1976)

Year	Defenses	Accumulated
1974	1	1
1975	1	2
1976	3	5
1977	0	5
1978	10	15
1979	7	22
1980	4	26

Not all the physicists who successfully defended their doctoral dissertations abroad went through the formalities required to homologate their titles at the National Commission for Scientific Degrees at the appropriate time. These figures may thus be taken to represent a lower amount

A.2 Tables Related to the First Decade of the Twenty-First Century (Sect. 6.8.1)

Table A.4 Physics class hours at junior high school (grades 7, 8 and 9) and senior high school (grades 10, 11 and 12) from 1967 to 2012. The hours marked with an asterisk correspond to the physics contents included in the subject named "natural sciences"

	Grades								
Periods	7	8	9	10	11	12	13	14	Total
1967–1974	80	80	110	110	110	110	110		710
1975–1986	80	80	80	120	120	120			600
1987–1989	80	80	80	117	117	117			591
1990–1993	80	80	80	110	160	126			636
1994–1999	80	80	80	110	140	104			594
2000–2004		120	120	110	140	104			594
2005–2008		120	120	92	126	172			630
2008–2009		41*	49*	84	138	196			508
2009–2010		41*	49*	82	123	90			485
2010–2011		41*	49*	78	117	58			343
2011–2012		74	111	78	117	58			438

Table A.5 Results of Cuban students in the physics Olympiad in the last decade

	Ibero American Olympics					International Olympics				
Year	Participants	Gold	Silver	Bronze	Mention	Participants	Gold	Silver	Bronze	Mention
2000	4	2	1	1	–	5	–	–	–	1
2001	4	2	2	–	–	5	–	–	–	–
2002	4	3	–	1	–	4	–	1	1	1
2003	4	4	–	–	–	1	–	–	.	1
2004	4	2	1	1	–	1	–	–	1	–
2005	4	1	2	1	–	4	–	–	–	2
2006	4	1	2	1	–	1	–	–	–	–
2007	4	–	1	2	1	1	–	–	–	–
2008	4	1	2	1	–	1	–	–	1	–
2009	4	1	1	1	1	1	–	1	–	–
Total	40	17	12	9	2	24	–	2	3	5

Table A.6 Physicists graduated in Cuban universities

Academic Year	UH	UO	INSTEC	UCLV	Total
2000–2001	22	9	7 LF+5 NE	–	43
2001–2002	9	8	8 LF+8 NE	–	33
2002–2003	20	7	12 LF+9 NE	–	48
2003–2004	9	6	8	–	23

(continued)

Table A.6 (continued)

Academic Year	UH	UO	INSTEC	UCLV	Total
2004–2005	10	5	4	–	19
2005–2006	11	6	8	–	25
2006–2007	4	6	13	–	22
2007–2008	14	5	12	–	31
2008–2009	14	8	12	–	34
2009–2010	14	5	18	–	37
Total	127	65	106 LF+18 NE	–	316
Accumulated	993	442	354	12	1,801

LF *licenciados* in physics, NE nuclear engineers

References

Alamino Ortega, D. 2005. Primeros pasos de las investigaciones en física del estado sólido en Cuba. *Rev. Cub. Fis.* 22(1): 81–88.

Alonso, M. 1958. *Física Atómica*, vol. 1. Havana: Cultural, S.A.

Altshuler, J. 1997. From shortwave and scatter to satellite: Cuba's international communications. In *Beyond the ionosphere: Fifty years of satellite communication*, ed. Andrew J. Butrica, 243–249. Washington, DC: NASA History Office.

Altshuler, J. 2006. *Para una historia de las ciencias físicas y técnicas en Cuba*. Havana: Editorial Científico-Técnica.

Altshuler, J., and K. Serafimov. 1991. *La hora del espacio*. Havana: Editorial Científico-Técnica.

Arias, O. 1997. *La microelectrónica: breve panorámica histórico sobre su desarrollo y estado actual en Cuba*. Unpublished manuscript.

Banco Nacional de Cuba. 2000. La Economía Cubana en el Período Especial: http://www.bc.gov.cu/Anteriores/Otros/economia%20cubana.pdf. Accessed 17 Oct 2013.

Baracca, A. 1999. El despegue de la física en Cuba desde 1959 hasta la década de los setenta. *Revista Española de Física* 13(4): 6–11.

Baracca, A. (ed.). 2005. *History of the development of physics in Cuba: The development of an advanced scientific system in an underdeveloped country*. Berlin: Max Planck Institute for the History of Science. Preprint 302: http://www.mpiwg-berlin.mpg.de/en/resources/preprints.html.

Baracca, A., and L. Méndez. 2001. Cincuenta años de física en la Universidad de Oriente, en Santiago de Cuba. *Rev. Cub. Fis.* 18(2): 146–154.

Baracca, A., V. Fajer, and B. Henríquez. 2001. The development of physics in Cuba during the sixties and seventies of the 20th century: An integrated approach. The formation of an advanced scientific system in an underdeveloped country. In *XXI international congress of History of Science*, México, July 8–14, 2001. HISTNEUR-L: The History of Neuroscience Internet Forum.

Baracca, A., V. Fajer, and B. Henríquez. 2004. El desarrollo de la física en Cuba. *Revista Iberoamericana de Física (Madrid)* 1: 54–61.

Baracca, A., V. Fajer, and C. Rodríguez. 2006. A look at physics in Cuba. *PhysicsToday* 59(9): 42–48.

Boletín. 1959. *Boletín Oficial Universitario 1959*. Havana: University of Havana.

Boletín. 1960. *Boletín Oficial Universitario 1960*. Havana: University of Havana.

Boletín. 1979. *Boletín Oficial Universitario Aug 1979*. Havana: University of Havana.

Cardero, E., M. Marrero, and J. Meitín. 1973. Aplicaciones de los equipos radioisotópicos en instrumentación y control y algunos resultados logrados en este campo en nuestro país. *Control, Cibernética y Automación* 1(2): 8–14.

Castaño, O., M. De Dios, and R. Pérez. 1975. The Beta-parameter and Velicky's approximation in the calculation of the optical properties of GaAs. *Physica Status Solidi* (b) 71: 111; Relaciones de dispersión: discusión del método de Velicky. *Comunicaciones* 17: 65–71.

Castro Díaz-Balart, Fidel. 1990. *Energía nuclear y desarrollo*. Havana: Editorial Ciencias Sociales.

CEDISAC. 1998. *Todo de Cuba*. Madrid: CEDISAC, Prensa Latina [Multimedia encyclopedia].

Cetto, Anna Marie, and H. Vessuri. 1998. *World science report*. Paris: UNESCO/Elsevier.

Clark Arxer, Ismael. 2010. Cuba. In *UNESCO science report 2010*, Chapter 6, 123–1331. http://www.unesco.org/new/en/natural-sciences/science-technology/prospective-studies/unesco-science-report/unesco-science-report-2010. Accessed 17 Oct 2013.

Comisión de Investigaciones. 1963. *La investigación científica: un panorama*. Havana: University of Havana.

Consejo Superior de Universidades. 1962. *La reforma de la enseñanza superior en Cuba*. Havana: Colección Documentos.

Crespo, F., E. Vigil, and D. Waisman. 1968. Sobre los primeros resultados en diodos de Ge obtenidos por aleación. In *Conferencia Química de Oriente,* Santiago de Cuba, February.

De Armas, R., E. Torres-Cuevas, and A. Cairo Ballester. 1984. *Historia de la Universidad de La Habana*, vol. 2. Havana: Editorial Ciencias Sociales.

Díaz, C., and J. Meitin. 1975. Determination of iron content in lateritic ores by radioisotope x-ray fluorescence using a ^{238}Pu source and a scintillation counter. *Revista CNIC, Ciencias Físicas* 6(1): 111–120.

Estrada, O.F. 2007. *Estudio sobre los Factores Determinantes de la Capacidad Nacional Innovativaen América Latina y El Caribe*, Programa de Doctorado en Economía. Havana: Universidad de La Habana/Universidad de Oviedo.

Estrada, M., A. Cerdeira, A. Fornés et al. 1972. Construcción de dispositivos semiconductores por un método de aleación. In *Memoria del 1er Congreso de Técnica Digital*. Havana.

Fernández Bertrán, J. 1962. *Informe sobre la visita a institutos de física italianos*. Santiago de Cuba: Universidad de Oriente.

Gálvez Taupier, L.O. 1986. *Ciencia, tecnología y desarrollo*. Havana: Editorial Científico-Técnica.

Gleijeses, Piero. 2002. *Conflicting missions: Havana, Washington and Africa: 1959–1976*. Chapel Hill: The University of North Carolina.

González, T. 1985. *Discurso por el XX Aniversario del Instituto de Meteorología*. Havana. (Report of the Instituto de Meteorología.)

Hardy, L., F. Martínez, O. Hoyos, and A. Núñez. 2006. Cuban strategy for reproducing, preserving and developing nuclear knowledge. *International Journal of Nuclear Management* 2(1): 31–37.

Hearnshaw, J. and J. Fierro. 2005. *Report by the executive committee of the International Astronomical Union and the President of the 46 IAU Commission on astronomy in the Republic of Cuba*, February 16, 2005. Commission 46 Work Group for the Worldwide Development of Astronomy (PGWWDA).

Historial de Geofísica n.d.. *Instituto de Geofísica y Astronomía: Sintesis del Historial del Centro.* (Document provided by Lourdes Palacios.)

Hurtado De Mendoza, Diego and Ana María Vara. 2007. Winding roads to big science: Experimental physics in Argentina and Brazil. *Science, Technology and Society*: http://sts.sagepub.com/content/12/1/27.

ININTEF. 1982. *ININTEF, XV Aniversario*. Havana: Academia de Ciencias de Cuba.

Jiménez Pozo, M.A., and C. Sánchez Fernández. 1993. Panorama de la Matemática en Cuba. *Revista Ciencias Matemáticas* 14(2–3): 99–123.

Levialdi, A. 1968: *Fotoconducción*. Lecture notes (unfinished, because of the author's passing), November 1968. University of Havana, Physics School.

López, A. 2009. Private communication obtained by Víctor Fajer.

Maggio, G., and L. Desdín. 2000. *Diseño y utilización de sistemas de control nucleónico*, 1st ed. New York: Organismo Internacional de Energía Atómica.

Marrero, M. and J. Meitín. 1976. *Analizador de hierro*. Patente Cubana No. 33625,12.12.76.

Marrero, M., et al. 1969. Instrumento isotópico de nuevo tipo para controlar la densidad y uniformidad de los cigarrillos. *Revista CNIC, Serie B, Ciencias Físicas* 1: 24–33.
Marshall, T.W. 1963. Random electrodynamics. *Proceedings of the Royal Society A* 275: 476–491.
Marshall, T.W. 1965. Statistical electrodynamics. *Proceedings of the Cambridge Philosphical Society* 61: 537–546.
Marshall, T.W. and P. Claverie. 1980. Brownian motion and quasi-Markov processes. *Physica A*, 103: 172–182; 104: 223–232.
Martí, R. and F. Auchet. 1962. *Informe general sobre la Escuela de Física.* Memorandum to the Rector of the University of Havana by the director of the Physics School and the head of the Department of General and Experimental Physics, December 14, 1962.
Martí, R.? 2009, private communication obtained by Hugo Pérez Rojas
Meitín, J., and J. Roig. 1971. Determinación del hierro total en el mineral laterítico por el método roentgenoespectral utilizando una fuente de bremmstrahlung de ^{3}H/Zr. *Revista CNIC, Ciencias Físicas* 2(1): 85–98.
Memoria 1966–67. *Memoria Anuario 1966–67.* Havana: University of Havana.
Memoria 1973–74. *Memoria Anuario 1973–74.* Havana: University of Havana.
Memoria 1976–77. *Memoria Anuario 1976–77.* Havana: University of Havana.
MES. 1997. *Catálogo.* Havana: Ministry of Higher Education.
Monet-Descombey, C. 1963. *Plan sobre una orientación posible del Laboratorio de Electrónica de la Escuela de Física.* Havana: University of Havana (Personal document of José Altshuler).
Montes, V. 1963. *Planificación y tendencias de la investigación: documento para la Comisión de Investigaciones.* Havana: University of Havana.
Morán-López, José Luis. 2000. Physics in Latin America Comes of Age. *Physics Today*, October, 38.
Ortiz, H. 1987. *Reseña Histórica de la Meteorología en Cuba. Conferencias y Estudios de Historia y Organización de la Ciencia, nº 53.* Havana: Academia de Ciencias de Cuba.
Pérez, M., M.E. Hernández-Diaz, and O. Diaz. 2011. Experience in medical physics and biomedical engineering for radiation in Cuba. In *Medical physics and engineering education and training, part 1*, ed. S. Tabakov et al., 185–194. Available at http://emerald2.eu/mep/e-book11/ETC_BOOK_2011_ebook_s.pdf. Accessed 18 Oct 2013.
Pérez Doval, J. 1991. *Datos Astronómicos para Cuba*, 29–33. Editorial: Instituto de Geofísica y Astronomía.
Pérez Rojas, H., D. Stolik, J. Fuentes, C. Rodríguez, A. D'Costa, R. Álvarez, B. Lazo, E. García, J. González, M. Ramos, and O. Arias. 1976. Estado actual de las Ciencias Físicas en Cuba. In *Las Ciencias Básicas: Examen preliminar de su situación actual en Cuba y a nivel mundial*, 35–61. Havana: Natural and Exact Sciences Section, National Council for Science and Technology.
Rodríguez Castellanos, Carlos. 1997. Universidad de La Habana: investigación científica y período especial. *Revista Cubana de Educación Superior* 18(3): 13–36.
Rodríguez Castellanos, Carlos. 1984–5. *La física teórica en Cuba.* From a report on physics in Cuba, prepared by the Cuban Physical Society.
Sáenz, Tirso, and Emilio García Capote. 1989. *Ciencia y Tecnología en Cuba.* Havana: Editorial de Ciencias Sociales.
Soto, H. 2000. Roberto Joaquín Soto del Rey (Biography). *Energía y Tú* 9: 36–37.
Tratado. 1962. Tratado sobre colaboración amistosa entre la Universidad de Oriente de Santiago de Cuba y la Universidad Técnica de Dresden.
UNDP. 1979. Project Introducción de las Técnicas Nucleares a la Economía Nacional. CUB/77/001/C/01/18, January 1979.
UNDP. 2003. *Human development report 2003. Millennium development goals: A compact among nations to end human poverty.* Oxford/New York: Oxford University Press.
Universidad. 1972. *Universidad de La Habana*, N. 196–197, 2–3, 377. Annual Report of the University of Havana.
Universidad de La Habana. 1963. La carrera de Ingeniero Geofísico. In *Riquezas que no vemos*, 20–24. Havana: University of Havana.

Universidad de Oriente. n.d. *Plan para la formación de físicos en la Universidad de Oriente*. (Probably written at the end of 1967 or the beginning of 1968).

Veltfort, T. 1998. *The beginning of semiconductor research in Cuba, a personal recollection*. Contribution to the second Congress of the Cuban Society on the History of Science and Technology, Havana, February 25–27, 1998. (Reproduced in this volume.)

Vida Universitaria 1968. N. 209. La Habana

Vida Universitaria, 1970, N. 222, 13.

Waisman, D. 1969. Andrea Levialdi (Obituary). *Vida Universitaria* 215: 40. (Reproduced in this volume.)

Chapter 7
Accomplishments in Cuban Physics (up to 1995)

Carlos R. Handy and Carlos Trallero-Giner

7.1 Introduction

In late October 1995, one of the authors (CRH) paid a personal visit to Cuba. While there, he took advantage of a longstanding invitation to present a research seminar to the Cuban Physics Society. With respect to the development of physics in Cuba, curiosity led to subsequent visits to various centers including the University of Havana (UH), the Pedagogical Institute, the Institute for Nuclear Science and Technology (INST), the Institute of Cybernetics, Mathematics, and Physics (ICIMAF), the Polytechnic Institute (ISPJAE), the Neuroscience Center, and the National Center for Scientific Research (CNIC).

Although Cuba's economy was on a positive swing compared to its bleakest period several years earlier, a first "American impression" was that no credible research could thrive in the midst of occasional electrical blackouts, gasoline and food rationing, little or no access to international journals, meager Internet access, unavailability of good PCs (and certainly no workstations), little support for scientific travel, and monthly academic salaries of \$20–\$30 (despite low-cost housing, free medicine and education). The truth is that although some excellent Cuban scientists had left the country for political or scientific reasons (i.e. lack of proper equipment to further their work), there still remains a first rate cadre of serious and dedicated scientists who publish in the top international physics journals (including *Physical Review Letters*), are welcomed in Europe, and who are in demand in Latin America. Indeed, as noted by Dr. F. García-Moliner, winner of the prestigious science prize in Spain (Premios Principe d'Asturias), member of the condensed matter group at the

C.R. Handy (✉)
Professor and Chair of the Physics Department, Texas Southern University, Houston, TX, USA
e-mail: handycr@TSU.EDU

C. Trallero-Giner
Department of Theoretical Physics, University of Havana, Havana, Cuba

A. Baracca et al. (eds.), *The History of Physics in Cuba*, Boston Studies in the Philosophy and History of Science 304, DOI 10.1007/978-94-017-8041-4_7,

International Center for Theoretical Physics (ICTP), and former Associate Professor (1961–1964) in the Physics Faculty at the University of Illinois:

> [...] twelve years ago I opened the door to the ICTP for Cuban physicists; they then took over and have impressed everyone with their high degree of competency and drive. They have one of the highest per capita numbers of Associate Memberships within the ICTP, and are eagerly welcomed. Never does politics enter into the professional interactions. Within Latin America, the Cuban semiconductor theoretical group is second only to that of the centre of excellence in Campinas, Brazil. American Nobel Prize winners have expressed strong interest in working with Cuban researchers.

The initial shock of seeing crumbling old buildings and riding on rusted out Mockovich and Lada Russian cars gave way to the recognition that despite Cuba's economic hardships, the people maintain their vitality and ingenuity. The ongoing embargo has forced Cubans to become self-reliant and develop the scientific and technical 'know-how' for addressing important problems to the society. The greatest irony is that Cuba has attained a remarkable level of scientific and technical independence that many other developing countries, politically favored by the United States, have yet to achieve, despite the best intentions and financial support of the US.

An important catalyst in Cuba's evolving physics capability has been the strong tradition for excellence in medicine. This has given added impetus and focus to the development of Cuban physics, emphasizing the need for diagnostic equipment and instrumentation, etc. In the course of pursuing this critical need, Cuban physics has simultaneously evolved to a level where all economically feasible forefront research areas are represented.

Prior to the 'Revolution' in 1959, there was no significant physics research in Cuba. Thirty-seven years after the Revolution, there is a highly evolved network of (at least) thirty-seven basic and applied research laboratories, all pursuing varying levels of forefront research problems. Some of these have demonstrated stunning achievements, such as the Medical Physics Laboratory, directed by Dr. C. Cabal, at the University of Oriente (UO) in Santiago de Cuba, one of the easternmost provinces. There, two tomographs were built and one of them is already being used for medical purposes.

Another physics achievement is the high level of expertise in ceramics, which enabled Cuba to synthesize high temperature YBCO superconductors in June of 1987 (the fourth country to do so), immediately after the publication of Chu's work. A third example is the neurophysics group within CNIC. Because of the ongoing economic blockade, and the prohibition on buying hardware and software with American made components, Cuban researchers built all the electronic hardware and software to visualize the brain. Indeed, this group is recognized as one of the first neurophysics groups in the world, leading in the stochastic analysis of nonlinear neural interactions. Representatives from this group were invited speakers at the 1995 Biomagnetism Conference in Santa Fe, New Mexico. American biophysicists have expressed a strong desire to recruit some of the Cuban graduate students for PhD/Postdoc training in the U.S.

In addition to the above, other important physics groups, laboratories and departments include those at the University of Havana (UH), involving 80 members of the Physics Faculty, specializing in optoelectronics, crystal growth, laser technology, solar cells, superconductivity, magnetism, ceramics, structural analysis, and theoretical solid state physics.

The Medical Physics group resides at University of Oriente, as previously noted. Coherent optics and microelectronics research is pursued at ISPJAE. The development of scientific instrumentation and laser technology is the responsibility of the Center for the Development of Scientific Equipment and Instruments (CEDEIC). Geophysics, seismology, and astronomy fall within the Institute for Geophysics and Astronomy (IGA). Additional areas are acoustics (ICIMAF), solar energy (IES), meteorology (IM), nuclear analysis (INST), nuclear techniques (CEADEN), nuclear medicine (INOR), radiometry (CNIC), nuclear reactors (NTC), metallurgy (SIME), metrology (CENMCC), electronic devices (CEE), telecommunications (LACETEL), analysis of minerals (MINBAS), and other industrial, academic, and governmental agencies (i.e. Ministry of Science, Technology, and the Environment).

Examples of the capabilities of these groups are the development and characterization of sensors, photo-detectors and electronic instrumentation for use in industry and/or medicine. Specifically, the CEDEIC produces advanced polarimeter lasers of relevance to the sugar industry, agriculture, and pharmaceutics (one of Cuba's more mature industries). In laser technology, the laboratory at UH designs and constructs neodymium YAG lasers for use in endoscopic and ophthalmological surgery.

One of the best multidisciplinary examples is the Medical Physics Institute at UO where NMR tomographs have been designed, built at a local factory, and put to medical use. All of this has been done under the "poorest working conditions," as noted by Dr. F. Garcia-Moliner. In addition they have built vibrational and torsional magnetometers, as well as spectroflurometers. Construction of these devices represents a considerable savings to the national economy and, where appropriate, resulted in increased medical resources to the general population.

The Acoustics Department within the ICIMAF, directed by Dr. G. Gonzalez, has manufactured Doppler detectors for use in analyzing moving structures, acoustic-driven sprays (nebulizers) for administering small medical dosages, ultrasound therapeutic equipment, etc. Piezoelectric materials for these devices are produced in the Ceramics Laboratory, under the direction of Dr. L. Fuentes. Their research is both basic and applied, for example, through their work on the optimization of electro-mechanical conversion processes, they have clarified the role of so-called ghost textures on piezoelectric phenomena.

The Ferroelectric Materials Group, directed by Dr. J. Portelles, in existence since 1983, studies ferroelectric, piezoelectric, and magnetostrictive materials. This research is directed towards the development of high efficiency piezoelectric energy conversion devices, as well as achieving low dielectric loss in ultrasound generation. Also, advanced materials research is pursued in the development of pyroelectric materials for infrared radiation detection, dielectric relaxation and diffuse phase transition, and thick films for Sol-Gel and laser ablation techniques.

An advanced technical capability is the Molecular Beam Epitaxy Laboratory directed by Dr. J. Fuentes. His group has developed nanostructures in III–V semiconductor compounds. In addition, Dr. P. Diaz's group conducts fundamental research on semiconductor lasers.

Another very important advanced technical capability is the Superconductivity Laboratory at UH, made up of very young (average age 30) researchers. They dedicated their first efforts to the fabrication of bulk ceramic RF SQUIDs soon after the first report in the scientific literature, and started to work on the transport properties of HT_c ceramics, leading to the first Cuban publications on the subject in 1990.

Besides the processing of HT_c ceramics and, later on, of laser ablated films, an important part of the work has concentrated in the study of the transport properties of ceramic superconductors, critical state models, relaxation phenomena, Mossbauer effect in HT_c s, etc. Most of the approximately 30 scientific papers published by this group, in international journals, involve a totally Cuban collaboration with very limited resources.

7.2 The PhD Program

Compared to many Central and South American countries, where no indigenous PhD physics programs exist, Cuba is more advanced. Out of a population of 11 million, approximately 160 hold a PhD in physics, and 75 % of these were earned at Cuban institutions. Approximately 1,500 individuals are *licenciados*, having completed a 5-year university undergraduate study.

In many developing countries, the low cost of maintaining a theoretical research thrust results in the total absence, or minimal presence, of an experimental capability vital to that society's growth. In Cuba, for every theorist there are at least ten experimentalists. Clearly, this is motivated by the extreme urgency in solving problems directly related to the national economy. For this reason as well, from 1985 through 1992, approximately, relatively little emphasis was placed on producing Cuban PhDs and peer-reviewed publications. The emphasis was on quickly producing a technically competent manpower base that could address the country's problems.

In the late 1980s and early 1990s, with the dissolution of the eastern European socialist block, Cuban scientists found that demand for them in Europe and other traditionally non-socialist countries depended on evidence of advanced training at PhD level and peer-reviewed publications. More specifically, with the end of the eastern socialists camp, Cuban scientists realized that to make themselves more attractive to outside collaborations and professional enhancement, they needed to make themselves more competitive. Since the early 1990s there has been a pronounced growth, year by year, in the production of PhDs and in the publication in top peer-reviewed international journals.

Many Cuban researchers and students collaborate or study in other countries in Latin America and Europe. From the American perspective of professional minority role models, the large black constituency within the Cuban population (approximately

61 % of the population has some African heritage) is well represented within the physics profession and student body. By comparison, the black and Hispanic American population is approximately 50 million. From 1973 through 1993, PhDs were awarded to 402 black/Hispanic US citizens. During the same time about 160 Cuban PhDs were produced, in a population one fifth that of the minority US population.

7.3 History of Cuban Physics

Before 1959, physics was taught at university level, but no research was done. There was a weak program in mathematical physics. In 1961–1962 the School of Physics was established, taught by Cuban professors with many good intentions but limited resources. The curriculum was modeled after the United States. During the same time, many other students studied in the eastern European countries as well. In 1966/1967, with the arrival of the new east-European trained Cuban graduates, there was a reorganization within the School of Physics transforming it into a more rigorous program with both Cuban and foreign visiting professors. Many of these non-PhD professors (with masters degrees) returned to Europe to complete their PhD studies. The first PhDs returned to Cuba in 1972–1973, with additional increases in 1976–1978. This period marks the beginning of Cuba's advanced physics capability. From this time until 1985, there was a steady flow of M.S. and PhD students to many European countries. The first Cuban-trained PhDs were produced in 1982. As indicated previously, with the dissolution of the eastern socialist sector, since 1992 there has been an increase in the production of PhDs.

7.4 Physics Education at the Pre-university Level

Cuban children from 8th through 12th grade are taught physics for at least 4 h per week. Topics in mechanics (grade 8) and electricity, acoustics, and optics (grade 9) are taught qualitatively, with little emphasis on integrating advanced mathematics. These topics are reintroduced at a higher level (i.e. incorporating algebra, trigonometry, and calculus) in grades 10–12. Specifically, mechanics is taught in grade 10 (oscillations and waves), molecular physics and electromagnetism in grade 11, and electromagnetic waves and atomic/nuclear physics in grade 12.

The above curriculum is for the standard high school. There are special science high schools ("Vocational High Schools") with about 15,000 of the best students (competitively selected at grade 9). At these schools, there is a greater level of training in science and mathematics. These schools exist in each of the provinces.

The excellence of the physics program in Cuba is further evidenced by its representation within the International Physics Olympiad (IPO), in which Cuba has regularly participated since 1983. By way of comparison, the only other Latin

American countries to participate have been Argentina (since 1994), Colombia (1988), and Mexico (1993).

With respects to Mexico's participation, Cuba helped to train their teams. IPO Cuban students are recruited from the special science high schools.

7.5 Undergraduate Physics

Students continuing in physics at university level may attend either UH, UO, the Pedagogical Institute (eleven campuses for training pre-university teachers), or INST. There are many other academic institutions in engineering, etc. (some with larger physics departments than those in the University of Havana) but the aforementioned four campuses are the only ones where degrees in physics are awarded. Entrance is based on general examinations and high school records.

The pure physics curricula at UH, UO, and INST are very rigorous. The undergraduate course of study lasts 5 years and students graduate with a *licenciatura* degree that involves thesis work. By the second year, an undergraduate physics student is expected to be actively working in research, either in an academic laboratory or industrial setting.

The student is placed on the basis of his/her interest and the available positions. The student pursues this research collaboration for 4 years, using it as the basis for his/her thesis. The minimum number of hours involved in research is 1,522 during the 4 years. No stipend is provided for this. The maximum GPA is five. Students advance from 1 year to the next only if they have earned three or more credits in each course. Students not performing at this level are allowed to repeat courses.

7.6 Advanced Studies

Upon completing the *licenciatura* degree program a student may elect to pursue a Masters Degree involving demonstrated independent thesis research and at least one international publication, or a PhD. The former may last up to 2 years. The PhD has a maximum time limit of 4 years. There are various ways of obtaining the PhD. The most basic is by directly going into such an academic program. Alternatively, someone working at an institute or in industry, after having completed various significant research projects, may seek the PhD by defending his/her research results. In each of these two routes towards the PhD the student must pass basic exams, have the research topic approved, pass a pre-defense process within a specified physics department (anywhere in the country), and pass the final defense in front of a national PhD committee.

Out of the 5,958 h of university study, the following list itemizes their distribution amongst the various physics and mathematics topics: general physics (906); mathematics (1050); theoretical physics (440); computation (124); electronics

(252); chemistry (100); thermodynamics (60); solid-state physics (86); experimental methods (186); history of physics (50). In addition, there are 144 h of electives in physics and mathematics, plus the 1,522 h of research.

The standard textbooks are a combination of Cuban, American, and Russian texts. The last two categories of texts are available in English. Some of these are the Halliday and Resnick text on *General Physics* (1992), Zemansky's *Thermodynamics and Heat* (1990), Landsberg's *Optics* (1976), Eisberg's *Modern Physics* (2000), Symon's *Mechanics* (1970), Goldstein's *Classical Mechanics* (1964), Jackson's *Classical Eletrodynamics* (1966), Davidov's *Quantum Mechanics* (1965), Landau's *Continuous Media Electrodynamics* (1987), Reif's *Statistical Physics/ Thermodynamics* (1968), Kittel's *Solid State Physics* (1971), Ashcroft and Mermin's *Solid State Physics* (1976), Millman's *Microelectronics* (1993), and others.

Because of the needs created by the embargo and the particular academic/research evolution of the Physics Faculty, many of the textbooks are by Cuban authors. Some of these are M. Ramos and R. Perez' *Theoretical Mechanics* (1982), R. Portuondo's *Introduction to General Physics* (1990), and R. Portuondo and M. Perez' *Mechanics* (1994), A. Martell and L. Fuentes's *Electromagnetism* (1989), R. Reguera's *Analytic Geometry* (2008), E. Puron's *Linear Algebra* (1987), C. Sanchez and C. Valdés' *Mathematical Analysis* I–II (1982/1983), J. Marín's *Complex Variable Functions* (1990). All of these books are lent to the students.

The "postgraduate studies" phase (i.e. Masters and PhDs) encounter the difficulties created by the economic situation, such as lack of current journals and modern texts. To this extent, students heavily rely on course notes and make use of the limited number of reference textbooks available in the library. Some of the advanced physics courses are: physics of magnetic phenomena; magnetic materials; material science; phase transformations; Mossbauer spectroscopy; X-ray diffraction; electronic microscopy; superconductivity; microcomputers in experimental research; photovoltaic phenomena; semiconductor devices; lasers, theory and application; Fourier optics; advanced computing; physics teaching; optical communication systems; solid-state theory; semiconductor quantum wells and super-lattices; quantum statistics.

The physics curriculum within the INST is similar to that at the University of Havana. Additional courses may be found in such topics as: nuclear theory, neutron spectroscopy and nuclear structure, nuclear physics, methods for materials analysis, and radiation detectors and measurements.

7.7 Physics Employment and Attrition Rate

In general, almost all physics graduates find physics-related jobs. Presently, approximately 60–70 first year university physics students enter the physics program at UH, UO, and INST. Before 1985, this number was in excess of 200; however only 30 % of the entering students finished the 5-year university program. During the early 1990s the selection criteria was raised, resulting in 100–150 entering physics students. Of these, approximately 50–60 % graduated. With the dissolution of the communist block and its negative impact on the Cuban economy, this number has dropped to the present level of 60–70. During the last 3 years, on average 33 graduates at the *licenciatura* level have been produced at UH. Of these, a growing number (19 last year) have continued with positions at degree granting research oriented institutions. A lesser number have gone to purely research institutes.

Only students with a GPA greater than four are allowed to take research-related jobs. The vast majority of graduates perform at this level. Approximately eight physics PhDs are produced, annually, throughout the country.

7.8 Interplay of Medicine and Physics

As previously noted, Cuba's excellent medical tradition has provided partial incentive to the growth and maturity of Cuban physics. Two exemplary areas are neurophysics and nuclear physics. We provide a brief overview of each.

7.8.1 Neurophysics

In the early 1970s Cuba designed and produced it's own computers. One of their earlier applications was to analyze the electroencephalogram in an attempt to provide a quantitative and objective evaluation of brain function. Successful field trials in the 1980s led to the establishment of a National Network of 42 totally computerized neurophysiology laboratories providing extensive health care, screening programs, and research. Amongst these, the CNIC carries out basic research in neurophysics, in addition to other research activities.

The neurophysics group within CNIC works on the solution of spatial and dynamic inverse problems involving the generation of tomographic images from the electrical and/or magnetic fields of the brain (i.e. determining neural currents from E&M field measurements on the external surface of the head). Their research into multimodal image tomography promises improved instruments for measuring normal and abnormal brain function.

In addition to the above, the neurophysics group also studies the effect of stochastic properties on nonlinear neural dynamics. Their methods are relevant in the evaluation of models for epileptic seizures.

7.8.2 Nuclear Physics

The first nuclear physics research in Cuba was in the fields of radiotherapy and nuclear medicine, in the early 1960s, at the Institute of Oncology and Radiobiology. In the beginning, the dosimetry control for X-ray units and other radiological services were performed there.

The first nuclear physics research at UH involved gamma ray spectroscopy and the construction of a Mossbauer spectrometer. A second Mossbauer spectrometer was acquired through a grant from the International Atomic Energy Agency (IAEA), enabling the study of lateritic minerals (containing Ni, Co and Fe components).

In 1969 the Nuclear Physics Institute (NPI) of the Academy of Sciences was founded. Approximately 50 physicists, with an equal number of chemists, electronics engineers and technical workers, were involved. The NPI had some experimental physics facilities, such as a subcritical reactor for neutron studies and a neutron generator for activation analysis with fast neutrons. In the NPI there were gamma spectrometric and dosimetric research, as well as condensed matter studies.

When the Cuban government decided to install a nuclear power station, a large program for educating nuclear professionals began simultaneously. This program started in 1979 with the selection of many of the best students from the "Vocational High Schools."

In 1980 the Faculty of Nuclear Sciences and Technology (FNST), within UH, was established. Therein also resided the radiochemistry and nuclear engineering groups.

The subcritical reactor of the NPI was moved to the FNST for training nuclear physicists and engineers. In addition, a second neutron generator was installed by an IAEA grant. Simultaneously, many students went to the Soviet Union to study nuclear physics and other nuclear specialties. After some years, the FNST separated from UH becoming the INST.

Several other nuclear institutions were founded as well. Since then we have the Center of Applied Studies for Nuclear Development, the Center for Radiation Protection, the Isotopes Center, the Center for Nuclear Technology, and the Radiation Safety Center. The latter is responsible for coordinating all adherences to nuclear safety regulations and guidelines.

The M.S. and PhD studies in nuclear physics began with the founding of the above institutions. As a result, many nuclear physicists obtained their PhDs in Cuba. Presently, nuclear physicists pursue academic duties, as well as theoretical research on nuclear structure and reactions, development and application of analytical techniques for new materials, geological sampling, environmental studies, and methods for non-destructive analyses. In addition, many nuclear professionals work in nuclear medicine at hospitals throughout the country. For example, there are presently about seven gamma-ray cameras throughout Cuba through which nuclear physicists perform single photon emission computed tomography (SPECT) and gamma transmission tomography diagnostic studies. At the various nuclear centers one can find research on laser spectrometry, nuclear detectors and instruments, materials science, radiation and environmental protection, reactor physics, etc.

There are several nuclear institutions for radiation security in the country, particularly in Cienfuegos, where a nuclear power station is being built. In order to improve the utilization of the anticipated nuclear power stations, the NTC develops strategies for nuclear core configurations in PWR and BWR reactors, including the use of advanced type fuels (i.e. *deep burn* types). A large part of the training and equipment for the aforementioned nuclear activities are obtained through the IAEA, which contributes about one million dollars per year in Cuban manpower training, projects, etc.

7.9 Effects of the Embargo on Scientific Progress

In the preceding narrative we have indicated many of the impressive achievements of Cuban Physics, despite the difficulties created by the embargo and the concomitant economical difficulties. Paramount amongst the latter is the prohibition on buying scientific equipment with American components.

Despite these considerable obstacles and notwithstanding the personal deprivations endured by all Cuban physicists, it is clear that within the context of Cuba's very limited resources, the progress of physics in Cuba has been stunning. Certainly a major motivation has been, for the society as a whole, survival. However, it would have been very easy for the Cuban intellect to surrender to the situation and sink to a quasi primitive subsistence culture. Instead, undaunted, the Cuban physicist has faced these challenges with exemplary intellectual fervor, focus, and professionalism, serving as a model to other developing countries as well as minority groups in the US. Beyond the political differences, one of the outcomes of Cubans having to build their technical and scientific infrastructure is an appreciation for the robustness and capability of the American industrial enterprise. From the other perspective, those Americans acquainted with Cuban society's evolution during the last four decades cannot help but admire their "pick yourself up by the boot-straps" attitude.

Only during the last 5 years has the American Physical Society (APS) permitted Cuban physicists to become members. There are now seven Cuban members of the APS, including one of the co-authors. These professional advances have been championed by Dr. Manuel Cardona, former member of the APS Committee on the International Freedom of Scientists (CIFS), and a founding director of the Max Planck Institute for Solid State Research in Germany.

This is a welcomed opportunity. The APS has, for many years, embraced and supported the professional right of physicists to pursue knowledge for its own sake. Part of this credo rests on the belief that the advance of scientific knowledge can come from any corner of the world, regardless of the local political environment. Given the plight of physicists all over the world, with shrinking resources, dwindling job opportunities, and an increasingly disinterested student population, the preservation of the profession depends on our defending and encouraging it wherever possible. Normal relations between the US and Cuba may not occur for some time, but perhaps with another generation. It would be a tremendous tragedy for the human intellect, particularly in the present physics context, if by turning our backs on the professional plight of Cuban physicists they might disappear. Certainly it is in everyone's long-term interest that a scientifically and technically competent workforce continues to exist in Cuba.

References

Ashcroft, Neil W., and N. David Mermin. 1976. *Solid state physics*. New York: Sounders College Publishing.

Davidov, A.S. 1965. *Quantum mechanics*. Havana: Instituto del Libro.

Eisberg, R., and R. Resnick. 2000. *Física Cuántica: átomos, moléculas, sólidos, núcleos y partículas*. Madrid: Limusa Grupo Noriega Editors.
Goldstein, Herbert. 1964. *Classical mechanics*. Cambridge, Mass.: Addison-Wesley Publishing Company.
Jackson, John David. 1966. *Classical electrodynamics*. New York: Wiley.
Kittel, Charles. 1971. *Introduction to solid state physics*. Havana: Instituto del Libro.
Landau, L.D., and E.M. Lifshitz. 1987. *The classical theory of fields*. Amsterdam: Butterworth-Heinemann.
Landsberg, G. 1976. *Optics*. Moscow: MIR.
Marín Antuña, J. 1990. *Teoría de funciones de variable compleja*. Havana: Editorial Pueblo y Educación.
Martell Arbelo, A., and L. Fuentes Cobas. 1989. *Electromagnetismo*. Santiago de Cuba: Editorial Ministerio de Educación Superior.
Millman, Jacob, and Arvin Grabel. 1993. *Microelectronic*. Madrid: Editorial Hispano-Europea S.A.
Portuondo Duany, R. 1990. *Introducción a la física general*. Havana: Editora Editorial Ministerio de Educación Superior.
Portuondo Duany, R., and M. Pérez Quintana. 1994. *Mecánica, Teórica*. Havana: Editorial Pueblo y Educación.
Purón Sopena, E. 1987. *Algebra lineal*. Havana: Universidad de La Habana.
Ramos Vázquez, M., and R. Pérez Álvarez. 1982. *Mecánica teórica*. Santiago de Cuba: Editorial Ministerio de Educación Superior.
Reguera Vilar, R., and M. Solana Sagarduy. 2008. *Havana: Geometría analítica*. Havana: Editorial F. Varela.
Reif, F. 1968. *Fundamentos de de física estadística y térmica*. Madrid: Ediciones del Castillo.
Resnick, R., and D. Halliday. 1992. *General physics I and II*. New York: Wiley.
Sánchez Fernández, C. 1982. *Análisis matemático T. I.* Havana: Editorial Pueblo y Educación; C. Valdés Castro. 1983. *Análisis matemático T. II.* 1983. Havana: Editorial Pueblo y Educación.
Symon, K.R. 1970. *Mecánica*. Madrid: Aguilar, S.A. de Impresiones.
Zemansky, H.W., and R.H. Dittman. 1990. *Thermodynamics and heat*. New York: McGraw-Hill.

Chapter 8
Physics at the University of Oriente

Luis M. Méndez Pérez and Carlos A. Cabal Mirabal

8.1 Introduction

This chapter will complement the information given in the previous chapter by Baracca, Fajer and Ródriguez on the development of physics in Cuba from 1959 to the present. Attention will focus here on the onset and development of physics teaching and research carried out at the University of Oriente, located in Santiago de Cuba, generally regarded as the second most important city of the country from an economical, social and cultural point of view. Its Physics School was formally established in 1970, replacing the physics chair that has existed since the founding of the university in 1947.

8.2 Initial Steps (1947–1961)

The inauguration of the University of Oriente on 10 October 1947 was reported in the Cuban press as "a transcendental event in the history of Cuban education [...] with a most promising start [...]" (*Diario de Cuba* 1947). Its foundation took place after countless efforts over two decades by various independent social organizations, made up of professionals, traders and industrialists with a long-term vision for development, in particular the Society for Higher Education in Oriente and the

L.M. Méndez Pérez (✉)
Departamento de Fisica, Facultad de Ciencias Naturales, Universidad de Oriente,
Santiago de Cuba, Cuba
e-mail: lmendezp@cnt.uo.edu.cu

C.A. Cabal Mirabal
Head of Images Group, Chemical Physics Division,
Center for Genetic Engineering and Biotechnology, Havana, Cuba
e-mail: carlos.cabal@cigb.edu.cu

A. Baracca et al. (eds.), *The History of Physics in Cuba*, Boston Studies in the Philosophy and History of Science 304, DOI 10.1007/978-94-017-8041-4_8,

Executive Council for the University of Oriente, one of whose most active members was the prestigious physics teacher, Roberto Soto del Rey (1913–1995). The establishment of the new institution was especially welcomed by the lower income middle class families of the former province of Oriente, since it gave their offspring the chance to access higher education, something that was beyond their means when their sole option before 1947 was the University of Havana. The local socioeconomic and historical context as well as the popular backing and energy of the new educational center added to the fact that its original staff included several well-known Spanish intellectuals exiled in Cuba after their involvement in the Spanish Civil War on the Republican side made the University of Oriente a popular and progressive institution.

It may be said that upon its foundation, the university incorporated the best of the national pedagogical tradition, as expressed by the eminent Spanish-born professor, Francisco Prat, in an interview with a local newspaper (*Sierra Maestra* 1982). According to him, "The University of Oriente was born with the purpose of renewing teaching in Cuba."

From the very beginning, the Engineering Faculty at the university offered a degree course in industrial chemical engineering, which included higher physics in its study plan for its first course. This subject was taught by Roberto Soto del Rey, who became the founding head of the first physics chair (Consejo Directivo de la UO 1947).

The working style of the university's newly created physics chair echoed the basic principles put forward by the best of the Cuban pedagogical tradition developed between the final decade of the eighteenth century and the first third of the nineteenth century. It was inspired by the Cuban philosopher and teacher, Father José Agustín Caballero, who fought for "the teaching in Cuba of Copernican Physics and that of the Englishman Newton; he wanted experimental and applied physics as a support for young people to help national development" (Augustín Caballero et al. 1944). Inspiration came also from Caballero's disciple, priest and philosopher Félix Varela, whose confidence in the principle that "knowledge can be arrived at from experience and reason, led him to inaugurate [in Cuba] the teaching of physics and chemistry through laboratory experiments" (*Granma* 1997). All this supports the assertion by Spanish-born professor Francisco Prat that "the University of Oriente was born with the purpose of renewing teaching in Cuba" (*Sierra Maestra* 1982).

Soto del Rey was a faithful follower of the above ideas, especially those propounded by Varela about "experience and reason," and had always been a devout follower of Cartesian rationalism and Galilean experimentalism. At the university, he pioneered activities aimed at establishing international contacts with prestigious scientific and academic institutions. He was the first professor at the university to benefit from sabbatical years in the early 1950s,[1] which he spent at the Sorbonne

[1]According to a letter to Soto del Rey, dated 26 February 1952 and signed by Ernesto Pujals Fernández, Secretary General of the University of Oriente, the University Board had decided on 30 January "to send two professors abroad every year to carry out special graduate studies" and chose him to be "the first one to enjoy the benefit of this decision."

and other European centers of higher education in search of a deeper understanding of relativity and quantum theories, statistical physics and tensor calculus. At the same time, he managed to have close social contact with local professionals, businessmen and industrialists who kept him informed of current technological and economical developments. He also profited from having been a student of Manuel Gran, his outstanding physics teacher at the University of Havana where he had done a degree course in physical-mathematical sciences from 1934 to 1939.

Soon after the creation of the university, Soto del Rey began to set up the first undergraduate physics laboratories, while in other university areas a well-stocked mechanical workshop and pilot plant were set up to help provide the students with solid practical training. According to Prat, they would "try to make the university a technological center dedicated to applied sciences, since right from the beginning emphasis was placed on chemical engineering studies, with the set up of the laboratories as soon as a budget was available" (*Sierra Maestra* 1982).

In 1951, a project was drawn up for the organization of the School of Sciences and the degree courses in the natural sciences and chemical physics within the Faculty of Sciences and Engineering. In 1956, the School of Sciences was changed into the Faculty of Natural Sciences which offered, among other degree courses, one in physical-mathematical sciences and another in chemical-physical sciences. These attracted very few students, to the extent that the only student to enroll in the former, transferred to the latter in her third course.

As far as the undergraduate courses in engineering and the sciences were concerned, this was a stage of definition and establishment of the institution's characteristic pedagogical approach. The sciences were given an applied profile, to the extent that the idea arose to offer a degree course in industrial physics, similar to the one on industrial chemistry—including industrial research—that already existed at the university. While this idea did not come to fruition, the movement—led by Soto del Rey—to avoid privatization of the university did succeed.[2]

8.3 The Period 1961–1967

Soon after the Cuban Revolution in 1959, Soto del Rey became active in implementing an in-depth curricular and organizational reform of the university, firstly as a member of the Statutory Assembly that was created at his educational center, and later as a representative of the university in the National Council of Universities created by the new government. During the early 1960s, he exchanged ideas and proposals with his Italian counterparts, and at the same time initiated official contact with Soviet educational authorities to promote collaboration for the development of

[2] Soto del Rey resigned from the staff and the Board of Directors of the university in protest of the approval of a proposal to privatize the university presented by the Rector to the Board on February 1, 1948. His tough critical stance strengthened the strong movement against privatizing, which was finally defeated.

national scientific activities. He visited the Soviet Union, Poland, Bulgaria and other Eastern European countries with the goal to obtain technical and scientific support from their higher education institutions (Soto 2000, 33).

Needless to say, an academic entity begins to achieve its characteristic personality when it is not absolutely dependent on externally generated knowledge and its carriers (curriculums, textbooks, equipment, etc.), and arrives at a stage in which it can generate these so that it can contribute, however modestly, to the preservation and generation of culture. In line with this ideal, a textbook entitled "Surface Tension and Liquid Solutions," was produced by Soto del Rey and Luis Aguilar Salcedo. "With this work—they stated in their prologue—the Physics Department begins publication of the physics course taught to engineering students at the Universidad de Oriente" (Soto del Rey and Aguilar Salcedo 1961, 1). It was followed by successive volumes on mechanics, hydrodynamics and wave motion.

The above-mentioned work bore witness to the particular personality of the university's Physics Department. They were purposely written so that the exposition of a new subject started with short paragraphs, apparently isolated from one another. These were followed by higher level texts that finally became book chapters where the most general approaches were dealt with. The style of writing was terse, giving just the information required for a clear grasp of fundamental concepts. This was characteristic of Soto del Rey's teaching style, which combined rationality with synthesis and began with a description of his many experiences whose results were generalized to arrive at theoretical conclusions through rigorous mathematics, while always striving to point out any practical applications derived from general principles that came up in the process.

At the time, the staff of the Physics Department was strongly linked to the teaching of high school physics, despite the requirements of their regular duties at the university. Soto del Rey and Aguilar were active physics teachers at the Santiago de Cuba secondary education institute, which enabled them to exert considerable influence on the rest of its teaching staff and on the mass of its high school students.

Soto del Rey had a strong belief in the motivational and cultural impact of the history of science on the training of students. A permanent exhibition of old instruments and equipment on the premises of the Physics Department and a mini portrait gallery of famous physicists along the corridor leading to the laboratories, helped Soto del Rey to create an intellectual atmosphere that was attractive to both students and visitors.

At this stage, sight was not lost of the understanding that scientific research combined with international academic exchange were essential to the progress of higher educational institutions. Accordingly, in May 1962, the Dean of the Faculty of Sciences, José Fernández Bertrán, paid a visit to East Germany and signed a collaboration agreement with Dresden Technical University (Treaty 1962). He also reported the results of his search for further collaboration possibilities with various Italian institutions and physicists in nuclear and solid state physics, and remarked that this "might contribute to the planning of physical research in Cuba" (Fernández Bertrán 1962).

Following Fernández Bertrán's visit to Italy, two Italian specialists arrived at the university: the nuclear physicist Piero Basso and Mario Chirco, an electrical engineer with a good command of physics and mathematics, who stayed several years in Santiago de Cuba.

The Mathematics and Physics Departments were organically incorporated into the Basic School for Technology and Sciences, created in 1966. In addition to its regular staff, the Physics Department at times counted on the temporary collaboration of qualified teachers, like Suárez Soto and Olivares, and the Salvadorian, Ricardo Arrieta Salazar, who had come from California and upon his return trip to the United States, met a tragic fate at the hands of the Salvadorian dictatorship of the time.

8.4 The Physics School at the University of Oriente

8.4.1 Formation of the School (1967–1970)

Since 1965, two commissions had been created at the university to analyze the best ways to train physicists and mathematicians and to solve teaching problems at the institution. These were led by Luis Oliva and Miguel Matute, respectively.[3] However, they did not prosper because at the time it was thought that the University of Havana was capable of training all the physicists and mathematicians needed by the country. In conversations with F.D. Kochanov, a Soviet specialist who was working as a consultant on the teaching of general physics at the University of Havana, it was agreed that apart from elaborating proposals for the best possible teaching methods in physics, it was important for the development of the country to promote a closer link between physics and industry, especially mining and some other industries. This led to the creation of a degree course in physical engineering within the Faculty of Technology, in which printed matter elaborated by Kochanov dealing with industrial applications and possibilities of physical science were used as text and guidance material and contributed to the consolidation of a favorable environment for the further development of physics at the university.

In 1967–1968, Jorge González Alonso and Homero Fuentes González (who had obtained their physics degrees at a German institute of higher education and at the University of Havana, respectively) joined the staff of the Physics Department, which was then under great strain because of the insufficient number of teachers taking care of the relatively large number of undergraduate physics students (González and Fuentes 1968). The staff members had little time to engage in research projects, and to take or give graduate courses. This state of affairs became increasingly acute because of the sudden increase of students in engineering, agricultural, medical and pedagogical degree courses. To provide an emergency

[3] Interviews undertaken by Miguel Matute and Ramón Pomés.

solution to this problem in the absence of previously organized degree courses for the specialized training of physicists, a "student assistant" movement took place at the university (and at all other Cuban higher education centers as well), to voluntarily engage the advanced students in the teaching of physics to students enrolled in lower-level courses. The proposed procedure was to:

Create a group, taken from most highly qualified second and third year students of the engineering or chemistry degree courses, with the purpose of training them as physicists.

Engage this group of students to simultaneously dedicate themselves to the teaching of general physics to students of various degree courses, and to the setting-up of student laboratories, where the existing teaching staff could tutor with the help of foreign technical assistance.

Accordingly, in 1967 authorization was requested from the Ministry of Education to implement a special plan to train 20 physical engineers who were to graduate in 1971 "with the purpose of solving technological problems and to engage in the physical analysis of minerals" (*Sierra Maestra* 1968). This in fact amounted to the first step in the formation of the Physics School, the goals of the plan being to provide the personnel needed for academic work and to train an initial nucleus of physical engineers who would work on various branches of applied physics at the university and in industry. These engineers would be capable of doing useful research work and introduce the results of advanced physical research into technology.[4]

This was, in effect, a plan for training industrial physicists. J. González and H. Fuentes, in collaboration with Soto del Rey and Aguilar Salcedo designed the pertinent curriculum to include such subjects as electrotechnology, basic electronics and industrial electronics (González and Fuentes 1968). The Ministry of Education not only agreed with the proposal, but also added that the university had "the possibility of developing scientific research work in various branches for it counted on specialists of international scientific stature (such as José Fernández Bertrán) who could lead the way to graduate research work of national interest."[5]

In 1967, 20 years after the foundation of the university, the first steps were taken to create its Physics School, which finally acquired official status in March 1970, and on December 8th of the same year, produced its first 19 graduates in physical engineering. Their graduation took place one course in advance, for the academic semesters had been compressed in time and the vacation period was cancelled by mutual agreement between students and staff, who made a special effort for the purpose. Still, the idea of training such physicists at the university was subject to much criticism, both from outside and from within. It was necessary to repeatedly explain and defend the concept that there was a real need to train on a regular basis physicists capable of performing well within a matrix of non-physics professionals,

---

[4]A plan for the training of physicists at the university. (Undated document, probably written by the end of 1967 or the beginning of 1968.)

[5]Report to the Ministry of Education's Vice Minister for Higher Education on the physical engineering degree course.

especially in view of the fact that in the Eastern part of the country, institutions did not invite physicists to join their staff; a newly trained professional in this field would have to prove himself by fruitfully linking physics to different areas. In short, it was claimed that the new professional would be something like a "physics cat," always capable of landing on its 4 ft, from whatever position it was flung.

At the time, several government officials and members of the teaching staff came to the defense of the new degree course, among them, the engineer Miguel Torres, first in his role as Dean of the Faculty of Technology and then later as Vice Rector; Luis Estévez, José Borges and Rosina Hing, members of the staff of the Mathematics Department; and the engineers Miguel Matute and Arístides Berenguer from the Electrical Engineering School, who also belonged to the staff of the Special Plan for Physics Engineering. Valery Smirnov, from the Leningrad State University, had a marked influence on the conception of the new degree course, on the modernization and extension of the undergraduate laboratories and on the actual training of the first generation of students. Johann Monecke, staff member of the Dresden Technical University, taught a model course in quantum mechanics in 1969.

After one academic course following the inauguration of the Physics School, the Mathematics School was inaugurated with a similar conception and also with the active participation of its students. Later on, the Biology School was created, which, in addition to the Chemistry School, completed the basic organizational structure for the fundamental sciences at the University of Oriente.

8.4.2 Consolidation of the School (1970–1976)

The foundation and organization of the Physics School in March 1970, together with the first and only graduation of physical engineers (at the time) in the following December, define the end of one stage and the beginning of the next, which now had the well-earned right to develop physics in the province of Oriente.

As soon as the Physics School had been officially established, it was decided that it would specialize in physical methods of analysis. Jorge González Alonso and Homero Fuentes were appointed Director and Vice Director, respectively, and the following five departments were created: Nuclear Physics, Optics and Spectroscopy, X-Ray Physics and Metals, General and Theoretical Physics, and Electronic Physics, headed by Luis Pérez Tamayo, Miguel Catasús Portuondo, Manuel García Ramos, Luis Aguilar Salcedo, and Carlos Cabal Mirabal, respectively. Following the Soviet educational pattern, the first three departments were supposed to offer specialized training to the students and engage them in applied research in their respective fields, while the task of the last two was to take care of the basic education of the physics and other students, and to cooperate with other departments.

Apart from Aguilar, the heads of the other departments were still students. Moreover, since the organization of the university was still in a state of flux, a few months later, in August 1970, the formerly appointed Director of the Physics School, the graduate Jorge González Alonso was promoted to Vice Dean of the Faculty of

Sciences, and Homero Fuentes and Carlos Cabal were appointed Director and Vice Director of the School, respectively. Before the end of the same year, another shift took place when Cabal became Director of the School (until 1973) in place of Fuentes, who had been appointed as Vice Dean of the Faculty of Sciences.

The School was considerably strengthened when the first 19 graduates in physical engineering joined its teaching staff. These included González, Fuentes, Soto del Rey and Aguilar Salcedo, plus other recently graduated physicists from the University of Havana. On the other hand, its scientific level was considerably enhanced thanks to the periodic working visits of a number of staff members from the Leningrad University (Smirnov, Petrov, Niementz, Braun, Lavzovsky, Borodin, Molchanov and Zanadvorov). Their collaboration at this stage (it extended up to 1976) was crucial for the launching of applied scientific research, which initially was certainly not at an advanced level, but did confirm the need for the further development of physics work at the university and its importance in the provincial vicinity.

Further development of the Physics School called for strong international ties, including visits by the staff members—either short or long-term—to appropriate institutions abroad. Accordingly, between 1971 and 1972 around ten members of the school staff visited Leningrad University and the technical universities in Dresden and Stockholm to become acquainted with their teaching methods, taking short special courses or taking up graduate courses leading to a doctoral degree. In 1970, Matute visited Italy for 6 months.[6] In September 1971, Homero Fuentes and Carlos Cabal traveled to East Germany and the USSR to establish or strengthen formal agreements with the Dresden Technical University and the Leningrad University, which was visited again in 1972 by Luis Aguilar and Miguel Catasús. French specialists, headed by Henry Pezerat, gave graduate courses on materials characterization in 1970, while from 1971 to 1972 some physics graduates (G. Lucambio and L. Méndez) took summer school courses from French professors at the University of Havana.

Up until 1976, 40 physicists had obtained their physics degree at the university,[7] some of them in nuclear physics (that year the Physics School had about 50 students). The main research subjects up to 1976 were those mentioned in (Pérez Rojas et al. 1976). Though the Department of Optics and Spectroscopy took the main load of thesis tutoring, the Department of Nuclear Physics began to catch up. Spectroscopy research developed in connection with the nickel industry in Oriente, with the application of atomic emission and absorption techniques, following two working lines: one on the plant product (M. Catasús), and another one on lateritic minerals (J. Ricardo). Work on X-ray analysis and metals dealt mainly with phase transformations.

In 1968, when Soto del Rey began teaching physics to first year medicine students, he arrived at the conclusion that the subject should be taught differently to correspond to the interests of his students. This led him to develop medical physics as a new subject. In the end, this was so successful that members of the

[6] Interviews granted by Miguel Matute and Ramón Pomés.

[7] Graduates Register, General Secretariat of the university.

teaching staff asked to take the new subject as a graduate course. It was later introduced as a standard course within the biology curriculum. As a by-product, in 1988 Soto del Rey published a four-volume textbook entitled *Introduction to Biophysics* (Soto del Rey 1968). This corresponded with his former view that given the national priorities, especially those particular to the province of Oriente, physics students should be trained in two main areas: technical and biological physics, as he called it. This idea preceded his involvement in the actual teaching of students of medicine and biology.

From 1976 to 1977, the first physicists graduates of the OU returned to Cuba after taking part in specialized and doctoral courses abroad. Margarita Cobas Aranda (optics) and Ramón Pomés Hernández (X ray crystallography) were the first to successfully defend their doctorates at Leningrad University. This made the Physics School of the University of Oriente the country's pioneer institution in the field of optics and spectroscopy. The achievements in the development of nuclear physics and X-rays and metals were not trivial, in both the experimental and the applied context.

8.5 A Period of Development (1976–1985)

When the Ministry of Higher Education was created in 1976, the Physics School and its Department for Electronic Physics was officially dissolved, while the other departments were absorbed into the Faculty of Physical, Chemical and Mathematical Sciences. In 1980, these were transferred to the newly created Faculty of Physics and Mathematics, whose first dean was Carlos Cabal Mirabal, who had obtained his doctoral degree from Leningrad University.

In addition to the international academic exchanges, especially with East Germany and the USSR, the improved qualifications of the staff trained abroad meant that training offered to physics students progressively improved. By 1985, the faculty included eight physicists with a doctoral degree (or candidate degree, as it was called at the time), one of whom had obtained this at the university, while 14 physicists out of 35 staff members had attained the rank of senior or auxiliary professor.

In 1983, some thirteen years after the creation of specialized studies in physics, the Laboratory for Physical Methods of Analysis summarized its main achievements: application of neutron techniques to the determination of the humidity of Cuban soils; determination of the structure of sucrose crystals experimentally grown in orbit during the 1980 Soviet-Cuban space flight (experiment Zona;[8]) determination of the composition of national minerals for inclusion in the 1:100,000 Cuban geological map; publication of more than a hundred scientific papers in national and international specialized journals; and participation in more than 80 scientific gatherings (ten of them of international nature) in the USSR, East Germany, The

[8] See the chapter by Ernesto Altshuler et al. on the Cuban technophysical experiments in space.

Netherlands and Canada. To this, it may be added that staff members of the Physics School were awarded prizes for their scientific work: in the 250th Anniversary of the USSR Academy of Sciences contest (R. Pomés, third place) and the Leningrad State University for Young Scientists Contest (Carlos Cabal, second place/1979, and first place/1980) (Project UCT 1983).

Several scientific monographs, textbooks and handbooks were published by members of the physics staff, among them: *Methods for the Determination of Gold in Minerals* (Cobas Aranda 1980), *NMR Studies of Paramagnetic Electrolytic Solutions* (Cabal Mirabal and Chizhik 1982), *Collection of Solved Problems in Physics* (Parera et al. 1984), *Electronic Methods in Experimental Physics* (Méndez Perez 1986), and *Introduction to Biophysics* (Soto del Rey 1988).

From 1980 to 1988, the Physics School strongly engaged in vocational guidance work for hundreds of senior high school students throughout the Eastern region of the country, which included its active support in the organization of Olympics for physics, chemistry and mathematics. Training courses were organized for high school physics teachers, and about a dozen popular scientific articles were published in national, provincial and university periodicals. Programs dealing with the relevance and possibilities of physics in the Cuban social context were also produced for TV.

The above scientific and educational work, which only emerged in the 1970s, had a beneficial influence not only on the quality of undergraduate teaching, but also on the development of graduate training (previously nonexistent) through the implementation of PhD-level studies as well various graduate courses and training programs. As early as 1979, the National Commission for the Peaceful Use of Atomic Energy deemed the school's Department of Nuclear Physics mature enough to entrust it with the national leadership of a PNUD project entitled "Introduction of Nuclear Techniques in the Country's Economy" (Project PNUD 1979). All of this contributed to the standing of the Physics School within the national physics community, to the extent that in 1985 the Cuban Physics Society held its Third Symposium not in Havana, but at the University of Oriente.

8.6 A Transitional Period (1985–1993)

At the beginning of the 1980s, an emigration wave of experienced staff members from the Physics School developed and grew with time. It originated from certain policy decisions at the national level, such as (a) the creation of a physics department for technical sciences that was attached to the creation in 1985 of the Julio Antonio Mella Higher Polytechnic Institute; (b) the elimination beginning in 1982 of specializations from the curricula; (c) the close-down between 1985 and 1986 of applied nuclear physics research by the Executive Secretariat of Nuclear Matters (SEAN), added to the impossibility of obtaining the financial resources and equipment for the two projects mentioned above. Consequently 17 senior staff members left the university for other institutions, with the result that the number of PhDs and professors (senior and auxiliary) was reduced to four and five, respectively, and the staff vacancies had to be covered by either inexperienced physics graduates or by graduates trained in other fields. By 1993, few of the original staff remained, especially after the retirement of Soto del Rey and Aguilar Salcedo and the passing of Arturo Guzmán, one of the first graduates of physical engineering, who was just finishing his doctoral thesis work on magnetic relaxation in rare earth paramagnetic systems.

Generally speaking, while the situation described above certainly brought about a decline in the scientific and academic standards as well as the philosophy that defined the early rise of the Physics School, the new staff managed from 1986–1987 to initiate the recovery of the school's former standards of excellence on new research lines established in place of those eliminated in 1982. After the purchase in 1987 of a suitable electron microscope, some of the remaining members of the former staff specialized in materials sciences and related techniques, while others dedicated themselves to the automation of experiments and computational physics, which were later combined for application in biophysics, medical physics and biotechnology specialties. A working group dedicated to nuclear magnetic resonance (NMR) was created in 1981, which from 1987 to 1990 bore its first relevant results that were to define the future path of this field.

A multidisciplinary Medical Physics Group, created in 1987, began its activities by establishing a successful partnership with the "Conrado Benítez" Oncology Hospital of Santiago de Cuba. It expanded to become the National Institute for Oncology and Radiobiology, carrying out automation and simulation tasks, and radio isotopic test calculations with computerized renographs. For the practical results it achieved with the digital photoplethysmograph UOangio-0, in collaboration with the "Saturnino Lora" hospital of Santiago de Cuba and the National Institute of Angiology and Cardiovascular Surgery, the Medical Physics Group was awarded a Prize for Scientific Merit. The results also had a high economic impact. In addition, the NMR Group was elected the most distinguished collective of the university for establishing a methodology for polymerization kinetics evaluation of hemoglobin in patients with drepanocytic anemia, and for producing an important preliminary study on the classification of lymphadenopathy useful for the quantification of breast cancer.

Project 35–26–7 NMR was started in December 1987, thus fulfilling a direct request from the country's leadership to develop technology in Cuba for the manufacture of nuclear magnetic resonance imaging equipment for medical purposes. This was necessary since it was impossible to acquire ready-made equipment of this kind for the country's public health system from abroad, especially since at the time the commercial firms involved were required to obtain a license from the US government to sell their products to Cuba. Initially the project was undertaken by five graduate physicists (only one of them with a PhD degree) and several students in the final year of their degree courses; they worked with a few engineers, whose goal was to calculate, design, set up, characterize, validate, register and apply to patients their NMR imaging equipment. The first machine went into regular operation in early 1991, at the beginning of the extremely difficult period the country endured after losing more than 80 % of its foreign trade with the former socialist countries in Europe. Despite these setbacks, three units for full body imaging were prepared in record time, together with several relaxometers and NMR magnetometers. A stable multidisciplinary collective was thus created that included physicists, engineers, chemists and computer scientists with a high professional standard recognized at national and international levels. The project was able to attract some valuable international cooperation, especially from the Institute of Physics of the University of Sao Paul, led by the late Horacio Carlos Panepucchi.

For the development of the "Giromag" NMR relaxometer and a methodology for the characterization HBS polymerization process, the collective received a national prize in 1994 from the Cuban Academy of Sciences awarded the collective responsible for above mentioned project a National Prize in 1994, and another one in 1995 for their development of the "Giromag" tomograph. The "Giorgio Albieri in Memoriam First Prize" was awarded to the NMR Cuban group led by Cabal at the 5th International Conference on Applications of Physics in Medicine in Trieste in 1996. Other national and international prizes followed. The results obtained stand among the most complex and significant technoscientific achievements accomplished in the country. In association with the development of equipment and related technologies, new research directions in the field of molecular and cellular biophysics were introduced that had a strong impact on Cuban biotechnology and the medical-pharmaceutical industries.

Rooted in the above achievements, which include the multidisciplinary collective, a Center for Medical Biophysics was set up on new premises in Santiago de Cuba, duly equipped for biomedical R+D work. In his 1993 inaugural speech, President Fidel Castro put forward an idea that was to become one of the conceptual pillars of contemporary scientific activity in Cuba: "Some day science and the productions of science must move into first place in the country's economy. We have to develop the production of human intelligence. This must be our place in the world, because there will be no other."

These thoughts were taken into account by the university authorities, who addressed this new approach by introducing successive structural changes within the university. One of these was the creation of the Faculty of Natural and Mathematical Sciences, which included a physics department in charge of the degree course (*licenciatura en física*).

8.7 Rebirth of the Physics Department

Starting in 1993, the physics department was reorganized so that the optics and spectroscopy group was transformed into a collective—led by Jorge Ricardo—dedicated to fluorescence and laser applications. Working groups were created for theoretical physics (led by Raúl Riera and backed by the University of Havana's Physics Faculty), materials sciences (led by José Anglada), computational fluid dynamics (led by Rafael Mut, backed by the University of Barcelona), and university physics didactics. Researchers in these groups worked on nanostructures, thermal and electric properties of ionic conductors, properties of ferroelectric materials and magnetic properties of soft magnetic materials. They also took part in the preparation of national programs for basic sciences, computational sciences and new materials, which were managed by the Ministry of Science, Technology and Environment (CITMA). Several national prizes and awards were won by members of these departments for the results of their work (among them, the Carlos J. Finlay and the Lázaro Peña orders awarded by the Council of State) and the Cabal was re-elected as a Senior Member of the Cuban Academy of Sciences for the period 2002–2006.

The Physics School continued to develop throughout the 1990s, despite the interruption in the previously close collaborative ties with the USSR and East European countries. The first "R. Soto del Rey In Memoriam" Physics Symposium held in 1997, was successfully revived in 2000 as an international conference attended by specialists from Brazil, Spain and other countries. Research and collaboration projects with Mexican, Brazilian, Spanish, and Venezuelan universities were later undertaken. By 2005, senior and auxiliary professors made up 80 % of the teaching staff, of which 60 % had a PhD and 67 % an MSc degree.

Fifty years after the foundation of the University of Oriente, its original physics chair had already developed into an academic complex which included a center for medical biophysics, a department for applied physics, and a physics department in charge of the degree courses in physics (awarded an "excellence certificate" in 2005 from the Ministry of Higher Education) and in physical engineering. More than 400 physicists, including some from African and Latin American countries, have graduated from the university in the five decades since its inauguration.

8.8 Conclusion

When the university was inaugurated in 1947, physics as a science was represented as a single basic subject in the chemical engineering curriculum. Later on, it was included in other engineering specializations and in agricultural and medical studies as well. Roberto Soto del Rey, head of the first physics chair, inspired the general orientation and further development of physics studies at the university. They experienced a very important turning point when a physical engineering degree course was established in 1967, especially since it supplied the country with a growing mass of well-trained specialists who were able to cover the teaching needs of higher education and applied research in important industrial sectors located in Eastern Cuba. Still, progress was not easy, due to serious difficulties that came at one time or another from radical changes in the university structure, migration of experienced specialists to other institutions and a limited material base available for experimental research. Still, professionals with a strong physics background who were trained at the university have been able to make important contributions to the educational, scientific, technical and economic development of the country, for which they have been amply recognized at the national level.

References

Augustín Caballero, José, Jenaro Artiles Rodríguez, Francisco González del Valle, and Roberto Daniel Agramonte. 1944. *Philosophia Electiva*, Biblioteca de Autores Cubanos, vol. 1. Habana: Universidad de la Habana.

Cabal Mirabal, C., and V.I. Chizhik. 1982. *Estudios de soluciones electrolíticas paramagnéticas por Resonancia Magnética Nuclear*, Serie Física Matemática, 1–84. Santiago de Cuba: Universidad de Oriente.

Cobas Aranda, Margarita. 1980. *Métodos de determinación de oro en minerales*. Santiago de Cuba: Dirección de Información Científico-Técnica, Universidad de Oriente.

Consejo Directivo de la Universidad de Oriente. 1947. Acta No. 13, 23 October 1947.

Diario de Cuba [newspaper]. 1947. Santiago de Cuba, 12 October 1947.

Fernández Bertrán, J. 1962. *Informe sobre la Visita a Institutos de Física Italianos*. Universidad de Oriente.

González, J., and H. Fuentes. 1968. *Ponencia al Encuentro de Graduados*. Universidad de Oriente.

Granma [newspaper]. 1997. 26 February.

Méndez Perez, Luis. 1986. Métodos electrónicos en la física experimental. Santiago de Cuba: Física-Matemática, Ediciones Universidad de Oriente.

Parera, J., E. Roca, L. Grave de Peralta, and R. López. 1984. Colección de problemas resueltos, I (Sobre tareas de la Física Matemática y aplicaciones a la Física).

Pérez Rojas, H., D. Stolik Novigrod, J. Fuentes Betancourt, C. Rodríguez Castellanos, A. D'Costa Méndez, R. Alvarez Morales, B. Lazo Orazábal, E. García Tarajano, J. González, M. Ramos Vázquez, and O. Arias Fuentes. 1976. Estado Actual de las Ciencias Físicas en Cuba. In *Las Ciencias Básicas: Examen Preliminar de su Situación Actual en Cuba y a Nivel Mundia*, ed. J. Altshuler. Havana: Consejo Nacional de Ciencia y Técnica, Dirección De Ciencias Exactas y Naturales.

Project PNUD. 1979. Introducción de las Técnicas Nucleares a la Economía Nacional. CUB/77/001/C/01/18, January 1979.

Project UCT. 1983. *Laboratorio de Métodos Físicos de Análisis*. Universidad de Oriente, Facultad de Ciencias Físico – Matemáticas.

Sierra Maestra [newspaper]. 1968. Santiago de Cuba, 3 July.

Sierra Maestra [newspaper]. 1982. Santiago de Cuba, 6 October.

Soto, H. 2000. Biografía: Roberto Joaquín Soto del Rey. *Energía y Tú* 9: 33.

Soto del Rey, R. 1968. Actividades docentes a desarrollar en este semestre. In *Memorandum by the Head of the Physics Department, dated 9 Nov 1968*.

Soto del Rey, R. 1988. Introducción a la biofísica. Santiago de Cuba: Editorial Oriente.

Soto del Rey, R., and L. Aguilar Salcedo. 1961. *Tensión superficial y soluciones líquidas*. Editora Universitara, Universidad de Oriente.

Treaty. 1962. *Tratado sobre colaboración amistosa entre la Universidad de Oriente de Santiago de Cuba y la Universidad Técnica de Dresden*. Dresden Technical University, May 1962.

Chapter 9
The Training of Physics Teachers in Cuba: A Historical Approach

Diego de Jesús Alamino Ortega

9.1 Introduction

The regular, systematic training of physics teachers in Cuba is quite recent when compared to the long history of physics itself. However, its development may serve to illustrate some interesting solutions to a long-standing question: How should a physics teacher be trained in agreement with a certain society at a given moment? In the Cuban context the answer to this question involves quite an original sequence of continuities and breaks, following perhaps the thoughts of Bolívar's teacher, Simón Rodríguez, who wrote in the nineteenth century: "Beware! The mania of slavishly imitating the enlightened nations may well make America play the role of an old lady in its infancy."

9.2 Background of Teaching and Teacher Training in Cuba

In order to seriously approach the history of the training of physics teachers in Cuba, it is necessary to first comment on the general evolution of teaching in the country.

While by 1492 several universities had long been created in Europe, and the place of the earth in the universe was being discussed, when Christopher Columbus arrived in America, he found the natives at the stage of civilization of a generally primitive community. The colonizer's unbounded thirst for wealth resulted in the decimation of the native population, while the education of the people was, needless to say, far removed from their main worries. It was the Catholic Church who took care of education, along with evangelization.

D. de J. Alamino Ortega (✉)
Exact Sciences Department, University of Pedagogic Sciences, Matanzas, Cuba
e-mail: diego.alamio@umcc.cu; diego.alamino@ucp.ma.rimed.cu

A. Baracca et al. (eds.), *The History of Physics in Cuba*, Boston Studies in the Philosophy and History of Science 304, DOI 10.1007/978-94-017-8041-4_9,

The Spanish educational system that was extrapolated to Cuba in colonial times favored the education of the nobility, the clergy, the rich, and high officials, while the education of the rest of the population lagged well behind: mid and higher studies were given greater importance than elementary schooling. In 1722, the San Basilio el Magno Seminar was founded in Santiago de Cuba, followed by the Real y Pontificia Universidad de San Jerónimo de La Habana (Real and Pontifical University of San Jerónimo of Havana), created in 1728 (Sosa and Penabad 2001). The Real y Conciliar Seminario de San Carlos y San Ambosio (San Carlos and San Ambrosio Real and Ecclesiastical Seminar)—a source of Cuban national consciousness and the cradle of some of the most advanced ideas in philosophy, science and pedagogy of its time in the country—was founded in 1773, also in Havana. These institutions prepared their students primarily for ecclesiastical careers and the humanities, and—with the exception of the San Carlos and San Ambrosio Seminar—were dominated by scholasticism.

The beginning of systematic elementary education in Cuba can be set in 1793 with the creation, in Havana, of the Real Sociedad Económica de Amigos del País (Royal Economic Society of the Country's Friends) which was supposed to "promote, supervise and lead public education" (García 1978). At this very moment, elementary schools began to be established in most Cuban cities and villages. The new teachers were poorly trained and most of them limited themselves to teaching the little they knew: religion, reading, writing, and the four basic arithmetical operations. However, in a few exceptional schools, the best contemporary trends in pedagogy and philosophy of education were assimilated, but they were only delivered to the children from the higher social classes.

Since its very beginning, the Economic Society made important efforts to regulate the quality of teaching by awarding teaching licenses that were given only after the fulfillment of various requirements. In this way, certain philosophical elements were introduced to support pedagogy and didactics.

In 1842, the Spanish Government promulgated a General Law for Public Education and began to take responsibility for its implementation. It established the requirement of being officially qualified to work as a teacher, and planned the creation of the Escuela Normal de Maestros (Teacher Training College), which first started operating in 1857 (Alamino Ortega 2002). It took 2 years for students to complete their training as elementary or higher education teachers. This training consisted mostly of religion studies, while training in the sciences were very limited. It ceased to operate in 1868 with the start of the Cuban "Ten Year War" of independence. Only 112 teachers graduated in this period. There were other educational institutions in the country at the time: the *Institutos de Segunda Enseñanza* (secondary education institutes) in Havana, Santiago de Cuba, Matanzas, and Puerto Príncipe, which admitted 9 year old children who had passed the fourth grade. Apart from some elements of mathematics, all they were taught belonged to the realm of the humanities.

In 1872 a teacher training school already existed and was supported by the Economic Society in accordance with its original goals. By 1880, the number of institutes of secondary education had increased to six, and some *Escuelas Primarias Superiores* (higher primary schools) had been created. In the light of the new Public

Education Plan of 1890, the *Escuelas Normales* (teacher training colleges) were reopened in 1892, but their work was interrupted by the 1895 war for independence.

Real Cuban independence was thwarted by the 1898 military intervention and subsequent occupation and control of national economy by the United States. The occupying power found the country economically devastated, but its people still had strong patriotic feelings. In this context the ruling authorities promoted the urgent training of elementary school teachers, which was not only required for a more efficient exploitation of Cuba's natural resources, but was done in a way that tended indirectly to erode feelings of national identity in favor of feelings propitious to an eventual annexation by the powerful neighbor—a goal which was only partially fulfilled among a sector of the population. There was a call to grant certifications as first, second and third grade teachers, depending on the candidate's performance in certain exams. Cuban youngsters traveled to the United States to study in summer schools for teachers. After the first Republic was founded in 1902, similar schools were created in Cuba aimed at the professional improvement of the already existing teachers. In any case, it must be said that while certain patterns typical of the education in the United States education strongly influenced teacher training, this led in general to a substantial improvement. Still, thanks to the efforts of Cuban teachers, a high dose of patriotic spirit, associated in particular with the national paradigmatic hero, José Martí, was never absent from the country's elementary schools.

The first teacher training colleges in the republican era were founded between 1916 and 1919: two in Havana and one in each provincial capital, which added up to seven schools. Admission was from the sixth grade, which was later extended to the eighth grade. When the teacher training college system was systematized, teachers formerly trained under different restraints were put on the same level as those whose had trained according to the new standards.

9.3 The First Physics Teachers

There are references from the beginning of the eighteenth century which evidence that physics was taught in religious (Catholic) institutions, including the University of Havana (Alamino Ortega and Simon 2004). The most prominent personality of this initial stage of physics teaching in Cuba was Félix Varela Morales (1787–1853), a young priest who taught physics subjects between 1811 and 1820 as part of his lectures in philosophy at the San Carlos and San Ambrosio Seminar (Alamino Ortega 2005).

By the middle of the twentieth century, Cuban universities were graduating "Doctors" in physico-mathematics, physico-chemistry and pedagogy, as well as engineers and architects, who devoted themselves to the teaching of physics, a compulsory subject in the natural sciences syllabuses. Relative to this stage, Professor Manuel Francisco Gran Gilledo (1893–1962) stood out due to his efforts to elevate the teaching of physics. He wrote some excellent textbooks and, during the 40 years between 1923 and a few months before his passing in 1962, taught at the University of Havana. Physics was also taught by graduates of the prestigious teacher training colleges, where Gran's *Elementos de Física* and White's *Descriptive Physics* were used well into the twentieth century as textbooks for the syllabus of the subject named natural sciences. But in point of fact, no physics teachers were specially trained as such in the country.

With the triumph of the Revolution of 1959, a reorganization of education at all levels took place (Fernández 1986). On September 18, 1959, the Ministry of Education issued a resolution establishing that "technical selection tests must be applied to cover the teaching staff of the general and professional secondary centers [hitherto] provisionally covered, as demanded by the proper organization of teaching." After this time, short courses for the training of "emergency" secondary teachers were offered. In provinces outside Havana, 6-month courses were given three times a week starting from June 1960. They were preceded, in April 1960, by the creation of the Instituto de Superación Educacional (Institute for Educational Improvement) aimed at the systematic improvement of the abilities of the teaching personnel. This institution was to award to the incumbents the academic qualification of "Physics Professor for Basic Secondary School" after fulfillment of certain requirements established by the Ministry of Education.

The physics syllabus for the 6-month courses—which were repeated in the following years—included laboratory experiments illustrating the properties of bodies, thermal phenomena, inertia, force and mass measurements, liquid pressure, atmospheric pressure, pulleys, the lever, inclined planes, light reflection and refraction, spherical mirrors, lenses, the construction of a telescope, the dispersion of light, sound and musical instruments, electrization, magnets, magnetic compass, construction of a battery and its insertion in a circuit, series and parallel circuits, etc. The textbooks most often used in these courses were Gran's *Elements of Physics* (in two volumes), Kleiber and Karsten's *A Popular Physics Treatise*, Alonso and Acosta's *Introduction to Physics*, Alonso's *Elementary Physics Course (Mechanics)*, Piorishkin's *Lectures for Everybody* (published in the central pages of the popular magazine *Bohemia*), and *Lectures sponsored by the National Association of Doctors in Pedagogy*. Practical activities could be undertaken in the laboratory thanks to the existence of modular equipment from the Spanish brand ENOSA, whose manuals contained detailed descriptions of the experiments. Class plan design was included in the methodological activities they had to comply with so that students emerging from the courses fulfilled the essential requirements for teaching the subject reasonably well.

9.4 The Pedagogical Institutes

In 1962, a thorough reform of higher education took place in Cuba. A couple of years later, three pedagogical institutes were founded, affiliated with the universities of Havana, Las Villas, and Oriente, respectively. Regular systematic training of physics teachers took place at those institutions. There were two levels: the basic level for training teachers for the *enseñanza secundaria básica* (junior high school, 7th–9th grades), and the higher level, to prepare teachers for the *enseñanza preuniversitaria* (senior high school level, 10th–12th grades). There were different syllabuses which included "double-specialization." Generally, students were boarders who profited from a vast scholarship system created for the purpose.

While the main emphasis at the pedagogical institutes was on mathematics and physics, their syllabuses included pedagogical and psychological subjects as well. Physics started with mechanics, and included all the typical subjects of general physics, ending with modern physics. Due to the experimental character of the subjects, there were a sizable number of demonstrations of experiments in class, as well as laboratory experiments. The textbooks most employed were Gran's *Elements of Physics*, Weber et al.'s *General Modern Physics*, Sears and Zemansky's *College Physics*, Resnick and Halliday's *Physics for Students of Science and Engineering*, and Frish and Timoreva's *General Physics*. Student training included a substantial component of practical school teaching.

Courses for workers started in 1971; these were based on a system of periodic meetings of student-workers with professors, in addition to a number of intensive class periods. This was an alternative way of attaining an official qualification, especially for those who were already working as 'unqualified' teachers, since it was not until the end of the 1960s that the country could count on the first physics teachers who graduated from the pedagogical institutes.

9.5 Plan for the Training of Teachers for General Intermediate Education

The extension of elementary education and its high permanence indices after 1959 produced a "matriculation explosion" at the secondary level by the early 1970s. To cope with this, in 1972 the Manuel Ascunce Domenech Pedagogical Detachment was created where graduates from secondary schools would be trained as intermediate education teachers by combining their study at the pedagogical institutes with actual practical teaching: students attended classes in the morning and taught in the afternoon, or vice-versa. Their teaching duties entailed quite a large number of hours and they were always coached by experienced professors. The Detachment was fed with students with a 9th-grade certificate, who received their official title after 5 years. The syllabus had ten semesters of mathematics and the same amount of physics—including experimental class demonstrations and laboratory experiments with good quality equipment, most of it imported from Sweden. There were two semesters of chemistry and Spanish language studies. Psycho-pedagogy took six semesters, followed by two semesters of special didactics. The syllabus also included six semesters of English language in view of its importance for handling most of the available bibliography—not least Soviet textbooks translated into English. The fundamentals of Marxist philosophy took two semesters from the fifth year, in addition to special seminars designed to go somewhat deeper into specific fields of physics and its teaching.

Since the input level was just the 9th grade, students had to tackle the contents of the pre-university level (10th–12th grades) alongside their training as teachers. Needless to say, the training of the Detachment graduate in specific topics such as literature was not as broad as what was expected from a pre-university 12th-grade graduate. The syllabus did not contain any scientific investigation—this was eventually performed by some advanced students in an extra-curricular fashion.

The official qualification attained by the graduates from this plan allowed them to teach physics at the intermediate level, and offered the possibility of studying two extra years in order to get the degree *licenciado en educación en física y astronomía*, equivalent to the regular courses. The syllabus for the extra years included mathematical methods of physics, theoretical mechanics, relativity theory and electrodynamics, atomic physics, nuclear physics, quantum mechanics and statistical physics, in addition to theory of education, elements of school organization, history of pedagogy, Marxist philosophy, and other socioeconomic subjects.

9.6 The *Licenciatura* in Education

In 1976, the pedagogical institutes were transformed into higher pedagogical institutes (ISP)—and became pedagogical universities themselves. They would produce *licenciados* (bachelor equivalent) in education, including specializations in physics and astronomy, with a 4-year study plan. To enter the HPI one should have finished high school (i.e., up to and including the 12th grade). The first syllabus was known as "Plan A," which, in this author's opinion, marked the coming of age of high school teacher training in Cuba—particularly in the field of physics. This plan attained a suitable balance between academic training and teaching practice, which was in harmony with the syllabus and not subordinate to the teaching needs of the country: the emphasis was now on academic training, especially physics and mathematics subjects—the syllabus included, for example, theoretical physics. The subject called methodology of teaching (with a total of 170 class hours) exposed the students to the actual textbooks and methodological rules that they would use in their future jobs as high school teachers. They also performed the corresponding experiments with the same equipment they were to use with their future students—all of this in addition to the theory and experiments they had to tackle as part of other subjects, of course. The experimental activities were performed with equipment imported mainly from Sweden, East Germany, and the Soviet Union. The *licenciados* in education degree was awarded after passing a national exam or, in the case of outstanding performance during the standard five study years, after successfully defending a diploma thesis containing original research work.

After the so-called Plan A, various successive syllabuses came into effect to adjust to the needs of modernization. Amongst the biggest modifications introduced in later syllabuses, one should mention the extension in the length of the studies from 4 to

5 years, and an increment in the time devoted to methodology of physics teaching, which eventually amounted to 268 h. As a rule, a 5-year period was allowed to elapse before introducing any important changes. The proposed modifications had to be examined in detail by the National Degree Course Commission and then submitted for evaluation by qualified reviewers. The new syllabuses that emerged would be introduced nationwide after reaching general assent. The so-called "Plan C," adopted in 1990 to produce *licenciados* in education specializing in physics and electronics was the most polished to date: this amounted to another turning point in the teaching of physics in Cuba, even if it was never fully implemented in practice (Ministerio de Educación 1990).

The textbooks used for the general physics subjects were—and still are—Resnick and Haliday's *Physics for Students of Science and Engineering*, Frish and Timoreva's *General Physics*, Kikoin and Kikoin's *Molecular Physics*, Savéliev's *A Course in General Physics*, Beiser's *Concepts of Modern Physics*, Strelkov's *Mechanics*, and Landsberg's *Optics*, among others. Some textbooks by Cuban authors were occasionally published for use in these courses. In the case of the theoretical physics subjects, Goldstein's *Classical Mechanics*, Jackson's *Classical Electrodynamics*, Bredov and co-authors' *Classical Electrodynamics*, Levich's *Theoretical Physics*, Landau and Lifshitz's *Theoretical Physics*, Blokhintsev's *Quantum Mechanics*, and Kompaneyets' *Theoretical Physics* were the main reference textbooks used. Since some of them were actually too complicated to follow in the regular courses, a few textbooks specially prepared by Cuban authors were introduced to match the syllabuses in a more workable way. It can be easily noticed that the books were typically written either by American or Soviet authors. Some Soviet physicists worked for years in Cuban higher pedagogical institutes, and a number of Cubans studied in pedagogical institutes in the former Soviet Union. The first Cuban PhDs in pedagogical sciences (specialized in physics) studied in the U.S.S.R or in East Germany, after having acquired a working knowledge of the corresponding languages in Cuban institutes specially set up for the purpose.

In later times, though, in response to the need for an urgent increase of practice in the school of the future professors, full-fledged theoretical physics subjects were dropped from the *licenciatura* in education (physics specialization) degree course, while the number of hours devoted to methodology of teaching and to direct school practice were increased.

During the 2002–2003 academic course, in order to match a number of changes made to the basic secondary syllabus (7th–9th grades), the new figure of 'general integral professors' was created. These would graduate from the pedagogical universities after 5 years of study and finish with a certificate of *licenciado* in education. Their job would be to take care of teaching a significant number of subjects at the basic secondary schools, including physics. In the 2003–2004 academic course, in order to match transformations at the pre-university level (10th–12th grades), the figure of professor in exact sciences emerged on an experimental basis. They also had to take a 5-year *licenciado* in education degree course, aimed at training staff to teach both general physics and mathematics at pre-university level. Both, in this and in the previously described case, syllabuses were designed for students to dedicate the greater part of their time to practical teaching after passing the second year of the degree course. The preparation of "general integral professors" was interrupted recently because in secondary schools (7th–9th grades), the professor returns to teach subjects in relation to areas of knowledge.

9.7 Conclusion

After 1959, with the avowed purpose of serving the needs of society, Cuba developed several projects for the training of physics teachers, all of which had the common feature of organically involving students in direct high school teaching practice. The urgent need of physics teachers at different times has been an important driving force in the modifications of syllabuses and the creation of new types of professionals: a process that has gone through several continuities and breaks.

Acknowledgement Syllabuses and textbooks used in physics teaching in Cuba have also been consulted. The author gratefully acknowledges his cooperation with former students and teachers, on whose testimonies much of the above is based.

References

Alamino Ortega, D. de J. 2002. Acerca de la enseñanza de la física en Cuba. *Boletín de la Sociedad Mexicana de Física* 16(3): 151–155.

Alamino Ortega, D. de J. 2005. Raíces históricas de la enseñanza de la física en Cuba: desde Varela a Gran; un material para maestros de física. In *Ensayos Históricos. Anuario del Instituto de Estudios Hispanoamericanos*, Universidad Central de Venezuela, segunda etapa No. 17.

Alamino Ortega, D. de J., and G. Simon. 2004. Physics education in Cuba. 129th American Association of Physics Teachers National Meeting, Sacramento, CA, July 3–August 4, 2004. *Announcer* 34(2), ISSN 1042-0851.

Fernández, J.R. 1986. Conferencia 'Desarrollo de la Educación en Cuba.' *Pedagogía 86, Encuentro de Educadores por un Mundo Mejor, 27 al 31 de enero de Palacio de las Convenciones, Cuba.* Empresa Impresoras Gráficas, MINED.

García, G.J. 1978. *Bosquejo Histórico de la Educación en Cuba*. Havana: Editorial de Libros para la Educación.

Ministerio de Educación. 1990. *Licenciatura en Educación. Carrera Física y Electrónica. Plan C: Programas*. Havana: Editorial Pueblo y Educación.

Sosa, E., and A. Penabad. 2001. *Historia de la Educación en Cuba*. Havana: Editorial Pueblo y Educación.

Chapter 10
Can Universities Develop Advanced Technology and Solve Social Problems?

Isarelis Pérez Ones and Jorge Núñez Jover

10.1 Introduction

This paper presents a case study of how the Cuban universities became directly involved with the economic and social development of the country.[1] It deals with the international debate on the "third mission" of the universities, on the basis of the unique Cuban experience.

The Cuban universities play an important role in national scientific development. Cuba currently devotes 0.49 % of GDP to R&D, a figure that is close to average in Latin America and the Caribbean. The number of researchers per 1,000 economically active persons is 1.06, higher than the Latin American average. Close to 80 % of the R&D is government-financed. The country has 211 public R&D institutes (Núñez et al. 2008b). There are 65 higher education R&D institutions located all over the island. In all municipalities nowadays there are higher education affiliates, and this has brought enrollment ratios up to 65 % of young people in the 18–25 age group (Núñez et al. 2008a).

The transformation of the Cuban university into an institution capable of producing, absorbing, disseminating and applying knowledge of economic and social importance is related to the processes that followed from the 1962 University Reform (Higher University Council 1962).

This paper will show how, from the late 1980s and early 1990s, the Cuban university experienced changes in the orientation and organization of its scientific research and started to be more directly and intensely involved with economic and

This paper was originally published in issue 36(2) (March 2009) of *Science and Public Policy*. It is reprinted here by kind permission of the authors.

[1] The general position of universities in the system of innovation of Cuba is described in Nuñez et al. 2008a, b.

I. Pérez Ones (✉) • J. Núñez Jover
University of Havana, Havana, Cuba
e-mail: isarelis@rect.uh.cu; jorgenjover@rect.uh.cu

A. Baracca et al. (eds.), *The History of Physics in Cuba*, Boston Studies in the Philosophy and History of Science 304, DOI 10.1007/978-94-017-8041-4_10,

social development. In this way, let us consider a case study related to a research group from the University of Havana.

10.2 From Research University to Innovation-Oriented University

The 1962 University Reform underscored the role of scientific research and high-level training, emphasizing sciences and engineering, as well as linking theory and practice in the training of students.

During the 1960s, against the background of economic, political, ideological and cultural transformation resulting from the Cuban Revolution, higher education became a key player in the social transformation of the country. In particular, universities became relevant players in building the emerging national science sector. Social relevance and commitment to society were taken up as key values of the new university and scientific institutionalization.

Practically all research and training programs developed by Cuban universities in the 1960s were oriented towards the economic and social development of the country. However, the relationship between the country's economic and social development did not always enjoy the same success in the following four decades. This success has largely depended on government strategies, and the more or less close relationship between the universities and the main leaders of the country, which, undoubtedly, is unique to the Cuban case.

By the late 1980s and vis-à-vis the evident crisis of European socialism (Cuba's main economically) and the need to generate new sources of wealth for the country, the highest level of the government held the idea that the Cuban scientific community should play a more direct and intense role in the economic and social development of the country.

Research centers were criticized for the scarce application of their results. Several transformations then took place in the scientific and technological policy with the aim to improve the use of scientific results (Rodríguez 1997). Probably the most brilliant result of those changes was the emergence of a robust, biotech-based medical-pharmaceutical industry that currently exports products at a value of US$300 million, that is supported by the West Scientific Pole of Havana (similar to clusters in other countries)[2] (Lage 2011).

In the 1980s, the universities received important financial and political support. Part of the university scientific community focused their work around the full cycle (from research to application of results). Groups were formed and they created production capabilities and in other cases established very close relations with the productive sectors. This innovation in the scientific practice of the Cuban universities was further strengthened by the creation in the mid-1990s of the Office for the

[2] The Scientific Pole of Biotechnology in Cuba includes over 40 institutions, more than 12,000 staff and 7,000 scientists, and has generated over 900 registered patents.

Transfer of Research Results (OTRI), which aimed at providing advice and commercial management to enterprises as well as the export of high value-added technology and products (Alonso and Rodríguez 2007).

However, these transformations affected only one sector of university research. Other groups continued to carry out their activities in traditional research. In the 1990s, the Cuban universities, while maintaining traditional graduate training (although with decreasing enrollment due to the economic difficulties of the country at the time), started expanding post-graduate programs. While some research groups continued with their customary programs, some research groups oriented their research towards innovation and its relation to economic and social development. These transformations are shown in the following case study involving the Laboratory of Synthetic Antigens (LAGS) of the University of Havana (UH).

10.3 The Laboratory of Synthetic Antigens and Quimi-Hib: Science and Society Meet

The LAGS is part of the Faculty of Chemistry of the UH. The UH is the most important institution of higher learning in Cuba, which was founded in 1728. It has 18 faculties and 21 research centers for exact and natural sciences, education, social sciences, economics, the humanities and to a lesser degree, engineering. It has campuses in all 15 municipalities of the capital city. It offers 32 university programs with some 60,000 students enrolled, of which 58 % are women. The faculty includes 1,400 professors and full-time researchers, 50 % of whom hold a PhD degree, and a little over 2,500 part-time researchers. There are 3,000 students (600 grads per year) in some 70 Master's programs. Some 800 students are pursuing their doctoral degrees and around 100 graduate every year. The approval of topics for doctoral thesis takes into account the priorities of the national scientific policy; 85 % of the new doctors studied for their doctoral dissertations in Cuba, although many conduct part of their training abroad. Scientific production in renowned journals, as recognized by the Institute for Scientific Information accounts for 25 % of the national product. From 1995 to 2004 the UH applications for patent protection for inventions amounted to 4.7 % of the total applications made by Cuban residents to the Cuban Office for Industrial Property (García et al. 2007).

LAGS has 19 researchers, nine technicians and seven staff; it groups professionals trained in the areas of chemical engineering and holders of bachelor's degrees in chemistry, biochemistry and pharmacy. Its origin dates back to November 1983, when the carbohydrates group (GC) was formed at the Faculty of Chemistry[3] of the UH, in the context of the above-mentioned institutional and policy changes.

[3] Subsequently Laboratory of Synthetic Antigens (1990).

In the early 1980s, the GC worked with two researchers from the Center of Biopreparations (BIOCEN)[4] on developing a synthetic antigen against mycobacterium leprae, which causes leprosy. Successful results were obtained in 1986, making it possible to diagnose, and this led to the disappearance of leprosy as an endemic disease in Cuba in the 1990s. This was the first successful encounter of the GC with the Cuban healthcare system. Later on, the group participated in the 1989 development of the antimeningococcal vaccine, VA-MENGOC-BC, led by the Finlay Institute.

In the 1980s the meningitis caused by haemophilus influenza Type B (Hib) was the bacterium with the biggest influence on meningitis and pneumonia in the country; consequently, it became one of the priorities of the Cuban public health system.

Vaccination in the world against Hib had begun in the 1970s with a vaccine that proved to be effective for children aged 18 months and older. It was followed by a new generation of vaccines called conjugates, where a process called conjugation was used to chemically bond the same capsular polysaccharide to a protein of bacterial origin. However, a decade after the introduction of the conjugation vaccines, only 38,000 out of the estimated 2.2 million cases every year are protected by vaccination; only 2 % of the children in the world at risk of catching the disease are protected. The introduction of the vaccine in developing countries has been slow, prices are relatively high and Hib kills half a million children every year with pneumonia. The import of the vaccine cost Cuba about US$2.5 million per year.

In 1987, Dutch scientists proved the scientific possibility of obtaining the vaccine through synthetic means. The challenge lay in turning the academic possibility of obtaining a small amount of synthetic antigen into a technology able to produce the antigen for millions of vaccine doses, and that such a process could compete economically with the existing one. In the 1990s, several universities and laboratories worked on alternatives with synthetic compounds, but failed to progress beyond the phase of clinical trials in humans. One of the reasons they had to abandon these efforts was that transnational pharmaceutical companies were not interested in a synthetic vaccine. A conventional bacteria-based vaccine existed; therefore, a second vaccine, although cheaper, was not needed and they were not about to generate competition to a product that was bringing in good earnings (Vérez 2008).

LAGS set out to make the process of chemical synthesis for the reproduction of capsular polysaccharide efficient. In this way, a close cooperation was established between LAGS and various research institutions in the West Pole of Havana and the Ministry of Public Health.

In the case of the cooperation between LAGS and West Pole, the role of the state as a promoter of cooperation networks was crucial. In 1999, the vaccine was given top priority in the Cuban biotech industry and following a decision of the Council

[4] In this case, a number of research-production centers from the West Scientific Pole of Havana were involved: National Biopreparations Center (BIOCEN), Finlay Institute, the Pedro Kouri Tropical Medicine Institute, Center for Genetic Engineering and Biotechnology (CIGB) and the Molecular Immunology Center.

of State, the CIGB, its staff and scientific and productive infrastructure were put to the task of developing the Hib vaccine.

Relevant foreign institutions and international organizations participated in the success of developing the vaccine. Also important was the participation of the University of Ottawa, Canada, through Professor Rene Roy, who is a co-author of the vaccine patent. Similarly, the World Health Organization (WHO) and the Pan-American Health Organization (PAHO) gave support to the Cuban researchers in terms of control methods for conjugates vaccines and the purchase of equipment and reagents that were difficult to get as a result of the US economic blockade of Cuba.[5]

Little by little, the technological process was optimized and the chemical synthesis was made efficient enough to compete with the conventional method.

Clinical trials were conducted in the province of Camaguey, with the decisive cooperation of the network of family doctor offices,[6] as well as the interaction of the educational sector, both primary schools and day care centers.

After 2 years of clinical trials, the vaccine proved to work with infants and induced a very high level of protection. Then the Center for State Control of Medicine Quality[7] issued the manufacturing license and the registration of the vaccine.

As a result of these 15 years of efforts, with the cooperation of several institutions, led by a small lab in the University of Havana, the study was completed and showed that the Quimi-Hib vaccine developed from a totally synthetic antigen is very safe and efficient. It was the first synthetic vaccine for human use approved in the world. Although at least 10 institutions and over 300 people were involved in obtaining it, the main author Vicente Vérez (2006) thinks that the vaccine is 'the first major product of the Cuban bio-tech industry with origins in university laboratories'. Until now four million doses have been produced (Vérez 2008).

The results achieved can be summarized as follows:

1. A world-level scientific and technical result was achieved, proving that talent and capabilities are not exclusively the domain of large companies and that much can be done by the countries of the South.
2. A health problem was solved in Cuba and similar opportunities were opened for other countries.
3. It proved that university science can be both at the forefront of human science and meet pressing human needs.
4. It proved that the motivation to make a relevant social and human contribution can become the main driver for researchers (professors, technicians, students, academic leaders) seeking to achieve major scientific results.

---

[5] Doctors Jose Luis DiFabio from PAHO and Edwin Griffith from the WHO very much facilitated work on the vaccine.

[6] National network at the level of primary healthcare that ensures one doctor and nurse for every 120 families. They are the first level of care in every disease prevention program in the country.

[7] The drug regulatory body of the Republic of Cuba, it performs the basic roles of access control to laboratories, registration of medicines and diagnostic kits, clinical trials, post-sales surveillance, good practices inspections, lot releases and issuance of licenses to establishments.

The results were published by the journal *Science* (2004: 305, 522). The United Nations Task Force that drafted the document, 'Innovation: applying knowledge in development' (2005), reflected on it extensively. The vaccine won the World Intellectual Property Organization Gold Medal Award for Best Invention (2005) and the Health Award from the Technical Museum of San Jose, California (2005). It also received several national and higher education awards.

Work is currently under way on the large-scale production of the vaccine. Over one million doses have been administered to Cuban children. The vaccine has patents in several countries and export agreements have been concluded. It is part of the world's only pentavalent vaccine against diphtheria, tetanus, whooping cough, hepatitis B and *haemophilus influenza* Type B.

According to the researchers themselves, the greatest prize is in having created a vaccine that can save the lives of many children.

10.4 Success Factors, Obstacles and Prospects

The success of the vaccine was based in a research project that was at the forefront of knowledge and also aimed to meet great needs of human health, both in Cuba and elsewhere. It was possible due to governmental support and the existence of a network of high-level centers devoted to research in bio-tech, as well as the support of international cooperation. The success was achieved due to the clear objectives and the perseverance of the team leader and the support received from a group of collaborators very much committed to the social objective sought. The Chemistry Faculty of the UH provided important human resources for research to advance and the university provided an environment of tolerance and understanding for their work. Success, however, is directly related to a number of policies, including:

1. A policy that gives top priority to public health, combining advanced services and own technology with free services. It should be noted that the project was conducted at a time of serious economic crisis in Cuba. Nonetheless, healthcare efforts remained a national priority.
2. The policy has favored the training of human resources, both inside and outside the university. Such a result requires a broad social distribution of training and capabilities.
3. The science and technology policy promotion of the bio-tech industry, which emphasized the health sector and began in the 1980s. This policy led to the creation of several institutions and groups focusing on these issues. The partnership with scientific institutions of the Scientific Pole of Havana proved to be crucial to obtaining the results.
4. The policy favored by higher education of promoting research institutions oriented towards innovation. The emphasis made in innovation did not prevent the understanding that strategic research can require a long time to produce results and it requires tolerance and support. The success of the vaccine was based on

the intelligent articulation of a research project that was at the forefront of knowledge and also aimed at meeting great needs of human of human health, both in Cuba and elsewhere.

Problems certainly abounded. At times, work was limited by resource constraints. Some younger members of the team abandoned the work, seeking economic improvement or less demanding academic work. At times, it seemed that the results would not be achieved; cutting-edge science and technology contains a dose of uncertainty. The groups working to produce papers have more freedom to rectify the course of their work, adjust pace, select results. Research groups aimed at a product or a technology, are required to observe very demanding requirements in the case of the vaccine, have to meet the expectations of the actors funding the project in a reasonable time and work under a lot of pressure. In the case study analyzed here, that demand was not associated with a better economic pay-off. The incentive was, above all, to solve a health problem of great importance. This is a case where ethics, social responsibility of the knowledge community and not the market, let them attain success. Inter-institutional cooperation—without excluding temporary tensions—and not competition between firms made it possible to achieve the targeted goal.

At present, the center works on several forefront issues seeking new vaccines against infectious diseases, cancer and Aids. Thus the LAGS learning process is being exploited for the creation of human vaccine and its experience in the organization of social and institutional networks. As a result of the prestige the center has gained, inter-institutional cooperation and government support has been expanded. LAGS is currently collaborating with the Finlay Institute to develop a vaccine against *pneumococcus*, which is the first cause of respiratory infectious diseases in children in the country. The vaccine was made a priority in the public health system in 2006. In tandem with the Molecular Immunology Center, they are working on a vaccine against breast cancer as well as other therapeutic vaccines for the treatment of several other types of cancer. Currently, new forms of integrating LAGS into the West Scientific Pole of Havana are being evaluated.

10.5 Final Remarks

Through its training and research agendas, Cuban higher education has played an important role in the economic and social development of the country, with contributions from programs related to health, housing, energy, and food production, to mention only some areas. Some universities have also proven to be capable of generating income through exports of goods and services. The orientation towards innovation and economic and social development is the result of the policies applied in higher education. The higher education system has facilitated the link between higher education research and development by emphasizing the importance of social relevance in the strategic planning of the institutions, as well as in evaluation and accreditation processes.

As we have seen, innovation-oriented university research became a priority area in Cuba in the 1980s. Several universities and research groups began orienting their research towards the 'third mission,' each one with its own potentials and particularities. Success has varied, but the examples are very interesting. In this context we could, for instance, mention the Central University of Las Villas, the University of the Orient and various research centers linked to the agricultural sector.

The success with the synthetic vaccine against *haemophilus influenza* Type B (Hib) can be considered an exceptional achievement, considering its worldwide reach, and its scientific, social and economic value. This result, however, was possible only in the context of government policies that depended on support from higher education that was orienting its innovation research towards the economic and social development of the country.

From this perspective, the synthetic vaccine against *haemophilus influenza* Type B (Hib) is *not* an exceptional case. It is the result of accumulated scientific tradition, the ethical commitment of the scientific community, and the applied policies that encouraged the orientation of universities towards economics and social development, or the 'third mission.'

References

Alonso, N., and D. Rodríguez. 2007. *Recuento de la géstion de innovación de la OTRI de la Universidad de La Habana*. Unpublished.

Garcia Delgado, B., R. Arencibia, and M. Sánchez. 2007. *La Universidad de La Habana, indicadores de desempeño. Su impacto en la generación de conocimientos*. Unpublished.

Higher University Council. 1962. La reforma de la enseñanza superior en Cuba, *colección de documentos*. La Habana.

Lage, A. 2011. *La economía del conocimiento y el socialismo: ¿hay una oportunidad para el desarrollo?* Available at http://www.cubasocialista.cu/index.php?q=la-economia-del-conocimiento-y-el%20socialismo. Accessed March 2014.

Núñez, J., G. Benítez, et al. 2008a. Universal higher education and sustainable social development: the Cuban model. *International Journal of Cuban Studies* 1(1): 102–113.

Núñez, J., L. Montalvo, et al. 2008b. Universidad, innovación y sociedad: la Universidad cubana en el sistema nacional de innovación. In *Cátedras de Integración Andrés Bello*, no. 4, 123–184. Bogotá: Convenio Andrés Bello.

Núñez Jover, J., L.F. Montalvo Arriete, I. Pérez Ones, and A. Fernández González. 2008. *University, innovation and society: the Cuban University in the national innovation system*. UniDev Discussion Paper Series paper no. 10. Published online at http://unidevnetwork.org/2006/10/25/cuban-note-on-biotech-2/ by Research Policy Institute, Lund, Sweden.

Rodríguez, C. 1997. Universidad de La Habana: Investigación cientifica y período especial. *Revista Cubana de Educacíon Superior* 17(3): 13–16. La Habana: CEPES.

UN Millennium Project. 2005. *Innovation: Applying knowledge in development*, Task Force on Science, Technology, and Innovation. London: Earthscan.

Vérez, V. 2006, 2007, 2008. Interviews given to the authors at UH.

Part II
Reflections from the Inside

Chapter 11
The Rise and Development of Physics in Cuba: An Interview with Hugo Pérez Rojas in May 2009

Angelo Baracca

Hugo Celso Pérez Rojas was born in 1938, and works as a senior researcher at the Institute of Cybernetics, Mathematics and Physics, at the Ministry of Science and Technology, Cuba. Pérez Rojas is emeritus member of the Academy of Sciences of Cuba, member of the Latin American Academy of Sciences and Fellow TWAS since 1994. He was one of the founders of the School of Physics in the University of Havana in 1962, and moved in 1971 to the Cuban Academy of Sciences. His national awards include the Rafael Maria Mendive and Carlos J. Finlay Medals. He was awarded in 2011 the National Prize in Physics from the Cuban Physical Society. His interests include quantum field theory and its applications to finite temperature problems in high-energy physics and condensed matter. Among these, Pérez Rojas has devoted special attention to quantum electrodynamics in matter and in vacuum in the presence of external fields, phase transitions in electroweak theory, relativistic quantum Hall effect, Bose-Einstein condensation in magnetic fields, and applications of physics to social sciences.

A. Baracca: I agree with you that a first piece of essential advice to the scientific-policy makers of any developing country is to attack the lack of scientific and technical development from its two extremes: by removal of illiteracy and by creating advanced scientific centers starting from the best of their universities. If they do not have enough good universities, they must start to build them. We also agree that this is a sound way of creating the "critical mass" able to start to do advanced research. That is why I believe that the experience of the School of Physics and the starting of international level research in some fields of physics in Cuba—which I have investigated for years in collaboration with several Cuban colleagues—is very important. You are one of the few witnesses of the birth of the School of Physics at the University of Havana, and you are leading at present a developed research group on theoretical

A. Baracca (✉)
Department of Physics, University of Florence, Florence, Italy
e-mail: baracca@fi.infn.it

A. Baracca et al. (eds.), *The History of Physics in Cuba*, Boston Studies in the Philosophy and History of Science 304, DOI 10.1007/978-94-017-8041-4_11,

physics. Thus, I would like to thank you for agreeing to tell me, however briefly, your impressions of those points and reviewing some historical moments and facts.

H. Pérez Rojas: The 1961–1962 University Reform was intended to modernize the old institutional organization, methods and degree courses. It soon faced a lack of teachers because the staff inherited from the old university was not enough, and much worse, quite a few had left the country. Also, as many students had taken an active part in the struggle against Batista's repressive regime, the new University started by sharing its governance with the Federation of University Students (co-governance). It started as a necessary step in 1961, but most of the students initially involved finished their studies pretty soon, and co-governance was dampened down after some six years.

Concerning the development of physics, I can see three distinguishable stages: the foundation, from 1961 to 1967; the consolidation, from 1968 to 1980; and the last stage, characterized by more development, leading to the rise of independent scientific research in several institutions, around 1979–1981. The first two stages were mainly concentrated at the School of Physics (which later became a faculty) of the University of Havana. I will start by talking about them.

I believe that Professor Jose Altshuler, appointed Vice-Rector of the University of Havana early in 1962, played a fundamental role mainly in the first (and later also in the third) stage. An electrical engineer by training, he had been a private tutor of university subjects for a long time and had started doing original research on network theory before 1959, which at the time was quite unique in the very discouraging scientific environment then prevailing in the country. He had a substantial share in shaping and implementing the degree courses, not only in Electrical Engineering, but also in the newly founded schools of Physics and Mathematics.

It is fair to say that the School of Physics had inherited a tradition of high quality – if somewhat limited in scope – teaching from the old School of Sciences, established by professors Enrique Badell, Miguel Angel Maseda, Manuel F. Gran and Marcelo Alonso. Its newly appointed Director, Dr. Ruben Marti, was a talented person, who had graduated in physical-mathematical sciences from the same school. He was strongly interested in theoretical physics, though he had a job as an actuary in the National Bank. Later on, after being appointed adjunct professor at the University of Havana, he started teaching theoretical physics subjects corresponding to the degree course of physical-mathematical sciences, which after Dr. Maseda passed away, had started to be taught by Dr. Marcelo Alonso. Dr. Alonso was a former student of Badell's, who had taken graduate courses at Yale University on modern physics, especially quantum mechanics and nuclear physics. Early in 1959 he faced a rather unpleasant situation at the University of Havana because he formerly held a leading post in an institution created by the Batista regime, which was much hated by the students. But he was duly cleared and kept on teaching as usual. Soon after, however, he accepted a post as a scientific advisor to the Organization of American States and left the country for the United States, where he became a professor of physics and a noted author/co-author of widely used physics textbooks. At this point, it must be said that by the year 2000 he visited Cuba twice, the first time invited to participate in a workshop on physics teaching, and the second time, invited by myself to participate in another physics workshop. He was highly impressed by the country's achievements in physics, and established connections with several Cuban physicists. In 2002 he wrote a short paper in *APS News*[1] urging the American Physical Society to take steps to promote interactions between Cuban and US physicists.

A. Baracca: This a very interesting fact: to return to his native country after nearly four decades, finding unexpectedly how science had developed to an impressive level despite many obstacles. I suppose this impressed him very much. But let us continue your talk about the initial steps.

H. Pérez Rojas: Dr. Marti started with great enthusiasm and good will in his new position as director of the new school, but faced, however, some challenging problems. First of all, a lack of suitable laboratories, books and journals, but above all a lack of national, highly qualified professors. Still, I remember that soon after a group of foreign teachers and technicians became available for various reasons, such as Claude Monet, a French electrical engineer, Amanda Blanco, a young Spanish-Russian physicist, Richard Bourret, an American mathematical physicist, Theodore Veltfort, an American electrical engineer, and Dina Waisman, an Argentinian solid state physicist. Trevor Marshall, a noted British physicist who came as visitor from Kingston University, started to teach quantum mechanics. I joined the School of Physics staff in 1961 as instructor of theoretical physics and mathematical methods of physics. But the staff was incomplete. For instance, the first group of new students could not take a course on statistical mechanics because no suitable teacher was available for the task. Hypothetically, the new School of Physics should have provided for specialization in radio wave propagation and fluid mechanics, but this proved to be an unrealistic dream. Students finished their studies with a generally good basis, but without some of the advanced courses and with minimal specialization. They had been taught excellent courses on classical mechanics and electrodynamics by following Goldstein's *Classical Mechanics*, parts of Sommerfeld's *Electrodynamics*, and other good textbooks. But the set of basic physics courses was not at all complete. Around 1966, Dr. Vladimir K. Grishin, an excellent professor of quantum mechanics, came from Moscow University.

From 1963, Dr. Marti had moved to the position of head of the Department of Theoretical Physics and Methods of Mathematical Physics. The new director was Dr. Juan Prohias, a professor of chemistry. Dr. Marti wanted to go farther and applied for a scholarship at the University of Göttingen. He got it, but the Rector of the University did not agree to back it, and instead proposed a similar scholarship for an East German University, which had collaboration arrangements with the University of Havana, however he did not agree. Interest in applying for PhD courses abroad could have been understood by anybody teaching at the School of Physics a few years later. By then the evolution of the School would require its

[1] Marcelo Alonso, "The Current state of Physics in Cuba: a personal perspective": reproduced in this volume.

professors to have a PhD degree and become independent researchers capable of teaching life sciences, not only textbook science, to advise students in research tasks, and even to lead new scientific groups.

A. Baracca: I heard that Ché Guevara visited the School of Physics at the University of Havana. Is that true?

H. Pérez Rojas: Yes, it is. In 1963, while Dr. Prohias was director, the School of Physics was visited by Ernesto Ché Guevara, at that time Minister of Industries. He suggested among other things working on semiconductor research and development, and its connection with the large-scale development of automation equipment and devices. Guevara had very advanced ideas, as he explained: "We are arriving at the automation and electronics era … electronics is a fundamental political task for our country." He had an extraordinary intuition about what then became the present era of informatics and communication systems, where physics has played an essential role.

A. Baracca: I heard also that you taught mathematics to Ché Guevara?

H. Pérez Rojas: Actually I did have such an honour, to which I should add that I gave him the invitation to visit the School of Physics. The story is as follows: Dr. Marti agreed with the Director of Automation of the Ministry of Industries to offer a course on information theory. It required an introduction on probability theory, which was preceded by a review of basic calculus, differential equations and complex variables. I was in charge of that review of mathematical topics. It took about 1 year, from 1962 to 1963. Before starting my teaching I was told that my students were a group of engineers. I was very surprised when I saw that Ché Guevara joined the group. After I finished my course, Dr. Marti continued with the teaching of probability and information theory.

In 1965 I was appointed Director of the School of Physics, after Dr. Prohías left the post. While at the time some people used to say that "the school was deformed in the theoretical sense," the truth was that we had weaknesses in both theoretical and experimental teaching, and many more regarding research. After this moment, the School underwent hard times. Dr. Marti resigned, along with some of the other teachers. This reduced our national teaching staff to a minimum. The role of the School of Physics was not properly understood by some University authorities. For instance, we were asked to start working on a draining project of the Zapata swamp. Since we did not have experts in soil physics, to embark on such a task would have diverted the School completely from the study of basic physics.

A. Baracca: And what was the students' attitude? I have known that students taking advanced courses often collaborated in teaching the younger students, due to the lack of teachers. How successful was this experience?

H. Pérez Rojas: Actually, this started to be a solution to a problem and ended as a problem requiring a solution. Such duality of student-teacher became a challenging source of difficulties, which conspired against the improvement of the teaching quality. There is a story about the visit of a professor from Czechoslovakia, who was an expert in magneto-hydrodynamics, and the School invited him to teach a short course on his field of expertise for the students' benefit. The reaction of some teacher-students was negative: their lack of maturity was a barrier to understanding that it should be useful to take such a course from an expert in that field. On the other hand, for students, teaching was a drain and a waste of time and energy, and for them it was more profitable to concentrate on their own duties.

Thus, I soon realized that three basic points needed to be changed: to improve the basic training of students in modern physics, to complete the teaching staff, and to start research activities. These three tasks were intertwining. Accordingly, I concentrated my efforts firstly in attracting young physicists who had recently finished their M.Sc. at Soviet Universities and were already in Havana, secondly, in improving the curriculum, and thirdly, in establishing collaboration agreements with Moscow University and the Ioffe Physical-Technical Institute in Leningrad.

After a few years the School of Physics, which later became the Faculty of Physics, started to be recognized as a stable and mature institution, having a capable staff who had established a tradition of serious teaching and research. I left as head of the School in 1968, though kept on teaching some statistical physics and thermodynamics courses for a while. In 1971 I moved to the Department of Electronics of the Academy of Sciences. In 1975, the Department of Electronics, led by Professor Altshuler, became a physical-technical institute, the Institute for Fundamental Technical Research (ININTEF), where several research groups were organized, some of which remain active up to the present time. From 1970 to 1980 a large number of the staff members of the Faculty of Physics and from ININTEF, among other institutions, started work on their Ph.D. degrees, either by staying in Eastern European academic institutes for 3–4 years, or by shorter yearly stays.

I believe that by 1980–1981, our country had achieved a critical mass of highly qualified physicists, able to do advanced research. By then, research was being done on semiconductor materials, magnetism and later on superconductivity, nuclear reactions, quantum field theory and on quantum statistics, optics, and other fields.

In addition to the above-mentioned institutions, new ones under the leadership of the Secretary of Nuclear Affairs were founded. The participation of a Cuban cosmonaut in the Soviet-Cuban space flight also stimulated the development of several interesting experiments to be performed under zero gravity conditions for the first time.

A. Baracca: You have mentioned the support given to the development of Cuban physics by the old Soviet Union and some other ex-socialist countries. But didn't Cuban physics also benefit from collaboration with West European countries?

H. Pérez Rojas: Certainly. I would like to mention the collaboration with institutions and scientists from France and Italy after 1968. Several summer schools were organized and were very fruitful. A systematic collaboration with the University of Parma worked for several years. We must emphasize the valuable support given to Cuban physics in the last two decades by the Abdus Salam International Centre for Theoretical Physics (ICTP) and the Third World Academy of Sciences (TWAS), both of them in Trieste.

One can say that at present the Faculty of Physics of the University of Havana, the Institute of Cybernetics, Mathematics and Physics (ICIMAF), various institutions in Santiago de Cuba and Santa Clara, and some others throughout the country are tuned in to international science, and at the same time contribute to the scientific, technical and educational development of Cuba. Nobel Prize winner Abdus Salam said that physics is the science of riches for a country.... Perhaps it has not yet given enough riches to Cuba, but I believe it will.... I believe that Salam was right!

Chapter 12
An Interview with Professor Melquíades de Dios Leyva, December 2008

Olimpia Arias de Fuentes

12.1 Introduction

When writing about the history of physics in Cuba, this remarkable professor of quantum mechanics must be mentioned, for he embodies a most genuine example of the turn taken by national educational policy after 1959: Education for all, at all levels, with no discrimination or elitism. The following is an interview granted by Dr. Melquíades de Dios Leyva, Outstanding Full Professor of the Physics Faculty of the University of Havana, to Dr. Olimpia Arias de Fuentes, Associate Professor at the same, and Senior Researcher of the Institute of Materials Science and Technology (IMRE) of the University of Havana.

12.2 Biographical and Professional Note

Melquíades de Dios Leyva was born on December 9, 1938 in the province of Santiago de Cuba. By 1959, although he had barely completed primary school, he had participated in the underground struggle that led to the January 1959 victory of the revolutionary forces, which fought against the tyranny that had seized power in the country, and was a member of the Rebel Army at the Sierra Maestra's Third Front. Soon afterward, the social changes that were being promoted in Cuba gave him the opportunity to complete his basic education. After successfully passing the entrance exams to the University of Havana in 1963, he was included in the vast scholarship program implemented by the Revolution, and completed his 5-years *licenciatura* in physics (created by the 1962 University Reform). Immediately after his graduation in 1968, he was awarded a junior teaching post at the Physics School

O. Arias de Fuentes (✉)
Institute of Materials Science and Technology, University of Havana, Havana, Cuba
e-mail: oarias@fisica.uh.cu

A. Baracca et al. (eds.), *The History of Physics in Cuba*, Boston Studies in the Philosophy and History of Science 304, DOI 10.1007/978-94-017-8041-4_12,

of the University of Havana. He was the first member of a higher education teaching staff in the country to obtain a MSc degree. In 1979, after a 4-years stay at the Moscow State University (Lomonosov University) in the former Soviet Union, he was awarded a PhD in Physico-Mathematical Sciences.

After more than four decades of teaching experience, he has contributed outstandingly to the training of the vast majority of Cuban physicists, who hold his pedagogic skills in high regard. He has written textbooks and published many scientific papers in important specialized journals, both national and international. A former Associate Member of the International Center for Theoretical Physics (ICTP, Trieste), he has been given various awards, including the Frank Pais Order, by the Council of State at the request of the Minister of Higher Education, for his long and brilliant teaching career and his contribution to the development of Cuban education. He has also held the Carlos J. Finlay Order, the highest honor awarded by the State to scientific personalities in recognition of their contribution to national science and the National Prize of Physics awarded by the Cuban Physical Society for his important role in the development of Physics in Cuba.

12.3 Interview

O. Arias de Fuentes: We know that you were born and lived until your early twenties in a small country village which belonged to the municipality of Palma Soriano in the province of Santiago de Cuba. How was it possible for you to gain access to higher studies in the country's capital?

M. de Dios Leyva: True, being of peasant origin I spent the first 20 years of my life in a humble country village in the former Oriente province. There was only a small primary school in the village and the majority of the people were illiterate, either totally or functionally. At the time, nobody in the village had any possibility or hope of accessing higher levels of education; we were sentenced to permanent ignorance. However, the victory of the Cuban Revolution on January 1st, 1959 completely changed the social outlook of the country, and our destiny in particular. Right from the very beginning there was a general call for the people to step up their educational level and I immediately jumped on the bandwagon. I arrived in Havana in January 1959 as a member of the Rebel Army and began right away to make the most of the new chances to enroll in a variety of courses.

O. Arias de Fuentes: Did you ever think of becoming a university professor?

M. de Dios Leyva: Never. In my small country village it was impossible to imagine that, not even in a dream. But I must tell you one thing: when in January 1959 I saw for the first time the grand entrance steps of the University of Havana and its imposing Alma Mater statue at the top, I felt an inner voice telling me: "You can do it, it all depends on you." And so it was, for a few years later I became a student at this beloved University.

O. Arias de Fuentes: Why did you choose a career in physics?

M. de Dios Leyva: That idea came up in a chat that I had with my high school physics teacher. The conversation had started when he casually said that his family was of very humble origins, and that to be able to complete his studies he had to make a lot of personal sacrifices. Among other things, he told me that he had to work very hard to pay his own way at college, and that most of the time he was forced to study late into the night by candlelight. Aware of my equally humble origins, he asked me what university degree course I wanted to take. I said I didn't know, but added that I liked very much the geometry and geometrical optics stuff I had come across in high school. Then, he said, "your future is in physics." And so it was.

O. Arias de Fuentes: How did you develop in your special area during the past years, and what activities did you undertake? What do you think this has contributed to physics in Cuba, not only from a purely scientific point of view, but also relative to the training of other professionals?

M. de Dios Leyva: Well, the educational and employment policies implemented by the Cuban Revolution from its very beginning allowed me to pursue a continuous line of development in my profession. Immediately after graduation in 1968 I got a teaching post at the then Physics School of the University of Havana, which in the end made me a teacher of the vast majority of the physics community in our country, thus being instrumental in exerting what I believe was a positive influence on their professional and human development. At least, that is what many of my former students told me after their graduation.

Quite a few foreign scientists from different countries who were sympathetic to the revolutionary process that was taking place here at the time visited us in the 1960s and 1970s to give summer courses in science. These had an important influence on the country's scientific development in general, and in particular speeded up my own scientific development. As a result, in 1972 I successfully defended my MSc dissertation in the physical sciences. Two years later I was chosen to do a PhD in the former USSR, where in 1979 I obtained my doctoral degree in Physico-Mathematical Sciences at the prestigious Moscow State University. Thereafter I continued doing research work, having published more than 90 scientific papers in specialized journals, and two textbooks related to the courses I teach. Some of my results have been especially valued by different scientific institutions; I have been awarded the rank of Outstanding Full Professor of the University of Havana.

O. Arias de Fuentes: What are you feelings regarding your contribution to the training of physicists abroad?

M. de Dios Leyva: Yes, I have also helped train physicists and chemists abroad by teaching basic and graduate courses. This I did voluntarily during my 2 years in Angola at the University of Luanda. It is comforting to come across someone in a foreign country once in a while who thanks you for your contribution to his professional training. For me it is quite important to say that several of my former students who have shared that feeling are now teachers and scientists who have earned national and international scientific recognition.

Fig. 12.1 Professor Melquíades de Dios Leyva teaching at the University of Havana. Photo courtesy of Dr. Olimpia Arias de Fuentes

O. Arias de Fuentes: According to your experience, what would be your advice to future professionals specializing in physics and other natural and exact sciences?

M. de Dios Leyva: I would tell them that teaching and scientific work is very important for the sustained development of any country. This is especially true for Cuba, compelled as it is to boost as much as possible those activities to successfully overcome the handicap presented by the limitations in our natural resources. Teaching and scientific work requires, of course, lots of dedication and persistence. Those teachers and scientists who have made important contributions to the progress of their countries and of humanity have been noted for their strong and sustained commitment to their work. Einstein had to devote more than 10 years of his very productive life to developing general relativity, even though he was an extraordinary genius. To those who want to become good professional scientists, I would summarize my advice in three words: earnestness, dedication, and perseverance. Add to this that I wholeheartedly adhere to what our great José Martí wrote in the nineteenth century: "Every person has the right to be educated, and then should repay for it by contributing to the education of other people."

Chapter 13
Experimental Semiconductor Physics: The Will to Contribute to the Country's Economic Development

Elena Vigil Santos

During the second half of the 1960s the Physics School staff increased sharply.[1] Before this, it had experienced near depletion due to the emigration of many of its members shortly after the triumph of the Revolution. This renaissance took place thanks to the incorporation of the first physicists to graduate from the Physics School itself and to the return of those who had graduated abroad, mainly from the former Soviet Union. This group of young physicists—with a heavy schedule of teaching duties in spite of their lack of experience—gave strong support to the development of physics in Cuba.

The group worked intensively on the redesign of departments and programs related to various subjects and devoted themselves to the introduction of duly oriented scientific research. These young professionals were very conscious of the need to become seriously involved in research work in order to become truly effective professors. In this way, they would be able to teach not only standard textbook knowledge, but also to offer the students new knowledge derived from original research. Involvement in research work was considered not just as the basic premise for achieving high quality teaching, but also as a way of contributing to the scientific development of the country.

This young but dedicated staff identified the development of research at the Physics School as their contribution to the new national goal to leave behind underdevelopment through the creation of a knowledge- and technology-based industry. The environment of change in social and economical structures, of "revolutionary effervescence" and of collective euphoria helped to undertake tasks that would be difficult to imagine in other countries with similar social and economical problems.

[1] The Physics School, specifically dedicated to the training of physicists, was created in 1962 as part of the Faculty of Sciences of the University of Havana. Before that, the only responsibility of the Faculty was the teaching of three four-year degree programs: physico-mathematical sciences, physico-chemical sciences and natural sciences, essentially designed to train high school teachers.

E. Vigil Santos (✉)
Faculty of Physics, Havana University, Havana, Cuba
e-mail: evigil@fisica.uh.cu

A. Baracca et al. (eds.), *The History of Physics in Cuba*, Boston Studies in the Philosophy and History of Science 304, DOI 10.1007/978-94-017-8041-4_13,

The Physics School was just another example of the same scenario. Joining efforts toward the development of physics research at the University of Havana and the rest of the country was also a principle of action in those years.

Some of the young graduates turned into professors by the circumstances, who had already acquired some research experience related to their graduation thesis work, participated in an intense debate about which research subjects were the most convenient to pursue, and about which lines would contribute most to the country's development. Solid-state physics and, in particular, semiconductor physics, were favored due to their novelty, accelerated progress, and popularity in those days. Quite a few foreign professors visiting the Physics School at the time also supported this point of view and some of them came at different times to help the scientific development. Moreover, on the occasion of the International Cultural Congress held in Havana, a number of distinguished physicists visited the country at the end of 1967. By coincidence, during their brief stay the first semiconductor diode fabricated by the alloying technique at the Physics School was obtained. All this catalyzed the decision to join efforts around solid-state physics as the main scientific target. After this congress, from 1968 to 1972, the Physics Summer School Programs organized by European—mainly French—scientists were quite valuable in developing research on solid-state physics at the University of Havana.

Groups were dedicated to experimental research on semiconductors, metals, and magnetic materials. This decision, aimed at some foreseeable emerging industries, clearly illustrates the ideal of Cuban physicists at the time to achieve not only better academic performance, but also to contribute to the economic development of the country as far as they could. It is worth mentioning that until the second half of the 1970s, the structure of the Physics School was designed to promote research development because of the potential economic impact of its results, and also because of its methodological value for training students. A double structure was created: one for the teaching departments, and a parallel one for the research departments. Therefore each professor had a double subordination and function. Research work did not adapt itself to the "traditional" subjects of the teaching departments but tended to concentrate on those subjects considered to have a strategic interest.

Due to its particular importance in the above context, what follows will provide an overview of the School's experimental work on semiconductor physics from the late 1960s to the end of the twentieth century.

The initial tasks taken on by the semiconductors group created within the Physics School of the University of Havana were related to the development of new microelectronic and optoelectronic devices, as well as to crystal growth techniques. The initial work consisted in the development of technological and characterization set-ups. Foreign advice was instrumental in the establishment of such an infrastructure. The newly created mechanics, electronics and glassware workshops were also important. The "First Seminar on Scientific Research" organized by the Faculty of Sciences was held in 1971. The results presented there included only the development of experimental set-ups and techniques, i.e., they reflected the creation of an experimental infrastructure for future research. It is worth mentioning that this ability—which continued to progress—was crucial in the development of science and technology in spite of a scarcity of resources.

Some significant results in semiconductor physics were rapidly achieved in the initial period up to 1975. For example, silicon technology for microelectronics was developed based on experimental set-ups built entirely by researchers themselves. This enabled the fabrication of CMOS capacitors and, later on, the development of integrated circuits based on metal-oxide-semiconductor (MOS) technology. In addition, the first monocrystals were grown, and gallium arsenide light-emitting diodes were produced. In addition, compound semiconductors were optically and optoelectronically characterized, as well as devices based on them.

However, the first publications by Cuban experimental physicists in "Web of Science" journals did not appear until the late 1970s. This may be explained by the efforts demanded by the difficult task of attaining rapid results of both scientific and socio-economical interest. Lack of experience of these physicists in publishing was also a significant handicap at the time. During the IV Latin American Symposium on Solid State Physics (IV SLAFES), held in Havana in 1975, Cuban physicists presented a series of quite impressive research results, showing a completely different situation from the one at the time of the Cultural Congress 7 years earlier.

In the second half of the 1970s a true contribution of solid-state physics to the development of the country materialized: the Cuban government decided to buy and set up the "Ernesto Ché Guevara" Semiconductor Factory, conceived for the production of discrete semiconductor components and integrated circuits. The idea was strongly motivated by the achievements of the Physics School at University of Havana in the field of microelectronics and of the Microelectronics Laboratory belonging to the "José Antonio Echeverría" Higher Polytechnical Institute (ISPJAE). The Physics School staff very seriously collaborated with the project. At the same time, it was decided that research in microelectronics should be continued outside the Physics School. This resulted in the movement of an important group of professors to the ISPJAE and to the Semiconductor Factory in the western province of Pinar del Río, where they continued their work.

In spite of the above, during the second half of the 1970s other research lines in crystal growth and optoelectronics flourished at the University of Havana. Different types of monocrystals (silicon, germanium and InSb) were grown and studied, as well as planar $ZnIn_2S_4$ crystals and CdS films. The Semiconductor Factory was technically advised on the technology for growing silicon crystals. The first red light-emitting diodes (LEDs) were obtained, and even integrated light digits were fabricated. At the same time, publications in the "Web of Science" journals began to appear (Hernández et al. 1976; Vigil et al. 1978), and papers in the proceedings of scientific meetings and national publications became more and more frequent.

Research on optoelectronic materials and devices at the University's Semiconductor Laboratory found very strong support from the laboratory then led by Dr. Z.I. Alfeorov at the Ioffe Institute of the USSR Academy of Sciences in what was formerly Leningrad (Dr. Alfeorov won the Nobel Prize for Physics in 2000). This collaboration was essential for the training of Cuban PhDs and continued from the mid-1970s up until the collapse of the Eastern Bloc. Besides joint publications of Soviet and Cuban physicists, there were various achievements that had real social and economic impact. For example, a solar panel based on AlGaAs with light concentration and sun-tracking was made, while new optoelectronic device technologies were developed. Semiconductor lasers were fabricated and studied in the University laboratories, as well as LEDs, solar cells and silicon photodiodes.

The team working on semiconductors was responsible at the University of Havana of the preparation of experiments in orbit for the Soviet-Cuban space flight which took place in September 1980. Because of their importance they are described in a separate article of this volume.[2]

The *Revista Cubana de Física* (Cuban Physics Journal) appeared in 1981. In its first year, 13 out of a total of 27 papers submitted from the whole country dealt with the physics of semiconductors, four of them produced by the Physics School's Theoretical Physics Group.[3]

Research maturity in semiconductor materials, structures and optoelectronic devices was achieved in the first half of the 1980s. The groups managed to master techniques of structure growth, and those needed to make semiconductor lasers, LEDs, solar cells and photodiodes. Among the techniques mastered, it is worth mentioning, for example, the liquid phase epitaxy for the growth of GaAs/AlGaAs-based heterostructures. Several techniques—some of them original—were developed to characterize and study the structures and devices obtained. Scientific publications based on those results were authored predominantly by research teams of Cuban physicists.

In 1985 the Institute of Materials and Reagents for Electronics (IMRE) was created at the University of Havana with the purpose of providing technical backing for the country's developing electronics industry. The development of technologies for obtaining LEDs and silicon solar cells in the Physics Faculty (formerly known as the Physics School) were among the relevant scientific results shown in order to justify its creation at the University of Havana. The experimental groups working on semiconductors were transferred to the IMRE, though not all at the beginning. Professors were incorporated as "adjuncts" and the number of full-time researchers increased with time. Again—but with small differences—the Physics Faculty had a research structure parallel to the teaching structure to promote the development of the country.

In the late 1980s, LEDs technologies were transferred to the Semiconductor Factory in Pinar del Río (diffusion-based first, epitaxially grown later). In fact, in the long run COMECON intended Cuba to be entrusted with the responsibility of developing and producing optoelectronic devices for all the socialist countries belonging to it. There was also a strong collaboration between IMRE and the Pinar del Río Semiconductor Factory for the transfer of photodiodes and silicon solar cells technologies.

The collapse of the Soviet block brought the Cuban electronics industry to a grinding halt. Nevertheless, the existing research groups at the University of

[2] For more information on these experiments, see the chapter in this volume by José Altshuler et al. on Cuban technophysical experiments in space.

[3] The Theoretical Physics Group of the Physics School concentrated since the beginning (at the end of the 1960s) basically on semiconductor physics. Although there were efforts toward joint experimental and theoretical work and a few examples of this, theoretical Cuban physicists proceeded independently. They have created an abundant scientific production published in high-impact scientific journals.

Havana managed to continue work on semiconductor materials physics and devices. It should be underlined that in the period 1993–1994, ultra-thin GaAs films were grown with a Russian Molecular Beam Epitaxy (MBE) device, but the work could not be developed further. The 1990s was an extremely harsh period for the Cuban economy, which affected every aspect of life. Due to the critical lack of financing there was a severe depression in the experimental activity based at IMRE. On the other hand, a strong accent was placed on international collaboration, especially with Latin-American laboratories, where in many cases the maturity reached by Cuban physicists was instrumental in optimizing the use of the "high-tech" experimental equipment existing in those laboratories. This resulted in joint publications and training of undergraduate and graduate students in Cuba and abroad.

Simultaneously, every effort was made to keep alive as far as possible the Cuban-based experimental activities. Keeping in mind that doing research was the only way to produce better physicists and new knowledge, and to contribute to the social and economic development of the country, laboratory staff concentrated on developing semiconductor materials and structures; particularly those potentially useful for the production of solar cell devices on an industrial scale. Throughout the 1990s, the collaboration with the semiconductors factory was kept alive by giving technical support to the manufacturing of solar panels. Collaboration expressed itself in developing appropriate technologies for producing silicon solar cells, and in providing the required systematic training for the technicians and specialists in photovoltaic energy.

In conclusion, it should be mentioned that the number of Cuban papers appearing in impact journals increased sharply in the difficult 1990s[4] thanks to the experience accumulated over the years and to a sizable dose of determination, as well as to the friendly collaboration achieved with foreign institutions. Such collaborations not only allowed Cuban-based research to be complemented by gaining access to structural and other techniques, which required expensive equipment, but also promoted a fruitful exchange of ideas and friendship with many esteemed colleagues.

Acknowledgments The author thanks José Folgueras, Adriana Fornés, Juan Fuentes, Osvaldo de Melo, María Sánchez-Colina, Daniel Stolik, and Julio Vidal for their observations, most of which have been incorporated into the present text.

References

Hernández, L., O. Vigil, and F. Gonzalez. 1976. Optical transitions and distribution of localized levels in $ZnIn_2S_4$. *Physica Status Solidi (a)* 36: 33–37.

Vigil, E., J.A. Rodríguez, and R. Pérez Álvarez. 1978. Optical constants of *n* and *p* type GaAs between 2.5 and 3.5 eV. *Physica Status Solidi* (b), 90: 409–414.

[4] See the chapter by Ernesto Altshuler in this volume on the impact of Cuban physics through scientific publications.

Chapter 14
Cuban Techno-physical Experiments in Space

José Altshuler, Ocatvio Calzadilla Amaya, Federico Falcon, Juan E. Fuentes, Jorge Lodos, and Elena Vigil Santos

14.1 Introduction

When Cuba joined the Intercosmos Program of the socialist countries in the mid-1960s, the great educational and scientific reform taking place at that time in the country had hardly begun to bear fruit. But when, a decade later, the Soviet Union offered all the participant countries the chance to make use of its space vehicles and related installations so that their cosmonauts could carry out original scientific experiments in space, the situation had changed radically in Cuba. In a short time around 200 people already involved in scientific and technological activities succeeded in designing and setting up—in close collaboration with various Soviet, East German and Bulgarian institutions—some 20 scientific experiments that were to be carried out in orbit around the earth during the joint Soviet-Cuban space flight of September 18–26, 1980 (Altshuler 1984). Those experiments, and a further one that was also set up for the same space flight—but carried out during a later flight, as mentioned below—are historically important since they were the first in their class to be carried out by humans in space under microgravity conditions.

J. Altshuler
Cuban Society for the History of Science and Technology, Havana, Cuba
e-mail: jea@infomed.sld.cu

F. Falcon
Institute of Sciences and Technology of Materials (IMRE), University of Havana, Havana, Cuba

J. Lodos
Faculty of Chemistry, University of Havana, Havana, Cuba
e-mail: jorge.lodos@zerus.azcuba.cu

O. Calzadilla Amaya • J.E. Fuentes (✉) • E. Vigil Santos
Faculty of Physics, University of Havana, Havana, Cuba
e-mail: calza@fisica.uh.cu; jfuentes@fisica.uh.cu; evigil@fisica.uh.cu

A. Baracca et al. (eds.), *The History of Physics in Cuba*, Boston Studies in the Philosophy and History of Science 304, DOI 10.1007/978-94-017-8041-4_14,

That the Cuban cosmonaut, Arnaldo Tamayo Méndez, on this flight became the first Latin American and Caribbean national to reach outer space and successfully work in orbit is a milestone in the history of space exploration. In the present context, however, it is worth recalling in particular that the contribution of Cuban scientists to the mission accomplished was of sufficient interest for the host country to strengthen its scientific collaboration in space research with Cuba. This continued to be successfully carried out until the closing of the Intercosmos Program for other non-scientific reasons.

The Cuban experiments set up for the above space flight can be grouped into four categories: biomedical, psychometrical, earth exploration, and techno-physical. Those in which the country's physicists played a central role were included in the last category under the names *Caribe*, *Azúcar*, *Zona*, and *Holograma* (Altshuler 1984; Avduyevsky 1984, 13–17). A brief description of each of these is given below.

14.2 Experiment Caribe

Developed by the Solid State Electronics Laboratory (LIEES) of the University of Havana's Physics Faculty, this experiment was designed as a follow-up to the experiments on obtaining crystals and alloys under microgravity conditions previously carried out at the *Skylab* space laboratory and at the *Soyuz-Apollo* orbital complex. It comprised two independent sub-experiments, the Caribe K-1 and the Caribe C-1 or SK-1, which made use of the *Saliut 6* orbital station ovens *Kristall* and *Splav*, respectively. They can be briefly characterized as follows:

14.2.1 Caribe K-1: Growth of a Ge-In Alloy

Its purpose was to study the degree of perfection obtained in crystals grown in orbit as compared to that obtained in crystals grown under analogous ground conditions, and to study how microgravity conditions affect the distribution of impurities during the recrystalization of strongly In doped Ge. Special graphite parts and a quartz spring were introduced to fix the samples in place inside the quartz ampoules employed, a novel technique to avoid the possible breakage of the ampoules due to the vibrations of the orbital station (Calzadilla et al. 1984). By applying different characterization techniques to the Ge-In samples the following results were obtained (Calzadilla 1994; Calzadilla et al. 1985, 1986):

- Though an air bubble became occluded in the space-grown sample, it was much more perfect than the twin sample grown on the ground.
- Induced by the temperature profile of the Kristall oven, two circulation cells appeared within the melt, which gave rise to an unexpected variation in the impurity distribution curve in the crystal grown in orbit. This confirmed that under microgravity conditions, circulation cells may be formed due to the temperature profile of the oven, and that thermocapilar currents and Soret thermal diffusion may appear.

- Evidence of convection within the melt under microgravity conditions was confirmed by means of experiment modeling and numerically solving the corresponding hydrodynamics equations.

The results of the Ge-In crystal growth in the *CARIBE* experiment can be found in NASA report CR-208314 (Riegel and Wilcox 1996).

14.2.2 Caribe C-1 or SK-1: Growth of Epitaxial GaAs and AlGaAs Layers

Monocrystalline GaAs and AlGaAs layers on GaAs substrates were obtained for the first time under microgravity conditions (Díaz et al. 1983, 1984; Vigil 1984; Vigil and Díaz 1981; Vigil et al. 1984). These epitaxial layers were grown from a 1.5 mm thick gallium solution by forced cooling. The experimental set up developed allowed to grow 16 layers simultaneously in four different temperature regimes. Substrates and growth solutions were placed inside small graphite containers of novel design and these were included in a vacuum-sealed quartz ampoule. This allowed for comparative analyses among epitaxial layers obtained under microgravity, as well as with similar samples grown at ground level. Zn doping of the layers during the growth process was studied, and, additionally, for the first time light-emitting diodes were obtained from the structures thus formed. Other results were the following (Vigil 1984):

- Comparison of the morphology of the layers grown under microgravity conditions with the morphology of those grown at ground level showed incomplete homogenization of the growth solution in orbit due to the absence of convection, and also to the relatively greater influence of surface tension.
- Dispersive X-ray analysis (EDAX) applied to aluminum concentration profiles measured in different points of the AlGaAs layers also pointed to the non-homogeneity of the solution of Al in Ga due to the absence of convection. It was determined that the diffusion coefficient of Al in Ga was less than $5.10^{-5}cm^2/s$.
- Photoluminescence analysis showed that under microgravity conditions, greater Zn concentrations were incorporated to the grown epitaxial layer. Modeling of these spectra showed that higher values of diffusion length of the electrons in GaAs:Zn were obtained, which indicated a lower concentration of structural defects. A high incorporation of Zn was confirmed by electroluminescent emission analysis.

14.3 Experiment Azúcar

The purpose of this experiment was to determine the macroscopic effect of microgravity on the crystallization of sucrose from its aqueous solution. This was to be the first time that the crystallization of an organic material was studied in space under microgravity conditions. The study was undertaken by specialists from the Cuban

Institute for Sugar Research (ICINAZ), who, with the experienced advice of Soviet space specialists, supplied for the purpose an optical bench, a crystallizer, injectors and extractors, plus the primary monocrystals grown, and the solutions to be used. The photo sequence for the kinetic study was taken with the permanent photographic equipment of the space station. It was found that the growth rate of sucrose monocrystals in space is eight times greater than that under the usual Earth conditions, a phenomenon whose mechanism was assumed to be related to the mechanism of the incorporation of sucrose molecules to the crystal lattice (Morera et al. 1983, 1984).

14.4 Experiment Zona

This experiment was designed and implemented by specialists from the Cuban Institute for Sugar Research (ICINAZ). It was aimed at determining the effect of microgravity on the microtopography of a crystal surface and on the molecular kinetics during the crystallization process using the technique known as zonal fusion with temperature gradient. The experience gained suggested the possibility of simulating at ground level the crystallization process in microgravity when the crystallization speed of the crystal face under study is antiparallel to the force of gravity (Falcón Rodríguez et al. 1984, 1985). Further elaboration of this idea was the central subject of a subsequent doctoral thesis (Falcón Rodríguez 1987).

Information from Azucar and Zona experiments has been useful in working out new models in materials science research.

14.5 Experiment Holograma

The original idea of this experiment can be traced to Roberto Homs, at the Institute for Fundamental Technical Research (ININTEF) of the Cuban Academy of Sciences, and was further elaborated and equipped by the scientific and engineering staffs of ININTEF, the Cuban Technical Military Institute (ITM) and the Ioffe Physical-Technical Institute of the USSR Academy of Sciences. It was designed as a two-part experiment. The purpose of the first one (*Holograma-1*) aimed at the evaluation of the degree of immunity to noise that the redundance typical of a hologram could offer to a TV transmission of two-dimensional images from space. The second part of the experiment (*Holograma-2*) aimed at demonstrating the practical possibility that specialists at ground level could count on full three-dimensional images for studying the evolution of certain phenomena produced in orbit.

Though Holograma had been originally prepared as one of the experiments to be executed in orbit during the space flight of the Cuban cosmonaut, it had to be postponed because of logistical transportation problems. Still, it was actually carried out in March 1981, during the next Intercosmos joint flight which included the Mongolian cosmonaut, Jugderdemidiin Gurragcha.

Holograma-1 was a relatively simple experiment consisting, firstly, in using the spaceship's standard television cameras for taking the images of ready-made holograms and transmitting them to earth so that they could be compared after reception with an exact duplicate of the holograms in space. Then, from the space control center to the orbiting spaceship a hologram formed with coherent light was transmitted through the standard television channel normally in use so that the holograms appearing in the TV monitors were photographed for later comparison. The images transmitted were specially designed for the determination of the spatial resolution in black and white without semitones and for the determination of semitone gradation. They included the Cuban Intercosmos logo and the Soviet USSR-Cuba logo (Gurragcha et al. 1982).

To carry out Holograma-2, a three-dimensional body and a dissolving salt in a liquid were recorded in orbit using a holographic instrument specially designed for use in space. Series of successive photographs were taken that made it easy to follow every detail of the dissolution process (which under microgravity conditions dissolved about 20 times slower than on earth). This showed that holography can be used for the control and testing of experiments in space and may eventually show its usefulness in space manufacturing processes (Avduyevsky 1984, Riegel and Wilcox 1996; Gansherli et al. 1982).

While the nature of the experiments briefly described above can be broadly classified as fundamental, it can be seen that they were generally aimed at eventual technological applications. They were also the first in their particular kind to be successfully carried out in space under microgravity conditions.

References

Altshuler, J. 1984. Space activities in Cuba. In *Study week on the impact of space exploration on mankind*, October 1–5, 1984, ed. C. Chagas and V. Canuto. Vatican City: Pontificia Academia Scientiarum.

Avduyevsky, V.S. (ed.). 1984. *Scientific foundations of space manufacturing*. Moscow: Mir Publishers.

Calzadilla, O. 1994. *Estudio de la influencia de la convección en los crecimientos por el método de Bridgman*. Doctoral thesis, University of Havana.

Calzadilla, O., J. Fuentes, J. Vidal, P. Diaz, R. Romero, C. Arencibia, E. Hernández, L.M. Sorokin, I.I. Shulpina, and A.S. Trebugova. 1984. Influencia de la microgravitación en la estructura y distribución del elemento indio en una aleación de Ge-In. *Ciencias Técnicas, Físicas y Matemáticas* 4: 17–21.

Calzadilla, O., J. Fuentes, I.L. Shulpina, and L. Sorokin. 1985. Estudio de los cristales de germanio dopados con indio en el experimento "Caribe". In *IX SLAFES, Latin American Seminar on Solid State Physics*. Mar del Plata, Argentina.

Calzadilla, O., J. Fuentes, I.L. Shulpina, and L. Sorokin. 1986. Caracterización estructural de una aleación de Ge-In crecida en el cosmos. *Rev. Cub. Fis.* 6(2): 47–52.

Díaz, P., E. Vigil, R. Romero, E. Purón, S. de Roux, F. Sánchez, V.M. Andreev, S.G. Konnikov, and T.B. Papova. 1983. Experiment "Caribe": Growth of GaAs and AlGaAs epitaxial layers and p-n junctions in space. In *International conference of socialist countries about materials science results in space program*, Riga, May 18–22, 1983.

Díaz, P., E. Vigil, R. Romero, E. Purón, S. de Roux, F. Sánchez, V.M. Andreev, S.G. Konnikov, and T.B. Papova. 1984. Homo y heteroepitaxia de GaAlAs-GaAs en condiciones de ingravidez. *Ciencias Técnicas, Físicas y Matemáticas* 4: 23–28.

Falcón Rodríguez, F.L. 1987. *Utilización de la fusión zonal con gradiente de temperatura (FGZT) en el estudio de los mecanismos de cristalización de la sacarosa.* Doctoral thesis, University of Oriente.

Falcón Rodríguez, F.L., P.V. Pérez, and G.F. Zhukov. 1984. Fusión zonal con gradiente de temperatura en el sistema H2O-sacarosa en condiciones de microgravidez. *Ciencias Técnicas, Físicas y Matemáticas* 4: 11–16.

Falcón Rodríguez, F.L., P.V. Pérez, S. Ameneiro, and G.F. Zhukov. 1985. Zonal fusion with temperature gradient in the H2O-Sucrose under microgravity conditions. Space experiment "Zona.". *Kosmicheskaia Issledovanie* 23(3): 488–492 (In Russian).

Gansherli, N.M., S.B. Gurevich, V.V. Kovalenok, V.B. Konstantinov, M. Cordero, I.A. Maurer, S. Mesa, R. Homs, S.A. Pisarevskaia, V. Rivera, et al. 1982. Holographic recording on board the Salyut-6 space station. *Zhurnal Tekhnicheskoi Fiziki* 52(11): 2192–2195 (In Russian).

Gurragcha, Z., S.B. Gurevich, V.A. Dzhanibekov, B.E. Kashonov, V.V. Kovalenok, V.B. Konstantinov, M. Cordero, S. Mesa, A.V. Militsin, R. Homs, et al. 1982. Exchange of holographaic information between Saliut-6 space station and flight control center. *Tekhn., Kino i Televideniya* 2: 5–11 (In Russian).

Morera R., J. Lodos, and E. Casanova. 1983. Study of sucrose monocrystal growth in space: experiment "Sugar". In *International conference of socialist countries about materials science results in space program*, Riga, May 18–22, 1983. Published as "Saliut-6"- "Soyuz" *Materials science and technology* (1984).

Morera, R., J. Lodos, and E. Casanova. 1984. Crecimiento de monocristales de sacarosa en el espacio cósmico. *Ciencias Técnicas, Físicas y Matemáticas* 4: 3–9.

Riegel, L., and W. Wilcox. 1996. *Modeling of detached solidification.* Final report, NASA-CR-208314. USA.

Vigil, E. 1984. Chapter 3. In *Estudio y aplicación del GaAs y sus soluciones sólidas con GaP y AlAs para la confección de diodos emisores de luz.* Doctoral thesis, University of Havana.

Vigil, E., and P. Díaz. 1981. Capas epitaxiales de GaAs y AlGaAs obtenidas en condiciones de ingravidez: características y propiedades. *University of Havana third scientific conference.* Havana, November 1981.

Vigil, E., P. Díaz, V.M. Andreev, and S.G. Konnikov. 1984. Capas epitaxiales de GaAs y AlGaAs obtenidas en condiciones de microgravidez. In *Fourth Mexican seminar of electronics physics. CINVESTAV*, México, D. F., May 16–18, 1984.

Chapter 15
Superconductivity in Cuba: Reaching the Frontline

Oscar Arés Muzio and Ernesto Altshuler

The start of experimental research in the field of superconductivity was a very special moment for Cuban Physics: Cuban scientists at the Physics Faculty, University of Havana, synthesized the first Cuban superconductor (a 123-YBCO ceramic sample) just 2 months after the publication of the famous paper by Wu and co-workers (Wu et al. 1987) that triggered the frantic race of High T_c superconductors all over the world. We timely joined the world's frontline in superconductor research.

The achievement had not been fortuitous: its roots can be found in our decades-long experience in the fabrication and characterization of magnetic materials, and the maturation of the scientific leadership at the Physics Faculty, University of Havana. Driven by Leonel Pérez-Marín (director of IMRE)[1] and by Carlos Rodríguez-Castellanos (dean of the Physics Faculty), Oscar Arés-Muzio (Head of the Magnetism Laboratory) "capitalized" the existing human potential to produce the new materials. The key was that Sergio García had the necessary knowledge on perovskites (the basic structure of the new superconductor YBCO reported by Wu and co-workers) due to the fact that he had been studying a perovskite-based compound as an intermediate phase in the formation of M-type ferrites (García and

[1] The Institute of Materials Science of Technology, originally Institute of Materials and Reagents for Electronics (IMRE) was created in May 1985 to back the National Program for Electronics. The two original driving forces were the work to obtain ferric oxide from Cuban natural resources headed by Leonel Pérez-Marín (Chemistry Faculty, University of Havana) and the experience in the fabrication of magnetic ferrites based on ferric oxide headed by Oscar Arés (Physics Faculty, University of Havana).

O. Arés Muzio
Center for Research and Advanced Studies of the National Polytechnic Institute, CINVESTAV, Mérida, Mexico
e-mail: oares@mda.cinvestav.mx

E. Altshuler (✉)
Physics Faculty, University of Havana, Havana, Cuba
e-mail: ealtshuler@fisica.uh.cu

A. Baracca et al. (eds.), *The History of Physics in Cuba*, Boston Studies in the Philosophy and History of Science 304, DOI 10.1007/978-94-017-8041-4_15,

Altshuler 1985). The first ceramic was synthesized by the end of April, 1987, by the expert hands of technician Armando Aguilar guided by professor García.[2] No X-rays or resistive transition were performed to check the material, but the appearance of the Meissner levitation—witnessed in Cuba for the first time—was a compelling reason to think that they had made it.[3]

After the appearance of this news in the newspaper *Granma* in May 1987, and following an invitation from Leonel Pérez-Marín, president Fidel Castro soon visited the Physics Faculty. This event catalyzed support for the development of IMRE and the idea to create a Superconductivity Laboratory, which we will call SCUH from now on. The Laboratory was officially established under the leadership of Oscar Arés Muzio as part of IMRE, but in strong collaboration with the Physics Faculty, University of Havana. Other original members, besides Oscar Arés Muzio, Sergio García, and Armando Aguilar were the young researchers Celia Hart-Santamaría, Reinaldo Aguilar, Daniel Carrillo, and Victoria Venegas. A few hundred thousand USD were assigned to the laboratory, which were invested in creating sample preparation infrastructure (furnaces, milling machines, an excimer laser, reagents, etc.), and some measuring instruments, prominently a VSM magnetometer—which was an old dream of our Magnetism community. At the same time, related investments where made at the IMRE, especially the installation of an industrial plant to produce liquid helium, aimed at feeding the RMN tomography machine at the "Hermanos Aimeijeiras" hospital. Some part of the liquid helium, of course, could be used by SCUH.

Oscar Arés Muzio immediately focused on the construction of a radio-frequency SQUID device[4] based on High T_c superconductors aimed at biomedical applications. In fact, he had systematic contacts with the Neuroscience Center headed by P. Valdés-Sosa (at the Western Scientific Pole of Havana), who was interested in incorporating a SQUID-based magneto-encephalogram to complement the conventional electro-encephalogram for the diagnosis of brain disorders. In order to build the SQUID device, Oscar maintained collaboration with the group of B.V. Vasiliev in Dubna, creators of one of the most celebrated ceramic-based SQUID at the time (Bobrakov et al. 1988). While his ceramic RF-SQUIDs based on hand-carved

[2] Curiously, the synthesis was not based on Wu et al. 1987, which provides the "whole recipe," since the paper was not available at the time. Instead, a less detailed paper, which appeared in *New Scientist*, was used as a main reference.

[3] The experiment consisted in making a small super-magnet float on a superconducting pellet (which rested in a shallow reservoir of liquid nitrogen improvised from a piece of white foam used for glassware transportation). Levitation is a standard qualitative test for the Meissner effect, that is, the expulsion of flux lines from a material when it changes from the normal to the superconducting states. It is worth noticing that not only the superconductor, but also the super-magnet, had been fabricated at the Magnetism Laboratory, Physics Faculty, University of Havana.

[4] Superconducting Quantum Interference Devices (SQUID) are very sensitive magnetometers used to measure extremely small magnetic fields, based on superconducting loops containing Josephson junctions. They are sensitive enough to measure fields smaller than a fT (10^{-15} T).

Fig. 15.1 Early hand-made devices at SCUH. *Left panel*: A magnetic shield for the characterization of ceramic SQUIDs. The walls were made out of hundreds of metal sheets accommodated in an oil can, surrounding a glass dewar which was inserted into the hole. Notice the "brand name" written on *top*: "Tropical SQUIDs, Inc." *Right panel*: assorted parts to study transport properties of superconductors at low magnetic fields (the long solenoid allowed the temperature to be controlled by moving the sample along its axis, without losing the uniformity of the magnetic field)

micro-bridges, were able to detect fairly small magnetic fields,[5] he devoted considerable energy to other tasks related to the project. One was the construction of a 30-cubic-meter Faraday cage consisting in a double layer of thick aluminum sheets soldered edge-to-edge *in situ*. Inside, an ingenious wooden frame supported a non-metallic dewar containing the SQUID sensor. Another challenge was the attempt to construct a dewar for biomagnetic measurements using an epoxy resin composite—an extremely difficult task due to the lack of appropriate materials and technical expertise (Fig. 15.1).

In May 1989 Ernesto Altshuler joined the original members of the group, contributing to the creation of a "critical mass" to start basic research in the field of transport properties of superconductors in order to complement the application efforts in the area of SQUID devices. The collaboration with his former scientific advisor Sergio García resulted in the first international paper in the field of experimental superconductivity undertaken exclusively in Cuba (Altshuler et al. 1990). This was the first in a series of more than 60 papers to appear in international

[5] A typical "microbridge" was a thin bridge carved between the edge of a superconducting pellet and one hole around 1-mm in diameter, near the edge of the pellet. The idea was to isolate one or a small number of Josephson junctions that occur naturally between superconducting grains in a typical YBCO ceramic sample. The current at the superconducting loop around the hole, interrupted by the "weak link" junction, is very sensitive to variations in the magnetic field through the hole.

journals over a period of almost 20 years—the overwhelming majority of them including experiments performed at SCUH.[6] In this long episode, collaborations took place between young researchers and professors who joined the laboratory later on, such as Andrés R. R. Papa, Jorge Luis González, Jorge Barroso, Carlos Abascal, Luis Eduardo Flores, Claro Noda, Carlos Martínez, Roberto Mulet, Milenys Acosta and Alexander Hernández, as well as other colleagues who came from the eastern regions of the country to accomplish their PhDs (especially Profs. Pedro Muné and Alfo José Batista-Leyva). The rhythm of publications in international journals of the SCUH did not decline during the golden days of High T_c Superconductivity, including the period of hardship in the worst of the Cuban economical crisis of 1992–1994. The papers were mostly "endogenous" in the sense that they were generally achieved exclusively in Cuba, including experiments, paper writing and editorial exchanges. It was perhaps a unique experience at the time for Cuban experimental physics, and it is fair to say that it was seen as the ideal of excellence by many young Cuban researchers. Viewed from a historical perspective, the Superconductivity Laboratory can be regarded as one of the world leaders in the understanding of granular superconductors at low fields and temperatures over 77 K, that is, the subfield we were forced to inhabit due to experimental shortages.[7]

In spite of its "endogenous" character, in some respects the SCUH maintained important exchanges with several institutions, such as the Superconductivity Laboratory headed by Francesco Cino Matacotta at the International Center for Theoretical Physics, Italy (ICTP), the Low Temperature Laboratory headed by Francisco (Paco) de la Cruz at Centro Atómico Bariloche (Argentina), the Unidad Centro de Investigación y de Estudios Avanzados (CINVESTAV), Mérida (Mexico), and, after the mid-1990s, the Centro Brasileiro de Pesquisas Físicas (CBPF), the Texas Center for Superconductivity, headed by C.W. Chu (USA) and the Superconductivity Laboratory headed by Tom. H. Johansen at the Physics Department, University of Oslo (Norway).

The "High T_c Revolution" not only triggered science at the SCUH. Shortly after its creation, two other groups that began to fabricate samples and/or performed experiments in the field of superconductivity, appeared in Cuba—one at the Center of Technological Applications and Nuclear Development (CEADEN), headed by Carlos Cruz Incán, and the other at the University of Oriente, headed by Pedro Muné, who completed his PhD at the University of Havana in 1995. The CEADEN's group concentrated on the effects of the irradiation of polycrystalline samples with gamma rays, a topic connected to the expected areas of interest of an institute originally devoted to nuclear research. In addition, the discovery of High T_cs triggered at an early stage theoretical research in the microscopic mechanisms of superconductivity among some Cuban physicists. Two important names related to these

[6] Many of them can be consulted at www.complexperiments.net. Accessed October 15, 2013.

[7] If one searches the very general subject "superconducting polycrystals" in *Google*, five out of the first ten hits are directly related to papers authored by members of the SCUH (sample search performed on February 18, 2011).

studies are Carlos Trallero-Giner at the Physics Department, University of Havana, and Alejandro Cabo Montes de Oca, at the Institudo de Cibernetica, Matemática y Física (ICIMAF).

A good overall indicator of the maturity of Cuban research in superconductivity is the participation of Cuban scientists in two of the most prestigious world conferences on High T_c superconductors. In Houston's 2000 M2S-HTSC-VI, five papers where authored by researchers from SCUH, and the Department of Theoretical Physics, University of Havana;[8] at least four of the Cuban authors attended the conference. In Rio de Janeiro's 2004 M2S-HTSC-VII, Cuban researchers "took over": they were represented both in the publication and in the Latin American Committee, and presented seven research papers, including one talk.[9] The authors were affiliated with SCUH, The University of Oriente, and CEADEN. At least four of them attended the event.

While the critical current temperature of the newly discovered superconductors increased exponentially during the late 1980s and early 1990s, promising a true technological revolution in power transmission and storage technologies, the golden dream of room-temperature superconductivity was never fulfilled. Moreover, theoreticians had been unable to produce a consistent microscopic theory of High T_c superconductors. These facts implied drastic cuts in support to superconductivity at many institutions worldwide. By 2004, SCUH had been officially closed as an independent laboratory, and incorporated into the Magnetism Laboratory, IMRE.

The intense scientific environment of SCUH, described by one of its visitors as "a wonderful place where what mattered was what people said and not who said it" had catalyzed the diversification of the scientific interests of its members since the early 1990s. For example, after years working in weak-link modeling of superconducting bridges,[10] Oscar Arés became interested in the new giant magnetoresistance materials discovered in 1988 (whose perovskite-based structure resembled those of the High T_c superconductors) and also started working in that direction. Later on, he performed experiments in magnetic effects on fluids and became interested in other materials—perhaps a *déjà vu* of the previous experience in transport properties in superconductors. Other members of the laboratory evolved towards the study of "complex systems" and statistical physics. To understand transport properties of superconductors, they needed to introduce ideas from percolation theory[11] and to understand vortex dynamics. But vortices could be modeled as grains, which opened the field of granular matter. From granular avalanches, those ideas fed back into superconductivity, resulting from years of work in the field of vortex avalanches.[12] If grains were given some level of "free will," they resembled ants… and some of us

[8] See conference proceedings published by *Physica C*, volume 341–348 (2000).

[9] See conference proceedings published by *Physica C*, volume 408–410 (2004) (one of the seven Cuban presentations was not published).

[10] See, for example, Arés et al. (1995).

[11] See, for example, Mulet et al. (1997).

[12] See, for example, Altshuler and Johansen (2004).

are now also working in the dynamics of social insects.[13] In fact, the merging of our research with the independent work of Oscar Sotolongo (Theoretical Physics Department, Physics Faculty, University of Havana) resulted in the creation in 2002 of the "Henri Poincaré" Cathedra of Complex Systems at the Physics Faculty, University of Havana.[14]

Today, the experimental research at the former SCUH survives against the odds. It concentrates in the transport properties of composite superconducting tapes patterned by laser ablation, and, in collaboration with the Superconductivity Laboratory at the University of Oslo, on the study of hot spots and vortex avalanches in superconductors. Meanwhile, the group of Pedro Muné at the University of Oriente works in the field of granular and nano-granular superconductors performing experiments primarily during visits at the Physics Department, University of Sao Paulo (Brazil).

Superconductivity in Cuba, as in many places in the world, is in a state of partial hibernation, quietly preparing for the next superconducting revolution, which, in the words of C.W. Chu, can explode "at any unknown laboratory in the World."[15]

Acknowledgements This paper would have been impossible without data given by Sergio García and also by Carlos Cruz-Inclán. Elena Vigil gave important insights relative to the early times of IMRE.

References

Altshuler, E., and T.H. Johansen. 2004. Experiments in vortex avalanches. *Reviews of Modern Physics* 76: 471.

Altshuler, E., S. García, and A. Aguilar. 1990. Hysteretic critical currents in Y-Ba-Cu-O superconductors: A microstructural approach. *Physica Status Solidi (a)* 120: K169.

Arés, O., C. Hart, and M. Acosta. 1995. Transition from magnetic Fraunhofer-like to interferometric behavior in YBCO bridges with decreasing width. *Physica C* 242: 191.

Bobrakov, V.F., B.V. Vasiliev, and V.N. Polushkin. 1988. SQUID operating at liquid nitrogen temperatures. *JINR Rapid Communication* 30: 101.

García, S., and E. Altshuler. 1985. Mössbauer study of the reaction kinetics of M-phase ferrites. *Physica Status Solidi (a)* 89: 427.

Mulet, R., O. Díaz, and E. Altshuler. 1997. Choice of sample size for high transport critical current density in a granular superconductor: Percolation versus self-field effects. *Superconductor Science and Technology* 10: 758.

Wu, M.K., J.R. Ashburn, C.J. Torng, P.H. Hor, R.L. Meng, L. Gao, Z.J. Huang, Y.Q. Wang, and C.W. Chu. 1987. Superconductivity at 93 K in a new mixed-phase Y-Ba-Cu-O compound system at ambient pressure. *Physical Review Letters* 58: 908–910.

---

[13] Interestingly enough, other groups in the world, such as Heinrich Jaeger's (James Franck Institute, University of Chicago), have followed an analogous evolution regarding scientific interests—at least up to the stage of granular matter.

[14] Sometimes referred to in scientific papers as the "Henri Poincaré" Group of Complex Systems (see the chapter by Sotolongo-Costa in this volume).

[15] Approximate quotation from a conversation of E. A. with C. W. Chu, Houston, Texas, 2000.

Chapter 16
The Physics of Complex Systems in Cuba

Oscar Sotolongo-Costa

16.1 Introduction

In establishing the circumstances that led to the birth and development of the physics of complex systems in Cuba, it is difficult to avoid being anecdotal—particularly because of the difficult times during which this research started. Cuban eclecticism, whose spectrum extends from religious syncretism to world-class medicine, seems quite coherent with the field of complex systems, characterized by the synergy of diverse fields. Such a combination, however, in the beginning seemed to be quite removed from the physicists' standard research dogmas.

Complexity as a concept emerges from daily experience and is actually dealt with in diverse contexts. For example, while the free fall of an object or the oscillatory motion of a pendulum would be considered "simple" phenomena by a physicist, economy, language and so on, would generally be assumed to be "complex" phenomena. Complex systems are characterized by several degrees of freedom, and also by correlations among different parts of the system, resulting in what we could call "coordinated dynamics." Such dynamics typically results in the emergence of spatial and temporal "structures."

Societies, for example, are organized in structures according to certain political rules. Such structures also appear in the purely spiritual domain, in language, in music, and even in science. Sometimes we are so used to those structures that we are no longer conscious of the miracle of their existence—our ancestors assumed they had a divine origin. Perhaps due to this, science has long dealt with the configuration or shape of complex structures, but not with their formation dynamics. This is a relatively new interest.

Starting in the 1960s, thanks to the increasing development of computers, we have witnessed the birth of entirely new methods for studying nature. As a consequence,

O. Sotolongo-Costa (✉)
President of the "Henri Poincare" group of Complex Systems,
University of Havana, Havana, Cuba
e-mail: osotolongo@fisica.uh.cu; sotolon@ual.es

A. Baracca et al. (eds.), *The History of Physics in Cuba*, Boston Studies in the Philosophy and History of Science 304, DOI 10.1007/978-94-017-8041-4_16,

insights that formerly used to develop separately within different branches of physics finally converge and show that the gap between "simple" and "complex" and between "order" and "disorder" is much smaller than has generally been assumed until now.

Thanks to the virtually unbounded calculation power of computers, we know, for example, that simple physical examples taken from undergraduate curricula can indeed show a complex behavior. If a simple pendulum is "kicked" by a periodically pulsed force, it can eventually show a rich variety of motion, including "casual," even "turbulent" separations from the equilibrium position. Conventional systems such as fluid layers and chemical reactions can, under certain conditions, result in "self-organization" phenomena of macroscopic dimensions producing spatial and temporal structures.

16.2 The Requirements

In 2002, in the frame of the celebration of the 40th anniversary of the *licenciatura en física* degree course, the Physics Faculty of the University of Havana inaugurated the "Henri Poincaré" Complex Systems Chair.[1] For the time being, the members of the group—who basically came from the Theoretical Physics Department and the Superconductivity Laboratory, Physics Faculty, and IMRE, University of Havana—had already contributed over 50 scientific papers to the field, and had cooperated with colleagues from several universities around the world.

But, why complex systems? Why in Cuba?

Perhaps the true catalyst for studying complex systems in Cuba was the extremely difficult economic situation in the country which derived from the collapse of the European socialist-bloc, and was substantially aggravated by an increase in U.S. embargos. These events certainly compelled the group to squeeze creativity to its limits. In addition, world opinion was clearly in favor of an engagement in the field of complex systems as front line research, in spite of the scant economic resources at disposal.

A particularly deleterious effect of the crisis was the scarcity of fuel, which resulted in the virtual disappearance of most motorized transportation, and the massive use of bicycles for any single activity in life. At the beginning, of course, it was difficult to imagine the many ways that bicycles would influence the development of a "tropical" style in the study of complex systems. The following examples will illustrate the point.

16.3 Breaking and Criticality

Given the situation, it was of central importance to find more efficient systems for combustion as a way to save fuel, so water-fuel emulsions started to be used in different places. This is based on the fact that tiny drops of water inside fuel drops

[1] Generally, members of the group explicitly state their affiliation to it in the papers they author or coauthor.

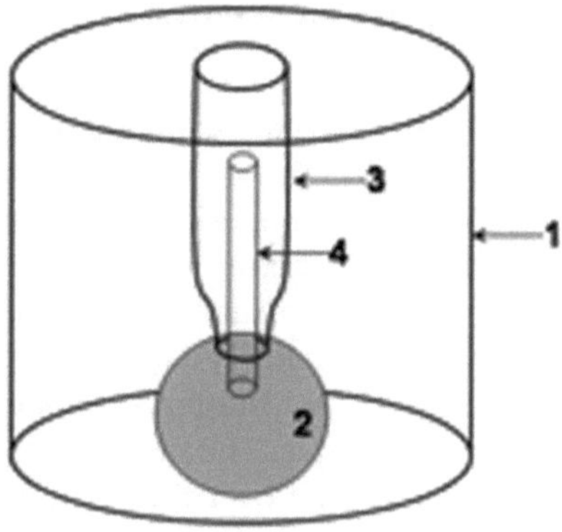

Fig. 16.1 Setup used to study the micro-explosion of fuel oil drops (*1*) Overhead projector sheet (strongly downsized in the figure), (*2*) Drop of fuel-oil, (*3*) External capillary tube, (*4*) Internal capillary tube

produce a fast nucleation of bubbles which burst the fuel drops during atomization within the combustion chamber of an engine, thus increasing the combustion effective surface—a process known as "micro-explosion." The subject motivated some Cuban physicists, who began to study the problem by imitating the "micro-explosion" by injecting air into drops of fuel oil. The air was injected into the drops using capillary tubes, and the resulting explosion was studied in detail. The external container used to control the air for injection was, of course a bike's inner tube.

Figure 16.1 shows the experimental setup for studying drop fragmentation. A drop (2) hangs from a capillary tube (3) and is penetrated by a second, thinner capillary tube (4), through which air is injected. The air comes from a reservoir consisting of the inner tube of a bike, whose pressure is controlled in order to make the drop burst. The drop fragments get stuck to the surface of a cylindrical screen made from an overhead projector transparency (1). The screen is then unfolded on a flat surface, and the sizes and number of the stains are measured under the microscope. It was found that, for bursting pressures that are large enough, the distribution of stain sizes (proportional to the fragment sizes) spread over a very wide range, i.e., the distribution had no "characteristic size" or "characteristics length."[2] This property means that the distribution is fractal, or scale-invariant, a quite popular subject in statistical physics at the moment. Moreover, the results suggested the existence of a close relation between combustion efficiency and the fractal properties of the distribution of fragments. Hence, it became clear that fractal geometry was an important tool to describe some of the technical problems where "classical" geometry is unthinkable. In the subsequent publication (Sotolongo et al. 1994), the authors underlined the use of fractal concepts in the field of emulsion combustion.

Fragmentation studies were later refined (Sotolongo-Costa et al. 1996) and extended theoretically to the field of earthquake dynamics (Sotolongo-Costa and Posadas 2004).

[2] Gaussian or other peaked distributions imply that there exists a characteristic size.

16.4 Piles of Grains

Shortly after the publication of a famous and controversial paper by Bak, Tang and Wiesenfeld (BTW) on "self-organized criticality" (SOC) (Bak et al. 1987), several scientists and laboratories throughout the world momentarily forgot about sophisticated lasers, sputtering machines and particle accelerators to "play with sand." We can mention, for example, an experiment designed by IBM scientists in 1990 in which sand was slowly added to the pan of a digital scale, so that the fluctuations in the weight of the resulting sand pile were monitored through a PC connected to the output of the scale. The fluctuations originated in avalanches—defined here as sudden leaks of sand over the borders of the circular pan—causing a decrease in the weight of the pile. The distribution of avalanche sizes could thus be deduced from the distribution of fluctuations of the weight of the entire pile.[3] The size distribution of avalanches was identified as a power law (after appropriate scaling), but only for pan sizes below a certain threshold (Held et al. 1990). The experiment suggested that SOC depended on some parameters such as the size of the pile—which apparently contradicted the "robustness" of power laws assumed in the original BTW paper (Bak et al. 1987). Of course, in the experiment only the avalanches including grains that abandoned the pile ("off the edge avalanches") were measured. The "internal" avalanches (grains suddenly moving inside the pile without abandoning it) were not accounted for due to technical limitations.

Another important experiment on pile avalanches was published in *Nature* by a group in Oslo in 1996 (Frette et al. 1996). This time the pile was made of grains of rice confined between two vertical glass plates, and the grains were added from the top using a commercial device. Instead of measuring weight loss, the Oslo team took digital videos of the pile, so it was possible to evaluate not only avalanches abandoning the pile by the lower edges, but also those associated with grain movements on the free surface of the pile. The Oslo rice pile experiment suggested that SOC behavior depends on the geometry of the grains involved in the pile. The image-analysis method was later tackled by a Cuban group in order to study avalanches in piles of steel beads.

After trying many materials, Cuban "sand piles" were made of ball bearings imported in massive amounts by the country for the maintenance of the millions of Chinese bicycles roaming the Cuban streets and roads during the tough 1990s (Fig. 16.2).[4]

Steel beads were added one by one from the top in a completely controlled way using a PC-controlled device (somewhat jokingly) called "chicharotrón."[5] The pile formed between two vertical glass plates separated by a bit more than one bead

[3] One clever detail of the setup was an inclined, rotating glass tube able to deliver grains of sand to "feed" the pile from the apex in a very slow way—not grain by grain, but almost.

[4] Some of them systematically transporting members of the "Henri Poincaré" Group, but this is not the kind of bicycle role we want to underline in this paper.

[5] This name comes from the Spanish word "chícharo," meaning "pea" in English. In fact, the experiment was originally conceived for peas, but these got bugs, so the authors decided to switch to the much less edible ball bearings.

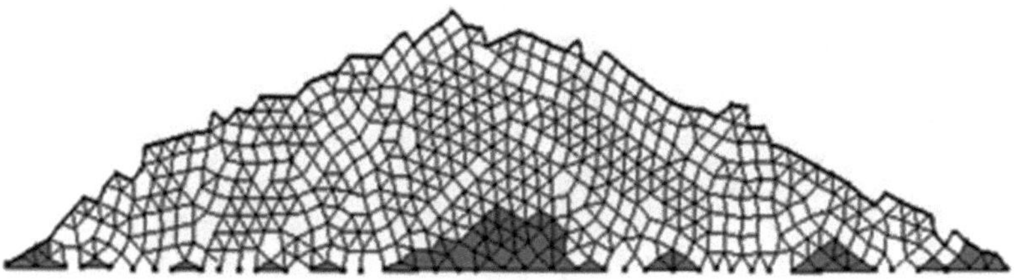

Fig. 16.2 Digitalized image of the front view of a two-dimensional pile of steel beads confined between two vertical glass plates, grown on a row of randomly spaced fixed beads (*bottom line*). The nodes correspond to the centers of the beads, and the lines represent a Voronoi triangulation. *Gray areas* correspond to regions that never changed after 50,000 beads were added one by one from the top. Off-the-edge avalanche size distributions suggest SOC behavior (Modified image taken from Altshuler et al. 2003)

diameter onto a horizontal row of fixed, randomly spaced beads. The idea was to study how the distribution of fixed beads in the base influenced the avalanche dynamics. The system worked like this: one bead was delivered by the "chicharotrón." An avalanche was then eventually triggered and the number of beads involved was obtained by measuring the fluctuation of the weight of the pile with a computer-controlled scale. After the event, a digital picture was taken to study the structure of the pile. Then, a computer ordered the "chicharotrón" to add a new bead to the pile. To get good statistics, this process was repeated around 50,000 times for each base type and each base size. After using a "scaling ansatz," the results indicated that scale-free distributions of beads were only obtained for random bases, and not for more "ordered" bases. Once again, SOC failed to be as robust as originally conceived: it demanded a certain level of "disorder" in the piles (Altshuler et al. 2001).

Very recent experiments in the same system with larger-sized piles and higher image resolution have allowed the authors to obtain, for the first time, the size distribution of avalanches involving the displacements of *all* the beads (i.e., not only those abandoning the pile or moving along the surface), (Ramos, Altshuler and Måløy 2009). This has produced two surprising outputs. One is the direct experimental demonstration that power-law distributed avalanches are possible in real piles, without the use of any "scaling ansatz." The second is the first experimental evidence of the possibility of predicting large avalanches based on the structural dynamics of a SOC-like system. This result is potentially relevant in many scenarios, particularly earthquake prediction, and has been featured recently in *New Scientist* (Ananthaswamy 2009).

While sand piles constitute the paradigm of SOC, power law distributions of avalanches have been investigated in many other systems. Due to the "superconducting roots" of some of the members of the "Henri Poincaré" Complex Systems Group, avalanches in superconducting vortices have been a popular subject of study in Cuba, resulting—for example—in the only review on the subject published in *Reviews of Modern Physics* (Altshuler and Johansen 2004).

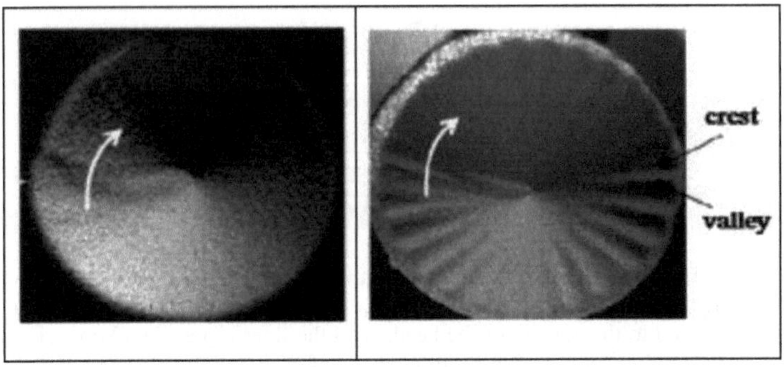

Fig. 16.3 *Top views* of sand piles showing the "scars" of revolving rivers. The rivers are streams of sand flowing down the hill from the apex to the base of the pile that revolve or "sweep" its surface as sand is added at the apex of the pile (sand addition was stopped to take the pictures). The so-called continuous (*left*) and intermittent (*right*) regimes are illustrated

The search for SOC behavior in natural sand led to the discovery of two new phenomena in sand flows: "revolving rivers" (Altshuler et al. 2003) and "uphill solitary waves" (Martínez et al. 2007). These phenomena have been found in rather mysterious sand from a particular location in Pinar del Río province (Cuba) and in a handful of other types of sand from different places around the world. Both phenomena were quite far from the SOC paradigm (Fig. 16.3).

16.5 Further Activities

The above examples are just a sample of the wide variety of subjects tackled by Cuban physicists in the—not very well defined—field of complex systems. Other areas of interest have been game theory (Marsili et al. 2001), and networks and algorithmic complexity, which have produced high-level results related to the problem of the coloring of random graphs (Mulet et al. 2002). Biologically-inspired problems have also been very popular, such as theoretical modeling of tumors, modeling of the immune system (in collaboration with the Cuban Center of Molecular Immunology), and the experimental study of collective behavior of social insects (Altshuler et al. 2005).

For the last decade or so, the Physics Faculty at the University of Havana has incorporated into its undergraduate and graduate curricula various subjects related to the Physics of Complex Systems, including laboratory experiments. Several dozens of students and young scientists have passed those courses. International meetings such as "Fractal Structures and Self Organization" (Satellite meeting of STATPHYS 2001) and the "First Latin American School and Conference on Statistical Physics and Interdisciplinary Applications" (2005) have also been organized by the Cuban complexity community.

16.6 Conclusion

The physics of complex systems emerged in Cuba at a most difficult time for the development of scientific research in the country. Nevertheless, several relevant scientific results have been obtained, some of them published in prestigious physical journals. Expanding the researchers' present horizons and establishing a solid material and human framework to systematize the corresponding original research, teaching and even applications are some of the goals aimed at by the small but enthusiastic national community now working on the physics of complex systems.

References

Altshuler, Ernesto, and T.H. Johansen. 2004. Experiments in vortex avalanches. *Reviews of Modern Physics* 76: 471–487.

Altshuler, Ernesto, O. Ramos, C. Martínez, L.E. Flores, and C. Noda. 2001. Avalanches in one-dimensional piles with different types of bases. *Physical Review Letters* 86: 5490–5493.

Altshuler, Ernesto, O. Ramos, E. Martínez, A.J. Batista-Leyva, A. Rivera, and K.E. Bassler. 2003. Sandpile formation by revolving rivers. *Physical Review Letters* 91: 014501.

Altshuler, Ernesto, O. Ramos, Y. Núñez, J. Fernández, A.J. Batista-Leyva, and C. Noda. 2005. Symmetry breaking in escaping ants. *American Naturalist* 166: 643–649.

Ananthaswamy, Anil. 2009. Beads get ball rolling on avalanche prediction. *New Scientist* 2698 (7 March).

Bak, Per, Chao Tang, and Kurt Wiesenfeld. 1987. Self-organized criticality: An explanation of 1/f noise. *Physical Review Letters* 59: 381–384.

Frette, V., K. Christensen, V. Malthe-Sørensen, J. Feder, T. Jøssang, and P. Meakin. 1996. Avalanches in a pile of rice. *Nature* 379: 49.

Held, G.A., D.H. Solina II, D.T. Keane, W.J. Haag, P.M. Horn, and G. Grinstein. 1990. Experimental study of critical mass fluctuations in an evolving sandpile. *Physical Review Letters* 65: 1120–1123.

Marsili, M., R. Mulet, F. Ricci-Tersenghi, and R. Zecchina. 2001. Learning to coordinate in a complex and nonstationary world. *Physical Review Letters* 87: 208701.

Martínez, E., C. Pérez-Penichet, O. Sotolongo-Costa, O. Ramos, K.J. Måløy, S. Douady, and Ernesto Altshuler. 2007. Uphill solitary waves in granular flows. *Physical Review E* 75: 031303.

Mulet, R., A. Pagnani, M. Weigt, and R. Zecchina. 2002. Coloring random graphs. *Physical Review Letters* 89: 268701.

Ramos, O., Ernesto Altshuler, and K.J. Måløy. 2009. Avalanche prediction in a self-organized pile of beads. *Physical Review Letters* 102: 078701.

Sotolongo-Costa, Oscar, and A. Posadas. 2004. A fragment-asperity interaction model for earthquakes. *Physical Review Letters* 92: 048501.

Sotolongo-Costa, Oscar, Enrique López-Pages, Felix Barreras-Toledo, and Jose Marín-Antuña. 1994. Scaling in drop distributions: Application in combustion. *Physical Review E* 49: 4027–4030.

Sotolongo-Costa, Oscar, Yamir Moreno-Vega, Juan J. Lloveras-Gonzáles, and J.C. Antoranz. 1996. Criticality in droplet fragmentation. *Physical Review Letters* 76(1): 42–45.

Chapter 17
Magnetic Resonance Project 35-26-7: A Cuban Case of Engineering Physics and Biophysics

Carlos A. Cabal Mirabal

The Magnetic Resonance Project 35-26-7 started in December 1987, commissioned by the [then] Cuban Prime Minister, Fidel Castro, who—concerned about introducing technological advancement into the Cuban health [system]—had for some months taken an interest in the possibility of building magnetic resonance imaging (MRI) equipment for medical diagnosis in Cuba (Zito et al. 1999, 56–66; Cabal 2002). Many of the companies producing MRI equipment were unable to deliver this technology to Cuba due to the *bloqueo*, the United States embargo against Cuba. Those who were later to advance the project's progress in scientific technology initially regarded the implementation of such a project in a developing country as unfeasible due to its complexity. But Fidel's belief and confidence and in turn the Cuban scientists' commitment to him and to Cuban science proved to be an undeniable factors for its success.

The project had been preceded by the nuclear magnetic resonance (NMR) group of the Physics Mathematics Faculty at the University Oriente in Santiago de Cuba, which emerged in 1975 following a [rewarding] post-graduate course given by Pyotr Mikhaylovich Borodin, a radiophysicist from the Faculty of Physics at Leningrad State University. Since that year, the NMR group have defined two lines of work: first, the maturity study of sugar cane with NMR in the Earth's magnetic field, searching for a generalizable procedure for the Cuban sugar industry. And secondly, the investigation of paramagnetic complexes in Nickel (II) and Cobalt (II) in solutions, a subject associated with the Nickel industry in which Cuba plays a prominent role. Thus, from the beginning, the NMR looked for a solution for the country's practical economic problems. On September 18, 1980,

Founding director of the Center of Medical Biophysics of Santiago de Cuba and research director of NMR project 35-26-7 until 2006, currently working in molecular MRI at the Center for Genetic Engineering and Biotechnology in Havana.

C.A. Cabal Mirabal (✉)
Head of Images Group, Chemical Physics Division,
Center for Genetic Engineering and Biotechnology, Havana, Cuba
e-mail: carlos.cabal@cigb.edu.cu

A. Baracca et al. (eds.), *The History of Physics in Cuba*, Boston Studies in the Philosophy and History of Science 304, DOI 10.1007/978-94-017-8041-4_17,

the very same day the first Cuban cosmonaut travelled in space,[1] Carlos Cabal Mirabal [the author] defended in this line of work his PhD thesis on the *Mecanismos de relajación magnética en soluciones acuosas paramagnéticas de Ni (II) y Co (II)* (Magnetic relaxation mechanism in paramagnetic aqueous solution of Ni (II) and Co (II)) in quantum radiophysics at the Faculty of Physics at Leningrad University, under the supervision of Vladimir Ivánovich Chizhik (Cabal 1980). The work, developed within the framework of this major thesis, was awarded with second place in 1979 and first place in 1980 respectively in the general Youth Science Competition of Leningrad University.

Between 1980 and 1987, the NMR laboratory at the Oriente University in Santiago was consolidated with experimental NMR installations, designed and built by the faculty (Chizhik et al. 1983). More specialists were trained with the support of the Soviet Union. MR investigations were undertaken of the microstructure and magnetic relaxation process in paramagnetic solutions of transition ion metal as well as lanthanides like Cerium (III), Samarium (III), Praseodymium (III), Terbium (III) and Europium (III) (Chizhik et al. 1983; Cabal and Chizhik 1982, 1983; Guzmán et al. 1987). Additional scientific work was conducted to contribute to issues in the nickel and cement industry (Gilart et al. 1987; Cabal et al. 1984). Despite its meager funding, the laboratory distinguished itself in the field of physics, as is proven by its publications and by the annual award *Al Merito Científico Técnico* it received for the courses 86/87 and 87/88 from the University Oriente.

Upon the emergence of the NMR project 35 26 7, conceived in various stages, a change in the direction of work (NMR geared toward biomedicine) and in the concept of scientific creation took place, namely its purpose, the rhythm of investigation, the role of science, the concept of how to achieve a scientific result as well as the extent to which the process can be taken (Cabal et al. 1999; Cabal 2002). Essentially the project went far beyond physics. It was a physical-technical program and thus its original group of 12 professionals and technicians (most of them young people and students) were likewise physicists. There were also a significant number of electronic engineers, as well as two mechanical engineers from other institutions who cooperated with the project: *Centro Nacional de Investigaciones Científicas de Ciudad Habana* and Retomed, a medical equipment company in Santiago de Cuba. The objective of the project's first stage was dedicated to:

- the design, construction and identification of parameters for an universal NMR relaxometer of 0.1 T for the measurement of spin-lattice relaxation time (T1) and spin-spin relaxation time (T2) with the different pulse sequences in radiofrequency existing at that time.
- the identification of high impact biomedical problems that could be addressed with magnetic resonance.

[1] Arnaldo Tamayo Méndez was the first Cuban citizen to travel into earth orbit. He was also the first Hispanic and first black person in space. Tamayo, along with Soviet cosmonaut Yuri Romanenko, was launched into space aboard Soyuz 38 from Baikonur Cosmodrome on September 18, 1980, at 19:11 UTC.

- finding out how to obtain a magnetic resonance imaging (MRI) equipment to study the human body in detail. Those initial three tasks were basically achieved within 7 months in July 1988.

The second stage consisted in calculating, designing, constructing, putting into operation and characterizing each and everyone of the systems constituting MRI equipment:

- Magnetic system including the electromagnet, its alimentation, refrigeration, shim coils and passive shim and devices for the intensity, homogeneity (some part per million), stability, magnetic measurement and correction in a volume sphere with a diameter of 40 cm.
- Magnetic field gradients with 120 V and 120 A and rise time of 10 milliseconds and duration of various milliseconds on the corresponding coils with inductance of hundreds of microhenries to guarantee the special linear spatial coding of the orthogonal axes in the significant area.
- Generation and modulation of radio frequency pulses allowing for kilowatt powers and millisecond durations as well as RF coils to produce the excitation for each part of the human body at the required frequency.
- Reception and detection of radiofrequency with RF coils where electromotive force (EMF) of some dozens of microvolts is induced in a wide frequency range in the presence of elevated noise levels.
- Detected signal digitalization with two-dimension Fast Fourier Transform (2DFFT). All this was undertaken with the existing computer technology available to Cubans at that time, based upon personal computers (PCs – Intel 8086). Software development (high and low level), the control and communication problems of the PCs with the other systems as well as managing information resources in order to deal with that, calculation and MR image formation were certainly rather complex tasks.
- Control and synchronization systems of all events guaranteeing the programming of the pulse sequences necessary to obtain the coded, spatial and temporary, information in the object under sampling and the stability phase between all systems mentioned and between them.
- Post-processing, evaluation and images presentation MRI system based upon software developed entirely within the project.

A profound understanding of physics and engineering was a prerequisite to achieve the electromagnetic calculations (direct and inverse problems of electromagnetism), mechanics, electronics, thermodynamics, and to implement a viable design with the technology available in Cuba. In order to prove its parameters and to determine the transfer functions within the distinct sections and the system as a whole, it was also necessary to calculate, design and construct measurement equipment and procedures to meet international regulations and norms. A magnetometer was constructed and calibrated together with various phantoms (patron object) and electronic devices to measure the gradient pulses, radio frequency, etc., allowing in part to evaluate the parameters of the constructed sections and thereby guaranteeing the quality of the whole MR machine (Cabal et al. 1997).

Hence, in August 1991, 3 years and 1 month after the start of the second phase, the first images of various parts of the human body were obtained with exclusively Cuban magnetic resonance equipment. This project received support and cooperation from numerous Cuban scientific institutions and from Horacio Carlos Panepucchi and his laboratory at the Physics Institute of the University of Sao Paulo in Brazil. From the very beginning, right from the conception of project 35-26-7RMN, it was considered that its success might mean the genesis of a research centre for applied physics, engineering physics and biophysics within the University Oriente. This was an idea that had been considered for decades by Roberto Soto del Rey, founder of the university and its physics department (Méndez-Pérez et al. 2012). Four general factors were considered to be vital for the emergence of this center:

- Factor 1 (fundamental): Clear definition of the line of scientific work in keeping with the national strategy, and in harmony with the local context as well as global scientific development.
- Factor 2 (organizational): Correlations among the researchers at the center and between them and the environment; work style and methods, discipline, collective and individual persistence; local, national and international links.
- Factor 3 (human): Disposition, qualification, customs, work capacity, identification with factors 1 and 2.
- Factor 4 (material): Scientific information, location, installations, transport, telephone, and finance.

In October 1991, some months after the MRI machine had gone into operation, the Cuban government made two quite transcendent decisions, as suggested by a group of experts:

- To relocate, install and put into operation the equipment built for the Hospital General Juan Bruno Zayas in Santiago de Cuba.
- To plan and construct the Centro de Biofísica Médica de la Universidad de Oriente, which was built in exactly 1 year and inaugurated on February 10, 1993 by Fidel Castro.

That day Fidel Castro wrote the following dedication in his book *History Will Absolve Me*: "To the Center of Medical Biophysics of Santiago de Cuba, which is not only the pride of this heroic city, but also of our whole nation. Congratulations and to the forging of new dreams and beautiful realities of tomorrow" and presented this to the center.

However, the laudatory burden of these words committed the [scientific] collective to a specific idea that represented a major challenge for Cuban science and technology. As Castro put it: "One day, science and its production must occupy the first place in the national economy. We must advance the production of results of the intelligentsia. This is our place in the world; there will be no other. Here we are on equal footing with the Japanese and the Germans."

Several renowned Cuban scientists have described the project as one of the most complex projects ever undertaken. Its main impacts can be distinguished into three areas: technology, science and society.

At the technological level: The principal contribution has been to prove the feasibility of the construction of diverse types of magnetic resonance equipment, including the images of the Giroimag series, and to improve quality to meet with international standards (Cabal et al. 1997). This is first world technology for one of the five most complex medical diagnoses. The project was realized in such a way that the entire expenditure used for the creation of technological premises (calculation, design, construction, identification, validation, registration in the National Health System and assembly in hospitals) as well as the primary equipment equalled the price of one single piece of equipment purchased on the global market.

Surrounding this Cuban project, and as part of the technological assimilation, multiple technologies have been propelled that allow us to compete with companies on the same or bigger scale and to provide consistency to the development that is taking place in the world of this and other related specializations.

In addition to detecting new and autonomous lines of investigation and development, this creative process of technological assimilation also features some original contributions, above all, regarding the use of microcomputers (hardware and software) to manage the MR equipment and to process the information obtained with them in some incompatibilities with new technologies, for instance, in the proceedings and algorithms of electromagnetic calculations of complex systems (Cabal et al. 1997; Sánchez et al. 2004).

At the scientific level: The contributions made in the field of MR applications, especially the study with the systemic biophysical approach (from molecular level to the organism) of the Sickle-cell disease (SCD) also known as sickle-cell anaemia (SCA) or drepanocytosis.[2] This genetic blood disorder stems primarily from Africa and occurs due to a mutation in the haemoglobin gene. It is characterized by red blood cells that assume an abnormal, rigid, sickle shape causing modifications in the blood rheology (blood circulation) and multiple physiological processes, which lead to an intense and progressive deterioration of the patients' health and can ultimately lead to death. Life expectancy is shortened, often to less than 40 years. According to the World Health Authority, the estimated number of people carrying this disease exceeds 250 million and affects all countries where African emigration has taken place at some time.

In this regard, the Cuban contributions have made a strong scientific and social impact. Based on the Cuban equipment of MR relaxometry, a new diagnostic methodology of the condition of patients with sickle cell disease has been established, which in addition to predicting their condition allows for the evaluation of the effectiveness of treatment schemes. It is based on studies of the kinetics of polymerisation of Haemoglobin S (HbS) under conditions of low oxygenation. New physical-mathematical models for the analytical description of those processes and prediction of the conditions under which the polimerization can be slowed have been published owing to the investigations developed within the scope of NMR project 35-26-7. Moreover, a new non-toxic drug is available in a test phase, which

[2] See (Cabal et al. 1998; Pérez Delfin et al. 2004; Lores and Cabal 2005; Lores et al. 2006; Losada et al. 1988; Fernández et al. 2005, 2009; Cabal and Ruiz 2008a, b; Noda et al. 2004).

has proved its efficacy as a palliative in the treatment of the disease. Both results will surely benefit in the near future the quality of life for patients.

The results of the project have received various important acknowledgments, including two awards from the *Academia de Ciencias de Cuba* in 1994 and 1995. In September 1996, it was awarded the Giorgio Alberi in Memoriam First Prize at the V. International Conference on Applications of Physics in Medicine and Biology in Trieste, Italy.

At the social level: During the more than 10 years of operation in various Cuban hospitals, contributing to saving the lives of some of the tens of thousands patients diagnosed using Cuban MRI equipment has been the most obvious social achievement (Noda et al. 1999).

The impetus of the Cuban MR provided the basis for the creation of the Center of Medical Biophysics in 1993. The center has ramified its investigations in more than a dozen directions, ranging from the technological assimilation to the basic studies in the fields of electromagnetism, signal processing, molecular biophysics and neurophysics. The high point of MR and the creation of the Center of Medical Biophysics fell victim to the darkest hours of the Cuban economy, yet national interest and the determination of its scientists has enabled modest achievements that planted the seeds for new scientific and technological developments.

The formation of human resources is impressive. In the early 1980s, only two MR research laboratories existed. The impetus of magnetic resonance caused without question an explosion in the training of experts in diverse branches such as physics, electronics, computation, chemistry, biochemistry, mechanics, industrial design, medicine and many others. Dozens of graduate and post-doc students have been involved with these laboratories. The researchers publish and participate in prominent international contexts and maintain an increasing cooperation with renowned institutions in Canada, England, Spain, Germany and France.

At the center as well as in Cuba, the interrelation between basic investigations and technologies are tested on a daily basis, as is the effectiveness of the links between academic activities and production and the implementation of scientific results. It should be emphasized how important it is for developing countries to seek new and alternative ways to establish actual and efficient connections between the diverse branches of physics and current scientific and technological work. While the entire global scientific and technological repertoire is intelligently digested via transfer and technological assimilation, the different fields of basic sciences should also be advanced (Cabal 2002, 2011; Cabal et al. 2009). They should be coherent in strategy and lay the groundwork for a native physics and at the same time contribute to a universal scientific community that can respond to the progress of all of our countries.

References

Cabal, C. 1980. *Nuclear magnetic resonance relaxation mechanism in aqueous solution of Ni(II) and Co(II)*. PhD thesis, Leningrad University.

Cabal, C. 2002. Biofísica médica. In *Cuba. Amanecer del Tercer Milenio. Ciencia, Sociedad y Tecnología: Biofísica Médica*, ed. Fidel Castro Dìas-Balart, 31–48. Madrid: Debate Editorial.

Cabal, C. 2011. A 20 Años de la Primera máquina de resonancia magnética de imágenes cubana: Un proyecto de Ingeniería Física y Biofísica. *Rev. Cub. Fis.* 28(1): 68–72.

Cabal, C., and V.I. Chizhik. 1982. Study of molecular motion and microstructure of hydrate shells of nickel(II) and cobalt(II) ions by the NMR relaxation method. *Theoretical and Experimental Chemistry* 17(3): 322–326.

Cabal, C., and V.I. Chizhik. 1983. Frequency dependence of ^{1}H and ^{7}Li relaxation times in concentrated solution of Ni (II). *Vestnik Leningrad University* 1(4): 20–24.

Cabal, C., L. Fiffe, and A. Guzmán. 1984. Posibilidades de la Relajación Magnética Nuclear de Protones para el Estudio de la Reaccion de Oxidación de Cobalto (II) en Medio Amoniacal. *Rev. Cub. Fis.* 4(3): 9–14.

Cabal, C., E. Gonzalez, H. Sánchez, A. Bordelois, A. Frómeta, J. Parra, and M. Sánchez. 1997. Spatial frequency response of Cuban MRI machine GIROIMAG. *Physica Medica* XIII: 211–213.

Cabal, C., A. Fernández, M. Lores, E. Alvarez, J. Losada, C. Soler, and E. Pérez. 1998. Magnetic relaxation in the kinetics of the polymerization of hemoglobin S. Clinical diagnosis and treatment with vanillin. Paper presented at the ISMRM Sixth Meeting and exhibition in Sydney, Australia. *Medicine* 3: 1027–1030.

Cabal, C., E. González, and J.E. Paz. 1999. Resonancia magnética de imágenes cubana: tres años de servicio en el sistema nacional de salud. *Revista Cubana de Salud Pública* 25(2): 123–128.

Cabal, C., E. González, Y. Torne, and A. Rojas. 2009. Micro magnetic and molecular magnetic resonance imaging in modern biotechnology and pharmacy. *Biotecnologia Aplicada* 26(4): 304–314.

Cabal Mirabal, C., and I. Ruiz Chaveco. 2008a. A model of the molecular aggregate process of haemoglobin S. Absence of crystallization. *Revista Integración, Universidad de Santander* 26(1): 13–22.

Cabal Mirabal, C., and I. Ruiz Chaveco. 2008b. A mechanism of crystallization processes of haemoglobin S. *Revista Integración, Universidad Industrial de Santander* 26(2): 1–9.

Chizhik, V.I., C. Cabal, A. Guzmán, A. Miranda, and R. Novas. 1983. Relaxómetro de Resonancia Magnética Nuclear. *Rev. Cub. Fis.* III(3): 33–42.

Fernández, A., C. Cabal, J. Losada, E. Alvarez, C. Soler, and J. Otero. 2005. In vivo action of Vanillin on delay time determined by magnetic relaxation. *Hemoglobin* 29(3): 181–187.

Fernandez, A., C. Cabal, M. Lores, J. Losada, and E. Perez. 2009. Sickle cell diseasc painful crisis and steady state differentiation by proton magnetic resonance. *Hemoglobin* 33(3–4): 206–213.

Gilart, F., N. Fleitas, and C. Cabal. 1987. Estudio del Cemento Pórtland-350 de la Fábrica 'José Mercerón' con el Método de Relajación Magnética. *Rev. Cub. Fis.* 7(1): 53–58.

Guzmán, A., C. Cabal, L. Molina, A. Fernández, and V.I. Chizhik. 1987. Procesos de Relajación Protónica en Soluciones Acuosas de Lantánidos. *Rev. Cub. Fis.* 7(1): 47–52.

Lores, M., and C. Cabal. 2005. Proton magnetic relaxation process during the polymerization of hemoglobin S. *Applied Magnetic Resonance* 28(1–2): 79–84.

Lores, M., C. Cabal, O. Nascimento, and A.M. Gennaro. 2006. EPR study of hemoglobin rotational correlation time and microviscosity during the polymerization of hemoglobin S. *Applied Magnetic Resonance* 30(1): 121–128.

Losada, J., F. Guilart, C. Cabal, et al. 1988. NMR relaxation study of sickle cell disease. In *Proceedings of the XXIV congress AMPERE magnetic resonance and related phenomena*, Poznan, 1027–1030.

Méndez-Pérez, L.M., E.J. Roca-Oria, and C. Cabal. 2012. Roberto Soto del Rey, fundador de la Universidad de Oriente y de su Departamento de Física. *Rev. Cub. Fis.* 29(1): 33–36.

Noda, M., C. Cabal, E. González, and J. Paz. 1999. Resonancia Magnética de Imágenes Cubana: tres años de servicio en el Sistema Nacional de Salud. *Revista Cubana de Salud Pública* 25(2): 123–128.

Noda, M., J. Losada, E. Pérez, E. González, A. Govin, I. González, and C. Cabal. 2004. Evaluación de Afectaciones Morfológicas en el Cerebro de pacientes con Anemia de células falciformes con el uso de Imágenes de Resonancia Magnética. *Revista de neurología* 38(1): 17–19.

Pérez Delfin, E., I. González Ferro, M. Noda Guerra, E. González Dalmau, A. Govin Cid, C. Cabal Mirabal, and J. Losada Gómez. 2004. Evaluación de Afectaciones Morfológicas en el Cerebro de pacientes con Anemia de células falciformes con el uso de Imágenes de Resonancia Magnética. *Revista de Neurologia* 38(1): 17–19.

Sánchez, H., C. Garrido, C. Cabal, and H. Saint Jalmes. 2004. Designing an efficient resistive magnet for magnetic resonance imaging. *IEEE Transactions on Magnetics* 40(5): 3378–3381.

Zito, M., M.M. Argüelles, et al. 1999. *Y sin embargo–: ciencia: hablan 30 investigadores cubanos*. Habana: Editoria April.

Chapter 18
Nanotechnologies in Cuba: Popularization and Training

Carlos Rodríguez Castellanos

18.1 Introduction

In Cuba, as in other countries, activities in the field of nanotechnology emerged from the converging development of research in materials physics and chemistry, microelectronics, supramolecular chemistry and molecular biology.

Their precursors go back to the 1980s and are linked to theoretical studies concerning low-dimensional semiconductor structures (quantum wells, superlattices and others), developed at the department of theoretical physics of the physics faculty of the University of Havana. During the 1990s, theoretical and experimental work on semiconductor nanostructures gained in importance. Cuban physicists led the organization of the *Red CYTED* (Network CYTED) to "study fabrication and characterization of semiconductor nanostructures for micro and optoelectronics" which functioned between 1998 and 2003 with the participation of eight Spanish-American countries. The network organized various courses and scientific meetings, edited a book and supported the scientific collaboration among the participant institutions.

During these decades, the research and training of doctors was extended, for instance, to nanobiology, nanomagnetism, nanocolloids and other dispersed systems, nanostructured polymers, drugs and biomolecules nanoencapsulation, nanoporous materials, carbon nanostructures, nanostructured photovoltaic cells, nanocrystals structure, synthesis and functionalization of metals, semiconductor and magnetic nanoparticles for medical applications, nanotoxicology, nanosensors, nems, computational simulation and modeling on nanoscale, fabrication of systems for obtaining nanolayers and equipment of nanometric resolution.

The experimental capacities of the Cuban institutions in this field are quite limited, causing the aforementioned research to lean heavily on international

C. Rodríguez Castellanos (✉)
Physics Faculty, University of Havana, Havana, Cuba
e-mail: crc@fisica.uh.cu

A. Baracca et al. (eds.), *The History of Physics in Cuba*, Boston Studies in the Philosophy and History of Science 304, DOI 10.1007/978-94-017-8041-4_18,

scientific collaboration. This has allowed hundreds of papers to be published in well-known journals and for more than 30 doctors to be trained in various specialties. In 2009, the *Revista Cubana de Física* dedicated an issue with an editorial and 15 papers to a review of part of the investigation on nanosciences and nanotechnologies developed in Cuba (RCF 2009). An analysis of the Cuban scientific output in the nanosciences and nanotechnology is given by Aguiar et al. (2012). A predominance in the field of physics can be observed as well as the importance of international cooperation. Papers that deal with economic and social problems linked to the development of nanotechnologies in the Cuban context have also been published by authors who, curiously, are physicists too (Rodríguez Castellanos 2009; Castro 2011; Estévez Rams and Aragón Fernández 2010a, b). Cuban specialists, most of them physicists, participate in the "Red Latinoamericana de Nanotecnología y Sociedad" (Latin American Network of Nanotechnology and Society) and the *Red CYTED* for Popularization and Training in Nanotechnologies.

Although a national research program in nanotechnology does not properly exist, research programs in this field have being financed in the context of other national programs. In order to support the activities in this and other related areas, the Ministry of Higher Education has financed the creation of the analytic laboratory LUCES with over one million US$. In 2008, the Council of State announced the creation of the *Centro de Estudios Avanzados de Cuba* (CEAC, Cuban Centre for Advanced Studies), a multidisciplinary institution devoted mainly to nanobiotechnology and nanomedicine. The project was triggered by a proposal made by a team of physicists convened by the Scientific Advisor of the State Council, Fidel Castro Díaz-Balart. The first phase of this investment is essentially finished. Within the framework of this project, work is done to educate and train young Cuban scientists at European universities.

International scientific events and courses related to nanoscience and nanotechnologies are regularly organized. These began in 2001 with an international event on nanoelectronics organized by the Centro de Investigaciones en Microelectrónica (CIME) of the Instituto Superior Politécnico "José Antonio Echevarría" (ISPJAE, Polytechnic University of Havana), and continued with the yearly summer schools organized by the Instituto de Ciencia y Tecnología de Materiales (IMRE) of the University of Havana (UH). Moreover, the Taller Internacional de Nanomagnetismo (2004), the XVII Simposio Latinoamericano de Física del Estado Sólido (2004), the conference La Ciencia de Materiales en la Era Nano (2009) and three international seminars on nanoscience and nanotechnologies (2006, 2008, 2010, 2012) stand out as examples of international scientific events. Binational meetings have been organized with representatives from Mexico (2003, 2009), the United Kingdom (2004), Brazil (2007, 2010) South Africa (2010), and a meeting of the *Red de Macrouniversidades de América Latina* (2006, Network of Macro Universities of Latin America). Many outstanding visitors have attended these events during the last decade, for example, the Nobel laureates Zhores Ivanovich Alfërov (2007 and 2010), and Robert Curl (2009). In addition, Christoph Gerber (2009), co-inventor of the atomic force microscope (AFM), has also been in attendance.

18.2 Popularization

Beginning in 2001, the dissemination and popularization of nanoscience and nanotechnologies for various audiences increased. One aspect that stands out is work devoted to inform the government and other decision makers about the content and possible consequences of the development of nanotechnologies. The latter started in 2002 when the *Observatorio de Ciencia y Tecnología* of the *Ministerio de Ciencia Tecnología y Medio Ambiente* (CITMA) organized a team of specialists to prepare the document "Elementos iniciales para el análisis de la nanotecnología en Cuba" (Initial elements for the analysis of nanotechnology in Cuba) which was directed at the government and the directors of various scientific institutions. A second project on "Nanomaterials"was prepared between 2006 and 2007 in the context of the *Programa Nacional de Ciencia e Innovación Tecnológica "Nuevos Materiales y Materiales de Avanzada"* (National Program of Science and Technological Innovation "New Materials and Advanced Materials").

Between 2005 and 2007 a team of experts, commissioned by the office of the scientific advisor of the Council of State, prepared a study and a proposal out of which the decision to create the previously cited CEAC arose. Moreover, the *Academia de Ciencias de Cuba* convened a group of experts that proposed changes to the ethical code of the scientific workers in Cuba to include some new problems posed from the development of nanotechnologies. In addition, work has been performed inside the scientific community in order to identify common interests and to promote multidisciplinary collaboration in this field.

In 2001, with the support of the *Centro Nacional de Investigaciones Científicas* (CNIC) and the IMRE, the national workshop "Las Nanotecnologías en la Biotecnología y la Industria Médico–Farmacéutica" (Nanotechnologies in Biotechnology and the Medical-Pharmaceutical Industry) was organized. Various organizations in the field participated in the workshop.

In 2002, the *Red de Nanotecnologías* of the Ministry of Higher Education, consisting of 12 institutions, was established. It aims at "promoting national and international cooperation in nanoscience and nanotechnologies." This network has organized several national conferences and coordinated the participation of Cuban specialists in the following meetings: Cuba-Mexico (2003, 2009), Cuba-United Kingdom (2004), Cuba–Brazil (2007) and the *Red de Macrouniversidades de América Latina* (2006). In 2009, the University of Havana created a *Colegio de Nanotecnologías*, which brings together specialists from the natural, economic and social sciences.

For the wider public, several initiatives have been developed, including conferences in schools and other institutions, press articles and TV lectures, among others, covering scientific events related to nanotechnologies that took place in the country. Several Cuban journalists as well as journals, for example *Juventud Técnica,* have shown interest in widening the popularization of nanoscience and nanotechnologies. A dozen illustrated articles on nanotechnologies have been published in periodicals and journals, emphasizing the practical applications and possible benefits of "nanoproducts." However, Cuban TV cannot acquire documentary or film material of sufficient quality to carry out a more attractive popularization that transmits to the public, especially to young people, the wonder of atom manipulation, the technological opportunities that it offers and the marvels of matter behavior on a nano scale.

18.3 Training

The majority of the training activities that have been developed are concentrated in postgraduate courses and in the final years of some degrees of sciences and engineering. Although a postgraduate program especially dedicated to nanotechnologies does not exist in Cuban universities, more than 30 PhD theses have been defended in the country, as well as an undetermined number of master's theses in the fields of physics, chemistry, biology, materials science and electronics, which are all directly related to nanotechnologies. As part of the project for the creation of the CEAC, a training program for young graduates of science and engineering careers in European laboratories has been created. Although the programs of physics, chemistry and biology given in Cuban secondary schools include the basic skills needed for illustrating typical phenomena and their applications on the nano scale, these are not explicitly mentioned in these courses. Moreover, no topic related to nanoscience and nanotechnologies is taught in primary schools. It is likely that didactic materials containing illustrative examples, simple problems, film experiments or simulations are lacking, as well as the corresponding training for the teachers and professors.

18.4 Conclusion

In a small, underdeveloped and embargoed country like Cuba, modest efforts are made to develop the skills needed to take advantage of the opportunities offered by nanotechnology. This becomes especially apparent when one takes into account the experience in training high-level personnel in the basic sciences, the development reached by biotechnology and the medical pharmaceutical industry, all of which are important elements for the technology sector in the Cuban economy.

This effort, training and popularization, directed to all sectors of society, but specially to young people, all play a central role since the main strength will always be the training and motivation of the human element. There is great potential both in the education system and in communication media to increase training and popularization in nanoscience and nanotechnologies. However, it is necessary to acquire a larger quantity and better quality informational material for teachers, professors and communicators.

Training in nanotechnology and the popularization of its use must not be simply aimed at impressing with new "nanoproducts" and emphasizing their practical usefulness, but also to stimulate the fantasy and creativity of young people, to presenting to them the wonders of the nanoworld and the new possibilities of fabricating and modifying matter provided by nanotechnology. At the same time, one has to consider the economic, social and environmental aspects related to the development of nanotechnologies, including the risks and dangers that it poses.

Activities related to the investigation, training in and popularization of nanotechnology needs both national support and strong international collaboration. This collaboration should increase the resources available in Cuba. By the same token, the modest achievements made within Cuba can and should be made available to others.

There is no doubt that the principal initiators for the take-off of nanosciences and nanotechnologies in Cuba have been the physicists, both in research and popularization, as well as in the promotion of advancing initiatives. Yet the modest Cuban investments in that field have rather concentrated on nanobiotechnology and nanomedicine, given that Cuba has better options in this area to convert scientific advancements into economic value. Although this policy does not seem to be fair to some physicists, the real situation is that the Centre for Research and Development of Medicaments announced in 2012 the start of clinical testing of the first Cuban "nanotechnological" medicament. It is a treatment produced by the Cuban pharmaceutical industry that uses nanoparticles for immunosuppression in the treatment of patients undergoing organ transplantation. Presumably, the new treatment will allow a reduction of the dose and thus minimize the adverse effects of the medicament.

This is a clear example of the vital, albeit indirect role of basic research in general and physics, in particular, in the emergency situation of the new scientific and technological paradigms, when the specialists linked to the application of the emerging technologies do not yet exist. Something similar happened in Cuba during the 1970s with the development of computer sciences, and again in the 1980s with bioengineering. It is possible that the physicist may not play such a central role in the final development of nanotechnologies in Cuba, but new technologies will certainly emerge—and the physicists will be there.

References

Aguiar, J.D., R. Arencibia, J. Araujo, and D. Alba. 2012. Producción Científica Cubana sobre Nanociencias y Nanotecnología. *Ciencias de la Información* 43(1): 5–14.

Castro Díaz-Balart F. 2011. La nanotecnología y el desarrollo. Oportunidades e incertidumbres. *Anales de la Academia de Ciencias* 1(1).

Estévez Rams, E., and B. Aragón Fernández. 2010a. Deshojando margaritas: ¿nanociencia o nanotecnología? *Revista Universidad de la Habana* 271: 8–33.

Estévez Rams, E., and B. Aragón Fernández. 2010b. La fábula de los tres hermanos: Las nanociencias y las nanotecnologías en el contexto cubano. *Temas: estudios de la cultura* 61: 107–117.

RCF. 2009. Número especial: nanociencia cubana 2009. *Rev. Cub. Fis.* 1: 26. See http://www.fisica.uh.cu/biblioteca/revcubfi/2009/vol.26-No.1/. Accessed 10 Oct 2013.

Rodríguez Castellanos, C. 2009. *Nanoamenazas y nanoportunidades*. Universidad 2010. http://revistas.mes.edu.cu. Accessed March 2014.

Chapter 19
Physics Studies at the University of Havana

Osvaldo de Melo Pereira and María Sánchez Colina

19.1 Introduction

The *licenciatura en física*[1] degree course was created as part of the 1962 University Reform. It started at the Physics School within the Science Faculty of the University of Havana, also including the Schools of Mathematics, Chemistry, Biological Sciences, Geography and Psychology (Henriques Rodríguez 2001). The degree of *licenciado* had replaced that of *bachiller* since 1880, but only the physico-mathematical sciences and physico-chemical sciences degree courses existed prior to the 1962 university reform.[2] At the time of its creation, the School of Physics had two departments: general and experimental physics, and theoretical physics and structure of matter.

In this paper, we will analyze some data concerning the undergraduate and graduate studies during the 46 years elapsed since the creation of the physics degree course at the University of Havana. Several related issues, such as the development of scientific research and the influence of international collaboration, are dealt with in other contributions to this volume.

19.2 Undergraduate Studies

Just after the 1962 reform, the physics syllabus included four semesters of higher physics (each semester took nearly 5 months), two semesters of rational (theoretical) mechanics, two semesters of atomic and nuclear physics, five semesters of physics

---

[1] We will refer to it just as "physics" from now on.

[2] See the contribution to this volume by Altshuler and Baracca.

O. de Melo Pereira (✉) • M. Sánchez Colina
Faculty of Physics, University of Havana, Havana, Cuba
e-mail: omelo@fisica.uh.cu

A. Baracca et al. (eds.), *The History of Physics in Cuba*, Boston Studies in the Philosophy and History of Science 304, DOI 10.1007/978-94-017-8041-4_19,

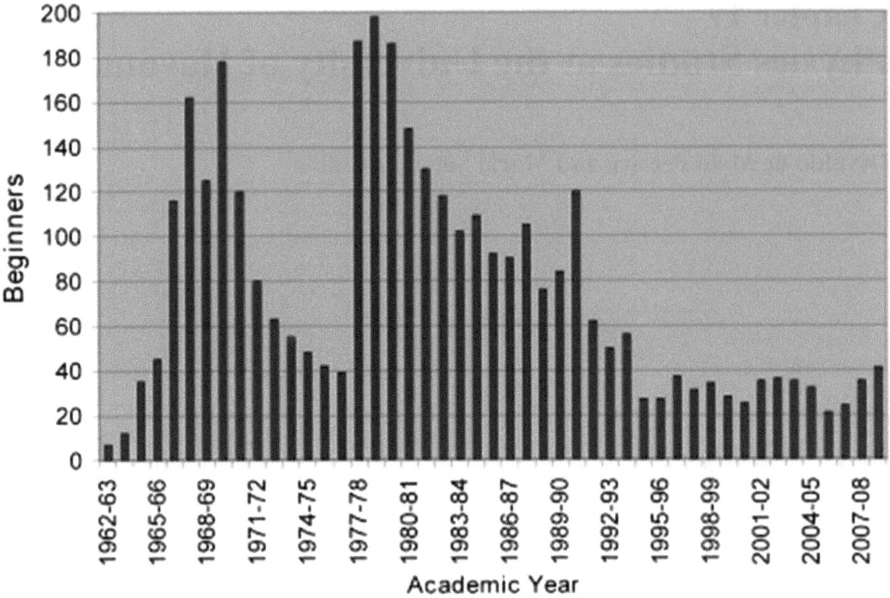

Fig. 19.1 Students entering each of the 46 editions of the physics degree course since 1962

laboratories, two semesters of atomic and nuclear physics laboratories, two semesters of electronics (electric circuits and radio-electronics), and six semesters of theoretical physics. The plan also included five semesters of philosophy studies, three semesters of a foreign language, and two semesters of general chemistry.

The above syllabus was later modified on six occasions, the last version to date (known as "Plan D") started in the academic year 2007–2008. Its structure is quite similar to that of the 1962 syllabus. Electronics and computer science have a stronger presence now, of course, but perhaps the most specific characteristics of the present syllabus are: (i) the increase in the number of hours devoted to scientific work, (ii) its higher flexibility, materialized in a larger number of optional subjects and (iii) a higher diversification of subjects in the last 2 years, implying a broader profile training. The degree course is still 5 years long.

In the last 46 years, 3,508 students entered the physics degree course, but only 906 managed to graduate. After eliminating the number of students in the last 4 years (189) and the number of those that graduated in the first 4 years (10), we get an "efficiency index" of 27 % graduating students out of the total input. This averaged figure, however, has had significant ups and downs, which require a more detailed exploration of input/output behavior during the relevant period.

Figure 19.1 shows the entering students for each of the 46 editions of the physics degree course since 1962. The dramatic increase corresponding to the 1977–1978 academic course coincides with the country's interest in the development of nuclear physics, and also with the additional graduation of an accelerated high school course to the standard one. After that, there was a steady decrease in the number of

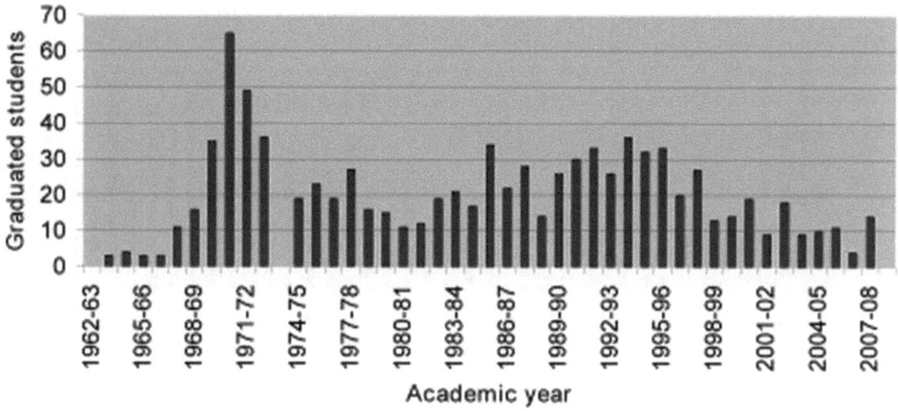

Fig. 19.2 Behavior of the graduating students

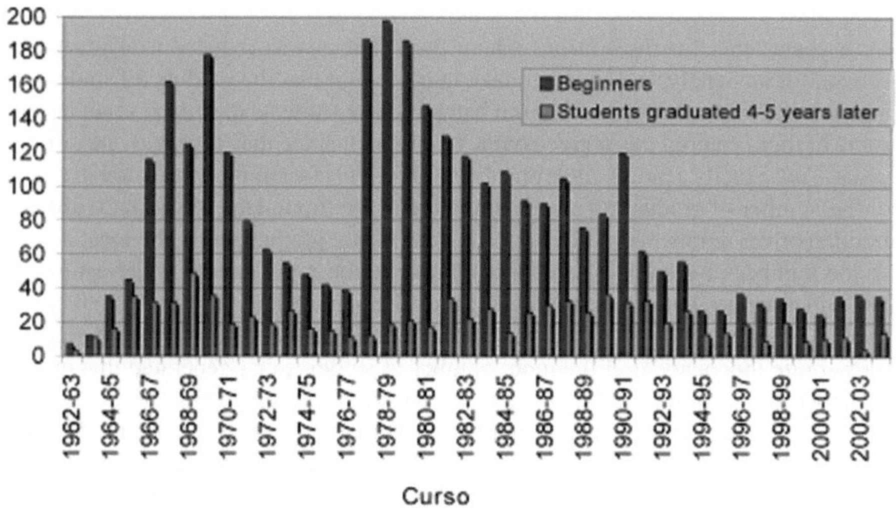

Fig. 19.3 A comparison between the number of new students and the number of graduating students

new students, until it stabilized at about 30 students, starting in the 1994–1995 academic course.

Figure 19.2 shows the behavior of the graduating students. The void in the 1973–1974 course is explained by the fact that, during 3 years, the degree course had a duration of 4 years—the 5-year regime was resumed just in the 1973–1974 course. By comparing Figs. 19.1 and 19.2 it can be seen that there is no trivial relationship between the number of graduating students and the number of new students.

To shed some light on this point, Fig. 19.3 displays a comparison between the number of new students and the number of graduating students 5 years later

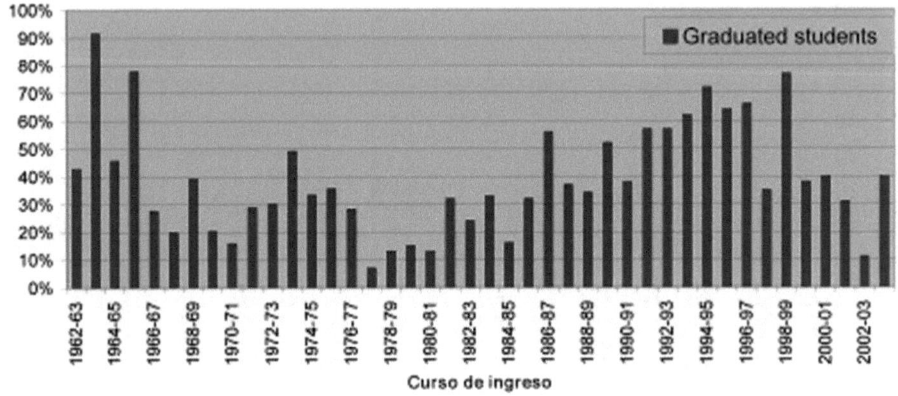

Fig. 19.4 The percentage of graduating students relative to the corresponding number of new students

(or 4 years later, for the editions where the syllabus was reduced to 4 years). Of course, for any analysis we have to take into account that the graduating students in a given year not only come from the batch of new students exactly 5 years before: some of them entered the degree course before that, since they repeated one or more years—not exactly a rare event in physics studies. This fact implies an over-estimation of the number of graduating students relative to the input. Due to similar reasons, a fraction of the number of students in any year would graduate after the ideal 5 years so the number of graduating students relative to the input can be under-estimated. We will assume that the two effects are of the same order, so the data shown in Fig. 19.3 is taken as good enough. At any rate, the number of graduating students looks quite independent of the input number of students that generated that output: while we see variations in the number of new students as many as 150 individuals, the variation in the number of graduating students is no larger than 15. From Fig. 19.3 we can calculate the percentage of graduating students relative to the corresponding number of new students 5 (or 4) years before, which is shown in Fig. 19.4. If we assume that this figure represents the efficiency, we see that the highest efficiency was attained in the 1990s.

An interesting result observed in Fig. 19.3 is the sharp decrease in the number of students in the period 1970–76. This is quite strange, since the "take-off" of Cuban physics actually took place by the end of the 1960s (as suggested in other contributions to this volume). There may be various reasons for this. The policy of sending young people to study physics to (then) socialist countries probably gained strength in those years; the physics degree course was opened at the University of Las Villas at the time, so that a number of potential students from the central region of the country was lost to the University of Havana (those from the eastern part of the country would study at the University of Oriente). Further reasons could have been the lack of attractiveness of the available employment offers, and the feeling among high-school students that physics was too difficult to follow.

By the mid-1970s, the number of students finishing the senior high school level was much higher, and the physics degree course was again given priority by the educational planners. In addition, the collaboration with the socialist block started to concentrate on the graduate level instead of the under-graduate level.

In our opinion, the input-output relationship in physics studies suggests that the number of high school students with real interest and abilities in physics has always been relatively small: any artificial "boost" in the input of physics students has only served to fill up the classrooms of freshmen, without a substantial contribution to the number of graduating students. This conclusion suggests that any attempt to increase the total number of students in physics must include a strong motivational campaign plus some kind of input selection process.

In fact, two important measures were recently taken to improve the quality and motivation of the physics students in Cuba. One was to establish a specific input system for the studies of physics (as well as chemistry and mathematics). Another was the creation of a new degree course: physics engineering.

The new entry system to the physics degree course started in 2003–2004. The main core of the new system was the application of a special exam to choose suitable candidates for the physics degree course. The ability of the new system to actually increase the number of students generated some controversy at the beginning, since it was argued that it might rather discourage some students instead of motivating them. Still, it showed itself to be attractive to the students with a previous inclination towards physics, who viewed the special exam as a challenge: if they passed it, their place at the University of Havana was guaranteed well before the rest of the students aiming at other degree courses. In addition, the exams were designed in such a way that there was an opportunity to pass even for students with poor formal preparation in physics, but with good reasoning abilities and physical intuition. In any case, the application of the new system prevented a drop in the input of students relative to previous years—as shown in Fig. 19.2 (only four students graduated in the 2006–2007 academic course). Since only two groups who passed the special exam have graduated from their physics studies to date, it is difficult to evaluate the effectiveness of the system.[3] By its own nature, the new system stimulated vocational work in physics in high schools since it was deemed necessary to visit them in order to inform them about the details of the examination process, the exam dates and so on. Many professors in the Physics Faculty at the University of Havana share the feeling that the new system has improved the quality of the physics students, but we insist that as yet there is not enough data to derive a reliable conclusion. We will have to wait a few more years to see the tendencies in the number of graduating students, and add enough bars in Fig. 19.4. Due to the typically small number of physics students, it is important to take data fluctuations into account.

The physics engineering degree course was inaugurated in September 2007. It is supposed to generate an engineer with a strong background in physics and mathematics aimed at applications. An important difference with the *licenciatura en física*

[3] Fifteen students graduated in each of these two courses on July 2008 and July 2009, respectively (note added during the editorial process).

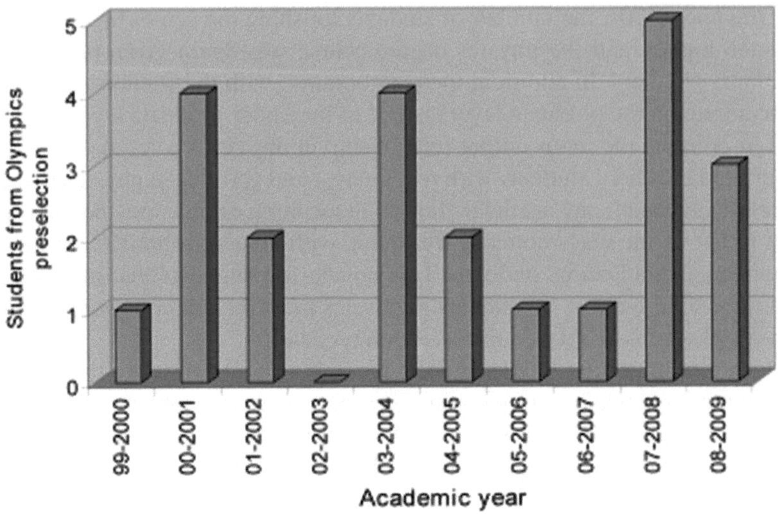

Fig. 19.5 Number of students entering the Physics Faculty from the "concentrate"

is an increase in the number of hours devoted to computer science, electronics and chemistry, at the expense of some of the time devoted to theoretical physics and mathematics. The creation of the new degree course implied a risk: the already modest number of incoming students could be split into two groups. But this has not been the case up to now: if one looks at the behavior of the new students in Fig. 19.2, an increment is detected in the last 2 years—the bars in the last two courses actually include both physics and physics engineering students. There were only five physics engineering students in the 2007–2008 course, but it can be seen that there are already 21 in the 2008–2009 course. Based on the data corresponding to these 2 years, we suggest that the creation of the new degree has led to a net increment in the total number of incoming students to the Physics Faculty, which rises in the course 2008–2009 to its highest value in the last 15 years. Still, it must be remembered that fluctuations are quite relevant to the conclusions that may be drawn from the analysis of data.

Cuban participation in the international physics olympics at the pre-university level has a long tradition. Cuban teams have had relevant performances either at the International Physics Olympiads (IPHO), or at the Ibero-American Physics Olympiads (Baracca et al. 2006). The preparation of these teams includes the recruiting of a national "concentrate" of gifted students from different pre-university centers all over the country, who are engaged in systematic problem-solving training (both theoretical and experimental). This "concentrate" can be thought of as a natural source of quality students for the physics and physics engineering degree courses. However, this has not always been the case. Figure 19.5 shows the number of students entering the Physics Faculty, University of Havana, from the above-mentioned "concentrate." If one takes into account that it includes nearly 20 students, and amongst them, no less than six to seven are in the final high school year (12th grade) we conclude that this source has not been efficiently exploited by the Physics Faculty: instead of physics (or physics engineering) they have in the end preferred other careers.

Relative to the undergraduate studies in physics, it should be underlined that, apart from the Physics Faculty at the University of Havana, there are other centers in Cuba producing physics-related professionals. The *licenciatura en Física* has also been studied at the University of Oriente since 1970—up to the year 2000, 334 physicists completed their studies there (Méndez Pérez and Baracca 2001). On the other hand, 12 physicists graduated from the Universidad Central de Las Villas before 1976[4] (after being closed, the physics degree course has re-opened there in the course 2006–2007). Also, an important number of nuclear physicists have graduated at the Higher Institute of Nuclear Science and Technology. Finally, a few tens of physicists studied abroad, mainly in the former Soviet Union. Considering all those sources, the total number of physics graduates in Cuba can be estimated (with a large margin of error) as 2000.

19.3 Graduate Studies

An important step in the development of physics in Cuba was the creation of graduate programs. A physics master program started in the 1970s at the University of Havana, and graduated nearly forty masters in physics. After an interruption of several years, the program re-started in 1994. Nearly 135 students have graduated since, in the areas of condensed matter theory, physics of semiconducting devices, materials science and, more recently, the physics of complex systems. The program has been given the category of "excellence" by a national board. Table 19.1 reports the incoming and graduated students from the Master in Physics program, year by year.

PhD theses in physics have been defended in Cuba since 1974. Figures 19.6 and 19.7 show updated data based on a 2001 report by the Cuban National Commission of Scientific Degrees (Rodríguez 2001). The data—comprising 214 PhDs—consist of the number of PhD defenses per year, and their subjects. If the PhDs defended abroad are included, the total figure may reach 250 or 300 PhDs in physics. In most

Table 19.1 Some statistics of the master in physics

Edition	Year	Number of students	Graduating students
1	1994	42	42
2	1996	24	24
3	1998	26	25
4	2000	23	18
5	2003	40	26
Total		155	135

[4] See the contribution to this volume by Baracca, Fajer and C. Rodríguez.

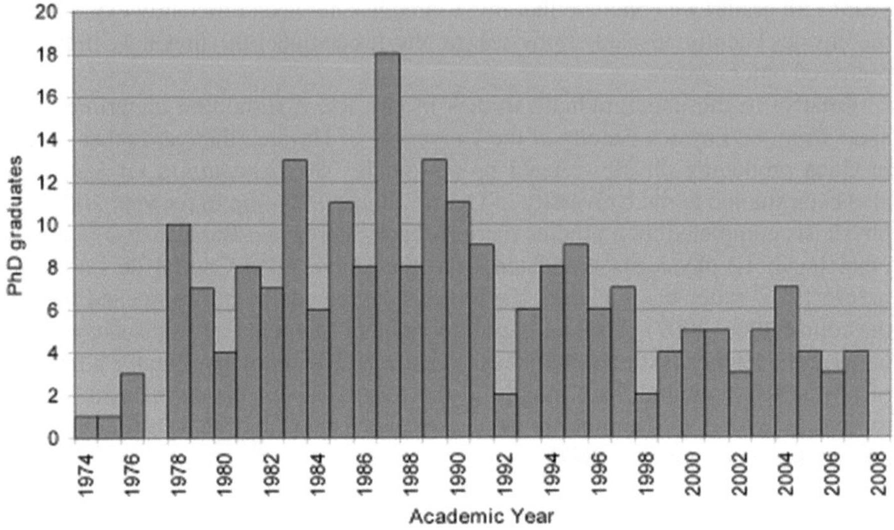

Fig. 19.6 Updated data based on a 2001 report by the Cuban National commission of Scientific Degrees

Fig. 19.7 Updated data based on a 2001 report by the Cuban National commission of Scientific Degrees

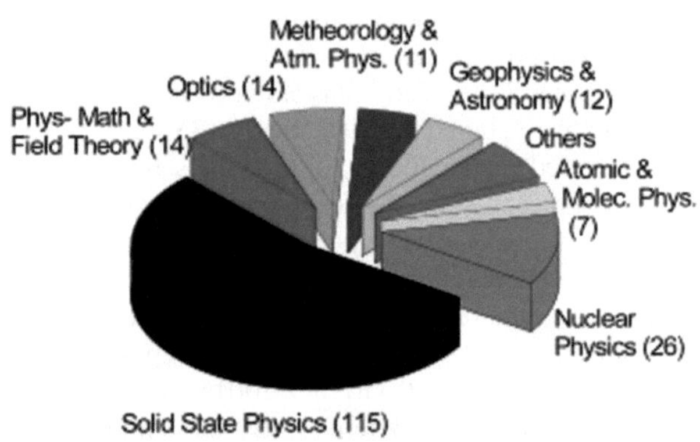

cases, a sizable contribution from foreign programs, institutions and laboratories has been required, basically due to the scarcity of equipment at home—which has also increased the duration of the PhD studies in many cases.

19.4 Conclusions

While it is true that the University of Havana is not the only Cuban source of physics graduates in Cuba, it has certainly produced the majority of Cuban physicists. Over 46 years, the undergraduate physics studies have gone through six successive undergraduate syllabuses, each one designed to match the international standards of its time. Generally, undergraduate studies kept a 5-year format to get the *licenciado en física* certificate, just after the defense of a research thesis before a board of experts. A decrease in the number of undergraduate students has been observed since the early 1990s, which has motivated a number of measures starting in the course 2003–2004: the increment of vocational activities in high schools, a new system to enter physics studies, and the creation of the new degree course in physics engineering. Due to the still small amount of undergraduate students, it is premature to evaluate the effectiveness of those measures. The available data hints, perhaps, at a discrete increment in the quantity and quality of the undergraduates.

Graduate programs in physics started in Cuba in the 1970s, and have now produced several hundred physicists with a scientific degree.

In this paper we have attempted to give insights based on hard data extracted from the archives of the Physics Faculty at the University of Havana, and from the National Commission for Scientific Degrees. However, analysis of further data sources might improve some of the numerical estimations given here.

Now, we would like to add some qualitative insights that might help round off a general view of the quality of physics studies in Cuba. Cuban physicists usually succeed in their graduate studies at foreign universities, including high-standard Latin American and European ones. The opinion of the employers on the preparation and abilities of Cuban physicists working in other Cuban institutions is generally very good. The following are two excerpts from statements made by representatives of important Cuban institutions who were asked to contribute with their criteria to the accreditation of physics as a "degree course of excellence" in 2005:

> The very frequent excellent performance of physicists in biological research has made us think about the differences between "training" and "information." Physicists get a solid training, expressed in their thinking structure, their ability to go from model to experiment and vice versa, their ability to orient themselves in the scientific literature and identify the essential regularities in specific phenomena associated to a concrete experimental model, and their ability to "recombine" information from different fields of science.[5]

> [Physicists] have shown an excellent theoretical formation, and a right approach to the research activity connected, in my view, to a ample and solid vision of the structure of matter. More importantly, they have a high standard of scientific ethics that distinguishes them from other young professionals. ... If I were required to nominate a "Scientific School" from the University of Havana, I would not hesitate to choose the Physics Faculty, which constitutes a source of pride to the University of Havana thanks to its rigor, scientific level and the human quality of the professionals it produces.[6]

Acknowledgments We acknowledge several colleagues from the Physics Faculty at University of Havana for their useful suggestions and comments, especially Dr. Carlos Rodríguez-Castellanos for the critical reading of the manuscript and providing some of the data. We also acknowledge J. Almaral, A. Cedeño, O. González and M. Coderch for their help in the search for information.

---

[5] From a letter provided by Dr. A. Lage Dávila, Director of the Center for Molecular Immunology, Havana, Cuba.

[6] From a letter provided by Dr. R. López Cordero, Director of the Unit of Catalysis, Center for Oil Research, Havana, Cuba.

References

Baracca, Angelo, Víctor Fajer, and Carlos Rodríguez. 2006. A look at physics in Cuba. *Physics Today* 59(9): 42–48.

Henriques Rodríguez, Ruth Daisy. 2001. Reforma Universitaria y la Facultad de Ciencias: memorias y reflexiones. *Revista Cubana de Educación Superior* 21(3): 8–17.

Méndez Pérez, Luis, and Angelo Baracca. 2001. Cincuenta años de física en la Universidad de Oriente, en Santiago de Cuba. *Rev. Cub. Fis.* 18(2): 146–154.

Rodríguez, C. 2001. Diagnóstico del estado de la Física en Cuba. Unpublished paper.

Chapter 20
Physics and Women: A Challenge Being Successfully Met in Cuba

Olimpia Arias de Fuentes

20.1 A Necessary Introduction

The history of physics in Cuba, like all the country's educational and scientific development, cannot be understood without taking into account its close relationship with the social changes that took place in Cuba during the five decades elapsed since 1959. This should include due consideration to the role played by women in this process, all the more since the link between science and gender is now generally regarded as a subject of growing special interest (Fernández Rius 2000).

It is quite a challenge for us even to outline the role of women in Cuban physics in the present context, however an overall view is attempted at the most relevant events related to the development of this science in the country.

While various approaches to the subject are possible, in any case it is not feasible to give in the few pages at our disposal a full and detailed enumeration of the many events and data deemed appropriate to the subject. Still, being proud of considering herself a part of the community of Cuban women physicists, the author feels it important to give some thought to various facts that emphasize the remarkable role played in the matter by the feminine gender. One must be aware that changes in individual subjectivity are slow to develop and that the relevant end results are long term, especially since we are dealing with ways of thinking deeply rooted through hundreds of years. It then becomes a great challenge for women to make themselves known within a predominantly masculine community with an androcentric cultural heritage, where certain values, behaviors and social barriers still prevail, in spite of the great efforts to revert them, made by the country, in the last five decades.

O. Arias de Fuentes (✉)
Institute of Materials Science and Technology, University of Havana, Havana, Cuba
e-mail: oarias@fisica.uh.cu

A. Baracca et al. (eds.), *The History of Physics in Cuba*, Boston Studies in the Philosophy and History of Science 304, DOI 10.1007/978-94-017-8041-4_20,

The above challenges are even greater for Cuban women physicists as far as academic media are concerned, because of the dedication and time called for by work in the field of physics, and the great deal of time required by the roles ordinarily played by women in society, especially in the familial environment, and the related way of thinking among many male physicists. A female scholar in the area of gender and science in Cuba recalled:

> A male friend, devoted to physics – a traditionally male science – told me once: 'women cannot do well in scientific work because they cannot concentrate as we do, since they have one neuron in the computer and the rest in domestic affairs and children, they don't have at their disposal the same time as we do to engage in work for hours and hours.' (Fernández Rius 2000)

Similar barriers on women working in physics are reported around the world (Ivie et al. 2002). Basically, such barriers are related to the existing conflict between the long time required to take care of children and the equally long time required by physics work. Associated with this is the fact that the field of Physical Sciences is one of the least populated by women on a global level and Cuba is not an exception. For the same reasons, the achievements of women physicists are in direct relationship not only to their intelligence, but also to the intensity of the work they have to do to compensate for the more limited time they can dedicate to scientific work, as compared to that available to their male counterparts.

Still, while the number of women physicists in Cuba is considerably lower than that of men who have graduated in physics, like in the great majority of countries—or perhaps all of them, the presence of women physicists in academic and scientific leading positions in our country shows prominent singularities (Arias de Fuentes 2008).

In the absence of a case study to find out the causes responsible for the accession of Cuban women physicists to such positions, this author advances her opinion that it has been the result of the increased prestige, social and professional, acquired by women in Cuba since the revolutionary changes that took place in the country after 1959. It shows how important it is for the Government to have the will to promote women to academic functions. The insertion of Cuban women in the development process of the country both as a leading actor and as a beneficiary, should be evaluated as one of the most successful social developments that have taken place in Cuba in the last 50 years, derived from the fact that women have had ample possibility to access public life nationwide just as men have.

Nowadays male and female students are graduated as physicists at the University of Havana—the oldest one in the country, founded in 1728—at the University of Oriente in Santiago de Cuba, and at the Higher Institute for Applied Science and Technology. In this last one, physics professionals receive specialized training in nuclear and earth sciences. Some 50 physicists per year obtained their degrees from these three universities in the 1990s (Baracca et al. 2006). High school physics teachers were trained in other higher education institutions whose profile is essentially pedagogical.

20.2 Some Interesting Figures

As an interesting piece of information it shall be noted that early in 1959 women made up 55 % of the illiterate Cuban population and about 17 % of the country's work force, with household jobs being their main employment source,[1] while presently women make up 46 % of the active labor force of the nation in the civilian state sector, and 65.6 % of the country's professionals and technicians.[2]

In Cuba nowadays women make up 51.6 % of the research personnel[3] and 60.4 % of the scientific research reserve of the country (González Bermúdez 2008) (i.e. young professionals with less than 2 years engagement in research work), which may be taken to predict an increase in the role played by women in this important sector of society. On the other hand, of the 3087 projects that compose the National, Branch, and Territorial Programs in the Science and Technology Innovation National System, 23 % are led by women (González Bermúdez 2008), which should be added to the recognized fact that the country is the second one in the region with the largest number of women incorporated to science (Padilla 2007). It should be noted, however, that while women represent 65 % of the university graduates in Cuba,[4] their proportion is much less in physics—about 20 %.

Figure 20.1 shows the percentage of women that have completed their physics degree courses in successive decades at the University of Havana, which is the main source of physicists in the country, from 1964 on. The first physics graduates in the

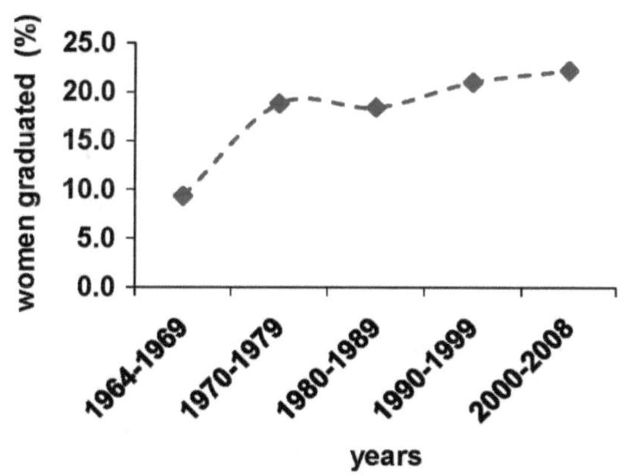

Fig. 20.1 Percentage per decade of women who have successfully finished physics degree courses taught at the University of Havana from 1964 to 2008

[1] See *Presencia de la mujer en la Salud Pública Cubana – Infomed*: http://bvs.sld.cu/revistas/spu/vol35_1_09/spu10109.htm. Accessed March 2014.

[2] See Aniversario 50 del Triunfo de la Revolución. Mujeres Cubanas en Cifras (2008): http://www.mujeres.co.cu/50%20aniversario/textos/Mujeres%20cubanas%20en%20cifras%20%282008%29.html. Accessed March 2014.

[3] See footnote 2.

[4] See footnote 2.

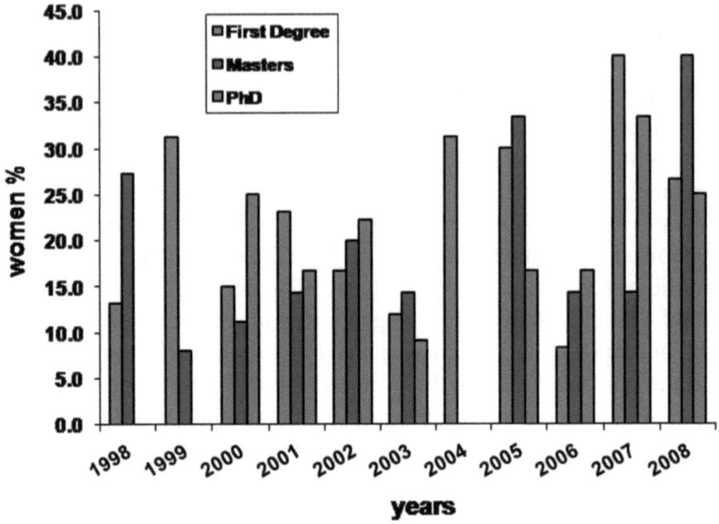

Fig. 20.2 Percentage of women graduated in all physics levels (1998–2008)

country finished their studies in 1964. It can be seen from the graph shown that the percentage of female graduates in physics has generally increased with time, having reached a maximum of 22.2 % in 2000–2008.

In Cuba, 72 % of the labor force in the Education sector is composed of women.[5] They are 61 % of the teaching staff at the University of Havana as a whole (Fernández Rius 2000). Curiously enough, while that percentage goes down to only 27.3 % at the Faculty of Physics, up to 66 % of the executives (deans and vice-deans) were female at certain times. An important fact in the history of physics in Cuba is the nomination for the first time of a woman physicist as Dean of the Physics Faculty at the University of Havana.

Figure 20.2 shows the percentage of women that obtained a *licenciatura*, MSc or PhD degree in physics in 1998–2008. Generally speaking, in this decade 20 % of the first level graduates (*licenciados*) were women, while the percentage of the MSc and PhD graduates was 16 % in both cases. This information refers to those who finished their corresponding studies in the country. On the other hand, we find out that of all those who obtained their MSc in physics after the initial and ephemeral existence of this degree in the 1970s (Baracca et al. 2006), 16 % were women, while of the total PhDs in physics defended in Cuba since it began in 1974, only 11 % went to women.

However, it should be noted that in spite of being such a minority, women were chosen as "best all around students" of many academic courses in physics at the University of Havana. In the last three, that condition was bestowed upon two women, one of whom was also elected the University's best all around student. Among the physicists that have been awarded to date the "Carlos J. Finlay"

[5] See footnote 2.

Order—the highest distinction awarded by the Cuban State to those professionals who have made important contributions to science and technology in the country—, 33 % are women. They have also quite a good representation in the yearly National Prizes awarded by the Cuban Academy of Sciences as exemplified by the fact that in three out of the seven dealing with physics subjects that were awarded in 2005, the leading author was a woman.

Generally speaking, the names of several women physicists regularly appear in the list of those who are awarded medals and national prizes for their scientific and educational work among them, the "José Tey" Medal awarded by the Council of State of the Cuban Republic the Distinction for Cuban Education, the Special Distinction of the Minister of Higher Education, the diplomas and medals awarded for participation in the preparatory tasks for the realization of the first Soviet-Cuban joint space flight, and the prizes for outstanding work awarded yearly by the National Science and Technology Forum, among others. All this, added to the fact that more than 35 % of the physicists who have been lately elected members of the Cuban Academy of Sciences are women, is a clear indication of national appreciation of the remarkable role played by women physicists in Cuba.

Two women physicists were nominated by Cuba for membership in the ICSU Regional Committee for Latin American and the Caribbean, and one of them was also elected by the International Council for Science (ICSU) as a member of its board of directors. A Cuban physicist was also elected member of the Academy of Sciences for the Developing World (TWAS), which brings together research workers of some 90 countries chosen through a highly competitive process, since less than 25 % of the nominees are finally promoted after taking into account their accumulated merits of their trajectory as professionals.[6] Cuba has eight members in this organization three of whom are women. Two of the members are physicists, a man and a woman.

A Cuban woman physicist was elected a member of the Executive Council of the Third World Organization for Women in Science (TWOWS), where she held the position of vice president for Latin America and the Caribbean for the period 1999–2006. She was also the president of the Cuban Chapter of TWOWS, within the framework of the Commission for Woman Scientists of the Cuban Academy of Sciences, which every year awards a prize to those women who are the principal authors of works that have won an Academy prize. In 2010 two Cuban female scientists, one of them a physicist, were awarded the TWOWS-TWAS international prize for scientific excellence. All of these facts are indicative that also at an international level Cuban women physicists have played an important role.

Among the prizes instituted by the Latin American Association of Biology and Nuclear Medicine Societies (ALASBIMN) with a view to stimulate research and development of nuclear medicine and related areas, only one bears the name of a woman, the "Mercedes Borrón" prize for technicians. This award was established

[6] Cuba News Headlines; Cuban Daily News (21.11.2008); "Eligen a científica cubana miembro de academia internacional": http://www.cubaheadlines.com/es/2008/11/21/14528/eligen_a_cientifica_cubana_miembro_de_academia_internacional.html. Accessed March 2014.

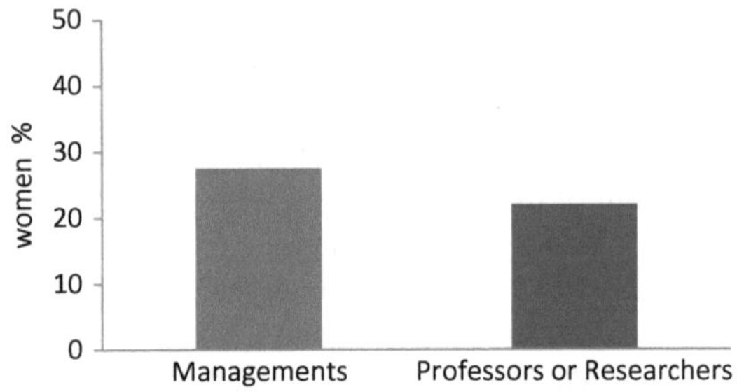

Fig. 20.3 Percentage of women female physicists holding posts as managers, professors or researchers referred to the total number of physicists in universities and scientific centers with a large number of physicists

and so named by ALASBIMN in honor of a distinguished Cuban physicist that passed away in 2002. She worked in the same area and had been awarded prizes by the Association. The "Sofia Kovalevskaya" international prize, sponsored by the organization of the same name, with the purpose of stimulating an increasing presence of women in the science and technology sectors of developing nations, has been awarded, since 2003, to 15 Cuban women scientists, 6 of whom were physicists. The Russian Cosmonautics Federation awarded another Cuban woman physicist the "Valentina Tereshkova" medal for her outstanding participation in the preparation of the scientific experiments which were performed during the joint Soviet-Cuban space flight.

Figure 20.3 shows the percentage of women physicists holding managerial posts as well of those who are involved in teaching and research work in the country's four universities that employ the largest number of physicists, and in two research centers with a sizable number of these professionals (Arias de Fuentes et al. 2009).

Interestingly enough, the percentage of women physicists who are scientific managers is larger than that of women physicists who do teaching and/or research. This can be taken as an indication of the academic and scientific leadership reached by women physicists in the country, and emphasizes the progress made by women in Cuba and the increase in their social prestige as far as national academic activities are concerned (Arias de Fuentes 2008). While still a minority, women physicists presently hold or have held in the past important managerial posts in the country's scientific institutions, such as:

- Director of Sciences at the Ministry of Science, Technology and Environment (CITMA). National Representative of the Ibero American Program of Science and Technology for Development (CYTED).
- President of the Commission of Women Scientists of the Cuban Academy of Sciences.
- Scientific Advisor at the Ministry of Science, Technology and Environment (CITMA).
- Director of the Managing Center for the Programs of the Ministry of Science, Technology and Environment (CITMA).
- Rector of the Higher Institute for Applied Science and Technology, one of the three universities in the country where it is possible to study for a career in physics.

- Advisor of the Rector of the University of Havana, the largest and most important university in the country.
- Director of the Integrated Directorship for Projects of the Agency for Nuclear Energy and Advanced Technologies.
- Vice rector of the "José A. Echeverría" Higher Polytechnic Institute (ISPJAE), one of the country's four higher education centers whose academic personnel includes the largest number of physicists.
- Vice director for Research of the Physics Faculty, University of Havana.
- Dean and vice dean of the Faculty of Physics of the University of Havana, whose academic personnel includes the highest number of physicists in the country. [Present]
- Vice director of the Institute for Materials Science and Technology (IMRE), University of Havana, one of the centers of the country with the largest number of physicists.
- Member of the managing group that founded the Cuban Physics Society.
- President and vice president of the Cuban Physics Society.
- Editor of the journal *Revista Cubana de Física.*
- Director of the Institute for Research and Projects for the Mining-Metallurgical Industry (CIPIMM), the leading center for technological research related to mining in Cuba.
- Vice director of the Institute for Cybernetics, Mathematics and Physics (ICIMAF), one of the scientific centers with the largest number of physicists in the country.
- Head of the Physics Department of the Institute for Cybernetics, Mathematics and Physics (ICIMAF).
- Director of the Institute for Geophysics and Astronomy (IGA).
- Director and vice director of the Research Center for Microelectronics (CIME) (Arias de Fuentes 2001).

20.3 The First Women to Graduate as Physicists

The Law for the Reform of Higher Education in Cuba was enacted on January 10, 1962, the 33rd anniversary of the assassination by government henchmen of Julio Antonio Mella, who lead the first movement for a general reform at the University of Havana. New degree courses, faculties and research centers were established.[7] Specialized Schools (the Physics School among them) with their associated departments, were created within the Faculty of Sciences to replace the former independent chairs dedicated to special areas of mathematics, physics, chemistry, and natural sciences (Pérez Rojas et al. 1976). An independent (5-year) degree course in physics was inaugurated in this context, since up to then the nearest thing to it in the country were two 4-year degree courses entitled "physical-mathematical

[7] Informe al Claustro Universitario convocado en conmemoración del 280 aniversario de la fundación de la Universidad de la Habana. 2008. Teatro Astral, Ciudad de la Habana.

sciences" and "physical-chemical sciences", both taught only at the University of Havana and essentially dedicated to the training of high-school teachers. The Physics School became the Faculty of Physics of the same university in 1984.

The first four *licenciados* in physics graduated in 1964. It should be noted that right from the beginning there was an important female presence, since two of the new graduates were women (50 %), one of whom is presently a senior professor of the Faculty of Physics of the University of Havana. In 1965 there were, again, four graduates, one of them a woman (25 %). In 1966 new graduates reinforced the staff of the Physics School of the University of Havana. Six of these were physics graduates from universities in the former Soviet Union (Pérez Rojas et al. 1976), two women among them (33.3 %).

20.4 The Presence of Women Physicists at Some Important Moments for Physics in Cuba

Women physicists were prominent in various events that took place in the 1960s and 1970s and had an important bearing on the subsequent development of physics in Cuba. Not only did they actively participate in relevant research groups, but in many cases women physicists were the principal authors of the first results obtained in the period, such as:

- The first semiconductor alloy diodes obtained in 1967, a milestone that marked the birth of solid-state physics research in the country (Baracca et al. 2006).
- The creation of the planar technology laboratory of the University of Havana (colloquially referred to at the time as "the little house") and all the results associated to the research work done in it.
- The manufacture at laboratory level, in 1969, of the first transistors and integrated circuits (Arias de Fuentes 1993, 2001; Arias de Fuentes and Martínez Morell 1997).
- The development of light emitting diodes (LEDs) in the 1970s.
- The first research papers on semiconductor physics published in indexed journals.

It is worthy of mention that in the period 1970–1975 a woman physicist served as vice director for research at the then Physics School of the University of Havana, while other women physicists served as heads of educational laboratories and departments. A woman physicist was in charge of the design, setting up and organization of the first educational laboratories of the Physics School of the same university.

Among other significant moments for the development of Physics in Cuba in which women physicists had outstanding participation, the following can be mentioned:

- First laboratory made solar cells (Arias de Fuentes 2001)
- Creation and development of the Research Center for Microelectronics at the "José Antonio Echeverría" Higher Polytechnic Institute (ISPJAE) (Arias de Fuentes 2001).
- Foundation of the Institute for Nuclear Physics, which included five women physicists
- Foundations at the Cuban Academy of Sciences of the Institute for Fundamental Technical Research (ININTEF), which later was expanded into the Institute for Cybernetics, Mathematics and Physics (ICIMAF), with a female presence in its group for theoretical physics. A woman physicist served as the vice director of ICIMAF for 9 years
- Foundation in 1985 of the Institute for Materials and Reagents for Electronics (IMRE) (Arias de Fuentes 1993, 2001; Arias de Fuentes and Martínez Morell 1997), now called Institute for Materials Science and Technology. Women physicists have served as vice directors.
- Design and development of the first integrated circuits to order (ASIC) (Arias de Fuentes 2001).
- Epitaxial growth while orbiting around the Earth, included in the Cuban experiments made in outer space during the Soviet-Cuban Space Flight in September 1980
- LED technology transferred to Cuban semiconductor device industry during the 1980s.
- Development of photovoltaic solar cells.

One of the promoters and organizers of the Cuban Physics Society in 1979 was a woman physicist, who was later elected its executive secretary for two successive periods and its vice president afterward. Later on a woman was elected president of the society. There is a female presence in its current executive committee. The editor of the *Cuban Journal of Physics* was also, for a long time, a woman.

Finally, it should be pointed out that, with the wholehearted support of the Cuban Physics Society, a Chair for Women Studies was created at the University of Havana, as well as a Working Group of Women Physicists at the Ministry for Science, Technology and Environment (CITMA). Since its inauguration, in 2002, this working group has actively participated in international gatherings of women physicists (Alvarez et al. 2002, 2005; Arias de Fuentes 2009; Vigil 2009), and has promoted the participation of women in all sorts of scientific activities, like conferences, round tables and panels. In particular, female scientific workers belonging to the ICIMAF have promoted debates on the place of women in physics and mathematics, while their colleagues at the Institute of Meteorology have systematically organized similar gatherings of women meteorologists.

By way of resumé it can be said that, in spite of the fact that women physicists make up only a small part of the Cuban physicists community, they have played an important role in the development of physics in the country, and have enjoyed a strong presence in academic and scientific leading posts that made them outstanding in comparison with the situation in other places abroad where the role of women in the area is still quite limited.

Acknowledgments The author wants to express her gratitude to all those colleagues who have provided invaluable information on the subject dealt with in this paper, in particular, to Lic. Odalys González Cruz, present secretary for Teaching Affairs at the Faculty of Physics of the University of Havana, and to Drs. Elena Vigil and Lilliam Álvarez, two important women physicists in Cuba, who have contributed to improve this document by providing relevant information, and useful comments and observations. Last but not least, the author wants to thank the editors of this volume for giving her the opportunity to set forth the role of women in the development of physics in Cuba.

References

Alvarez, Lilliam, Aurora Pérez Martínez, and Magarita Cobas. 2002. Women doing hard sciences in the Caribbean. In *Women in physics: The IUPAP international conference*, ed. B. Hartline. Melville: American Institute of Physics.

Alvarez, Lilliam, Magarita Cobas, and Aurora Pérez. 2005. Cuban women in sciences: The physicists speak. In *Women in physics: 2nd IUPAP international conference on women in physics*, ed. B. Hartline. Melville: American Institute of Physics.

Arias de Fuentes, Olimpia. 1993. *La microelectrónica en Cuba*. Paper read at Primer Taller sobre Historia de la Tecnología Eléctrica en Cuba, 26–28 January, La Habana, Cuba.

Arias de Fuentes, Olimpia. 2001. *La microelectrónica y el protagonismo de la mujer en esta rama en Cuba*. Paper read at IV Taller Internacional "Mujeres en el siglo XXI", 13–16 Novembre, at Ciudad de la Habana.

Arias de Fuentes, Olimpia. 2008. Mujeres Físicas y su presencia en la vida académica en Cuba: Algunas cifras y datos interesantes. In *VII Congreso Iberoamericano de Ciencia, Tecnología y Género*. La Habana, Cuba.

Arias de Fuentes, Olimpia. 2009. Some interesting data about women physicists in Cuba. In *AIP conference proceedings*. Melville: American Institute of Physics.

Arias de Fuentes, Olimpia, and Ofelia Martínez Morell (eds.). 1997. La Microelectrónica: Breve panorámica histórica sobre su desarrollo y estado actual en Cuba. In *100 Aniversario del Electrón*. ed. A. Jerez Mendez. (UNED) Madrid: Lerko Print.

Arias de Fuentes, Olimpia, Lilliam Alvarez, and Elena Vigil. 2009. Women physicists in Cuba. In *Women in physics: 3rd IUPAP international conference on women in physics*, ed. B. Hartline and R. Horton. Melville: American Institute of Physics.

Baracca, Angelo, Victor Fajer, and Carlos Rodríguez. 2006. A look at physics in Cuba. *Physics Today* 59(9): 42–48.

Fernández Rius, Lourdes. 2000. Mujeres académicas: ¿Conflictos de roles? In *Feminismo: Del pasado al presente*, ed. M.T. López de la Vieja, 156–159. Salamanca: Ediciones Universidad de Salamanca.

González Bermúdez, Fernando Mario. 2008. *Palabras de apertura de VII Congreso Iberoamericano de Ciencia, Tecnologia y Género*. La Habana: Ministro en funciones de Ciencia, Tecnología y Medio Ambiente de Cuba.

Informe al Claustro Universitario convocado en conmemoración del 280 aniversario de la fundación de la Universidad de la Habana. 2008. Teatro Astral, Ciudad de la Habana.

Ivie, Rachel, Roman Czujko, and Katie Store. 2002. Women physicists speak. In *Women in physics: The IUPAP international conference*, ed. B. Hartline. Melville: American Institute of Physics.

Padilla, Maria Esther. 2007. Verdadero caudal humano. *Juventad Técnica* 335: 2–3.

Pérez Rojas, Hugo et al. 1976. Estado Actual de las Ciencias Físicas en Cuba: Presidente Ing. José Altshuler.

Vigil, Elena. 2009. Women's presence in the development of semiconductor physics and semiconductor devices research in Cuba. In *Women in physics: 3rd IUPAP international conference on women in physics*. Melville: American Institute of Physics.

Part III
Reflections from the Outside

Chapter 21
The Beginning of Semiconductor Research in Cuba

Theodore Veltfort

I was invited to Cuba in 1962 to initiate some efforts in semiconductor development. I had been a physicist and senior research engineer with various electronic companies of the "Silicon Valley" of California, south of San Francisco. I had heard of the efforts made by the new revolutionary government of Cuba to advance the level of science and technology, and I was anxious to see what I could do to help.

I found that Cuba had very advanced development in several areas, including communications technology. However, the means for producing the components, as well as most electronics apparatus were still lacking. After some initial exploratory efforts in Juceplan and in the Ministry of Industries, it was a brief conversation with Ché Guevara (the Minister of Industries) that led me to the University of Havana to initiate the development and production of electronic components in Cuba.

Ché had a way of answering questions by asking questions of his own, usually very much to the point. In this case his reply to my query of what could be done to accelerate the development of semiconductors, his reply was: "don't you need a basis (plataforma) in physics before you can go ahead with the engineering aspects?" The important implication was that the place to start should be the School of Physics at the University of Havana, or possibly the Academy of Sciences.[1]

Dr. José Altshuler, the Vice Rector of the University and (among other things) an important communications engineer, confirmed Ché's view and welcomed me into the School of Physics. Dr. Dina Weisman, visiting professor of physics from Argentina, also joined and became essential in the developments which were to follow.

Theodore Veltfort (Author was deceased at the time of publication) (1915–2008)

[1] An elaboration of Ché Guevara's point of view is given in the article "Tareas Industriales en los Anos Venideros", in *Cuba Socialista* 7, p. 28, March 1962. See as well the discussion with Ché in *Cuba Socialista* #17, p. 67, January 1963.

T. Veltfort (✉)
School of Physics, University of Havana, Havana, Cuba

A. Baracca et al. (eds.), *The History of Physics in Cuba*, Boston Studies in the Philosophy and History of Science 304, DOI 10.1007/978-94-017-8041-4_21,

An excellent program in theoretical physics had been initiated prior to the Revolution by Professor Gran, and was being carried out by others. Experimental physics, however, was almost completely lacking. We could not help but notice that the physics laboratories resembled those of the developed world at the time of World War I. With the impulse given by Dr. Altshuler and visiting faculty advisors from the USSR the situation was to change rapidly.

Our first steps were to establish embryonic teaching laboratories and institute a program of training for graduate and senior level students. An appeal to Dr. Villaseca, then Rector of the University, resulted in our obtaining the building vacated by the Faculty of Technology, which had moved to more extensive quarters in a newly established campus in Cujal. The electrical facilities as well as the space were essential. We proceeded as rapidly as possible to take advantage of our new location. The teaching program included a seminar in solid-state physics given by Dr. Weisman, and the courses that I gave in Electronic Circuits and in Applications of Semiconductor Devices. Those attending the courses included professors and auxiliary professors of the School of Physics and students of the upper undergraduate years.

At this time we were fortunate in obtaining the interest of the Academy of Sciences in our program. I met with Dr. Nuñez Jimenez, who assured me of the interest of the Academy in our projects. Shortly afterwards Dr. Lopez Sanchez, Director of the Academy, said that it would be very important for us to obtain direct advice and assistance from the Academy of Sciences of the Soviet Union. Dr. Lopez Sanchez then arranged for Dr. Weisman and me to make a 2-months trip to the Soviet Union. The (for us!) bitter climate of the northland in November of 1964 was greatly alleviated by the friendly welcome of the scientists at the Joffe Semiconductor Institute in Leningrad. There we were able to obtain invaluable assistance in planning our own program for teaching and research.

In addition to our observations and studies at the Joffe Institute we visited other laboratories in Leningrad as well as several in Moscow and Tallin. We learned much about the most recent advances in solid-state physics, especially about semiconductors. We were able to receive literature and information about articles and books essential for our work in Cuba. Summaries of courses given in Leningrad and Moscow, as well as laboratory manuals proved very helpful in developing our teaching programs at the University of Havana. An added bonus was their very "material" help: I refer to semiconductor materials!

This "material" help came in the following way. I recall that I was asked by the very helpful scientists in one laboratory: "What sort of things do you need most?" I pointed to the array of sundry apparatus and materials which filled every available crevice in the crowded laboratory, and answered: "Junk!" Their amusement changed rapidly to complete understanding as I pointed out that when one needs to assemble a piece of apparatus in the developed world, one has an array of discarded equipment and components which have accumulated from past work. This accumulation is just as essential for the construction of new apparatus as brand-new parts which may take month to arrive. However, if there has been no previous laboratory work, there is no "junk!" The Joffe scientists followed through. A month after Dina and I returned, a tremendous box from the USSR arrived at the Havana docks addressed to our laboratory. When we opened it, we found everything from pieces of plastic and metal to all kinds of semiconductor devices and even sections of electronic instruments. This established our laboratory!

Of course, various complete equipment was necessary too. Essential equipment started to arrive. A crystal-growing furnace came from England, a Zeiss vacuum chamber apparatus for measuring semiconductor characteristics came from the GDR, and a 20 KW air conditioner arrived from Havana! This equipment and some medium high purity silicon and germanium enabled our (at least partial) completion of a "clean room" laboratory for growing semiconductor crystals and the assembly and measurement of semiconductor devices. A minor hitch developed in the construction of the "clean room." The carpenters, unfamiliar with the purpose, altered the plans to put louvers in the doors of the access chamber. The chamber, of course, had to be completely airtight to maintain a sterile atmosphere within the laboratory to avoid any possible contamination. Some sheets of plywood and some epoxy solved the problem.

It was still some time and many obstacles had to be overcome before we achieved some success in our experimental program. At last, Dina and I were able to activate the crystal-growing furnace and grew a crystal of electrically pure germanium, and from a slice of this made a germanium diode. This little piece, humble and primitive, was the first semiconductor device ever made in any Latin-American country with local talent and facilities, that is, without importing a completely assembled manufacturing plant.

All of this was long ago. In 1968, shortly after the above-described events, I returned to the United States. In subsequent visits I found that the Cubans I recall as students of junior scientists had followed through with very impressive advances in semiconductor research. Now Cubans design complete integrated circuits and much in the way of complete semiconductor apparatus. Each time I return to Cuba I am astonished to see the development made by a courageous people despite the severity of the blockade, which imperialists to the north attempt to impose.

Appendix

A Short Biography of Theodore Veltfort by Connie Veltfort

Ted Veltfort was born into a conservative upper-middle class family in Cambridge, Massachusetts, in 1915. He became radicalized as a college student at Princeton and Swarthmore in the early 1930s. When war broke out in Spain in 1936, he joined the International Abraham Lincoln Brigade and drove an ambulance on the war front. Ted developed a passionate admiration for the Soviet Union because of the support that it lent to the Republic and the struggle against fascism. After his return to the United States, and after the Second World War, he suffered the repression of the McCarthy era for his leftist politics and his history as a "premature anti-fascist."

Shortly after the Cuban Revolution in January of 1959, Ted saw in it the rebirth of his ideal, and offered his services as an electronics engineer and physicist to the new revolutionary government. His offer was accepted, and in 1962 he arrived in Havana with his family. With the same fire and conviction with which he fought against fascism in Spain, he embraced the Cuban Revolution and its quest to create a just and developed society. He worked first for JUCEPLAN, under the leadership of Ernesto Guevara, and then for 5 years at the University of Havana in the School of Physics, directed by Dr. José Altshuler.

Chapter 22
Andrea Levialdi in Memoriam

Dina Waisman

Professor Andrea Levialdi was born in Bologna Italy in 1911, son of a very modest scientist who at the time was active in the socialist ranks. From an early age Levialdi felt the contradictions between the bourgeois environment surrounding him and his family's deep antifascism. He earned a doctorate in mathematics and physics at the University of Rome in 1937 with a dissertation on *Photoelasticity, methods and applications*. Soon after, he was awarded a scholarship for specializing in optics at the Arcetri National Optics Institute (Florence).

Feeling uncomfortable with the political situation at home he applied for and obtained an academic scholarship to France. He emigrated to this country in 1938 and began to work at the Experimental Physics Laboratory of the Collège de France, led by Paul Langevin, where he dedicated himself to the study of the hydrodynamical and optical characteristics of certain amorphous substances. Later on, he became research assistant of Professor Edmond Bauer, with whom he collaborated on the study of infrared absorption spectra of CO.

During Levialdi's stay in Paris he cooperated with his country's resistance movement, which included secretly going in and out of Italy several times. As the war threat rapidly developed in Europe, he collaborated with Joliot-Curie, Bruno Pontecorvo and others to help the French military defense, and worked in underground laboratories in Paris to produce better explosives for use against an eventual Nazi invasion. Some of the devices he created were operative in the Maginot line.

In 1941 he emigrated to Argentina and got a job at the Córdoba National Astronomical Observatory from which he was later fired for having participated in an anti-fascist congress which was held in Montevideo in those years. He managed

This is a translation of a slightly rearranged version of the obituary published in the Cuban journal *Vida Universitaria* (20, 215/April–June 1969), 40.

D. Waisman (✉)
Reactor Physics, National Atomic Energy Commission (CNEA), Buenos Aires, Argentina
e-mail: dwaisman2@gmail.com

A. Baracca et al. (eds.), *The History of Physics in Cuba*, Boston Studies in the Philosophy and History of Science 304, DOI 10.1007/978-94-017-8041-4_22,

to keep working on both fundamental and applied physical research. In 1946 he became a faculty member of the La Plata University, where he continued his research work on luminescence.

He was an active supporter of Professor Grinfold, who had been arbitrarily expelled from the university for political reasons. After the fall of the Peron government in 1955, Levialdi returned to higher education together with the mass of professors and students who had been estranged from it. He was appointed delegate of the Buenos Aires Faculty of Sciences to the National Council of Scientific and Technical Research, a job that he combined with the teaching of solid-state physics. An active member of the Human Rights League, the Peace Movement, and other organizations, he took a definite stance on Argentinian social and political problems and his name was generally associated with the democratic and progressive movement that was developing in the country.

When a big rally was organized in Buenos Aires' Luna Park in 1959, on the occasion of the visit of a Cuban delegation to Argentina, he was invited to share the stage with the visitors. But when the police would not let them through, Professor Levialdi took the floor and began his speech by saying: "We welcome the heroic people of Cuba who have turned military barracks into schools." This can be taken as the starting point of his subsequent cooperation with the educational revolution which was taking place at the time in the Caribbean island, which he visited for the first time in 1960 on his way back home after participating in the Congress for the Peoples' Self-determination and Sovereignty that had just taken place in Mexico. As soon as he returned to Argentina, he started preparing a set of physics experiments for use by Cuban students, a project that in the end he was not able to carry out.

In 1961 he decided to settle in Italy after he was diagnosed with lung cancer. Still, he managed to continue doing scientific research for which results can be found in his 43 scientific publications, a large part of which appeared while he was undergoing medical treatment for his terminal illness. At the same time, he took advantage of the personal prestige he enjoyed in the academic world to help find appropriate jobs for the Argentinian professors that had resigned their posts when in 1966 the police forces stormed the Faculty of Sciences of Buenos Aires University. Consistent with his radical position, he took sides with the European students that had revolted against unchecked academic power, in addition to wholeheartedly engaging in the many activities that were organized against the genocidal war waged on the Vietnamese people.

He was deeply moved when he received an invitation to pay another visit to Cuba to collaborate with the Physics Faculty of the University of Havana, to the extent that he decided to accept it in spite of the contrary opinion held by his doctors and family members in view of his delicate state. A faculty member of the University of Parma, he gave himself enthusiastically to the task of organizing collaboration between his University and Cuba, aimed at training high level research workers, a collaboration that would last for many years. He died in Havana on December 8, 1968, only a few days after he gave his last seminar session on photoconductivity in solids. The intimacy and modesty that accompanied the burial of his remains were consistent with the modesty and austerity that preceded and accompanied all aspects of his life.

Acknowledgments The author is grateful to Prof. Levialdi's widow, née Veronica Kleiber, who was with him on his last trip to Cuba, for providing nearly all the information included in this obituary related to his European years.

Chapter 23
The Andrea Levialdi Fellowship

Roberto Fieschi

My first encounter with Cuba dates back to winter 1967–1968 at the Cultural Congress of La Havana, a very large international event to promote greater understanding of the reality of the Cuban Revolution. In fact the person invited was my friend and colleague Andrea Levialdi (Andrea already knew Cuba and loved it) who, unable to participate, allowed me to go in her place. So I landed at the airport of the "first free country in Latin America" with the delegation of the Italian Communist Party. In Havana I met other Italian physicists whom I already knew, among them Bruno Vitale and Daniele Amati. They, like me, were embarrassed by the generous hospitality of 'Havana Libre,' especially in a country which was going through such difficulties. Despite our best efforts we did not succeed in receiving a more modest welcome.

The event was great and exciting; present were the leading experts of the worldwide Left, from Sartre to Archbishop Guzman, to Zatopek. We visited the Escuela de Fisica, with a very poor supply of instruments and researchers, talked with Cuban physicists (among them Elena Vigil, Fernando Crespo, and Juan Fuentes) and with foreign guests such as Dina Waisman and other young, enthusiastic Argentinians who had been students of Levialdi. In our ingenuousness, we wrote a short report with some recommendations, such as developing research on magnetism, as we were told that Cuba had minerals suitable for use in magnetic materials. The naive attitude of "advisers" – foreign intellectuals and technicians enthusiastic about the Cuban revolution – was typical of that time; unfortunately some of these recommendations led to wrong choices, especially in agriculture and the breeding of animals.

On November 5, 1968 Levialdi departed for Cuba: the University of Havana had invited him to give a course on solid-state physics. His health was already irreparably undermined by a disease, which had progressed quickly. Despite his poor health

R. Fieschi (✉)
University of Parma, Parma, Italy
e-mail: roberto.fieschi@fis.unipr.it

A. Baracca et al. (eds.), *The History of Physics in Cuba*, Boston Studies in the Philosophy and History of Science 304, DOI 10.1007/978-94-017-8041-4_23,

he left Italy for Cuba with his beloved Veronica. In fact, after some weeks I received a telegram from Veronica: "Yesterday Andrea passed away." He was buried in the great cemetery of Havana. Years later, I went there to give a short course on statistical mechanics (after determining that the theoretical preparation of the students was sufficient), and was able to visit his grave.

To honor his memory and to follow his desire to promote constructive scientific ties with Cuba, I undertook fund raising to host young Cuban physicists in the laboratories at the Istituto MASPEC/CNR and at the Physics Department of Parma University. I wrote to all the comrades and physicist friends I knew in Italy and abroad – a difficult task – and received enough contributions to start the program "Beca Levialdi" (Levialdi Scholarship). After a few years, the funds were exhausted, but the collaboration continued for a long time thanks to funding from the IILA (Italian Latin American Institution).

The first young Cuban we hosted was Joaquin, kind and gentle; I remember that he sang Guantanamera for us although he was tone-deaf. Unfortunately he is no longer with us. Then came Osvaldo, then Luis, Diego, Fernando, Julito, and many others. The money was little and the young Cubans, who were aware of this, lived satisfied with the minimum, also helped by the generous hospitality of Fabrizio Leccabue.

The collaboration was beneficial and reciprocal. Our guests, while learning, worked hard and intelligently on the development of our research in the field of materials science: semiconductors, magnetic materials, electrical characterization, optical, structural, etc. Some of the techniques they learned in Italy were then successfully transferred to the University of Havana.

I know that many of them later became internationally recognized professors and researchers.

I have met some of them on successive occasions and with many I have maintained a long and fruitful relationship.

Appendix: Two Letters Written from Cuba by Andrea Levialdi to Roberto Fieschi

Letter 1

November 14, 1968 *Havana*

Dear Roberto,
Here quickly a list that I can finally send you:

1) meter of magnetic field and gaussimeter
2) 2 or 3 graphite bars of high density and purity: 5 cm diameter, minimal length 15 cm
3) chemically pure germanium oxide – 500 gr
4) chemically pure germanium – 500 gr

5) a synchronous electric motor of almost 1/10 HP (here there are 60 c.p.s.) if possible with a flexible extension of the axle (type Blentista) with the possibility of fixing at its extremes discs of pertinex or diamond coated cardboard for cutting
6) diamond coated disks – 2 dozen
7) 2 dozen Kover feedthroughs for fixing electrodes through glass in vacuum (*sent*)
8) a gas fluximeter for almost 1 bubble per second (Paorici is practical and precision is scant) (*sent 2*)
9) some meters of tungsten wire, 1 mm diameter
10) insulating pearl for the thermocouple of 1 mm internal diameter (*sent*)
11) chromel-alumell wire for thermocouple almost 1,5 diameter – some meters (*sent*)
12) a hairdryer hot-cold (*sent*)
13) a small synchronous electric motor of 1/16 HP and even less (60 c.p.m.)
14) a book about the technical practice of pyrometry

It's not all that is lacking, but before I end up with nothing, I have begun to make this first list of things without which I cannot work.

I have made a program of the kind we had discussed, i.e. Guerci during the summer for two or three months – two grant holders at Parma for one or two years. In March, or later on, during a visit from you and Gozzini – I will explain this to you in detail.

Dina [Waisman] leaves again for Argentina in January, the poor girl has really done very much in a very difficult environment.

Letter 2

Program Cuba

It is proposed that close contact is kept between the Institute of Physics of Havana and the one in Parma, establishing a certain continuity.

1st stage: Prof Fieschi goes to Havana in order to evaluate the situation, the possibilities, and to formulate a program together with the Cubans.

2nd stage: Prof. Levialdi will give two months of lectures and seminars and will try to begin some investigative work and to single out two young people to come to the Parma Institute for 1–2 years.

3rd stage: Placing in Parma of two grant holders with a working plan to learn techniques which can be used on their return in Cuba.

4th stage: At the same time that the 2 grant holders are sent to Italy, a chemical technologist should be sent to Cuba.

5th stage: Course to be given by Prof. Fieschi.[1]

[1] Note from the editors: the two letters, sent by Andrea Levialdi from Havana to his colleague Roberto Fieschi in Parma, were kindly provided by Levialdi's widow, Mrs. Veronica Kleiber, whom we warmly thank. The notes in italics were added in pencil by Roberto Fieschi.

Chapter 24
A Witness to French-Cuban Cooperation in Physics in the 1970s

Jacqueline Cernogora

In France in 1968 many lively discussions and debates took place at several universities and laboratories in which official authority was questioned. Very often in such debates someone would stand up and ask the previous speaker: "Who are you to assert such a thing?" or "From where are you speaking?" Forty years later, to avoid such questions, I will say right away "from where" I am writing this text, which is by no means an exhaustive study of French-Cuban collaboration in physics at that time, but rather a personal recollection.

Since 1969, I have been a member of the so-called "Comité de Liaison Scientifique et Universitaire France-Cuba."[1] This committee was created just after the "Cultural Congress of Havana," which took place precisely in 1968. At its creation, this committee declared its aim was to support the development of sciences and technology in Cuba (see Document 1 in the Appendix). I was a member of this committee, first as a treasurer for about 3 years, then as a lecturer in the 1970 summer school (*Escuela de Verano*), and later co-responsible for "solid-state physics." A bulletin has been published around three times a year since its creation; each scientific branch had its own report (Document 2). I will concentrate here in particular on the year 1970, the year I spent in Cuba.

Document 3 synthesizes the program of the "*cursos de superación para graduados de la Escuela de Física*" (improvement courses for graduates from the School of Physics) for the year 1970, which is taken from the general program of "Universidad de la Havana 1970." This includes no less than 20 different schools: from mathematics to geography, from pedagogy to history, and so on. The lecturers came mostly from France, but also from other countries such as Italy, and a few

[1] This committee was founded against the background of solidary initiatives supporting the anti-Imperialism movements in, for example, Cuba and Vietnam. Capdeville and Levesque (2011) show for the faculties of science at the University of Orsay that a French-Cuban committee was founded as early as 1968.

J. Cernogora (✉)
Centre National de la Recherche Scientifique (CNRS), Paris, France
e-mail: jacqueline.cernogora@wanadoo.fr

A. Baracca et al. (eds.), *The History of Physics in Cuba*, Boston Studies in the Philosophy and History of Science 304, DOI 10.1007/978-94-017-8041-4_24,

Fig. 24.1 Jacqueline Cernogora teaching at the Summer School in Havana, 1970 (see Document 3)

from Great Britain. The exchanges not only took place during the summer at the summer school, in certain cases they also lasted throughout the year. Some of the Cuban students also completed their theses in France: for example, Pascal Lederer told me that this was the case, for instance, in Orsay.

Returning to physics, I will remark that the professors came mostly from Paris and Orsay in France, but as ever, there were exceptions. The main reason for this is that those who had begun to form relations with Cuba came from laboratories in those places, and from there the relations grew. This proved to be very fruitful indeed since one of the main goals of the collaboration with Cuban scholars had been to maintain continuity over the course of the exchanges. This meant that the professor or technician involved in the collaboration could change quite often, from 1 year to the next, but that their successors should proceed from the same French laboratory or university, that is, from the same scientific culture. This requirement made it possible to fulfill, if possible without interruption, the very concrete requests of the Cuban colleagues: the programs could endure from one summer school to the next. Figure 24.3, translated in Document 4, gives a concrete example of this kind of precise request.

In 1969, for example, Jean-Marie Debever was in Cuba, and in 1970, when I was also there, we worked for a time in the same laboratory. Around the same period, Eric Gessler came several times to ensure the setting of magnetic resonance; Jean-Claude Chervin worked with the Cuban physicist Antonio Cerdeira building a furnace for silicon devices; Pollard had begun, or more precisely, had transported from his company, an epitaxial bank, and subsequently Daniel Bois continued this work. As a personal memory, I remember how worried I was about how to transport

Fig. 24.2 Lecturers and students at the University of Havana Summer School in 1970. From left to right: Eric Geissler and his wife (fourth and fifth from the left); Bernard Prum (lecturer in statistics); J.C. Chervin's wife; Jacqueline Cernogora; Jean-Claude Chervin; Claude Monet

such a large machine over such a long distance. In the end, the carpenter of our laboratory succeeded in securing the bank adequately for the trip.

It is of course impossible to do anything without money. Although the Committee had more than 100 persons who paid membership fees, this was not enough. To increase funding, the Committee was even involved in the quite successful release of an album of Cuban songs with a cover designed by Wolinsky. However, if I am not mistaken, a great part of the resources came from the Cubans themselves. When lecturers went to Cuba, the Cubans paid for their journeys, but when they came with their spouses (which was the most probable situation in those days), they had to pay the Committee for their journeys, and in this way money was collected to buy what was necessary for the smooth running of the Cuban laboratories (or the best possible with the modest means at hand).

Up to now, I have written about the Committee. Before this came into existence, however, some persons had already visited Cuba with the aim to help the new scientific institutions. In this respect, Claude Monet, who went to Cuba almost immediately after the revolution deserves a special mention.[2] Monet was a very resourceful man who knew how to make anything from nothing, or, better, where to find what was needed. With him, I learned that liquid nitrogen could be found on a farm where artificial insemination was practiced on cows, or that a pulse generator can be found on a military plant (I don't remember exactly where this was, so am not revealing any military secrets).

[2] See the contribution by Baracca, Fajer and Rodríguez to this volume.

In closing, I would like to add a couple of further personal recollections. One day I heard a Cuban student complaining: "because of Cuba's underdevelopment, we cannot get the latest modern laser that you have abroad." I answered that the true underdevelopment in the laboratory consisted rather in not finding the right nut for the bolt. I was also very surprised initially to find Spanish engines in Cuban laboratories. These were brand new and just waiting for the engineers to put them to work. I was surprised to learn of this relationship between Cuba and Spain because at that time Franco was still alive. Another time, two persons came from Cuba to Europe (and I apologize for not remembering who they were) with a little money in one pocket to buy a calculator and a lot of money in the other pocket to buy a component company in Spain. I thought then that it should be the aim of those summer schools not only to train people to teach or to perform research, but also to manage industrial-size projects.

I am aware that this paper is not precise or thorough enough to consider as a historical assessment since the focus, apart from documents included, is on personal recollections from around 1970. I hope the reader is able to enjoy this as a story, and remember that a story can also present history, and in this sense I must emphasize that this is a story of solidarity (not free of political conflicts) and joy.

Acknowledgements I would like to thank Jean-Claude Chervin, who has kept so many documents for such a long time and has kindly allowed them to be used here.

Appendix[3]

Document 1

Declaration of the Committee for French-Cuban Scientific and University Collaboration

Greetings to the Committee from his Excellency, Dr. Baudilio Castellanos Garcia, Cuban Ambassador to France

Paris, 10 October 1969

In recent years, French scientists and academics have had the opportunity to visit Cuba, notably in 1968 for the Cultural Congress organized in Havana, during which the problem of the underdevelopment of science and technology was studied.

The blockade of the island makes contact between Cuban universities and scientists and their West European and American counterparts difficult, impeding the free circulation of scientific information.

---

[3] The documents in this appendix were provided by the author. The excerpts featured here were selected and translated from French by Angelo Baracca.

It therefore seemed useful to create an organism charged with the establishment of academic and scientific collaboration between Cubans and French. It is our Committee that proposes to play this role. [...]

The goal of the Committee is to promote French-Cuban cooperation in the domains of teaching, research, technology, and culture.

Of course, the Committee hopes to see official relations between France and Cuba develop as much as possible through the signing of agreements for scientific, technical and cultural cooperation, through the exchange of delegations of experts and of grant-holders. [...]

Before the official creation of the Committee, a certain number of activities took place.

In the first place, in collaboration with the Havana University, we were able to organize Summer Schools in various fields.

In 1968: Mathematics (probability calculus), statistics (test planning in view of applications to agronomy), solid-state physics (basic course and technical course on the problems of transistors), molecular biology and immunology.

In 1969: Mathematics (functional analysis), statistics (following the 1968 course), programming, process planning, solid-state physics (following the 1968 course), molecular biology (following the 1968 course), genetics (directed at practical work), plant physiology (directed at specific applications for Cuba), geography and soil study, problems in forestry, organic chemistry (physical methods of separation), history (problems in methodology), child and learning psychology.

In 1968: thirteen teachers gave courses; there were forty this year. For next year, numerous courses have already been conceived and discussed, dealing at the same time with the problems of teaching the sciences and the basic humanistic disciplines, but also with specific technical problems. The number of courses in technological fields in particular will be considerably increased. [...]

For these summer schools, the Committee collaborated with some highly qualified foreign specialists, notably from America. England, Germany (Federal Republic), Italy. On the whole, these schools were a success. They enabled very useful discussions to take place on teaching (programs and methods), on techniques of practical work and the use of material, on the relation between theory and practice. Numerous personal and social contacts were established, and French teachers were able to experience first-hand the inherent difficulties of developing countries, and to observe entirely new experiences. For our Cuban colleagues, these human relations often help them to break out of their professional isolation.

The Committee's activities are not limited to the summer schools. We made great efforts to send French teachers to Cuba for periods ranging from six months to two years, as well as researchers who were capable of leading a team. We have benefited from the interest shown in this problem by the Directors of the Centre National de Recherche Scientifique. We try to favor Cuban grant holders coming to France and sending young French voluntary workers to Cuba by way of military service; a possibility that French students and young technicians are unaware of. Finally, we can facilitate the purchase of material for our Cuban colleagues by providing them the necessary documentation. [...]

We would like to associate with our efforts the highest number of scientists and academics interested in the problems of cooperation with a country that is making great sacrifices to develop its culture, as has been shown by past campaigns for literacy and mass education, and now shown by current efforts to introduce modern technology to agriculture, industry and the economy.

Cuban authorities and colleagues have always put great trust in us. The cordiality of their acceptance, the efforts and economic sacrifices they made in establishing a close collaboration among the scientists and academics of the two countries leave us in no doubt that it is necessary to vigorously develop the activity of our Committee and to associate to it the highest possible number of academics, researchers and engineers.

For the Committee officials:

Mr. HELLER, President of the Committee, Professor at the Parisian Faculty of Sciences
Mr. D. DACUNHA-CASTELLE, General Secretary, Master of Conferences at the Strasbourg Faculty of Sciences

and the honorary members of the Committee:

Mr. Raymond FEVRIER, General Inspector of the I.N.R.A.
Mr. POITOU, Dean of the Orsay Faculty of Sciences
Mr. Jean JACQUE, Research Director of C.N.R.S. at the Collège de France
Mr. Yves LACOSTE, Professor of Geography at the Institute of Geography
Mr. Adam KEPPES, Research Director of C.N.R.S. at the Collège de France
Mr. Pierre LEHMANN, Professor of Physics at the Orsay Faculty of Sciences
Mr. Laurent SCHWARTZ, Professor of Mathematics at the Parisian Faculty of Sciences
Mr. Pierre VILAR, Professor at Sorbonne
Mr. Charles THIBAULT, Director of the Station Centrale de Physiologie animale

[Greetings from Dr. Baudilio Castellanos Garcia, Cuban Ambassador to France, to the members of the Committee ...]

French university researchers and scholars, together with Cuban and other colleagues of many other nationalities, organized summer courses in 1968 and 1969 in Havana, Cuba. To launch and develop their plans, they organized themselves into a committee known as the Committee for French-Cuban Scientific and University Collaboration.

We consider these activities to be one of the most impressive human adventures currently unfolding in the formation and diffusion of science.

Professors from France, England, Germany, North America and Italy lectured at a summer school that they themselves organized. Their courses were offered during the last weeks of July, in August, and sometimes during the first week of September. The duration of the courses varied, in general, between fifteen and forty days.

With no intention of offending the modesty of these professors, we should point out that in giving these courses, they renounced their summer break, thus sacrificing time with their families during the traditional holiday period. In doing this inspiring deed that is worthy of recognition, we should point out as well that their services were offered for free, and moreover, that their work in Cuba involved costs for books, materials and air transport from Paris to Madrid.

The first summer school is devoted mainly to natural sciences. The second expands this essential activity with courses in the humanities and technology.

These courses were aimed at Cuban professors at our three universities, at students in their final years, and at technicians and researchers at various research institutes and ministries.

Once in Cuba, the professors were not satisfied with simply giving their respective courses, they helped, moreover, to revise our curricula, both for university teaching and for secondary, technical and primary education.

Furthermore, the professors visit our facilities and laboratories, discuss with our technicians, travel in the regions of our country, and finally offer their advice and criticisms.

The remarkable aid offered to our country by the Committee and by the professors supplement the assistance already received from international organizations such as UNESCO, WHO and others, as well as that granted by the French government and by other countries, in particular, the socialist countries, and that what we have received from the industrial and commercial firms with whom we have relations.

The leading members of the Committee have given themselves the objective of honouring the higher qualifications within the national framework of our country in order to launch during the next decade the creation of autonomous and high-quality schools in Cuba.

In the coming months, the Committee proposes to expand both horizontally and vertically. For the current scholarly year of the Cuban universities, several professors have already been sent. And young Cuban researchers are already in France in laboratories chosen by the Committee.

Making use of experiences from previous years, the Committee, in collaboration with Cuban universities, is now planning the activities for summer 1970. It is up to us to create, with farsighted and meticulous work, the summer school 1970 and, after Cuba has obtained a large crop of ten million sugar canes, a fruitful scientific event.

Three hundred men of science from twenty or thirty nations or disciplines could produce a chain reaction of human intellectual energy and generate a fusion of neurons that could shake up the international scientific community and release new forces so that summer schools may be created in other countries of the underdeveloped world.

The delay in the cultural, scientific and technical development of the peoples of the third world is more acute than the delay in its economic development and more noticeable since it denies any effective means of solving the problems of modern society.

It is agreed that the importance of the sacrifice and of the magnificent tasks carried out by the Committee and the researchers at universities in France and other nations are not only a service to Cuba, but also to the whole of humanity.

To them, the infinite gratitude of our people.

Document 2

Report on the Physics Courses Held at the Summer School of 1971, and Projects and Remarks for the Summer School of 1972 (excerpts).

[...]

1971 Summer Schools

During summer 1971, 6 courses took place in Havana:

- Detection of weak signals (Weisbuch – 27)[4]
- Hall effect (Jacquinot – 8)
- Planar technology (Pollard and Chervin – 14)
- Thin layers (Leger)
- Metals: electron microscopy, metallurgy (Gantois and Schmidt)

Who are the students? Where do the courses take place? Why does the number of students vary from one course to another? The answers to these questions enable a survey of more or less all the organizations where, let us say, physics "is done" in Cuba.

The students come:

- <u>from the School of Physics at the University of Havana</u>. These are either persons who already work at the School of Physics (where they teach and work in the laboratories) or students from the previous year
- <u>from CNIC</u> (Cuban CNRS)
- <u>from IFN</u> (Institute of Nuclear Physics)
- <u>from CUJAE</u> (Technological University, situated near Havana).

Some Remarks

- Among the 6 courses held this year, two were held for the first time: those on metals (the Cubans have been asking for these courses for a long time, but it is often difficult to find persons who can bridge the university and industrial sectors).
- The other courses were held in continuity with those held in previous years. The Summer Schools on the other hand are more effective due to this continuity, and as exchanges are established between the Cubans and French over the course of the year. This proved to be particularly true in the sector of planar technology

[4]The name of the person who held the course is given in parenthesis, as well as the number of students who took part, if known.

where progress from one year to the next is particularly spectacular (it is worth mentioning the dynamism and competence of the Cuban team involved).

- As well as a certain continuity of subjects (in certain aspects, at least corresponding to the experiences already made in the laboratory), there was up to now a real continuity from the side of the French teachers. This does not necessarily imply that it is the same person who guarantees a course from one year to the next, but rather persons who, in France, work in the same group with analogous subjects or techniques.

The Difficulties

- For the teachers, the difficulties often consist in the heterogeneity of the level of the students.
- Difficulties are often met in what concerns collaboration and coordination among the various systems. But this year all is going quite well: in planar technology a group exists that works at the School of Physics, and another one at the CUJAE, and one can say roughly that human resources came from the School of Physics and materials from the CIJAE. Pollard's influence was decisive in provoking the necessary collaboration between the two teams.
- Also, several groups from the School of Physics will be called upon to collaborate on the implementation and utilization of certain equipment (Lock In implemented by Weisbuch).
- Evidently, difficulties persist that are related to underdevelopment, at the level of absence of small materials (for instance screws), of the non-standardization of equipment, and of a sometimes not very efficient running of the mechanics workshop (the arrival of a French technician this year was on the other hand invaluable).

1972 Summer Schools

The Cubans have requested this year:

- A course on metallurgy (in continuation of the 1971 course – professors acquired with Schmidt and Gantois acting as intermediaries).
- A course on crystal growth.
- A course on MOS devices.
- A course on electronic devices which in principle must be held by an engineer.
- A course on magnetic resonance which could probably be held by the same person as in 1969 and 1970 (Geissler).
- A course on theoretical physics (still quite badly defined since it should be held in connection with experiences made in the laboratory and everyone knows the difficulties of bringing experimenters and theoreticians together).
- A course on thin layers.

In addition, we point out that several courses were requested by Santiago:

- A course on experimental spectroscopy (no one has been found yet).
- A course on X-rays (in connection with chemists).
- A course on the application of radioisotopes to agriculture (but according to the agronomists, the situation is not mature enough for such a course).

Document 3

University of Havana 1970
School of Physics, Specialization Courses for Graduates

PHYSICS OF TRANSITION METALS

1. **Electronic Structure of Metals:**
   - Calculation of band structure.
   - Analysis of bands of almost free electrons.
   - Band structure vs. localized states.
   - d-band and virtual d-level.
   - Width of the d-band.
   - Density of states.
   - s-d mixture.
   - Fermi surface of copper and nickel.
2. **Correlation. Electron-Electron. Magnetism:**
   - Effect of correlation and dependence on spin.
   - Large and small magnetic moments.
   - Finite temperature magnetism.
   - Sin-orbit effect.
3. **Introduction to Landau theory of Fermi liquids:**
   - Paramagnetic and ferromagnetic liquids.

   Professor:
   Jean Hanus (Appointed researcher at CNRS. Laboratory of Solid-State Physics)

p-n JUNCTION IN SEMICONDUCTORS

I. **General Properties of p-n Junctions**

1. Short revision of the general principles of deriving a p-n junction: diffusion, alloying, etc.
2. Elementary principle for deriving the insertion.
3. Control of p-n junctions.
4. p-n profile.
5. Electrical characteristics.
6. Experimental work
7. Experimental study of diffusion in silicon.
8. Voltage-current characteristics of samples.
9. Depth of diffusion.

II. **Technology of Diffusion Ovens**

1. Calculation of diffusion oven.
2. Criteria of selection for the constituent elements.
3. Method of calibration of thermocouples.

4. Measurements at high temperatures.
5. Systems of temperature regulation.
6. Experimental work:

   Construction of an oven at regulated and stable temperatures suitable for operating at temperatures up to 120°C.

III. **Planar Technology of Semiconductors**

1. Physics of surfaces in semiconductors.
2. General principles for deriving MOS devices.
3. Experimental work:
   - Implementation of planar diodes in silicon.
   - Preliminary implementation of diffused transistors.

Professors:
Christian Verie (Dr. of science, appointed researcher at CNRS)
Jean Claude Chervin (Research engineer, CNRS)
Gonzalo Velasco (Research engineer, CNRS)

MAGNETIC RESONANCE

1. Basic principles of magnetic resonance. Classic account.
2. Dilation dipole in solids.
3. Relaxation process in solids and in liquids.
4. Indirect coupling between nuclear spins: metals, molecules.
5. Experiments for multiple impulses.

Professor:
Erik Geissler (Master of Conferences at the Faculty of Sciences, Grenoble. Laboratory of Physical Spectrometry)

INTEGRATED CIRCUITS FOR COMPUTERS

1. **Introduction**
   - Description of a computer.
   - Principal parts.
   - Electronic functions used in computation.
2. **Evolution of integrated Circuits**
   - The large families of integrated circuits: ECL, TTL, DTL, RTL, MOS, LSI.
3. **Techniques used in the fabrication of integrated circuits**
4. **Specific technical problems in the use of integrated circuits**
   - Implementation, supply, adapted and non-adapted combinations, bugs.
5. **Theoretical considerations on the behavior of diodes and transistors in impulse regimes.**
   - Practical methods of calculation.

6. **Modeling (from the aspect of electrics and not logic) of the integrated circuits of different families.**
7. **Integrated circuits available on the market:**

   – Manufacturers, costs and predicted development.

8. **Practical work.**

   Professor:
   Victor Chantal de Chanteloup (Research Engineer at IRIA. Structure of Electrical Computers)

PROPERTIES OF GALLIUM ARSENIDE

1. **Generalities**

   Crystal structure
   Physical constants
   Methods of preparation

2. **Band spectrum**

   – Generalities

     Approximation methods
     Atomic orbitals, molecular orbitals in the case of one molecule
     Atomic orbitals, molecular orbitals in the case of the solid

   – Bloch theorem
   – Method k.r. Notion of effective mass

3. **Optical properties**

   – Summary of perturbation theory
   – Generalities, absorption, direct and indirect transitions
   – Spontaneous emission, induced emission
   – Theory of the "laser" effect

4. **Gunn effect**

   – Experimental description of the phenomenon
   – Theoretical description

5. **Practical work**

   Professor:
   Jacqueline Cernogora (Appointed researcher at CNRS, Group of Solid-State Physics)

MEASURE OF LIFETIMES

1. **Summary of the lifetimes of minority vectors**

   – Direct combination:
     Case of a direct forbidden band

Application to Ga As
Case of an indirect forbidden band
Application to Ge and Si

- Recombination through mediation of capture centres
  Application to Ge and Si
- Generalization of the notion of lifetime in the case of a very excited semiconductor, kinetics of recombination

2. **Experimental methods**

– Pulse method:
Light source
Light modulation
Detection
Application to Ge, Si, Ga As

– Resonance method:
Theory
Experimental device and applications to Ge and Si

Professor:
Jean Marie Debever (Master of Conferences of the Faculty of Sciences, Paris, Group of Solid State Physics, ENS.)

PLANAR TECHNOLOGY

1. **Presentation of planar technology**

Comparison with other techniques.
Microelectronics.

2. **Basic techniques in the context of planar technology**

Diffusion technology.
Oxidation:

Thermic oxidation.
Anodic oxidation.

Epitaxial growth. Different techniques of preparation. Theory and methods of study. Thin films. Realization. Physical characterization.
Auxiliary techniques: photogravure, micro welding, etc.

3. **Practical work**

Realization and characterization of silicon oxide.
Characterization and realization of a p-n junction diffused in silicon.
Physical characterization of thin films.

Professor:
Gonzalo Velasco (Research Engineer, CNRS.)

Document 4 (Fig. 24.3)

University of Havana, Materials for work on planar technology

1.-	Filter holder for Millipore pipe XX40 02500	4
2.-	Packet of Millipore filters standard 0.8 m type AA with diameter 25 mm	1 pack.
3.-	Pre-filters diameter 25 mm	1 pack.
4.-	Filter holder in PVC for clarification of 142mm type YY 40 142 00	1 pack.
5.-	Canalization tube in inox. (¾ inch)	10 m
6.-	Cock for 1” (1 inch) vacuum	2+4
7.-	Cement Desmarques 169–19	
8.-	Quartz thin slide.	
9.-	Flexible wire of compensation for Pt–Rh/Pt	
10.-	Spherical entrance 19/9 Pyrex.	
11.-	Spherical entrance 19/9 in inox.	
12.-	Pure platinum wire for thermocouple.	
13.-	Specimen holder for diffusion.	
14.-	Quartz tube 10 mm dia., external.	
	- with quartz crucible welded for diffusion.	1
	- with quartz leaf for holding aluminum crucible of 30 42 mm	1
15.-	Quartz tube 45 40 with conical entrance for diffusion total length 1,300 mm	
16.-	Quartz tubing 45 40 with male spherical jaws for oxidation. Total length 900 mm	
17.-	Conical entrances for previous tubing.	
18.-	3 liters of acetone for electronics [handwritten]	

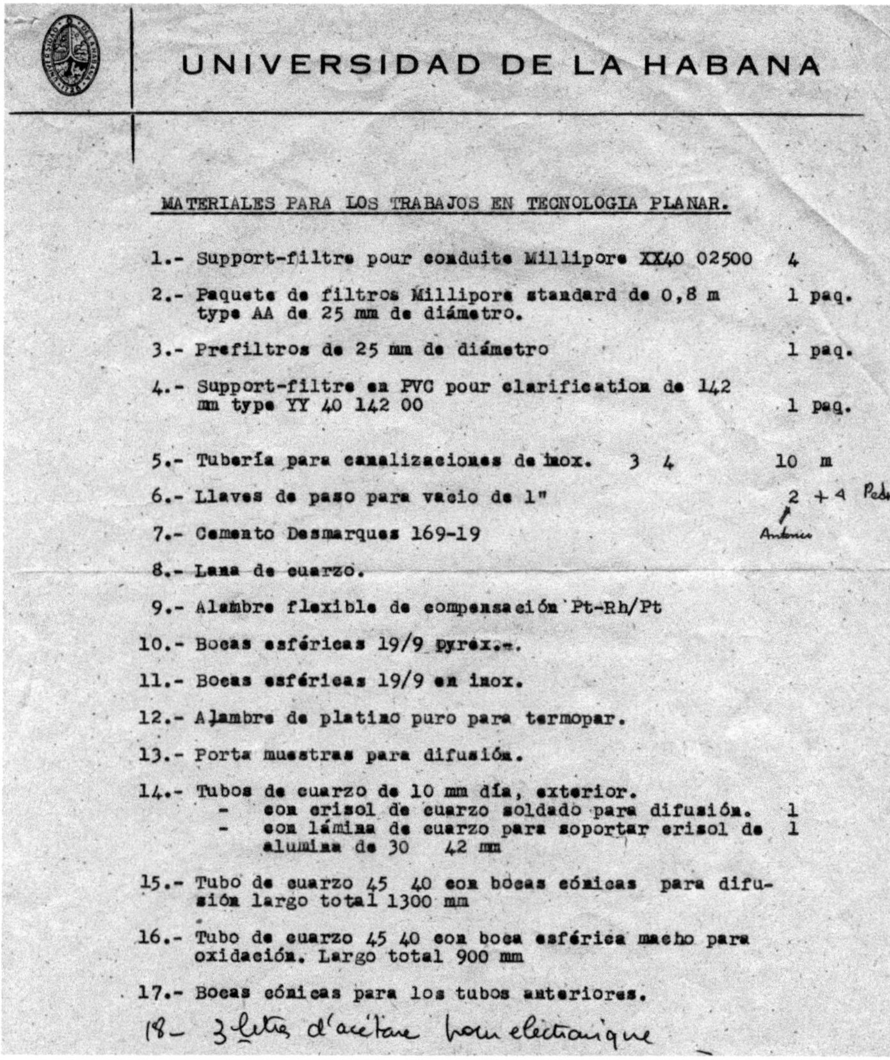

UNIVERSIDAD DE LA HABANA

MATERIALES PARA LOS TRABAJOS EN TECNOLOGIA PLANAR.

1.- Support-filtre pour conduite Millipore XX40 02500 4

2.- Paquete de filtros Millipore standard de 0,8 m type AA de 25 mm de diámetro. 1 paq.

3.- Prefiltros de 25 mm de diámetro 1 paq.

4.- Support-filtre en PVC pour clarification de 142 mm type YY 40 142 00 1 paq.

5.- Tubería para canalizaciones de inox. 3 4 10 m

6.- Llaves de paso para vacio de 1" 2 + 4 Pedro Antonio

7.- Cemento Desmarques 169-19

8.- Lana de cuarzo.

9.- Alambre flexible de compensación Pt-Rh/Pt

10.- Bocas esféricas 19/9 pyrex.

11.- Bocas esféricas 19/9 en inox.

12.- Alambre de platino puro para termopar.

13.- Porta muestras para difusión.

14.- Tubos de cuarzo de 10 mm día, exterior.
- con crisol de cuarzo soldado para difusión. 1
- con lámina de cuarzo para soportar crisol de alumina de 30 42 mm 1

15.- Tubo de cuarzo 45 40 con bocas cónicas para difusión largo total 1300 mm

16.- Tubo de cuarzo 45 40 con boca esférica macho para oxidación. Largo total 900 mm

17.- Bocas cónicas para los tubos anteriores.

18- 3 litres d'acétone pour électronique

Fig. 24.3 Original document from the University of Havana, Materials for work on planar technology (Document 4)

Reference

Capdeville, Yvonne and Dominique Levesque. 2011. *La Faculté des sciences d'Orsay et le Vietnam. De la solidarité militante à la coopération universitaire (1967–2010)*, Paris: L'Harmattan.

Chapter 25
My Collaboration with Cuban Physicists

Fabrizio Leccabue

25.1 Introduction

My first meeting with the scientific Cuban community was in 1969 when the first of four young Cuban physicists, Joaquín Torres Orosco†, came to the Physics Department of Parma University through the 'Andrea Levialdi Fellowship,' an Italian bursary promoted by Roberto Fieschi using a fund, subscribed to voluntarily by the Italian physics community.[1]

In 1974 I attended a summer school at the Physics Institute of Havana University and, during this period, came into closer contact with the issues, difficulties, problems and goals of research in Cuba. In 1981 I spent a sabbatical year at the Physics Institute of Havana University, mainly to promote experimental research in crystal growth, and more broadly in the area of materials science. Alongside Cuban researchers and technicians, we devoted our attention to two classes of materials: calcopyrites and hard magnetic materials. Even during these early collaborations the excellent level of young Cuban physicists and students and their aptitude towards research activity surprised me.

During this period, I met a new generation of physicists who grew up after the Revolution: Pedro Díaz†, Juan Fuentes, Luis Hernández, Osvaldo Vigil, Julio Vidal, Fernando González: a dynamic generation devoted to the new scientific and technological issues, but who also paid particular attention to teaching and tuition, with many ideas and with a strong vocational approach to what was needed to overcome the economic difficulties, due mainly to the US embargo. The embargo also affected the scientific community in various ways, for example, Cuban researchers were not allowed to publish in scientific journals edited in the USA. A second and third generation of researchers followed. Some of these reacted differently to the everyday

[1] See the contribution by Roberto Fieschi in this volume.

F. Leccabue (✉)
Institute IMEM of Consiglio Nazionale delle Ricerche, Parma, Italy
e-mail: mirella.pelizzoni@libero.it

A. Baracca et al. (eds.), *The History of Physics in Cuba*, Boston Studies in the Philosophy and History of Science 304, DOI 10.1007/978-94-017-8041-4_25,

economic difficulties of Cuban life and made different personal choices: Osvaldo de Melo, Diego Seuret, Francisco 'Chuby' Calderón, Celia Hart†, Oscar Ares, José Luis Sánchez, Sergio Díaz, Nelson Suárez, Enrique Vasco, Juan García Rodríguez, Daniel Carrillo, Humberto Rodríguez, Julio Rimada.

Within these collaborations and efforts, I had the impression—and perhaps so did my Cuban colleagues—that scientific exchanges with the USSR were useful, but could not cover the whole set of conditions for the development of this field of research. In fact, during our meetings, discussions and conversations the idea was developed to promote wider international collaborations, to boost the professional development of Cuban researchers in stages throughout Europe and in Latin America, and to favor the material conditions of research work through the fabrication of apparatus, reagents, bibliographies, etc.

It is in this direction that my efforts, and those of some colleagues from Parma and Europe and of first-generation Cuban researchers, was aimed: to organize international scientific events in Havana and to look for ways to promote fellowships and scholarships. This led to the advancement of a stable relationship with the Faculty of Physics in Havana and the other centers involved, both Latin American and European, as will be detailed below.

Cuban colleagues played a prominent role both during the elaboration of most projects and the organization of international or Latin American events, both in Cuba and in other Latin American countries. The main goal of these activities was to enable closer connections and consolidate scientific networks.

25.2 From 1970 to 1990

These efforts developed in several directions with the active participation of Cuban physicists. The wide network of scientific collaborations at Parma University was extended to include Cuban physicists and students and to promote effective scientific collaborations. Financial and material support was requested from all international scientific organizations in order to obtain scholarships or help in providing and shipping scientific equipment. Important international scientific appointments were promoted in Cuba, for example, meetings and conferences.

25.2.1 International Scientific Collaboration

Cuban fellows were progressively and successfully engaged in international collaborations. From 1983, an extensive network was created: with Spain, Antonio Hernando (Universidad Complutense de Madrid and RENFE), Manuel Vázquez (ICM/CSIC) in the field of magnetism, Fernando Briones (INM/CSIC, Madrid) in microelectronic and sensors; with France and Austria, in magnetism Dominique Givord

(CNRS, Grenoble), R. Grössinger and Josef Fidler (Technical University of Vienna); with Portugal, in piezo-ferro-electric materials, Maria de Matos Gomes (Universidade do Minho, Braga). In Latin America a network was established with México: Feliciano Sinencio, Isaac Hernández (CINVESTAV-IPN, Ciudad México), J.L. López Morán, Manuel Mirabal (Universidad Autónoma de San Luis Potosí), Juan Luis Peña, Román Castro (CINVESTAV-INP, Mérida) in magnetism and semiconductors; with Venezuela, Vicente Sagredo in magnetism (Universidad de los Andes, Mérida); with Brazil, Frank Missell in magnetism (Universidade de São Paulo).

In Italy, a close collaboration arose with the LAMEL/CNR Institute in Bologna and, more broadly, with the Physics Department and the Engineering Faculty of Parma University, and with all the groups working at the Istituto MASPEC/CNR in Parma. In particular, the collaboration with the MASPEC Institute (later called IMEM) of the CNR was possible thanks to the International Centre of Theoretical Physics of Trieste (ICTP) and to the Istituto Italo Latino Americano (IILA) of Rome, which promoted a regular exchange of researchers over many years. In the case of the ICTP conferences, scientific publications, the shipping of apparatus, travel expenses, and the participation of Cuban scientists in European conferences were funded.

25.2.2 International Events

Starting in the 1980s, in spite of the economical and political difficulties, strong efforts were devoted to the organization of schools, workshops and conferences, with the aim of exchanging, enlarging and transferring the scientific collaborations and experiences between Latin American and European groups of researchers. Some events could be organized in Cuba; in all cases the active participation of Cuban physicists was promoted.

The International School on "Crystal Growth and Characterization of Advanced Materials" (Christensen et al. 1988) was organized in Havana in 1988, sponsored by the IUCr, IUPAP, ICTP.

The First Latin American Workshop on "Magnetism, Magnetic Materials and Their Applications" (Leccabue and Sánchez 1992) was held in Havana in 1991, sponsored by IUPAP, ICTP, IUCr, Centro Latino Americano de Física, Centro Internacional de Física and CNR; followed by the second in San Luis Potosí 1993 (Mexico) (Lopéz Morán and Sánchez 1994), the third in Mérida 1995 (Venezuela) (Leccabue and Sagredo 1996), the fourth in São Paulo 1997 (Brazil) (Missell 1999); the fifth in Bariloche 2001 (Argentina). The proceedings of the latter were published in *Mater. Sci. Forum*. The sixth took place in Chihuahua 2003 (Mexico), published in *J. Alloys and Compounds*; the seventh in Temuco 2005 (Chile), published in *Physica B*; the eighth in Rio de Janeiro 2007 (Brazil), published in *J. of Magnetism and Magnetic Materials*.

In 1993 the International Workshop on "Optoelectronic Materials and Their Applications" (Leccabue et al. 1993) was held in Havana, followed by the second in 1998 including "Solar Cells" (Leccabue et al. 1999).

From 1995 to 1997 a network was constituted through the Action COST 514 EU, "Ferroelectric Thin Films," between the Centro Interdipartimentale di Ricerca-MTI (Parma, Italy), Depto. de Física de la Universidade de Braga (Portugal), NMRC of the University College of Cork (Ireland), Depto. de Física de la Universidade Federal de São Carlos (Brazil), Instituto de Física de la Universidad Nacional Autónoma de México (Ensenada, Mexico), Facultad de Física de la Universidad de La Habana (Cuba) and Depto. de Física de la Universidad Católica del Norte (Antofagasta, Chile), (Leccabue et al. 1997).

25.2.3 *Scientific Activity*

The investigation addressed different classes of materials and particular attention was devoted to those with technological applications: calcopyrites and related alloys for photovoltaics (Vigil et al. 1984; Leccabue et al. 1985); hard (hexaferrites, intermetallic alloys), (Leccabue et al. 1987) and soft (FeSiB based alloys) magnetic materials; sensors, magnetically coupled mechanical systems; piezoelectrics; ferroelectrics for non volatile memories (PZT, SBT); in the 1990s research focused on the preparation and characterization of ferroelectric thin films and oxides, obtained by pulsed laser ablation deposition and sol-gel processing (Watts et al. 1991; Díaz et al. 2000; Castro Rodríguez et al. 1998); amorphous, nano-micro-crystalline magnetic materials by melt spinning (Leccabue et al. 1991); more recently, scientific activities include the design and construction of MOCVD reactors to prepare silicon carbide films.

The research on hexagonal ferrites (obtained both through ceramic and chemical methods) was of paramount importance for the set-up of a small production of permanent magnets realized by the Physics Faculty of Havana University. In particular, this was made possible thanks to a systematic study of magnetic properties as a function of grain dimensions. In the years 1986–1987 several thousand permanent magnets were used in microphones for electrical guitars, devices for the magnetic treatment of water, reparation of mechanical equipment for beer-bottling and for magnetic door-locks. Starting from this pilot production two researchers from the Physics Faculty, Sergio Díaz and Francisco Calderón, managed to develop a factory as part of the Impresa de Componentes Electronicos in Pinar del Rio, where permanent magnets were produced for equipment dedicated to the treatment of water for the Cuban cane-sugar industry.

Supervision and coordination activities were developed at the Physics Faculty of the University of Havana. I was involved in several activities: as far as teaching is concerned, in the supervision of Masters and PhD theses (from 1985 to 2007); the supervision of fellowships funded by the CNR, by IILA in Rome, by ICTP in Trieste through the Training and Research in Italian Laboratories, and by the European Union.

It should be emphasized that the preparation of Cuban graduates or PhD students and young researchers was of a very high level with high degrees of autonomy and critical skills being developed for designing and carrying out research projects. I collaborated directly in research projects for 6–12 month periods with about 20 young Cuban graduates, and it is also thanks to them that the scientific results we obtained were published in international journals. They also enabled us to participate in numerous international conferences.[2] Cuban researchers demonstrated very good critical analysis skills and scientific quality.

25.3 Conclusion

These initiatives and activities have produced a complex network of (scientific and personal) experiences and results.

I believe that many aims and goals have been accomplished. Yet, there is still much work to be done, perhaps requiring substantial renewals following the changes and developments that are taking place internationally, and in particular in the world of science and research.

Finally I would like to mention three of our Cuban colleagues who, unfortunately, are no longer with us: Joaquin Torres, the first to design future perspectives for those who came after him; Pedro Díaz, for his scientific talent; and Celia Hart, for her enthusiasm and spirit of life.

Acknowledgements All of these activities would not have been possible without the contributions of Dr. B.E. Watts, R. Panizzieri, E. Melioli and other researchers and technicians at IMEM/CNR in Parma (Italy).

References[3]

Castro Rodríguez, R., M. Quadrelli, E. Vasco, F. Calderón, F. Leccabue, and B.E. Watts. 1998. Effects of particle size on the phase transition in $Pb(Zr,Ti)O_3$ grown by sol-gel. *Materials Letters* 34: 326.

Christensen, N., F. Leccabue, C. Paorici, and O. Vigil (eds.). 1988. *Proceedings of the international school on "Crystal growth and characterization of advanced materials"*. Singapore: World Scientific Publ. Co. Ltd (WSP).

Díaz, S., F. Leccabue, B.E. Watts, and R. Yapp. 2000. $PbFe_{12}O_{19}$ thin films prepared by pulsed laser ablation deposition on Si/SiO_2 substrates. *Journal of Magnetism and Magnetic Materials* 220: 79.

Leccabue, F., and V. Sagredo (eds.). 1996. *Proceedings of 3rd Latin American workshop on "Magnetism, magnetic materials and their applications"*. Singapore: World Scientific Publ. Co. Ltd (WSP).

[2] A selection of the articles published can be found in the references.

[3] The names of Cubans are underlined.

Leccabue, F., and J.L. Sánchez (eds.) 1992. *Proceedings of 1st Latin American workshop on "Magnetism, magnetic materials and their applications"*. Bristol: Institute of Physics Publ. Co. (IOP).

Leccabue, F., D. Seuret, and O. Vigil. 1985. Sintered n-$CuInSe_2$/Au Schottky diode. *Applied Physics Letters* 46: 853.

Leccabue, F., G. Albanese, and O. Ares. 1987. Magnetic and Mössbauer study of coprecipitated $Sr_2Zn_2Fe_{28}O_{46}$ (SrZn-X) hexagonal ferrite. *Journal of Applied Physics* 61: 2600.

Leccabue, F., B.E. Watts, G. Bocelli, G. Calestani, L. Dimesso, A. Deriu, J.L. Sánchez, and D. Carrillo. 1991. *Crystallization and Mössbauer study of the DyFe(B,C) alloys prepared by melt spinning.* International conference on magnetism (ICM-91), Edinburgh (UK), September 2–6, 1991.

Leccabue, F., O. de Melo, and I. Hernández (eds.). 1993. *Proceedings of international workshop on "Optoelectronic materials and their applications"*. Pisa: Edizioni Tecnico Scientifiche (E.T.S.).

Leccabue, F., B.E. Watts, and G. Bocelli (eds.). 1997. *Proceedings of EU-action COST 514 workshop on "Ceramic ferroelectric thin films"*. Pisa: Edizioni Tecnico Scientifiche (E.T.S.).

Leccabue, F., M. Sánchez, and A. Escobosa (eds.). 1999. *Proceedings of II workshop on "Optoelectronic materials and their applications (Including solar cells)"*. Pisa: Edizioni Tecnico Scientifiche (E.T.S.).

Lopéz Morán, J.L., and J.L. Sánchez (eds.). 1994. *Proceedings of 2nd Latin American workshop on "New trends in magnetic materials and their applications"*. Plenum Publ. Co. (IOP).

Missell, Frank P. (ed.). 1999. *Proceedings of 4th Latin American workshop on "Magnetism, magnetic materials and their applications". Materials Science Forum*, Vol. 302–303.

Watts, B.E., F. Leccabue, G. Bocelli, G. Calestani, F. Calderón, O. de Melo, P.P. González, J. Vidal, and D. Carrillo. 1991. On the preparation PZT thin film by laser ablation deposition. *Materials Letters* 11: 183.

Vigil, O., O. Calzadilla, D. Seuret, J. Vidal, and F. Leccabue. 1984. $ZnIn_2S_4$ as a window in heterojunction solar cells. *Solar Energy Materials* 10: 139.

Chapter 26
Scientific Cooperation Between the German Academy of Sciences in Berlin (DAW) and Cuba in the 1960s and 1970s

Helge Wendt

After the ratification of its constitution in 1959, the young Cuban Republic sought new cooperation partners in a number of different fields. One of these fields was scientific cooperation. It seems the Cubans quickly found partners in the academies of science of the USSR, Czechoslovakia and China, whereas the German Academy of Sciences in Berlin (DAW) was reluctant to engage in long-term cooperative projects. In the early 1960s, some universities of East Germany (GDR) began to send docents and scientists to Cuba where they participated in the summer schools, taught for one semester or more in one of the universities and undertook research that would be useful for their home institutions. However, the DAW carefully observed the reestablishment of Cuba's own academy of science before becoming involved in common projects with Cuban partners.

Early in the 1960s, only a minority of members of the DAW was in favor of initiating a collaborative program with Cuban partner organizations. The problem was that the DAW considered only national academies to be adequate partners for research cooperation. The Cuban Academy of Science was dissolved after the Revolution and reestablished only in 1962. The second problem was to find a suitable area of research in which both parts were equally interested. In addition, a highly evolved consciousness had developed within the Berlin academy concerning the neo-colonial effects of scientific cooperation, which they tactfully avoided.

The first connections between the Cuban scientific community and the DAW in Berlin were established during a trip to Cuba in February 1962 by the president of the East Berlin Humboldt University, Kurt Schröder. During the trip, Schröder met with Nuñez Jimenez who was working on the reestablishment of the Cuban Academy of Sciences. Jiminez expressed his interest in establishing contacts with

H. Wendt (✉)
Max Planck Institute for the History of Science, Berlin, Germany
e-mail: hwendt@mpiwg-berlin.mpg.de

A. Baracca et al. (eds.), *The History of Physics in Cuba*, Boston Studies in the Philosophy and History of Science 304, DOI 10.1007/978-94-017-8041-4_26,

the Alexander von Humboldt research group at the East German academy and planned to visit the institution during a trip to Europe the following year.[1]

In the autumn of 1962, the General Secretary of the DAW, Günther Rienäcker wrote to the Secretary for Science and Technology, Herbert Weiz, that no cooperation with Cuba should be undertaken before the Cubans had clarified their intentions.[2] The GDR representatives therefore waited for the Cubans to clarify in which matters they needed help and cooperation and for an institution to be established that was able to coordinate any further cooperation.

Early in 1963, progress began to be made and the Republic of Cuba officially asked the DAW to establish a research institute on the island. It was now up to the academy to decide which departments should take part in this "Caribbean venture." The Cubans suggested that the DAW should establish an institute in Cuba for tropical research, which would cover fields such as geography, geology, botany, oceanology, meteorology, zoology, tropical medicine, history and literature. The Cubans also sent a long list of scientific instruments that they wished to obtain from the DAW.

The president of the DAW then approached different departments within the DAW for their advice on how to determine the major areas of research required and how to move forward. The first research proposal involved testing machines and insulating materials in the tropical environment of Cuba. A second proposal involved the Institute for Astronomical Observation, which was thought to be useful for both academies.[3] A third proposal aimed to establish a seismic research unit in Cuba in order to register seismic activity and to test instruments built in the GDR.[4]

Next, the DAW looked for cooperation partners inside the GDR because the institution was aware of the high costs of such a transatlantic partnership. One cooperation partner was the German Bureau for Material Control and Product Testing (Deutsches Amt für Material- und Warenprüfung), whose director, Helmut Lilie wrote a favorable note to the General Secretary of the DAW. Lilie expressed his interest in experiments on corrosion and what he called "climate protection," a term that in the 1960s did not refer to protecting the climate, but to the protection of materials and machines from weather induced corrosion.[5] By March 1963, the

[1] Transcript. Mission der Deutschen Demokratischen Republik in der Republik Kuba. Havanna, 21.2.1962. Akademieleitung 1945–1968, 515. The information given in the footnotes is taken from archive material found at the Archive of the Berlin-Brandenburg Academy of Sciences.

[2] Der Generalsekretär Berlin, 18. September 1962. Akademieleitung 1945–1968, 515.

[3] Auszug aus dem Protokoll der Sitzung der Klasse für Mathematik, Physik und Technik vom 24. Januar 1963. Akademieleitung 1945–1968, 515.

[4] Prof. Dr. O. Meißer (Sekretär der Klasse für Bergbau, Hüttenwesen und Montangeologie) an Gen-Sekr. Rienäcker, Berlin 6.3.1963. Akademieleitung 1945–1968, 515.

[5] Dr. habil Lilie, Deutsches Amt für Material- und Warenprüfung, Berlin 12.3.1963, an Rienäcker. Akademieleitung 1945–1968, 515.

establishment of the Tropical Institute at Cuba was making good progress. Seven major fields of research were mentioned:

1. The seismic research station should be established by the Institute for Seismic Dynamics in Jena
2. An oceanic biological research unit
3. A regular exchange of information in questions of tropical medicine
4. Research on the question of morphological processes and economical geography
5. Cooperation in botanical and zoological fields of research
6. Support in writing a Cuban history of culture
7. Testing materials and products of the GDR under tropical conditions with a focus on corrosion and climate effects on materials.[6]

The problem was that in Cuba only a few cooperation partners could be found as most of the scientific institutions were only about of being founded. In addition, those who were able to do scientific research were already working on Soviet, Chinese, Czech and Hungarian led research projects.[7] These countries had established institutes for different fields of research much earlier than the DAW. As mentioned earlier, teaching staff from the GDR universities were already working in Cuba, but the DAW was not willing or not politically inclined to rely on those experiences.[8]

In a first cooperation agreement in June 1963 between the DAW and the National Commission of the Academy of Sciences of the Republic of Cuba, the two partners agreed that a GDR delegation should study the circumstances on the island under which scientific research could be realized. The second point of this agreement stipulated that a limited number of Cuban students should stay for a maximum of 1 year at different research institutes in the GDR.[9]

In 1964, the progress in founding any institute in Havana slowed down. One of the reasons for this was the predicted high costs. For the year 1964, the DAW calculated expenses of more than 900,000 East German marks. This was designated for equipment for the new institute, for travel costs and currency change taxes.[10] Another obstacle to the establishment of the institute was that some of the above-mentioned research fields were put on hold because the German partner had pushed a new research area in the field of chemistry to the forefront. The responsible person at the DAW, apart from the General Secretary, was now the chemist and leader of the

[6] Bisherige Vorstellungen zur Zusammenarbeit mit der Akademie der Wissenschaften der Rep. Kuba, Berlin, 22.3.1963. Akademieleitung 1945–1968, 515.

[7] Langer, Kulturabteilung der Botschaft der DDR, Havanna, 26.2.1965. Bestand Klassen, 208.

[8] Protokoll der Sitzung der Kommission zur Vorbereitung der wissenschaftlichen Zusammenarbeit mit der Akademie der Wissenschaften der Republik Kuba am 9. April 1963. Akademieleitung 1945–1968, 515.

[9] Übereinkommen über die wissenschaftliche Zusammenarbeit der Deutschen Akademie der Wissenschaften zu Berlin und der Nationalen Kommission für die Akademie der Wissenschaften der Republik Kuba, 7.6.1963. Akademieleitung 1945–1968, 515.

[10] Zur Entstehungsgeschichte des Tropenforschungsinstituts, 3.6.1964. AKL.1945–1968, 515.

Department of Chemistry, Horst Sinnecker. Sinnecker became involved in the process because the Cuban side had partly rejected the program proposed by the DAW. The National Commission asked instead for basic research, for a much more practical and economically useful cooperation. Sinnecker proposed to do chemical research on laterites,[11] a material found in secondary sediments. These sediments were rich in nickel, which Sinnecker and the Cuban partners considered to be of further use for the economy.[12] Sinnecker had worked on uranium chlorides in the 1950s (Sinnecker 1960), before becoming an expert on nickel extraction. During his stay in Cuba, he published at least two articles (Sinnecker 1968, 1969) and completed a second dissertation on that theme in 1972 (Sinnecker 1972).

Shortly before the inauguration of the Tropical Institute, the DAW partially withdrew further investment and the German Bureau of Material Control (DAMW) and the Academy of Agriculture took over its direction. Thereafter, the DAW planned to found a bureau for scientific relations only, with Sinnecker as its head.[13] In 1968, shortly before the entire DAW was reorganized and renamed, this form of engagement in Cuba was openly criticized. For some of the critics, it remained unclear who directed the institute: the DAW or the Academy of Agriculture. Additionally, Sinnecker, after his arrival in Cuba, had ordered a complete chemical laboratory to be transported from Rostock to Cuba. Its costs exceeded all estimates and forced the General Secretary to stop any further orders from Sinnecker in Cuba. Both for the general members of the DAW and those inside the DAW headquarters, it remained unclear who was in charge of the Cuban enterprise. This confusion caused some to demand a common commission for all Cuban affairs. The role of the Ministry for Science and Technology was also vague. In 1968, with no further explanation, the Ministry refused to reopen negotiations between the partners of the Tropical Institute (DAW, DLA and DAMW).[14] Sinnecker had to return to Germany in the years that followed; he became Director of the Central Institute for Inorganic Chemistry until 1974 and maintained his status of a member of the DAW. What happened to the laboratory and the equipment is an open question.

In the mid 1960s, the DAW maintained a wide range of international cooperation projects, reaching beyond those with academies of the close allies within the Warsaw Bloc. For example, a well-developed cooperation existed with India. A delegation had attended the annual meeting of the Council of Scientific and Industrial Research of India (CSIR) in 1965 and an exchange of researchers had already been agreed

[11] Sinnecker, Themenvorschlag für die Bearbeitung im Stützpunkt der DAW bei der Kubanischen Akademie der Wissenschaften in Havanna: Untersuchungen zur Analytik und Aufbereitung kubanischer Laterite, 5.1.1965. Bestand Klassen, 208.

Rienäcker an Liers (Vizepräsident des Deutschen Amtes für Meßwesen und Warenprüfung, Berlin 6.1.1966. Akademieleitung 1945–1968, 515.

[12] Sinnecker, Themenvorschlag für die Bearbeitung im Stützpunkt der DAW bei der Kubanischen Akademie der Wissenschaften in Havanna: Untersuchungen zur Analytik und Aufbereitung kubanischer Laterite, 5.1.1965. Bestand Klassen, 208.

[13] Zum Vereinbarungsentwurf des DAMW vom 24. Jan. 1966, Berlin, den 27. Jan 1966. Akademieleitung 1945–1968, 515.

[14] Akademieleitung 1945–1968, 515.

upon. Less extensive cooperative projects in different domains existed with Indonesia, the Sudan, Ghana, East Africa and Iraq.[15] The intention of the DAW to cooperate with Cuba in the 1960s has to therefore be considered in a wider framework. International cooperation was initially organized within the structure of COMECON member states. This cooperation was comprised of both bilateral and multilateral partnerships. As the working program of the DAW of 1970 shows, the bilateral cooperation with Soviet project partners was in a privileged position:

> The international cooperation in research with scientific institutions and institutions of the USSR and other socialist countries is of essential importance for developing the scientific potentials of the DAW. The importance of international research cooperation with these countries is growing according to its inner logic. Therefore, the ideal development and the enactment of this research cooperation have become a central issue of the international relations of the DAW.[16]

In 1970, the DAW was aware of the necessity to restructure its international engagement in a way that would increase its scientific output. The aim of this restructuring, which was implemented at the same time as the reorganization of the entire DAW, was to create larger research areas and to unite different institutes and departments of the DAW to encourage institutional and interdepartmental cooperation. In 1970, international cooperation projects were assembled into "Complexes of Work," "Departments of Research" or "Centers of Research," each one headed by a director who regularly reported to the president, the party and to the government.[17]

In the early and mid 1960s, Cuba itself was not involved in most of the structures for scientific cooperation within the COMECON. For example, there were no Cuban delegates in attendance at the meetings of the Bureaus for International Relations of the Academies of Socialist Countries between 1962 and 1965. During this period, the academies of China, North Korea and Vietnam had obtained the status of observers, even though these countries were not members of COMECON.[18] Cuba that became a full member of the COMECON in 1972, was also not invited to attend the socialist countries' Second Conference of the Academies of Science, held at the end of March 1963 in Berlin. In addition to the European socialist states, Vietnam

[15] Übersicht über die Beziehungen der Deutschen Akademie der Wissenschaften zu Berlin zu Nationalstaaten in Asien und Afrika (Stand 10.3.1965). AKL.1949–1968, 433.

[16] Arbeitsprogramm der Deutschen Akademie der Wissenschaften zu Berlin zur Durchsetzung der Beschlüsse des Politbüros des ZK der SED, des Staatsrates und des Ministerrates der DDR zur Gestaltung der sozialistischen Wissenschaftsorganisation der DAW (1970). Akademieleitung 1969–1991, 154.

[17] See point III of Konzeption z-u-r Gestaltung der internationalen Forschungskooperation und sonstigen Beziehungen zum Ausland als Bestandteil der Wissenschaftsorganisation der DAW, Entwurf, Berlin, den 21.8.1970. Akademieleitung 1969–1991, 154.

[18] Bericht über die Arbeitssitzung der Leiter der Auslandsabteilungen der Akademien der Wissenschaften sozialistischer Länder vom 29. Nov. bis 1. Dez. 1962 in Prag/CSSR; in Balaton-Szabadi vom 10.–13. Sept. 1963 and Information für das geschäftsführende Präsidium über die 3. Arbeitssitzung der Leiter der Auslandsabteilungen der Akademien der Wissenschaften sozialistischer Länder vom 6.–9. Juli 1965 in Bukarest und über den Stand der Vorbereitung der IV. Konferenz der Akademien der Wissenschaften sozialistischer Länder im Dezember 1965 in Moskau. Akademieleitung 1949–1968, 433.

and Mongolia were also present at the conference, while China and North Korea were absent. The following quote gives an explanation of why no other socialist country participated in the conference:

> Scientific institutions of other socialist countries were not invited because at this stage of cooperation, the intention was to confine participation in this conference to only those countries that had already received an invitation to the first conference.[19]

Equally, no Cuban delegate is mentioned to have participated in the April 1964 Third Conference of the Academies of Science at Sofia. But in the same year, it was agreed that Cuba would participate in the Soviet-coordinated commission on "The History of the Great Socialist October Revolution." This can be seen as the first official Cuban commitment to a scientific activity of COMECON.[20]

In 1971, a balance of international cooperation was drawn between the departments of cosmology and geosciences, amongst them those organized jointly with Cuba. The Institute for Geography led a 5-year research program, focusing on territorial economy and geoeconomics. The Central Institute for Terrestrial Physics hosted a common program on seismology. Another cooperation was conceived for the field of ocean sciences, namely a scientist exchange program in 1974/1975. Cuba was very active in the COMECON Intercosmos program in which, it seems, cooperation between GDR and Cuban scientists did occur.

One example of this cooperation is the 1976 visit to Cuba by the General Secretary of the DAW, Claus Grote, and his delegation. This visit had three major goals. The first was to sign a cooperation contract between the academies in Havana and in Berlin covering the years 1976–1980. Secondly, Grote and his colleagues inaugurated the "Days of Technology of the GDR at Cuba." The last larger event was the anniversary of what was left of the former Tropical Institute, known then as the Institute of Basic Research in Tropical Agriculture at Havana. The delegation visited different institutions of the DAW in Cuba; for example, the institutions for nuclear research, geophysics, astronomy and chemistry. Grote was especially interested in the Institute for Nuclear Research as the Cubans and Soviets working at this institute sought to develop it in a manner similar to the Rossendorf Institute for Nuclear Research near Dresden. Here, Grote may have seen a possibility for future cooperation between the two academies. He envisioned developing a plan to cooperate on a joint nuclear research project that would last until the end of 1976, a plan for further seismological investigation, and a program for physics in the higher ionosphere that would last until the end of January 1977.[21]

This short overview shows that among the COMECON academies, the DAW was a latecomer in engaging in Cuban scientific development. The projects that

---

[19] 2. Konferenz der Akademien der Wissenschaften sozialistischer Länder vom 24.3.–1.4. 1963 in Berlin. Akademieleitung 1949–1968, 434.

[20] Problemkommission "Geschichte der Großen Sozialistischen Oktoberrevolution". Akademieleitung 1949–1968, 435.

[21] Grote, Bericht der Delegation der DDR über den Aufenthalt in der Republik Kuba zu Fragen der Zusammenarbeit auf dem Gebiet von Wissenschaft und Technik in der Zeit vom 20.10.1976 bis 31.10.1976. FOB Geo- und Kosmoswissenschaften, 221.

Cuba and the GDR agreed to execute and the jointly founded Tropical Institute in the 1960s were only mildly successful. Further research should focus on those aspects of the history of the bilateral relations of both academies, which were continued into the 1970s. The multilateral cooperation in the frame of COMECON has also to be studied in depth. Furthermore, research is lacking on the relations between East German and Cuban universities.

References

Sinnecker, H. 1960. *Zur Kenntnis der Chemie einiger Uranchloride*. PhD thesis, Halle (Saale).

Sinnecker, H. 1968. *Sobre el componento manganeso-cobalto-niquel en las lateritas*. La Habana: Academia de Ciencias de Cuba, Instituto de Investigaciones Tropicales.

Sinnecker, H. 1969. *Acerca de la influenca del componente manganeso-niquel-cobalto de la laterita sobre las propiedades de sedimentación de ésta*. La Habana: Academia de Ciencias de Cuba, Instituto de Investigaciones Tropicales.

Sinnecker, H. 1972. *Zur Chemie der Nickelextraktion nach M.H. Caron*. PhD thesis, Berlin: DAW.

Chapter 27
A Beautiful Story

Federico García-Moliner

In 1981, a time when our science administrators still believed that small is beautiful, I was sent to Cuba with a very simple mission: To establish contact with the solid state physicists at the University of Havana in order to identify affinities and prepare the ground for possible collaborations. I count this as one of the most fortunate events in my professional and human experience. It marked the beginning of a long lasting and very fruitful collaboration as well as of many brotherly friendships that have significantly contributed to the enrichment of my life as a human being.

When I arrived I saw a community of hard working, highly motivated, solidly prepared young physicists (we were all young then) working under atrocious conditions, a situation I could easily understand, as it reminded me of our own conditions in Spain not so long ago. In spite of their many hardships, their work was very good and stimulating, the problems they were tackling were interesting and timely in their field and it was very easy to identify common interests and grounds for a collaboration which for many years to come shaped a substantial part of my own activity and that of my closest collaborators in Spain. To work with my colleagues at the University of Havana was a sheer pleasure. Besides the warmth of our personal relations and the many satisfactions I had, I benefited a great deal from this association regarding the actual research work. In spite of the many difficulties they had, they managed to maintain the quality of their work and their determination to carry on. The many years I managed to keep this association active were for me an endless source of happiness and benefit. In this respect my days of wine and roses ended only when my retirement inevitably arrived, but it is satisfying to see that some of my Spanish colleagues were attracted to participate in this process through the links I take pride in having initiated and part of these are still active, and will remain so, I hope, for a long time to come.

F. García-Moliner (✉)
University 'Jaume I', Castellón de la Plana, Spain
e-mail: garmol@aug-arquitectos.com

A. Baracca et al. (eds.), *The History of Physics in Cuba*, Boston Studies in the Philosophy and History of Science 304, DOI 10.1007/978-94-017-8041-4_27,

From the moment I established contact with my Cuban colleagues it was clear that their background in physics and mathematics was very strong. They had nothing to envy, in that respect, in anyone else. The rigour and objectivity in the "public defence" of a PhD thesis was admirable. My contacts were limited to physicists and a few mathematicians, but it is reasonable to assume that the scene was essentially the same in other fields. From the first moment, I recognized a familiar scene: a young generation, working under extremely poor conditions, with a firm determination to build up science in their country. We—some before us and then my generation—had the same experience in Spain after our civil war.

There was, however, a significant difference: in our case we simply worked hard and that was all; the young scientists I met in Havana and elsewhere, particularly in Santiago, felt concerned to find a meaning to their effort and asked themselves, in the words of one of them, "*key questions for our development: where to concentrate our efforts; relationship between theory, experimentation and application; which were the ethical principles specific of the scientific work; with what rigour had we to approach the publication of our results and many other*..." They were also more interested than we were in basic questions of education—should they favor the education of specialists or was it better to give a strong basis as a preparation for further individual development and adaptation to the world outside? In a way it was like an extension of my own education. There were important questions we, in my generation in Spain, had not asked ourselves and I found the experience of debating these ideas with them most stimulating. Still today I spend most of my time thinking, reading, writing and talking about similar and related questions.

Years passed, the world changed and these changes resulted in increased hardship for Cuban scientists. For a time I thought that in the turn of events, notwithstanding the added problems, there were for them some positive aspects in the new situation. Hardship is tough for everybody, but they were in this sense strong people who had survived difficulties under which many others would have collapsed. On the other hand, there was the cultural factor. They had had a first class education, largely based on the Soviet model and often from Soviet teachers in the times when Soviet science was first class, but that was not their natural cultural environment. And then the new situation meant for many of them a shift towards increased contacts with countries where the prevailing culture constituted a more natural environment for Cubans. I am referring to European countries like—mainly but not exclusively—France, Italy and Spain and some South American countries like Mexico and Brazil. Since these changes caught them when they were full-grown scientists and no longer needed tutoring but simply contacts and occasions for various collaborations, I thought they might end up by having the best of both sides, perhaps a blessing in disguise after all.

Meanwhile, admirably, they kept maintaining the quality of their work, but new problems were in store for them. Their economic problems grew worse and reached the stage in which survival, in this sense, became a compelling prosaic necessity. This forced a change for the worse: they were pushed into a policy of commercializing their scientists as teachers. Then, instead of seeking affine groups where they could carry on a mutually beneficial collaboration, they had to begin looking for places that lacked qualified teachers as a way of helping themselves and their home departments economically. Unfortunately this trend, if continued for a long time, will not help.

Furthermore, as the difficulties increased, Cuban science began to suffer from another problem, namely brain drain. Leaving one's own country is a very painful decision, as only those who have been through it know, but the real question is what do both sides involved do about it. Those remaining at home feel distressed and perhaps somewhat irritated, but those in self-exile are no happier. We had this problem in our difficult years in Spain; the difference is that the world has changed a great deal since then. Present circumstances allow an option to be considered which was not possible in our time, not to any extent remotely comparable to what is possible today. I am referring to the so-called *diaspora option*, an alternative now conceivably viable due to globalization, intense internationalization and efficient information and communication technologies. Urging intellectual emigrants to "stay home" or "come home" is not the only alternative, nor does it seem the most practical attitude. The question is whether all stakeholders involved—government and academic and scientific institutions, as well as national scientists at home and abroad—decide to come to terms with the reality of the present world and adopt policies and practices which have a chance of changing the *brain drain* into the *brain gain* by combining local efforts with the intellectual and material resources of self-exiled nationals. Some countries are trying to implement such policies as the only practical option in the world we live in.

I do not know whether some of my Cuban colleagues may disagree with this appraisal of their current situation. If such is the case I can only assure them that my comments are moved exclusively by one motivation: With my utmost sincerity, I only wish the best for them and their country. To a significant extent the help they can get, in one form or another, from their colleagues around thc world will be no doubt an important factor and in this, as in other cases, those who are in a position to help and enjoy a more fortunate situation should recognize that helping fellow colleagues in trouble is an ethical obligation, both as members of the scientific community and simply as decent human beings. But it is obvious that, in any case, it all depends mainly on their endurance, capacity, willpower to push forward and wisdom in their policy options. Knowing them, I dare hope that they will find their way.

Chapter 28
The Current State of Physics in Cuba: A Personal Perspective

Marcelo Alonso

After 40 years of absence I returned twice to Cuba, in January and December of 2000, to participate as a guest lecturer in two international scientific meetings. The first dealt with physics education, and the second with current issues related to quantum mechanics. In addition to a few participants from Europe, the US and Latin America, the two meetings were well attended by Cuban physicists.

International meetings are very useful for Cuban physicists, whose travel possibilities are limited unless financed by foreign sources, and thus offer them the opportunity to interact with foreign colleagues. For me the meetings were very helpful because I could talk at length with several Cuban physicists, allowing me to get first-hand information about physics education and research. Both have changed during my absence. Prior to 1959 there were three official universities, Habana, Central and Oriente, and one private, Villanueva. Now there are several official universities, polytechnic institutes and pedagogical institutes, so higher education is much more diversified. Only two universities in Cuba offer a degree in physics: the University of Habana, in Habana, and the University of Oriente, in Santiago, although other universities offer physics courses for students of chemistry, engineering, biology, and so on.

On both occasions I was able to visit the University of Habana, where I had been Professor of Theoretical Physics until 1960. The main campus, on a hill, with neoclassical architecture, remains the same except that the use of some buildings has changed because the academic structure of the University has also changed. Unfortunately the buildings are not well maintained, but that is a general problem in Cuba.

I found that since my time the physics curriculum in the University of Habana has been reorganized substantially and the academic staff expanded considerably. The Faculty of Physics, headed by a dean, consists of three departments: general

---

Marcelo Alonso (deceased) (February 6, 1921–November 11, 2005)

First published in 2002 in vol.11, No.8, Aug./Sept. of *APS News*.

M. Alonso (✉)
Graduate School and Florida Tech, Georgetown University, Washington, DC, USA

A. Baracca et al. (eds.), *The History of Physics in Cuba*, Boston Studies in the Philosophy and History of Science 304, DOI 10.1007/978-94-017-8041-4_28,

physics, theoretical physics, and applied physics, with a total academic staff of 69 persons with about 40 holding a PhD. The faculty offers a 5-year *licenciado* which corresponds to a level between a bachelors and masters degree in the US. Beginning with the third year, students must work in some laboratory, and at the end of the fifth year students must submit a thesis in order to obtain their diploma. Masters and PhD degrees are also offered which are to a great extent comparable to the US. At least a Masters degree is required to teach in a University. My general impression is that the physics students (currently about 100) and the staff are very well prepared, in spite of severe limitations in resources (equipment and library).

In many cases students can take graduate courses or do their Masters or PhD thesis in some of the research institutes that operate under the Academy of Sciences, such as the Institute for Cybernetics, Applied Mathematics and Physics (ICIMAF) and the Advanced Institute for Nuclear Science and Technology (ISCTN) that offers 5-year *licenciado* and PhD degrees in nuclear physics and in nuclear engineering.

In addition to the two universities and research centers offering advanced physics degrees, there are 16 higher pedagogical institutes that offer a 5-year *licenciado* degree in education with specialization in physics education. This degree is required to teach physics in secondary schools, although university physics students must take courses on the pedagogy of physics, just in case they decide to teach.

After graduation a student must work up to 2 years in some government research center or equivalent (social work). In addition to the physics courses, students must take courses with social and political content, a tradition inherited from former Soviet universities.

During the period of Soviet influence in Cuba, from the early 60s until the demise of the Soviet Union, many Cuban scientists were trained in Russian centers, mostly in Moscow and St. Petersburg (formerly Leningrad), as well as in some East European countries, such as Hungary. The Cuban scientific establishment was patterned after the Soviet organization of science, with universities and technological institutes providing mainly scientific and engineering education, and most of the research done in specialized governmental institutes operating under the Cuban Academy of Sciences, the Ministry of Science, Technology and Environment, or other government agencies. This structure still exists.

In a centrally planned and operated economy as is the Cuban system, all job opportunities are in governmental institutions. To be considered for a position (research and teaching) in a university, the *licenciado* in physics must have graduated with an average of at least 4.0 points out of 5.0, and must take advanced courses related to pedagogy in the areas in which they will teach. Cuban physicists work in research centers of the Ministry of Science, Technology and Environment and other government agencies, in hospitals and biomedical research centers, and in industrial and technical services. The main fields in which Cuban physicists work are (1) optics, lasers and spectroscopy, (2) condensed matter and materials physics, (3) electronics and computation, (4) non-conventional energies, mostly solar, (5) biophysics and medical sciences, (6) geosciences, (7) theoretical physics (complex systems, cybernetics, particle physics, field theory, etc.), (8) nuclear physics, (9) teaching, and (10) physics education research at all levels. In some instances it is a combination of fields.

Currently there are in Cuba about 1,600 physicists, of which about 180 are PhDs, and about 700 are engaged in research. The Cuban Physical Society has about 500 members, and publishes the *Cuban Journal of Physics* with three issues per year. Other technical journals are available, some of a popular nature such as *Energia y Tu* (Energy and You) published by CubaSolar and *Nucleus* published by the ISCIN. Aside from research, physics education at all levels receives special attention and several semi-popular journals have that orientation.

An important difference with the US is that *all* students when they finish secondary (high) school have taken physics. In elementary school students are taking science courses, with some physics content, in the third grade. However physics as an "obligatory" course for secondary (high) school students is taught in grades 7 through 12. All physics teachers in secondary schools must be *licenciados* in physics education, graduated from a higher pedagogical institute. Thus in spite of possible deficiencies in laboratory and computing equipment, secondary (high) school graduates are much better prepared in physics (as well as in mathematics and other subjects) than their counterparts in the US.

If I am asked what is the best way to help physicists in Cuba, I would recommend as the first priority to establish a modest fund to invite Cuban physicists to attend conferences and seminars in the US, and to teach one semester courses or work with a research group in US academic institutions. Considering how inexpensive travel is between Miami and Havana ($300 round trip) I assume that the amount needed per individual physicist would be of the order of $2,000–$5,000 depending on the place and length of stay. Organizing seminars in Cuba, in which US physicists would participate, is my other priority. I hope very much that funds for these two purposes can be found.

Chapter 29
Engaging Cuban Physicists Through the APS/CPS Partnership

Irving A. Lerch

In his reflections on Cuban physics below, Marcelo Alonso urges APS to take steps to promote interactions between Cuban and US physicists. As an introduction to Marcello's essay, this note will summarize past and current activities.

After a prolonged period of political estrangement, we have actively engaged colleagues in Cuba in a number of collaborations over the past 2 years. In many ways, this joint effort mirrors the APS policy of engagement pursued during the Cold War with the physics communities of the USSR and China. But scientific communications with the Soviet Union and China were not hampered by extraordinary legal impediments such as the economic embargo levied against Cuba in 1960. Nonetheless, this past April, more than 30 US medical physicists participated in an international congress in Havana (International Conference in Medical Physics, April 8–10, 2002) and many more are expected to attend the VIII Inter-American Conference on Physics Education to be convened July 7–11, 2003, in Havana.

The policy that underlies this relationship was enunciated by the APS Council in 1989 with a "Statement on the International Nature of Physics and International Cooperation," which, while advocating the rights of physicists, strongly promoted open international exchange (see http://www.aps.org/policy/statements/89_2.cfm).

With not much more than 11 million people on an island smaller than Pennsylvania, Cuban physicists were little in number, known to colleagues in Latin America and the Soviet Bloc, but practically unnoticed in the US.

In the early 1990s, during Ernest Henley's Presidency, the APS made a commitment to invigorate its ties with colleagues in Latin America and embarked on a series of initiatives to include the organization of joint Canadian-Mexican-US physics meetings called CAM (Canadian Association of Physicists, American Physical

First published in *APS News*, August/September 2002, 8.

I.A. Lerch (retired) (✉)
Professor New York University, Director International Programs, American Physical Society, Washington, DC
e-mail: ialerch@verizon.net

A. Baracca et al. (eds.), *The History of Physics in Cuba*, Boston Studies in the Philosophy and History of Science 304, DOI 10.1007/978-94-017-8041-4_29,

Society, Sociedad Mexicana de Fisica). These, in turn, led to regular meetings with the Federation of Latin American Physical Societies, a consortium of 17 national physical societies.

It rapidly became clear that notwithstanding Cuba's size, the island's intellectual community was a major presence in Latin American science. Cuban physicists often took up residence in the universities and labs of the larger countries of Latin America, Spain, France and Russia. And gradually, growing numbers of Cuban students and scholars began coming to the US.

At the April, 2000, APS meeting in Long Beach, California, the President of the Cuban Physical Society, Victor Luis Fajer Avila, was invited to attend a discussion on future APS/CPS collaborations with the officers of the Society, and an agreement was made to organize joint meetings in Cuba. In May, a Reciprocal Member Agreement was signed by the two societies exchanging some privileges. By the following year, the two societies had agreed to hold joint meetings on medical physics and physics education and both Bernd Crasemann (then Chair of the Committee on International Affairs), and I participated in an international conference in Havana in early June, 2000, and met with the Board of CPS and officers of the Cuban Academy of Sciences. At these meetings we finalized the procedures to be used to facilitate accelerated contacts between the two communities.

To limit the impact of the embargo, we agreed to exploit that portion of US law that promoted intellectual and cultural exchange sponsored by international organizations of which both Cuba and the US were members (although the law specifies that the organization may not have its headquarters in either country). Thus IUPAP sponsorship—to include the sponsorship of most of the other international disciplinary unions—was a viable means for promoting scientific relationships.

The success of the medical physics conference (supported by the International Union of Physics and Engineering Sciences in Medicine and the International Organization for Medical Physics) prompted the Brazilian physics community to offer to host a second meeting and it is likely that the series will continue. Since the Inter-American Conference was due to be convened in 2003, the Council of the conference readily accepted Cuba's offer to hold the next meeting in Havana (and the organizers have applied to IUPAP for sponsorship). But we have not been nearly as successful in meeting Marcelo's other demand that a way be found to increase Cuban participation in APS meetings. Funding continues to be a significant obstacle.

I was first scheduled to arrive in Cuba on October 28, 1962, by parachute from a C123 troop transport. History intervened and made me wait almost four decades.

It was well worth the wait.

Postscript in January 2014

In October of 1962, I was a junior officer in the 101st Airborne Division, one of two airborne divisions assigned to the initial assault on Cuba during the Cuban Missile Crisis. I had been involved in the planning for this operation and was very aware how close the US came to invading Cuba. Thankfully, I had to await almost 40 years before peacefully invading Cuba during a scientific congress that I helped organize. This occurred at the April, 2002 Congress on Medical Physics that I allude to in the article—almost 40 years after the threatened invasion. It is worth noting that as a part of the settlement that President Kennedy made with Premier Khrushev at the end of October, 1962, the US Administration pledged never to invade Cuba. On the basis of this tenuous beginning, the US physics community was able to engage in modest exchanges and to plan mutual undertakings.

This article was first published in APS News in 2002 in response to an essay by Marcelo Alonso (see previous chapter). In view of the fact that it was written 12 years ago, it is reasonable to assume that there have been many changes in fact and policy. Whereas US policy has not changed a great deal over the past decade, there have been some accommodations between the two nations that have altered the relationship between them. Although Raul Castro's assumption of the Cuban Presidency in 2008 along with the election of Obama in the US occasioned some optimism, little has changed. However, recent reforms by the Cuban government have resulted in some loosening of political controls over social mobility and the economy. Still, official government contacts remain largely frozen. During the Bush Administration, some economic restrictions were made more severe but under Obama, these have been eased. The major question thus is what changes, if any, have altered the relationship between the scientific and cultural communities of the two nations? There are some signs that cultural and scientific exchanges may be liberalized in the coming years and this may portend the harbinger of more fluid political exchanges. Time will tell.

Chapter 30
A Perspective on Physics in Cuba

Carlos R. Handy

The author is Cuban by birth: the son of an Afro-Cuban mother and an Afro-American father. He spent the first 8 years of his life attending schools in Havana and New York City. He left Cuba in 1958 so as to gain a more uninterrupted educational experience. This led to his PhD in theoretical physics from Columbia University in 1978, followed by a post-doctoral appointment at the Los Alamos National Laboratory in New Mexico. He returned to Cuba in 1980 as a member of an academic tour organized by the University of New Mexico in Albuquerque. This provided an introduction to various government-run scientific facilities in Havana and the easternmost city of "Santiago de Cuba" in the province of Oriente. The dim memories of that first visit do not conjure visions of any particularly impressive achievements except for one. It was revealed that within the former Soviet block, Cuba was responsible for building mid-sized computers. He specifically remembers asking to be taken to the facilities where they were designed. Fifteen years later, he returned to Cuba and received an eye-opening exposure to how its physics development had progressed. He had always known from his mother's experiences that Cuba had one of the best medical systems (particularly excellent doctors) in all of Latin America prior to "La Revolucion." It was this tradition, coupled with the effects of the American-led embargo in restricting the importation of vital scientific instruments that provided an impetus for the emergence of Cuba's talents in physics. The following remarks are a result of the observations and discussions initiated at that time.

Most individuals are unaware of Cuba's intellectual vitality, no doubt the result of a significant mix of world cultures (i.e. African, Middle Eastern, European, Asian, etc.). Their impressions of this island nation essentially evolve from its many accomplishments in sports and musical entertainment. Occasionally some news reports provide a glimpse of Cuba's emerging prowess in pharmaceutics and medicine. The 2002 "James Bond" thriller "Die Another Day" culminates within an

C.R. Handy (✉)
Professor and Chair of the Physics Department,
Texas Southern University, Houston, TX, USA
e-mail: handycr@TSU.EDU

A. Baracca et al. (eds.), *The History of Physics in Cuba*, Boston Studies in the Philosophy and History of Science 304, DOI 10.1007/978-94-017-8041-4_30,

ominous high tech Cuban bio-medical complex. This is an implicit acknowledgement of that country's ascent on the world stage of significant achievement in biotechnology. However, despite its dire economic situation, Cuba maintains a competitive posture in many other areas, not least in physics. A society's capability in this discipline is a measure of its intellectual maturity and resolve to become a "creator of knowledge" instead of a "consumer of knowledge." The importance of physics as the fundamental scientific foundation on which all scientific and engineering advances depend was enshrined in the resolution passed by the 23rd General Assembly of the International Union of Pure and Applied Physicists (IUPAP). Certainly the author's continued interactions with some Cuban scientists and institutions over the last 13 years have provided continued affirmation of the Cuban government's respect for, and dependence on, physics.

In 1995 the author returned to Cuba at the invitation of the Cuban Physics Society presided over by Dr. Maria Elena Montero. Years earlier (1985–1988) he had discovered a powerful and radically new theory for computing important physical parameters associated with certain "delicate" quantum operators (Handy and Bessis 1985; Handy et al. 1988). These methods relied on important theorems related to the classic "Moment Problem" in pure mathematics. Together with his collaborators, the author had pioneered the development of (what are now referred to as) Semi-Definite Programming (SDP) methods for dealing with singular perturbation strongly coupled systems. SDP methods are now important in many areas, particularly optimization theory (Lovasz 1992; Lasserre 2009), combinatorics analysis (Goemans 1997), and reduced density matrix theory for quantum many body calculations (Greenman and Mazziotti 2008). In addition, the author's "moment quantization" methods provided an alternate framework for understanding how continuous wavelets might be useful in quantum mechanics (Handy and Murenzi 1997). Common interest in these methods has sustained an ongoing collaboration between the author and some Cuban physicists.

At that time, Cuba was slowly recovering from the negative economic impact of the collapse of the former Soviet Union. Whereas the Cuba of 1980 manifested a modest well being for its citizens, the Cuba of 1995 was very different with severe electrical cuts, food limitations, flirtations with small private enterprise (i.e. individually owned restaurants called "paladares"), three currencies (i.e. the Cuban peso, the Cuban "divisa" or the equivalent to the U.S. dollar, and the U.S. dollar), and incredible public transportation resources (i.e. "Los Camellos" (camels)), or 18-wheelers pulling long trailers capable of carrying several hundred passengers). Although many everyday items were subsidized by the Cuban government (i.e. housing, education, medicine, etc.), the author was shocked to learn that many professionals, including surgeons, made around 30 dollars per month! He remembers that after presenting one of several talks, one Cuban physicist (author of a *Physical Review Letters* article) was embarrassed that he could not invite the author (understandably) to a \$2–3 lunch. The author witnessed many other bizarre situations involving highly skilled professionals plying their trade within poor facilities (i.e. dancers of the world famous "Ballet Nacional de Cuba" slipping on weak stage floors). If there was ever a time when the Cuban government could have "legitimately" eliminated its support for basic research, the arts, etc., it was then! Instead, Cubans persevered. Young couples walked briskly in the dark, content displayed on their faces, as they found their way to some night entertainment. Many fixed their cars (i.e. engine overhauls, etc.) right on the street. The author remembers learning that an international conference on "second hand" items was being hosted in Havana. How apropos!

The author visited many scientific centers including the former Capitol Building which had been converted into the main offices of the Cuban Academy of Sciences (with an impressive architecture). He toured the University of Havana with its many artistically stunning buildings and academic halls (i.e. "El Aula Magna"). He met the Director of the School of Physics, Dr. Carlos Trallero Giner, who eventually became a Fellow of the American Physical Society in 2003. The author has enjoyed several collaborative research endeavors with him and his students (Handy et al. 2001; Rodriguez et al. 2003, 2004). In addition, through his deep knowledge of the history of Cuban physics, many important achievements were revealed. The author conveys here some of the important points which are intended as a guide through the myriad of Cuban accomplishments and governmental scientific centers.

As previously suggested, an important catalyst in Cuba's evolving physics capability has always been the strong tradition for excellence in medicine. In the formative period, this gave added impetus and focus to the development of diagnostic equipment and instrumentation, etc., particularly in the face of a restrictive embargo that limited Cuba's access to technology.[1]

In the United States there is a special interest in encouraging the "underrepresented minorities" (i.e. particularly women, Black Americans, Hispanic Americans) to pursue careers in physics. As a Black American, the author had a special interest in documenting, as much as possible, the manifest ethnic diversity of Cuban physicists. Fortunately this was the case: there was a significant representation of visibly Black Cubans within the University of Havana. The most notable is Dr. Alejandro Cabo, a dynamic theoretical physicist. Having said this, it must be understood by all that the African heritage is extremely pervasive throughout all of Cuba. Many seemingly "white" Cubans have black ancestors. One good example is Dr. Valdes Sosa (M.D and Ph.D. in mathematics. A noted researcher in stochastic modeling of the brain, who was invited to a Los Alamos Bio-Physics conference in this area). The author was surprised to learn that he grew up in the United States and his family faced discrimination when it became clear that one of his grandparents was black. The common joke known to most Cubans (both inside and outside Cuba) is that if some Cuban says he is "white" one then demands to see their grandparents ("… y tus abuelos donde estan?")!

From a geopolitical standpoint, the author firmly believes that many Black Americans are unaware of the tremendous kinship that should exist between them and most Cubans. Whereas most Black Americans want to reach out to Africa, they overlook the "brothers and sisters," many highly educated, that exist 90 miles off American shores. The author is very proud that Cuba (including two of his cousins) fought to overcome apartheid in Southern Africa. Hopefully, with the arrival of a

[1] See the contribution by Handy and Trallero-Giner to this volume.

new, black, American administration, greater awareness of these ethnic ties may serve to remove many barriers between the United States and Cuba.

During the 1995 tour, the author had the pleasure of meeting an outstanding quantum chemist, Dr. Luis Alberto Montero Cabrera (who was on his way to visit Dr. Ronald Hoffmann, 1981 Nobel Prize winner from Cornell University). Despite the problems created by the embargo in securing computers for his research, his technical wizardry allowed him to circumvent these obstacles. Many more incredible intellectuals were met, too numerous to mention here. Suffice it to say that all were undaunted by their economic plight and steadfastly continued to undertake research vital to their nation's survival as well as to their own intellectual curiosity.

Upon returning to the United States, the author consulted with various U.S. scientists who confirmed that Cuba was producing outstanding students who would be easily accepted within any graduate program. He resolved to invite several Cuban scientists to his (former) institution in Atlanta, Georgia: the Center for Theoretical Studies of Physical Systems (CTSPS), at Clark Atlanta University, where he served as Co-Director. During 1996 no visa problems were encountered in inviting Drs. Trallero-Giner, Valdes-Sosa, and Montero-Cabrera. However, a subsequent visit proved more difficult since the author was told by a U.S. State Department official that "the United States does not want to facilitate Cuba's ascendancy in the sciences," this is a paraphrase of the actual remark. This sparked numerous inquiries by the American Physical Society and the American Institute of Physics, leading to several articles protesting travel restrictions on Cuban scientists (Kumagai 1997, 1998). Once these were overcome, the author invited several others including Dr. Alejandro Cabo.

During this exciting time CTSPS became highly visible as a U.S. melting pot for various Cuban physicists, nationals and exiles. We hired the irrepressibly brilliant, exiled political dissident, Dr. Rolando Roque-Malherbe. We hosted other physicists such as the (then newly arrived) married couple Drs. Efrain Ferrer and Vivian Incera (now at the University of Texas at El Paso). The author's objective was to do all he could to insure that the excellence of Cuban physics, wherever it was found, could be preserved for the future.

References

Goemans, Michel X. 1997. Semidefinite programming in combinatorial optimization. *Mathematical Programming* 79: 143–161.

Greenman, Loren, and David A. Mazziotti. 2008. Electronic excited-state energies from a linear response theory based on the ground-state two-electron reduced density matrix. *Journal of Chemical Physics* 128: 114109.

Handy, Carlos R., and Daniel Bessis. 1985. Rapidly convergent lower bounds for the Schrodinger equation ground state energy. *Physical Review Letters* 55: 931–934.

Handy, Carlos R., and Romain Murenzi. 1997. Continuous wavelet transform analysis of quantum systems with rational potentials. *Journal of Physics A* 30: 4709.

Handy, Carlos R., Daniel Bessis, Gabriel Sigismondi, and Thomas D. Morley. 1988. Rapidly converging bounds for the ground state energy of hydrogenic atoms in superstrong magnetic fields. *Physical Review Letters* 60: 253–256.

Handy, Carlos R., Carlos Trallero-Giner, and Arezky H. Rodriguez. 2001. Generating bounds for the discrete state energy values of the infinite quantum lens potential. *Journal of Physics A* 34: 10991.

Kumagai, Jean. 1997. US policy toward Cuba undermines scientific exchanges. *Physics Today* 50: 55.

Kumagai, Jean. 1998. As US-Cuba ties improve, scientists ask that travel restrictions be lifted. *Physics Today* 51: 56.

Lasserre, J.-B. 2009. *Positive polynomials and their applications*. London: Imperial College Press.

Lovasz, Laszlo. 1992. Combinatorial optimization: Some problems and trends. Technical Report 92–53, DIMACS, Rutgers University.

Rodriguez, Arezky, Carlos R. Handy, and Carlos Trallero-Giner. 2003. Excited states in the infinite quantum lens potential: Conformal mapping and moment quantization methods. *Journal of Physics: Condensed Matter* 15: 8465.

Rodriguez, Arezky, Carlos R. Handy, and Carlos Trallero-Giner. 2004. Reply to Comment on "Excited states in the infinite quantum lens potential: Conformal mapping and moment quantization methods". *Journal of Physics: Condensed Matter* 16: 2945.

Chapter 31
Cuban/US Research Interactions Since 1995

Maria C. Tamargo

Interactions between Cuban physicists and researchers in the United States are difficult, to say the least. The complexities associated with communication and travel between Cuba and the US greatly hamper these efforts. Nevertheless, scientific interactions are permitted within the limits of the US embargo, and travel to Cuba to attend international scientific conferences or for well-documented research and educational purposes is allowed.

My first interaction with Cuban physicists was in 1994 at the Latin American Congress of Surface Science and its Applications (CLACSA) meeting in Cancun, Mexico. There, I was introduced through a Mexican physicist and friend, Isaac Hernandez-Calderon, to a large contingent of Cubans that were participating in this event. My first trip to Cuba (first since 1962) followed soon, in 1995, after I expressed an interest to the Cuban participants in the Cancun conference. They quickly responded with the needed invitation and research plans. It was the tail end of the so-called "special period." As a Cuban American (my parents are Cuban) the opportunity to visit Cuba was truly an exceptional prospect for me. Suddenly a window had opened in the insurmountable wall that, since my youth, I felt existed between me and what I consider to be my homeland.

The purpose for the visit, which lasted about 1 week, was to interact with Juan Fuentes and his group at the University of Havana. He was attempting to build a molecular beam epitaxy (MBE) research program (a Russian-built MBE instrument had been purchased a few years earlier). I was struck by the lack of resources and dismal facilities, but at the same time I was fascinated by the enthusiasm, resourcefulness and sheer joy they seemed to exude toward their activities. I was also impressed by the level of academic training and the awareness of the recent developments in their fields. With almost nothing as far as laboratory equipment and supplies, the researchers in the Faculty of Physics had active research programs and published

M.C. Tamargo (✉)
Department of Chemistry, City College of CUNY, New York, USA
e-mail: tamar@sci.ccny.cuny.edu

A. Baracca et al. (eds.), *The History of Physics in Cuba*, Boston Studies in the Philosophy and History of Science 304, DOI 10.1007/978-94-017-8041-4_31,

their work in several leading areas, such as superconductivity, nanotechnology, lasers, and theoretical solid state physics.

The MBE effort turned out to be misguided. The costs associated with this kind of research were well above what the University of Havana could support, and although Fuentes persisted for a few years, it was finally abandoned. However, during my visit I also got to know and became friends with several other scientists, most importantly Pedro Diaz-Arencibia, who did optical characterization and physics of semiconductors, and Osvaldo de Melo, who had implemented in his lab a novel (and less costly than MBE) method of semiconductor epitaxial growth, with which he has been quite successful. Both of these became fruitful collaborations that resulted in jointly published work. Over the next several years we published several papers with Pedro Diaz on the physics of exciton localization in ZnCdSe quantum wells grown in our MBE system (Diaz-Arencibia et al. 2000a, b, c). Our exchanges with Osvaldo de Melo involved characterization of samples that we grew together during one of my visits using the isothermal closed space sublimation technique that he has developed (Larramendi et al. 2004).

At the time of my first visit, Carlos Trallero-Giner, an accomplished theoretical condensed matter physicist, was the dean of the Faculty of Physics at the University of Havana. Prior to its break-up, the Soviet Union was the primary avenue for scientific exchange with Cuba. Much was gained scientifically in Cuba from this interaction and the excellent training of the scientists at the University of Havana at that time was largely due to the strong foundation they had received at places like the Ioffe Institute, Moscow University, and others.

However, the end of the "special period" brought a new outlook to the Cuban scientific community which was expecting that new avenues of exchange would be built to replace the old ones. This included, at least in the minds of some, an opening of exchange with the US. As a result, some Cubans traveled from the University of Havana to the US, in search for these potential exchanges. Carlos Trallero-Giner, who visited CUNY and other places back in 1995, was one. Although strong research interactions were established, many of which still persist, I would safely state that these efforts, which were mostly intended to establish bridges for student training and scientists' exchanges, were not nearly as successful as he had hoped.

Even so, we began research collaborations with Carlos Trallero on understanding of the electroreflectance spectra of lens shaped self-assembled quantum dots that we were fabricating by MBE (Muñoz et al. 2003, 2004; Rodriguez et al. 2005). He also developed then a long-lasting and fruitful interaction with Joseph Birman of our university. Jointly, they addressed the problem of the theory of confined semiconductor excitons (such as in GaAs) trapped in quantum wells, and investigated the nature of the excitations in a Bose-Einstein condensed (ultra low temperature) atomic system such as Rb vapor (Trallero-Giner et al. 2002, 2008). He also established strong bonds with other scientists he visited elsewhere in the US. In part recognizing his efforts to forge alliances with US physicists, Trallero-Giner was elected Fellow of the American Physical Society in 2001, perhaps the only Cuban physicist to hold that distinction. The broader exchange programs, however, did not materialize.

Besides advancing common research interests, I have also tried to promote exchange with Cuba over the years. On several occasions I invited Cuban scientists to US conferences, such as the North American MBE conference in 1999, where Diaz presented a paper on our collaborative work, and the International Conference on II-VI Compounds, held in Niagara Falls in 2003, where Trallero was an invited speaker. I have traveled to Cuba five other times, always to attend conferences. I facilitated the travel of other US scientists to Cuba, notably Myriam Sarachik, professor at CCNY and past President of the American Physical Society, for whom the first trip was also a pilgrimage to her childhood places, Jacek Furdyna of Notre Dame University, Fred H. Pollak of Brooklyn College, and scientists from abroad, such as Takafumi Yao from Tohoku University in Japan. And most importantly, over time I have forged true friendships with the Cuban scientists.

My last trip was in January of 2009 to Santiago de Cuba to present a paper in the School of Nanophotonics and Photovoltaics, organized by Alexey Kavokin of Southampton University, UK. I was the only US scientist at that meeting. Still, these collaborations and interactions are the exception, and participation of Cuban physicists in conferences in the US, Europe and Japan is still unusual. Even more unusual is the presence of US physicists in conferences held in Cuba. Of course, US federal funding agencies do not readily provide financial support for travel to or from Cuba, which also limits the exchange possibilities.

Research on condensed matter physics in Cuba persists. Scientists find ways by which to interact with others around the world, primarily leveraging good relations that exist with a few countries, such as Mexico and Spain, but also Germany, Italy, Brazil, and others. They also maintain, with very limited resources, experimental facilities and computational capabilities of their own (in Cuba) that allow them to advance their work while at home. An example of a success story is the semiconductor effort of de Melo, which he carries out in Havana. Unfortunately, the physical infrastructure of their facilities has not progressed in noticeable ways, and the fact that many of their young scientists leave the country and establish themselves elsewhere (being very well trained, they are quickly picked up in other places) makes their efforts to build and grow their programs much harder. Nevertheless, to the benefit of the scientific community, their determination, combined with the very strong academic base and an unyielding desire to stay in the game, will likely keep them going for the foreseeable future. It is interesting to consider what their contributions would be if more open and favorable exchanges with the US, their closest geographical neighbor, were possible.

References

Diaz-Arencibia, P., I. Hernandez-Calderon, L.M. Hernandez-Ramirez, and M.C. Tamargo. 2000a. A two-level model for intra-well exciton migration in Zn1-xCdxSe quantum wells. *Physica Status Solidi B* 220: 27.

Diaz-Arencibia, P., I. Hernandez-Calderon, L.M. Hernandez-Ramirez, and M.C. Tamargo. 2000b. Photoluminescence properties of intra-well exciton migration in ZnCdSe quantum wells. *Microelectronics Journal* 31: 443.

Diaz-Arencibia, P., I. Hernandez-Calderon, L.M. Hernandez-Ramirez, and M.C. Tamargo. 2000c. Temperature dependence of exciton localization in ZnCdSe quantum wells. *Journal of Vacuum Science and Technology B* 18: 1526.

Larramendi, E.M., O. de Melo, M.H. Velez, and M.C. Tamargo. 2004. Absorption, desorption and interdiffusion in atomic layer epitaxy of CdTe and CdZnTe. *Journal of Applied Physics* 96: 7164.

Muñoz, Martín, Shiping Guo, Xuecong Zhou, Maria C. Tamargo, Y.S. Huang, C. Trallero-Giner, and A.H. Rodríguez. 2003. Contactless electroreflectance of CdSe/ZnSe quantum dots grown by molecular beam epitaxy. *Applied Physics Letters* 83: 4399.

Muñoz, Martín, Hong Lu, Shiping Guo, Xuecong Zhou, Maria C. Tamargo, Fred H. Pollak, Y.S. Huang, C. Trallero-Giner, and A.H. Rodríguez. 2004. Contactless electroreflectance studies of II-VI nanostructures grown by molecular beam epitaxy. *Physica Status Solidi B* 241: 546.

Rodriguez, A.H., C. Trallero-Giner, M. Muñoz, and M.C. Tamargo. 2005. Electroreflectance spectroscopy in self-assembled quantum dots: Lens symmetry. *Physical Review B* 72: 45304.

Trallero-Giner, C., F. de León-Pérez, Meng Lu, and J.L. Birman. 2002. Scattering of hot excitons due to optical phonons in quantum wells: Multiphonon resonant Raman process. *Physical Review B* 65: 115314.

Trallero-Giner, C., J.C. Drake-Perez, V. Lopez-Richard, and J.L. Birman. 2008. Formal analytical solutions for the Gross-Pitaevskii equation. *Physica D* 237: 2342.

Chapter 32
Viva La Ciencia: Cuba's Creative Scientists Aim to Make Knowledge Their Country's Sugar Substitute

Rosalind Reid and Brian Hayes

At first, peas served as particles in Ernesto Altshuler's experiment. A mechanical dispenser would drop the *chícharos* one by one into the space between two glass plates, forming a tidy two-dimensional approximation of a sand pile. Lattice structure appeared, then vanished, as the pile self-organized and went critical—avalanche!

But Havana's insects soon found the peas in Altshuler's physics lab. "I began to have too many degrees of freedom," he recalled with a smile.

For a physicist working under harsh economic conditions of Cuba in the early 1990s, options were few. Yet altshuler's solution came as a byproduct of the crisis: Because of fuel shortages, the country had begun importing Chinese bicycles, and ball bearings were available in abundance. Thus the peas have been replaced by steel beads, but Altshuler and his students still call their machine the chícharotron.

Forty-five years after Cuba's revolution, and a decade after the loss of economic support from the U.S.S.R., Altshuler and other scientists of his generation stubbornly and inventively carry on basic and applied science as well as the education of students from Cuba and neighboring countries. Their work is constrained by continuing personal hardship, the whims and troubles of a dictatorship, and a U.S. embargo that impedes access to computer software and laboratory equipment and supplies as well as meetings, fellowships, collaborations and opportunities to publish research. Just before the latest tightening of the U.S. embargo this summer, *American Scientist* visited Havana to talk with some Cuban scientists about their work, which because of these barriers is less well known in the U.S. than it might be.

First published in *American Scientist*, Vol. 92, September/October 2004, 416–18. Reprinted here by kind permission of the authors.

R. Reid (✉)
Executive Director, Council for the Advancement of Science Writing
e-mail: ros@casw.org

B. Hayes
Senior Writer, American Scientist magazine
e-mail: brian@bit-player.org

A. Baracca et al. (eds.), *The History of Physics in Cuba*, Boston Studies in the Philosophy and History of Science 304, DOI 10.1007/978-94-017-8041-4_32,

In proportion to the size of its national economy, Cuba invests more in science and engineering than its Latin American neighbors, except Brazil, and far more than developing countries in general, according to 2002 statistics from the Network on Science and Technology Indicators. A 2001 RAND report for the World Bank rated Cuba "scientifically proficient" on the basis of its high relative investment and number of scientists and engineers. Despite the economic and political constraints under which it operates, Cuban science has an unusual vibrancy. The country's high literacy rate and the Castro government's emphasis on science and investment in training feed bright students into a research-and-development enterprise that remained a source of pride during the island nation's worst years.

"The important thing is to have the people," said Sergio Jorge Pastrana, who oversees international relations for the Cuban Academy of Sciences. "Money comes afterward."

Altshuler is one of the human assets Cuba has succeeded in keeping. A decade ago, he was working on superconducting materials, studying "avalanches" of vortices moving through a crystal lattice. But experiments on superconductivity were impractical in Havana. (For one thing, there was no liquid helium for refrigeration.) To continue the same line of work would have entailed spending much of his time abroad, or else confining himself to theoretical studies.

He chose instead to make a career out of "experiments doable in Cuba," whatever they might be. Avalanches of various kinds remain a central theme, investigated not only with peas and ball bearings but also with Cuba's considerable variety of beach sands. (He has recently discovered a new phenomenon in sandpile physics, in which "rivers" of sand rotate around the pile.) He also has a long-running experiment on seismic instability, in which the apparatus consists of a salvaged stripchart recorder and a handful of tiny magnets from Radio Shack. Even the local abundance of insect wildlife, which plagued the chícharotron, has been turned to good effect: Altshuler has been looking into the "physics" of panicked ants in a Petri dish.

Osvaldo de Melo and María Sánchez are husband and wife as well as dean and vice-dean of the Havana physics faculty. Their interests lie in condensed-matter physics, in fabricating semiconductor devices, magnetic materials and the like—the kind of work often done in a gleaming white cleanroom by technicians in lint-free bunny suits. The laboratories in Havana never approached that standard, but through the 1980s it was still possible to do competitive solid-state research. Much of the apparatus was Russian-made, including the department's prize possession: a molecular-beam epitaxy (MBE) machine for depositing thin layers under precise control. The MBE device is still there in a basement room, its stainless steel vacuum vessel cradled amid ducts and tubes and wires, but inoperable without spare parts and supplies. Much other equipment is kept in working order only through improvisation.

Improvisation, in fact, is a theme of Cuban science and life. Sánchez has completed crucial steps in some of her experiments during visits to better-equipped laboratories overseas. She has also resorted to computer simulation in lieu of the lab bench. De Melo has adopted the "mother of invention" approach: lacking access to MBE equipment for making thin-film devices, he has turned to research on low-tech, low-cost ways of accomplishing the same thing. He reports promising results using sublimation in a simple quartz-lined oven. These varied strategies have kept Sánchez and de Melo active and publishing, but they have also limited the scope of their work. Studies of material properties are feasible, but not fabrication of complex devices.

In the U.S., debates about the level of support for science tend to focus on research grants and departmental budgets. But in Cuba it gets more personal. Sánchez points out that the monthly salary of a senior member of the faculty is 600 pesos, equivalent to $23. Although many basic services are supplied at little or no cost, this is not nearly enough to support a family. Thus the issue for many Cuban scientists is how to keep teaching and doing research while also scrambling to make a living in some other way. Fellowships and visiting professorships in Europe and Latin America have become increasingly important not just as opportunities to collaborate with foreign colleagues but simply as a means of acquiring hard currency.

Things are rather different out at the "Scientific Pole" on Havana's western edge, where modern facilities house the research, development and pilot-production facilities that fuel Cuba's biotechnology effort. Here Ernesto Moreno welcomes his visitors to the Center of Molecular Immunology (CIM) with a marketing video that touts the monoclonal antibodies, vaccines and other products developed at the Pole.

Educated in Moscow as a nuclear physicist when the country was attempting to launch a nuclear power program, Moreno was ready to apply his quantitative skills to genetics when the nuclear project was abandoned in 1992. He went to Sweden for a Ph.D. in structural biology and today works in molecular modeling at the CIM, where cleanrooms and bunny suits are indeed standard. Three decades into Cuba's ambitious biotechnology push, biomedical product sales fund national health care, Moreno notes; additional revenues, Pastrana said, are invested in education and the research effort itself. Joint ventures have been negotiated with China, India, Malaysia, Vietnam, Germany and other countries possessing the capital to build production plants and license production techniques and distribution rights from Cuba—or the ability to conduct clinical trials that will win acceptance for Cuban drugs in affluent nations.

Biology, Pastrana points out, can be done with more limited resources than physics. Among Cuba's available assets are its natural resources, not just sugar cane and tobacco but an array of wildlife said by some to be unusually well preserved because of the island's underdevelopment.

Twenty-five kilometers south of central Havana, botanist Rosalina Berazain has had the run of a grand laboratory since her student days more than 30 years ago. The National Botanical Garden has specimens of much of the nation's 6,500 vascular flora (half the species of the entire Caribbean). Berazain is one of the 10 professors from the university who train young conservationists at the National Botanical Garden for work in regional protected areas around the island. Although Berazain lacks facilities to do the genetic analysis that undergirds taxonomy and conservation today, she has field laboratories to offer to overseas collaborators. In return, a botanical garden in Berlin performs molecular analysis on her specimens.

Much of Cuban science has the strongly pragmatic emphasis typical of developing-country research programs. For some years talented students have been steered toward the biotechnology sector, and now they are being enticed by a new initiative in informatics. (Still more young people are lured away by the tourist industry, which has the important inducement of offering dollars rather than pesos.)

But basic science has survived in Cuba, and recently it has been singled out for encouragement and new financial support. Last January, the Ministry of Science, Technology and Environment announced a new series of competitive grants for research in basic sciences and mathematics. Although the amount of money is small—20 million pesos, or about $750,000, spread over 5 years—it represents "official recognition at the highest level," as one mathematician remarked, a phrase that all Cubans recognize as referring to Fidel. At least eight of the new grants will go to workers at the Institute for Cybernetics, Mathematics and Physics, which occupies several large, colonial-era houses scattered around a neighborhood near the university. The institute has programs in quantum physics, cosmology, linear algebra, statistics and geometry, among other areas.

American visitors tend to see Cuban affairs through the lens of American foreign policy. The long trade embargo and travel restrictions surely have had profound effects on everyday life in Havana, and yet the event seared most deeply into the memories of Cubans is the dissolution of the U.S.S.R., the subsequent withdrawal of Soviet support and the collapse of living standards that followed.

Still, the embargo does give an edge to the stories Cuban scientists tell visiting U.S. journalists. When Pastrana, a historian of science, received a shipment of journal issues on CDs from an American publisher, every disk had been broken. María Sánchez was the president of the Havana chapter of the Institute for Electrical and Electronic Engineers; in 2002 she received a letter from the IEEE severing connections with the chapter and denying support, services and privileges to individual members. The letter said the actions were taken to "comply with U.S. government restrictions." Altshuler had been planning to visit the Santa Fe Institute—his third trip to the U.S.—but his visa application was denied.

Science goes on regardless. On a Friday afternoon toward the end of the academic term, students filed into the lecture hall of the physics department for presentations of undergraduate research, followed by an awards ceremony. In many respects, the scene could have been at any American university. Students in the audience cheered friends and classmates; departmental rivalries flared up; at one point the computer running the PowerPoint presentations had to be rebooted.

But there was also an intensity to the proceedings that seemed distinctly Cuban. The young speakers were cross-examined, and a debate erupted in the audience, bouncing back and forth between students and faculty. Subtleties of quantum mechanics were argued with the same urgency and passion as the merits of contending baseball teams. These young men and women, already zealously engaged in the rituals and traditions of the scientific community, face uncertain career prospects. But for a moment it seemed possible to be optimistic about a vision of Sergio Pastrana's: "Eventually, Cuba could live off knowledge instead of sugar."

Part IV
Scientific Communication and Its Conditions

Chapter 33
Physics in Cuba from the Perspective of Bibliometrics

Werner Marx and Manuel Cardona

33.1 Introduction

Latin America is a large region involving 21 countries with a wide array of cultures, traditions, ethnicities, and economic performance. Its total population is over 500 million people, with an average yearly gross domestic product (GDP) of nearly US$ 6,000 per capita at current exchange rates, corresponding to an average purchasing power of US$ 10,000. The four most economically developed countries (Brazil, Mexico, Argentina, and Chile) only surpass these average figures by about 50 %. This economic performance per capita is an order of magnitude less than that of highly developed nations, usually leading scientific countries such as Canada (US$ 42,900 at exchange rates). Nevertheless, some of the more developed Latin American countries have made, and are still making, a significant contribution to science. Two Nobel prizes have been granted for scientific work performed in Argentina and three to natives of Latin America (Argentina, Mexico, Venezuela) for work performed abroad. Many citizens of Latin American nations have made brilliant careers in science in more advanced countries (Western Europe, USA, Canada), only a few having returned to their countries of origin. They have nevertheless contributed to the development of science in their countries and provided guidance to young countrymen/women for careers in science.

W. Marx
Information Service, Max Planck Institute for Solid State Research,
Stuttgart, Germany
e-mail: w.marx@fkf.mpg.de

M. Cardona (✉)
Emeritus Scientific Member, Max Planck Institute for Solid
State Research, Stuttgart, Germany
e-mail: M.Cardona@fkf.mpg.de

A. Baracca et al. (eds.), *The History of Physics in Cuba*, Boston Studies in the Philosophy and History of Science 304, DOI 10.1007/978-94-017-8041-4_33,

Cuba is a particularly interesting case, with a GDP of about US$ 5,000 in terms of purchasing power, well below the average for Latin America.[1] Moreover, it is the only country in the area with a socialist, centrally controlled economy. Punitive embargoes have been a heavy burden on the Cuban economy and a handicap to the free exchange of ideas with non-socialist countries. We present here an investigation of the development of the physical sciences in Cuba since the revolution, performed with bibliometric techniques involving the number of publications in international journals and the number of citations to these publications. The overall Cuban scientific articles have been analyzed in a former study (Araujo Ruiz et al. 2005). We have also investigated the patterns of collaboration of Cuban scientists with foreign colleagues and institutions. We have confirmed the suspicion that creative science was practically non-existent until the fall of the Batista dictatorship (end of 1958). But, even so, the revolution left the universities without teaching personnel as a result of massive emigration during the years 1959, 1960 and 1961. This personnel, in spite of its shortcomings, was instrumental in transmitting some of the basic tenets of science and scientific method. Their teaching was sorely missed after the Revolution. From 1962, the more fortunate young students obtained fellowships to study in Eastern Europe, and especially in the Soviet Union.

The first bachelor degrees in physics were granted at the University of Havana around 1964. At this time, research began to be encouraged and the first papers were written as internal, local publications. The first scientists with physics degrees returned from the Soviet Union in 1967, but without senior guidance, they did not know how to publish in international journals. In 1973 increasing numbers of PhD holders (*Kandidat* in Russian parlance) began to return. As we shall see, this resulted in the appearance in international journals (we shall call them 'source journals' following the Science Citation Index notation) of articles in the natural sciences with at least one author based in Cuba. Their numbers increased, especially during the 1985–1990 period, a period of relative economic stability. The collapse of the Soviet block (and its subsidies to Cuba) around 1990 put a stop to this growth. Consequently there was a small decrease in scientific production until around 1995, when the country had partly readjusted to the new political and economic realities. A new period of growth followed up to the year 2000 when internal labor pressure, a measure of liberalization and attractive offers from abroad enabled the emigration of many young graduates. As an example, as many as 50 competent young Cuban physicists are now working in Mexico. Our bibliometric studies revealed the ensuing mild decrease in output after the year 2000, leading to the present stagnation in a period of political uncertainties and expectations.

An analysis of collaboration with other Latin American countries, as given by the number of published articles involving at least one author based in Cuba, shows

[1] Because of the vagaries of the government set rates of exchange the GDP based on this rate is near meaningless. Also, the centrally controlled commodity prizes make the purchasing power based GDP difficult to compare with that of market economies. This number is an average of several data found in the web.

that such collaboration has been the strongest with Mexico, Brazil, Colombia, and Argentina (in this order). This fact reveals not only the ranking of these four countries according to scientific activity, but also their interest in establishing some collaboration with their sister nation Cuba, despite differences in ideologies and of periods of right-wing military dictatorships in which relations with collaborating countries cooled down. Significant collaboration has also taken place with countries outside the Latin American area, in particular Spain, Italy, Germany, the US, the USSR, and even small Switzerland (in this order). In terms of citations (they vary in number from 51 to 157), we also present a study of the ten most cited physics articles with at least one Cuban author. Some considerations based on the recently introduced Hirsch index h are also given.

Last but not least, the scientific performance of Cuba is compared with that of countries with similar output, using the *Essential Science Indicators* (ESI) of Thomson Reuters, an in-depth evaluation tool for ranking scientists, institutions, countries, and journals based on the WoS data of the last decade. Finally, the Cuban output and performance is split into 18 different fields of endeavor, again based on the ESI data.

33.2 Methodology

The data presented here are based on the databases relevant for measuring the productivity (output in terms of the number of papers) and the research performance (impact in terms of the number of citations) of scientists, institutions, and countries in the field of physics:

1. *The Thomson Reuters Science Citation Index* (SCI) is accessible under the *Web of Science* (WoS).[2] This database gives access to the articles (no books, no conference proceedings unless they appear in journals or, recently, in the *Conference Proceedings Citation Index*) published since 1900 that are covered by the so-called SCI source journals. Currently about 8,000 journals selected by the staff of Thomson Reuters are considered to contribute significantly to the progress of science. The other citation indexes available under the WoS are not relevant here or do not cover the full time range. The Cuban physics papers were selected using the following physics relevant sub-fields (WoS subject areas):
   - physics, applied
   - physics, atomic, molecular & chemical
   - physics, condensed matter
   - physics, fluids & plasmas
   - physics, mathematical
   - physics, multidisciplinary
   - physics, nuclear
   - physics, particles & field

[2] *Thomson Reuters* WoS: http://scientific.thomsonreuters.com/products/wos. Accessed October 11, 2013.

Note that this analysis does not include the research fields of chemistry and materials science (with the exception of about 100 papers which have been assigned to subject areas out of these two research fields resulting in some overlap). Physical chemistry is one of the WoS subject areas under the chemistry research field, and materials science is a separate WoS research field, which is also subdivided into several subject areas.

2. The INSPEC database for physics, electronics, and computing is accessible via the database provider *STN International*.[3] This database is the appropriate subject-specific literature database in the field of physics with the advantage of a larger coverage of the relevant literature (including books, conference proceedings, dissertations, and reports). It does not provide any citation-based data.
3. The *Thomson Reuters Essential Science Indicators* (ESI) is an in-depth analytical tool for ranking scientists, institutions, countries, and journals.[4] This tool has been used to compare the scientific performance of Cuba with that of countries with similar output and to reveal the trend analysis of output and impact.

33.3 The Output (Number of Articles)

The publications within the natural science disciplines are predominantly published as journal articles. These research fields are covered reasonably well by the source journals of the WoS citation indexes, in particular the *Science Citation Index* (SCI). The number of articles published in the WoS source journals has become a standard measure for scientific productivity (output in terms of the number of publications). The WoS based data set contains all papers (1) with at least one Cuban author (i.e., the term "Cuba" appearing within the author address field), (2) published in one of the source journals covered by the SCI, and (3) assigned to one of the physics subject areas mentioned above.

Note that publications are not even at present a priority among Cuban physicists. They are particularly for those who actively participate in scientific collaboration with Western colleagues and perform basic research: Universities, ICMAF, etc. In other cases, 'know-how' and practical contributions of economic or social kind are considered more important.

When measuring the visibility of the genuine Cuban contributions to physics, papers with the first author (or the corresponding author) having a Cuban affiliation should be selected. Unfortunately, author names and affiliations are not individually allocated in the WoS citation indexes. Therefore, the Cuban first author papers cannot be retrieved using the WoS search and analyze functions.

If only productivity (number of papers) and not their impact (number of citations) are considered, the subject-specific literature databases are a valuable additional (or alternative) information source. In contrast to the SCI, the INSPEC database does not include the affiliations (and countries of origin) of all authors, but only those of

[3] *STN International*: http://www.stn-international.de/index.php?id=123.

[4] *Essential Science Indicators* (ESI): http://thomsonreuters.com/essential-science-indicators/.

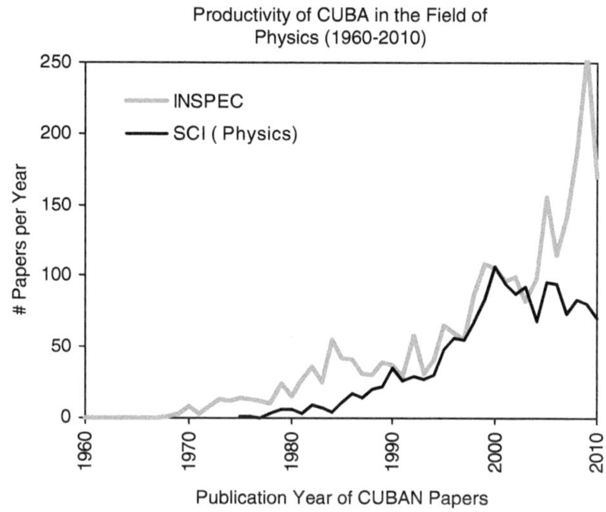

Fig. 33.1 Time evolution of the productivity of Cuba in the field of physics. The search was carried out by searching the term "Cuba" in the author address field (Source: SCI under the WoS and INSPEC under STN *International*)

either the first author or the corresponding author. Thus, this database has the advantage of delivering a better picture of the genuine Cuban contribution to physics.

At the date of search (14.02.2011) the SCI revealed altogether 1,526 papers with at least one author based in Cuba. That is about 10 % of the total Cuban productivity as covered by this database throughout all WoS subject areas (currently 14,448 papers). The overall number of papers in INSPEC is higher (2,576) than in the SCI. This is caused by the broader subject coverage: additional journals not included in the list of the SCI source journals and the inclusion of conference proceedings and other types of documents. The time evolution of the Cuban physics papers based both on the SCI and INSPEC is shown in Fig. 33.1.

The currently 329 physics related Cuban-based conference proceedings covered by the *Conference Proceedings Citation Index – Science* (CPCI-S) have not been included and analyzed here. The CPCI-S does not extend back prior to 1992.

Some benchmarks should be mentioned here:

1962: first students go to socialist countries.
1964: first graduates at the University of Havana.
1967: first graduates return from socialist countries.
1973: an increasing number of PhD holders return from the USSR.
1985–1990: period of relative economic stability.
1990–1995: stagnation due to collapse of the Soviet block.
1995–2000: stabilization and growth.
After 2000: decrease of SCI source journal articles (due to "braindrain"?).

At present, Cuban authors publish about 100 journal based research articles and another 100 conference proceedings per year in the field of physics. The number of physics papers compared to the overall number of Cuban papers covered by the SCI increased from about 2 % in 1975 to around 10 % in 2010. The different time

Table 33.1 Number of papers with at least one author from the countries of Latin America published since 1974 within the journals of the physics subject areas

Country	# Papers
Brazil	40,161
Mexico	18,495
Argentina	14,311
Chile	4,155
Venezuela	2,969
Colombia	2,246
Cuba	**1,526**
Uruguay	578
Ecuador	352
Peru	272
Costa Rica	142
Bolivia	74
Panama	25
Dominican Rep	12
French Guiana	12
Guatemala	10
Honduras	9
El Salvador	4
Paraguay	4
Nicaragua	1
Haiti	0

Source: SCI under STN International

evolution since 2000 of the Cuban productivity based on the SCI in comparison to INSPEC is caused by an increasing fraction of conference proceedings covered by INSPEC and not covered by the SCI (to a large extent not even by the *Conference Proceedings Citation Index*).

The productivity of all countries in Latin America within the field of physics has been analyzed and is shown in Table 33.1. At the date of search (14.02.2011) Cuba ranks seventh (behind Brazil, Mexico, Argentina, Chile, Venezuela, and Colombia). This ranking follows closely the scientific strength of the countries collaborating with Cuba (see below).

The 1,526 SCI based Cuban physics papers have been analyzed with regard to the co-author countries and the physics sub-fields (WoS subject areas) by using the WoS analyze function. Table 33.2 reveals that Germany ranks fifth as a cooperating country, (behind Mexico, Brazil, Spain, and Italy) yet still in front of the USA and the USSR (now Russia). Here, the differences in political and economic systems come to the fore.

Table 33.3 lists the preferred source journals of the Cuban physics publications and Table 33.4 shows that the most important subject areas of scientific activity within the field of physics are condensed matter physics and applied physics.

INSPEC performs a categorization with regard to the experimental and theoretical nature of the works via the so-called treatment codes. The analysis of the Physics publications of Cuba showed that almost half of the papers (48 %) are classified as theoretical. An analysis with respect to the document types results in 82 % journal

Table 33.2 Distribution of the Cuban physics papers throughout the contributing countries (only countries with at least ten papers)

Country	# Papers	% Papers
Cuba	1,526	100.0
Mexico	341	22.3
Brazil	304	19.9
Spain	230	15.1
Italy	140	9.2
Germany	83	5.4
USA	57	3.7
Colombia	53	3.5
USSR	38	2.5
England	32	2.1
France	31	2.0
Switzerland	29	1.9
Argentina	27	1.8
Finland	26	1.7
Japan	26	1.7
Chile	25	1.6
Portugal	23	1.5
Russia	20	1.3
Belgium	18	1.2
Poland	18	1.2
Venezuela	18	1.2
Canada	17	1.1
Norway	16	1.0
PR China	15	1.0
Austria	14	0.9
Slovakia	13	0.9
Hungary	10	0.7
Sweden	10	0.7

Source: SCI under the WoS

Table 33.3 Distribution of the Cuban physics papers throughout the journals (only journals with at least 20 papers)

Journal	# Papers	% Papers
Physical Review B	126	8.3
Revista Mexicana De Fisica	93	6.1
Physica Status Solidi B-Basic Research	75	4.9
Journal of Applied Physics	72	4.7
Journal of Magnetism and Magnetic Materials	48	3.1
Journal of Physics-Condensed Matter	38	2.5
Solid State Communications	33	2.2
Thin Solid Films	33	2.2
Materials Letters	32	2.1
Physical Review C	30	2.0
Physica Status Solidi A-Applied Research	27	1.8
Physica Status Solidi B-Basic Solid State Physics	26	1.7
Physica C	24	1.6

(continued)

Table 33.3 (continued)

Journal	# Papers	% Papers
Applied Physics Letters	23	1.5
Journal of Chemical Physics	23	1.5
Physical Review E	23	1.5
Journal of Physical Chemistry A	22	1.4
Physica C-Superconductivity and Its Applications	22	1.4
Ferroelectrics	21	1.4
Physical Review Letters	21	1.4
Chemical Physics Letters	20	1.3
Physical Review D	20	1.3
Physics Letters A	20	1.3

Source: SCI under the WoS

Table 33.4 Distribution of the Cuban physics papers throughout the physics subject areas (only subject areas with at least ten papers)

Subject area	# Papers	% Papers
Physics, condensed matter	644	42.2
Physics, applied	387	25.4
Physics, multidisciplinary	281	18.4
Materials science, multidisciplinary	221	14.5
Physics, atomic, molecular & chemical	169	11.1
Chemistry, physical	126	8.3
Physics, mathematical	102	6.7
Physics, nuclear	97	6.4
Physics, particles & fields	93	6.1
Materials science, coatings & films	51	3.3
Engineering, electrical & electronic	49	3.2
Instruments & instrumentation	36	2.4
Astronomy & astrophysics	34	2.2
Nuclear science & technology	28	1.8
Physics, fluids & plasmas	25	1.6
Mathematics, interdisciplinary applications	21	1.4
Nanoscience & nanotechnology	21	1.4
Spectroscopy	20	1.3
Mechanics	18	1.2
Mathematics, applied	16	1.0
Computer science, interdisciplinary applications	15	1.0
Optics	15	1.0
Chemistry, multidisciplinary	10	0.7

Source: SCI under the WoS
Note that a certain amount of papers assigned to physics is also assigned to other research fields (in particular *Chemistry* and *Materials Science*)

articles, 25 % conference proceedings (with many of them published as journal articles) and almost 3 % general reviews. The dominance of theory may be attributed to the high cost of experimental equipment which finds little use outside of physics. However, the only costly equipment required for theoretical work are computers which

Table 33.5 Distribution of the Cuban physics papers throughout the author affiliations (only affiliations with at least 20 papers)

Author affiliation	# Papers	% Papers
Univ Havana, Cuba	939	64.5
Inst Politec Nacl (IPN), Mexico	151	10.1
Inst Cibernet Matemat & FIS, Cuba	133	8.9
Univ Nacl Autonoma Mexico (UNAM), Mexico	107	6.8
Univ SAO Paulo, Brazil	95	6.0
Univ Oriente, Cuba	93	5.5
CSIC, Spain	82	5.4
CNR, Italy	55	4.1
Univ Estadual Campinas, Brazil	55	3.3
Inst Super Ciencias & Tecnol Nucl, Cuba	53	3.4
Univ Fed Fluminense, Brazil	50	3.4
Ceaden, Cuba	49	2.8
Univ Fed Sao Carlos, Brazil	45	2.7
Ctr Estudios Aplicados Desarrollo Nucl, Cuba	42	2.6
Univ Antioquia, Colombia	35	2.3
Univ Autonoma Madrid, Spain	35	2.0
Acad Sci Cuba, Cuba	33	2.0
Cinvestav, Mexico	33	1.9
Univ Fed Rio de Janeiro, Brazil	31	2.0
Inst Super Tecnol & Ciencias Aplicada, Cuba	23	1.7
Max Planck Inst Solid State Res, Germany	22	1.7
Ctr Brasileiro Pesquisas Fis, Brazil	21	1.5
Int Ctr Theoret Phys, Trieste, Italy	20	1.5

Source: SCI under the WoS

are extant in almost any country because of the needs of financial institutions and government statistics. These computers can be accessed in part to do scientific work.

Table 33.5 gives a more detailed picture of the collaboration and reveals some leading affiliations involved in physics research, both in Cuba and the major countries of cooperation. Note that the author affiliations including the countries of origin were scarcely covered by the WoS prior to around 1972 and that until present, institutional names are insufficiently standardized by the WoS producer. The name variants identified as belonging together have been given cumulatively.

33.4 The Impact (Number of Citations)

An increasingly popular way to estimate the impact of a scientist, a research institute or a country is to count the number of times that their articles have been cited. The number of citations is often taken as a measure of the attention scientific work has attracted and of the importance or usefulness within the scientific community. Although citation numbers are no ultimate scale of the final importance and quality of publications, they reflect strengths and shortcomings and hence, are frequently used

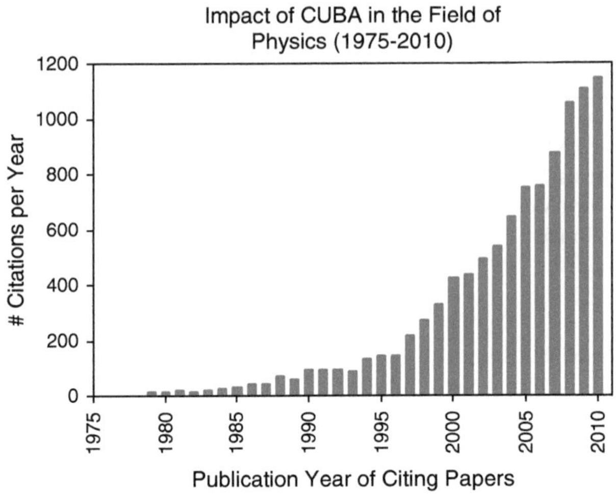

Fig. 33.2 Time evolution of the impact (citations per year) of the Cuban physics papers (Source: WoS citation report based on the SCI data)

for research evaluation (Van Noorden et al. 2010, Van Noorden 2010a, b). In contrast to expert advisory committees (peer review) where only a few scientists are involved, a large number of anonymous scientists within a field of research are involved when using such an evaluation procedure. The question of how suitable citation numbers are for the evaluation of research performance has been debated extensively and therefore will not be discussed here any further (Bornmann and Daniel 2008).

The WoS citation report function reveals the total number of citations per year of the 1,526 physics papers with at least one author based in Cuba and published since around 1972. The time evolution of the citations is given in Fig. 33.2. As the number of articles and the number of their citations accumulate, the number of citations in each year typically tends to increase with time. The slopes of such 'sales curves' can be taken as indicators of the citation impact of scientists and countries, respectively.

The plot in Fig. 33.3 shows the number of citations in each year of all articles from the year of appearance until the given publication year of the citing papers. The impact of the articles published by physicists from Cuba recently accelerated, reaching more than 1,100 citations at present. An acceleration of the number of citations with time may be generally expected due to the cumulative character of the WoS citation report graph but is not always to be found. A linear increase or even a bending-off may also be observed. For example, if a researcher towards the end of his/her career stops publishing, the impact curve shows a significant waning and eventually a bending-off.

The ten currently most highly cited articles in the field of physics with at least one author based in Cuba are listed in Table 33.6. Note that this listing implies no real impact ranking, as papers published in different years have accumulated their citations within different time periods and thus cannot be easily compared concerning their citation counts. It is interesting to note that nine out of the top ten

Table 33.6 The ten currently most highly cited articles in the field of physics with at least one author from Cuba

1. Polar optical vibrational modes in quantum dots
Roca E, Trallero-Giner C, Cardona M
Physical Review B 49(19) 13704–13711 (1994)
Times Cited: 157
2. Optical vibrons in CdSe dots and dispersion relation of the bulk material
Trallero-Giner C, Debernardi A, Cardona M, et al.
Physical Review B 57(8) 4664–4669 (1998)
Times Cited: 122
3. Theory of one-phonon Raman scattering in semiconductor microcrystallites
Chamberlain MP, Trallero-Giner C, Cardona M
Physical Review B 51(3) 1680–1693 (1995)
Times Cited: 115
4. Analysis of the phenomenological models for long-wavelength polar optical modes in semiconductor layered systems
Trallero-Giner C, Garcia-Moliner F, Velasco VR, et al.
Physical Review B 45(20) 11944–11948 (1992)
Times Cited: 85
5. Colloquium: Experiments in vortex avalanches
Altshuler E, Johansen TH
Reviews of Modern Physics 76(2) 471–487 (2004)
Times Cited: 75
6. Coloring random graphs
Mulet R, Pagnani A, Weigt M, et al.
Physical Review Letters 89(26) 268701 (2002)
Times Cited: 74
7. Fusion and elastic scattering of Be-9+Zn-64: A search of the breakup influence on these processes
Moraes SB, Gomes PRS, Lubian J, et al.
Physical Review C 61(6) 064608 (2000)
Times Cited: 72
8. Influence of the Li-6, Li-7 breakup process on the near barrier elastic scattering by heavy nuclei
Maciel AMM, Gomes PRS, Lubian J, et al.
Physical Review C 59(4) 2103–2107 (1999)
Times Cited: 64
9. The reaction Ca-48+U-238 ->(286)112 studied at the GSI-SHIP
Author(s): Hofmann S, Ackermann D, Antalic S, et al.
European Physical Journal A 32(3) 251–260 (2007)
Times Cited: 55
10. Efimov states for He-4 trimers?
Gonzalez-Lezana T, Rubayo-Soneira J, Miret-Artes S, et al.
Physical Review Letters 82(8) 1648–1651 (1999)
Times Cited: 51

Source: SCI under the WoS (date of search: 14.02.2011)

papers appeared in journals of the *American Physical Society* (APS) and one in the *European Physical Journal A*.

A new index has been introduced by Jorge Hirsch as a measure of the cumulative impact of a researcher's scientific work (Hirsch 2005). This so-called h-index is defined as the number of articles in WoS source journals that have had h citations or more. For example, a researcher with an h-index of 40 will have published 40 articles that have received at least 40 citations each. The index is at the same time a measure for both quantity *and* performance of research. Although developed for measuring the scientific contribution of researchers, the index may also be extended to journals, to research fields or even to countries as a whole. If the altogether 1,526 physics papers published by authors from Cuba were produced by a single author, it would correspond to a Hirsch number of h=35.

Different disciplines have different citation habits (i.e., different average numbers of references per article) resulting in different average citation rates (average citations per article) making it rather difficult to compare citation rates (and h-index values) out of different scientific disciplines. For example, the average citation rates of mathematics and of molecular biology & genetics differ by almost a factor of 10.[5]

The most cited Cuban physics article appeared in 1994 and has received 157 citations until present time (see rank one in Table 33.6). If all disciplines in the SCI are considered, the highest number of citations (1,848) belongs to a paper published in 1995 in the field of medicine (Bosch et al. 1995). However, articles in medicine are cited on average about 2.5 times as much as those in physics.

33.5 Essential Science Indicators: Cuba

Thomson Reuters offers an in-depth analytical tool for ranking scientists, institutions, countries, and journals based on the WoS data of the last decade (here: 01/2000 until 10/2010). The *Essential Science Indicators* (ESI) provides

> Internet access to a unique and comprehensive compilation of essential science performance statistics and science trends. The chief indicators of output, or productivity, are journal article publication counts. For influence and impact measures, *Essential Science Indicators* employs both total citation counts and cites per paper scores. The former reveals gross influence whereas the latter shows weighted influence, also called impact.[6]

We first compare the scientific performance of Cuba with that of countries with similar output. Table 33.7 shows the country ranking in physics (ranked by the number of papers). Cuba ranks 68 out of the 89 countries listed. A ranking by the overall number of citations again places Cuba in rank 68 and a ranking by the average number of citations per paper puts Cuba at number 73.

---

[5] ESI baselines: http://esi.webofknowledge.com/baselinesmenu.cgi.

[6] ESI product overview: http://esi.webofknowledge.com/help//h_whatis.htm.

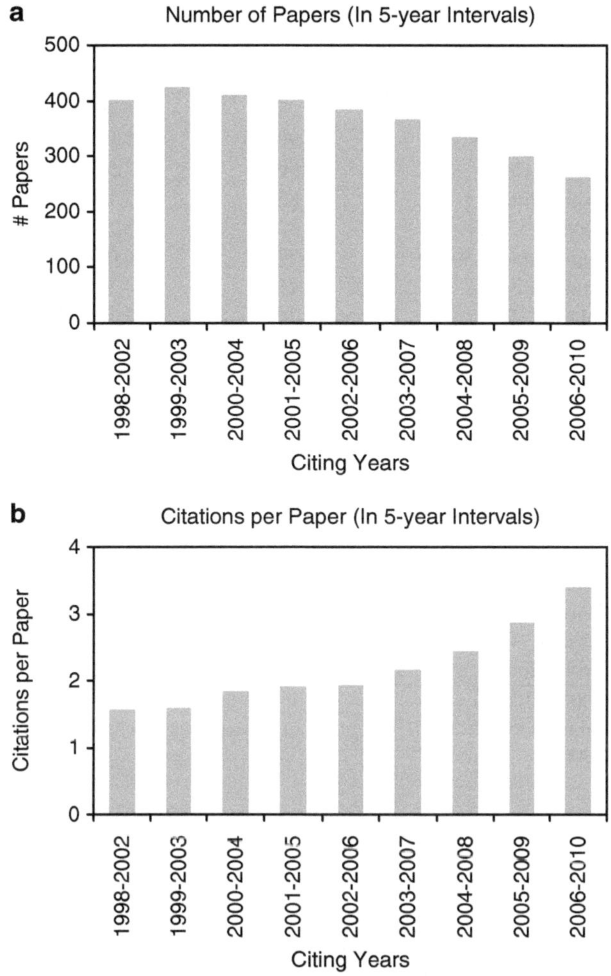

Fig. 33.3 (**a–b**): Time evolution (running average in 5-year intervals) of the number of papers and the average number of citations per paper of Cuban physics papers. (**a**) Number of papers (In 5-year intervals). (**b**) Average citations per paper (In 5-year intervals) (Source: ESI 01/1998–10/2010)

Finally, we have split the Cuban performance into attributions to 18 different fields of endeavor. The result of the field ranking for Cuba (again ranked by the number of papers): Physics ranks third, before agricultural sciences and behind chemistry, followed by clinical medicine. Note that due to the varying dependence on average number of citations within each discipline, a comparative ranking of the fields by the overall number of citations or by the average number of citations per paper is not very meaningful.

The running average of the citations of papers within the first 5 years after their publication, as given by the ESI, has been consulted for checking the impact trend of the physics papers from Cuba. Figure 33.3a–b show the time-dependent number of

Table 33.7 Overall country ranking in physics – ranked by the number of papers

Rank	Country	# Papers	# Citations	Citations per paper
61	Uzbekistan (27.8)	982	5,407	5.51
62	Rep of Georgia (4.4)	908	5,829	6.42
63	Vietnam (86.9)	861	4,514	5.24
64	Serbia (9.9)	859	2,195	2.56
65	Estonia (1.3)	825	5,262	6.38
66	Yugoslavia (22.9)	790	6,078	7.69
67	Latvia (2.2)	756	4,397	5.82
68	Cuba (11.2)	743	3,700	4.98
69	Moldova (3.6)	616	3,642	5.91
70	Azerbaijan (9.0)	607	1,943	3.20
71	Jordan (6.5)	491	2,396	4.88
72	Bangladesh (150.0)	482	1,860	3.86
73	Kazakhstan (16.4)	471	3,046	6.47
74	Cyprus (0.8)	414	4,087	9.87
75	Uruguay (3.4)	353	2,699	7.65
76	Cameroon (19.4)	329	1,552	4.72
77	Indonesia (237.6)	305	1,588	5.21
78	U Arab Emirates (4.7)	284	1,596	5.62
79	Lebanon (4.3)	276	1,712	6.20
80	Serbia Monteneg[a] (10.5)	241	1,702	7.06

Source: ESI 01/2000–10/2010
The population (in millions, date: 2010) is given in parentheses
[a]From 2003 till 2006

papers and the average number of citations per paper within the shifted publication and citation time window of 5 years each, since 1998. In contrast to Fig. 33.2, the plot shown in Fig. 33.3b is more suitable to detect impact changes and therefore has become a standard method in bibliometrics. The impacts of ensembles of papers from different publication time periods become comparable with each other, because all publications accumulate their citations within the same time window. Note that on average the maximum of the citation rate is about 3 years after publication, depending somewhat on the research discipline. The shifted 5 years window is also taken into account in order to smooth fluctuations and thereby to better visualize the output and impact trends.

According to the ESI data, the output of Cuba in terms of publications per year has decreased in the field of physics by about 35 % since the year 1999. The average number of citations per paper, however, increased significantly by a factor of 2.2, thus reflecting an impressive average increased impact of each individual publication. This fact would be laudable if it would reflect an increase in higher quality work at the expense of a decrease in support (and thus in the number of papers of lower quality). However, it may be too early to reach such a conclusion. Note that due to the different accumulation periods the data for the 5-year intervals given in Fig. 33.3b (3.4 for the 2006–2010 period) is correspondingly lower than for the 10-year interval given in Table 33.7 above (4.98).

33.6 Conclusions

The contribution of Cuba in the field of physics has been analyzed by bibliometric methods. Among the countries of Latin America, Cuba ranks seventh (with Brazil, Mexico, and Argentina at the top, followed by Chile, Venezuela, and Colombia) concerning the productivity (number of papers) within the physics research field. Spain, Italy, Germany, the USA, and the USSR have been identified as the main collaborative partners among the highly developed countries. Condensed matter physics and applied physics were found to be the most productive physics research fields. An in-depth analysis of the output and impact trends showed that the number of journal based research articles per year has decreased since the year 2000. Within the past few years, while probably necessary to keep full employment of scientists, emigration seems to have slowed scientific output. The average number of citations per paper, however, has significantly increased. The results reported here suggest that even countries with a weak economic base and a measure of political ostracism can make a significant contribution to science.

References

Araujo Ruiz, J.A., G. van Hooydonk, R.G. Torricella Morales, and R. Arencibia Jorge. 2005. Cuban scientific articles in ISI citation indexes and Cuba Ciencias databases (1988–2003). *Scientometrics* 65(2): 161–171.

Bornmann, L., and H.D. Daniel. 2008. What do citation counts measure? A review of studies on citing behavior. *Journal of Documentation* 64(1): 45–80.

Bosch, F.X., M.M. Manos, N. Muñoz, et al. 1995. Prevalence of human papillomavirus in cervical cancer – A worldwide perspective. *Journal of the National Cancer Institute* 87(11): 796–802.

Hirsch, J.E. 2005. An index to quantify an individual's scientific research output. *Proceedings of the National Academy of Sciences of the USA (PNAS)* 102(46): 16569–16572.

Van Noorden, R. 2010a. A profusion of measures. *Nature* 465: 864–866.

Van Noorden, R. 2010b. How to improve the use of metrics. *Nature* 465: 870–872.

Van Noorden, R., et al. 2010. Do metrics matter? *Nature* 465: 860–862.

Chapter 34
Contemporary Cuban Physics Through Scientific Publications: An Insider's View

Ernesto Altshuler

34.1 Introduction

The Cuban physics community is relatively small, but takes quite seriously the publication of research results in national and international journals, as compared to the situation in other disciplines. This bent is well illustrated by the fact that physicists have won nearly one-third of the annual prizes for the "Best Scientific Paper" given by the University of Havana since 1993. Numerically, however, physicists represent the smallest community at the university, as compared to those constituted by mathematicians, chemists, life scientists, geographers, and others.[1]

In a previous paper on the impact of Cuban physics based on international publications (Altshuler 2005), I made clear that I did not expect this to be an exhaustive bibliometric study of the impact of Cuban physics and that it "would be excellent if a real expert [engaged in] a definitive study on the subject, which would surely contribute to an evaluation of the international impact of Cuban physics and its perspectives in the short and medium terms." The paper included in this volume by Marx and Cardona nicely answers this call. It is the purpose of the present paper to provide some complementary data and points of view from the position of an 'insider.'

[1] After searching the Archives at the University of Havana's Scientific Council, I was able to obtain the information for all years from 1993 to 2007, except for the year 2001. An average of 14 papers from all areas of the University competed each year. Physics authors have won 9 out of a total of 30 prizes—I include here two papers authored by physicists who can be classified as belonging to the fields of pharmaceutical chemistry and quantitative biology.

E. Altshuler (✉)
Physics Faculty, University of Havana, Havana, Cuba
e-mail: ealtshuler@fisica.uh.cu

A. Baracca et al. (eds.), *The History of Physics in Cuba*, Boston Studies in the Philosophy and History of Science 304, DOI 10.1007/978-94-017-8041-4_34,

34.2 Production and Impact of Cuban Physics Papers up to 2000 and Beyond

Since the early 1960s, Cuban policies for the evaluation of scientific performance have always emphasized the "introduction into social practice" of scientific results much more than their publication. By 1978, the Cuban National Commission for Scientific Degrees, created the year before, established a minimum of two papers in peer-reviewed journals as a requirement to complete a PhD degree. In 1986, with the creation of the National Permanent Boards (NPBs) for the defense of PhDs dissertations, further emphasis was put on the quality of the journals in which these two papers should be published. It was not too long before the Physics NPBs established a standard of three papers as a prerequisite to defend a dissertation in that particular scientific field.[2]

The above requirements were instrumental in overcoming the publication inertia of Cuban scientists, but this proceeded at a slow pace until the second half of the 1990s, when a substantial acceleration took place. The economical crisis of the early 1990s (reaching a low by 1993–1994) fostered the exchange of Cuban scientists with 'western' countries. In the new situation, the well-known "publish or perish" motto suddenly became a reality for many Cuban physicists visiting foreign institutions. While a relatively small number of Cuban physicists had been systematically publishing the results of their work in international journals since the 1970s, only in the second half of the 1990s did many more physicists become exposed to the "culture" of international journals. They learned how to present scientific results in an attractive way, which journals were best suited for publishing a specific scientific output, the art of discussion with referees and editors, and the need for a good command of "technical" English. Some even became aware of non-trivial elements in the ethics of scientific publishing, such as who should appear in the authors' list, who should only go into the acknowledgements, and if the attainment of material support for your research should modulate some of these issues.

Against this background, the Annual Prize for the Best Scientific Paper award was created at the University of Havana in 1993, as stated above. In the case of the Physics Faculty, a strong policy of striving for excellence in publications as a prerequisite for promotion to higher teaching ranks was also established in the early 1990s.

These facts boosted the volume of Cuban physics publications, as is shown in Fig. 33.1 in Marx and Cardona (a graph of the papers/year with at least one Cuban-based author vs. year, based on SCI and INSPEC databases). The graph shows a modest increment from 1975 to 1994 with a slope of 1–2 papers/year^2, which increases dramatically to some 10.5 papers/year^2 from 1995 to 2000. The number then decreases with a slope of approximately −2.5 papers/year^2 following SCI, or keeps

[2] Personal communication by Dr. C. Peniche, Secretary of the National Commission for Scientific Degrees.

growing beyond the year 2000 following INSPEC.[3] All in all, the approximately 70 papers/year produced in 2008 (following SCI) is more than twice the production 15 years earlier using the same database. The possible decrease (or saturation) after the year 2000 is suggested by Marx and Cardona to be connected to the "brain drain." An important contribution to this is the fact that several young Cuban physicists, who by the early 1990s had departed for Mexico, Brazil, Spain, etc., to perform graduate studies, remained abroad once they had finished their studies around the late 1990s. While the brain drain is undoubtedly a very important element to consider, other factors may come into play: most of Cuban physics papers depend heavily on foreign collaboration, and the number of months/year spent by Cuban physicists abroad may have decreased since 2000, though unfortunately it is very difficult to obtain reliable data on the matter. Such a tendency might become more marked in coming years due to new regulations being enforced. It is worth saying that the long visits of Cuban physicists to foreign institutions tend to benefit their scientific productivity, not only because of their exposure to state-of-the-art equipment and first-rate scientific environments, but also because they have more time to do research, since teaching, administration duties and the complexities of everyday life are left behind.

Another factor that might have eroded to some extent the overall production of papers after 2000 is the continued increase in the quality of the papers contributed by Cuban physicists, as is explained below. Figure 33.2 in Marx and Cardona (Chap. 33) clearly indicates a dramatic increase in the number of citations of Cuban physics papers, starting around 1995: while the slope between 1985 and 1995 is of approximately 10 cited papers/year^2, the slope approaches 70 cited papers/year^2 in the period 1996–2008. A qualitatively similar tendency was predicted earlier (Altshuler 2005) using a different approach that is explained as follows: The number of citations is the clearest indication of the impact of an individual paper among the specific scientific community to which the paper is relevant. However, there are other elements that influence the visibility of scientific articles, such as the nature and impact factor of the journals where they are published. Figure 34.1 shows the evolution of the "total impact" of Cuban publications in APS journals in the period 1994–2004, calculated as

$$I_T = \sum_{i=1}^{N} I_i$$

where N is the total number of papers published during a certain year, and I_i is the average impact factor (Popescu 2002) of the journal where paper i was published.

Notice that a systematic increase of the total impact of Cuban publications in the journals under analysis begins quite suddenly in 1996. If we fit this to a straight line in the period 1996–2004, the slope is approximately 6 impact units per year^2. As mentioned earlier, this may be interpreted as an increase in the quality (or, at least, visibility) of Cuban physics starting from the mid 1990s.

---

[3] My earlier analysis (Altshuler 2005) using the APS database suggests a saturation after 2000.

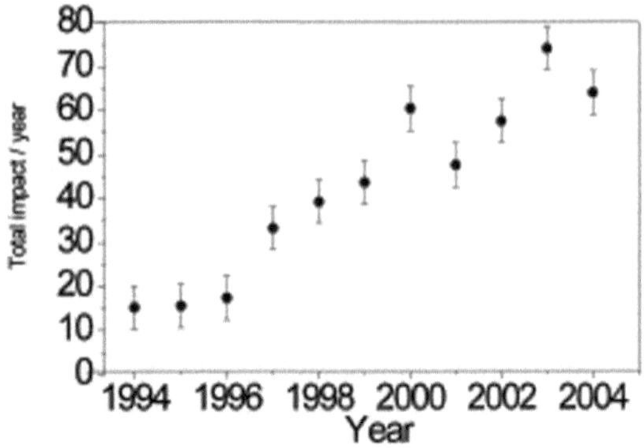

Fig. 34.1 "Total journal impact" per year of articles that appeared in APS and AIP journals, signed by physicists working in Cuban institutions (period 1994–2004)

34.3 Highly Cited Cuban Physics Is Quite Specialized, Very Concentrated, Mostly Theoretical and Internationally Collaborative

The overwhelming majority of the most cited Cuban papers in the field of physics deal with solid-state physics (or materials science)—this being a historical strength and, at the same time, a weakness of Cuban physics. This fact is well documented in Table 33.4 in Marx and Cardona's chapter where the Cuban papers are classified into subfields of physics using the SCI database: the condensed matter + materials science subfields contribute to more than 60 % of the total (and the 25 % attributed to applied physics surely contains much work related to materials science!). It comes as no surprise then to see that many of the most cited Cuban papers have been published in *Physical Review B*[4] and a sizable contribution to the total production of Cuban physics articles have appeared in journals such as *Physica Status Solidi*, *JMMM*, and others specialized in solid-state physics.[5]

Cuban-based authors of most physics papers tend to be highly concentrated in a few centers located in the city of Havana. Table 33.5 of Marx and Cardona's chapter clearly shows that concentration: approximately 65 % of the physics papers by Cuban-based scientists are produced by the University of Havana (which includes staff from the Physics Faculty and the Institute of Science and Technology of Materials, IMRE), 9 % correspond to the Institute of Cybernetics, Mathematics and Physics (ICIMAF), 5.5 % to the University of Oriente, 5 % to the Higher Institute of Technology and Applied Science (former Higher Institute of Nuclear Science and Technology), and nearly 3 % to the CEADEN. With the exception of the University of Oriente, the rest of the centers are located in the city of Havana –a

[4] See, for example, Table 33.6 in the previous chapter by Marx and Cardona.

[5] See Table 33.3 in the previous chapter by Marx and Cardona.

qualitative conclusion which corresponds to my previous analysis in (Altshuler 2005). The situation is due not only to the concentration of resources for research in Havana, but also to the internal "brain drain" from other provinces to Havana in the early stages of graduate studies. Curiously, a second important pole of Cuban physics could have been established at the "Nuclear City" (located in the central region of the country), but this project was aborted shortly after the collapse of the Soviet Union. Even within the University of Havana the most cited physicists concentrate on a few authors, which may be deduced from Table 33.6 in Marx and Cardona's chapter (a list of the 10 most cited papers with Cuban-based authors). In my view, such levels of geographic and individual concentration constitute an important weakness of contemporary Cuban physics.

Theoretical physics overwhelmingly dominates the most cited publications by Cuban physicists. If one examines Table 33.6 in Marx and Cardona's chapter, around half of the papers are purely theoretical and the rest are concerned with experimental work performed outside of Cuba. As in many developing countries, a sizable fraction of the most capable physics students choose to study theoretical physics, due, amongst other things, to its lesser dependence on experimental infrastructure. This results in a concentration of talent among theoreticians. In general, the lack of material infrastructure makes it very difficult to produce high-impact experimental work in Cuba. This is especially true in the field of nuclear physics, but actually affects all branches of physics. It is customary for Cuban scientists working in the field of materials science, for example, to systematically bring their sample materials to foreign facilities in order to characterize their structure and other physical properties. Fabrication of state-of-the-art materials—including nanomaterials—is also extremely difficult in national laboratories. One strategy used by a few Cuban experimentalists to produce significant materials science is to use low cost experimental set-ups in a creative way (like the fabrication of nanostructured materials using chemical methods). Another choice has been just to turn to the field of complex systems, where costly equipment is not the experimental standard in the world. However, several former experimentalists have become theoreticians (especially in the field of computer simulations)—a sort of "hibernation" while waiting for the experimental scenario to improve.

A final feature shared by many of the Cuban papers in physics—and particularly by those highly cited or published in high-impact journals—is the large participation of foreign co-authors. If we go back to the most cited papers by Cuban-based physicists (Table 33.6 from Marx and Cardona), we see that none are signed solely by Cuban physicists. It can be said that many of the most cited papers emerge from the contributions of visiting Cuban scientists to ongoing research projects at foreign institutions. This demonstrates the talent, adaptability and capacity for hard work of many visiting Cuban scientists, but also constitutes a weakness in Cuban-based physics.

It is worth analyzing the Cuban contribution to "wide-scope" journals, since these reach a wider scientific (and even non-scientific) audience, and help to foster interdisciplinary interactions. Papers by Cuban-based physicists in wide-scope physics journals are very scarce: there are only 18 papers in *Physical Review Letters*, and just one in *Reviews of Modern Physics* (data taken by December, 2009).[6] There are no contributions by Cuban-based physicists to very wide-scope journals such as *Nature*, *Nature Physics*, *Nature Materials*, *Science* or *PNAS*.

Let us now briefly examine the Cuban-based papers published in a high-impact, wide-scope journal such as *Physical Review Letters*, by Cuban-based authors.[7] First of all, it should be noted that the majority of these papers do not rank within the 10 most cited Cuban physics papers. This is partially due to the fact that a sizable fraction has been published recently (13 of them after the year 2000). Contrary to what happens with the most cited papers, in this case solid-state physics is represented by only 3 out of 18 papers, which might suggest a process of reorientation (or, at least, diversification) of Cuban work in physics. On the other hand, 6 out of 18 papers contain experimental results—a much higher proportion than is the case for most of the cited papers. Curiously enough, a larger fraction of the experiments were performed in Cuba.

An additional measure of the impact of research in society is the frequency with which scientific articles are featured in popular journals. In the last years, a number of papers authored by Cuban-based physicists have been featured in important journals such as *American Scientist*, *New Scientist* and *Discover*.[8]

34.4 Is It Particularly Difficult for Cuban Physicists to Publish in International Journals?

Aside from the difficulties emerging naturally from poor or badly managed resources—including an enervatingly slow Internet connection—I would say that Cuban physicists generally face the same conditions for publication as do the rest of the world.

As far as I am aware, only once have political issues influenced the normal course of the publishing process of Cuban papers submitted to international journals. This happened after February, 2004, when a governmental "reinterpretation" of the US trade embargo laws was announced which potentially affected a number of countries (Cuba amongst them). This stated that manuscripts submitted to US journals from those countries were not allowed to be subject to any sort of editing process, which meant that a journal editor would not be allowed to correct the grammar, or

[6] This can easily be checked for each of these journals by entering the keyword "Cuba" in the "affiliation" field of the open access search engine of *Physical Review Online Archive* (PROLA), http://prola.aps.org. Accessed October 11, 2013.

[7] See the previous footnote.

[8] See, for example, Hayes (2003) and Ruvinsky (2005).

even the spelling, of a paper submitted, say, by an Iranian author. In practice, this would mean freezing any paper from the embargoed countries submitted to the journal. Some Cuban manuscripts were reportedly frozen by indexed journals in the process. An obvious offense against scientific ethics, the decision was subject to protest by the international scientific community (Brumfiel 2004), resulting in its final dismissal, at least de facto. It should be recalled that, contrary to the wait-and-see attitude of some other scientific organizations, the *American Physical Society* (APS), the *American Institute of Physics*, and the *American Society for the Advancement of Science* refused from the very beginning to comply with the prohibition. My own experience is that APS never hesitated in pursuing the editing process of Cuban papers submitted to its top-visibility journals, even at the worst point of the "crisis."

34.5 Conclusion

Since the early 1990s, Cuban physicists opened up as never before to international journals, in spite of a decreasing supply of material resources—a convincing demonstration of the survival instinct of Cuban physics. While the volume of papers has increased up to the year 2000, and may have decreased or saturated afterwards, the international impact of Cuban articles has been growing sharply since 1995. I doubt that the total scientific production will ever decrease below the levels of the early 1990s. A wide community of Cuban physicists has systematically contributed to this situation, including a number of high-level theoreticians and several experimentalists, who fight against all odds to produce research of excellence—some eventually transforming into "hibernating computer physicists."

The majority of the most cited international publications by Cuban-based physicists have some common features: they often concentrate on solid state physics; they are authored by a reduced number of researchers affiliated with Havana-based institutions; they are predominantly theoretical; and they are produced in a context of international collaboration. At least the first and the fourth features are common to most Cuban physics articles. Cuban contributions to wide-scope physics journals, on the other hand, tend to depart from the solid-state physics scenario.

I believe that the renewal of Cuban physics needs a two-fold effort: the strengthening of the Cuban-based experimental and theoretical capabilities in solid state physics (especially those related to nanoscience and nanotechnology), and the maturation of emerging research in interdisciplinary fields. A "self-organized" renewal process is in progress.

Acknowledgements I thank the "Abdus Salam" ICTP for supporting the writing of the first version of this manuscript. M. Sánchez-Colina and A. Baracca must be acknowledged for encouraging me to write this paper and for the revision of the manuscript, and C. Trallero-Giner, for sending me an earlier version of Marx and Cardona's paper.

References

Altshuler, Ernesto. 2005. Quantifying the impact of physics through scientific publications. *Rev. Cub. Fis.* 22(2): 173–182.

Brumfiel, G. 2004. Publishers split over response to U.S. trade embargo ruling. *Nature* 427: 663, http://www.nature.com/nature/journal/v427/n6976/full/427663a.html. Accessed March 2014.

Hayes, Brian. 2003. On the threshold. *American Scientist* 91(1): 12.

Popescu, Ioan-Iovitz. 2002. *Science Journal Ranking by Average Impact Factors.* http://www.iipopescu.com/Jo_rankingb.htm. Accessed March 2014.

Ruvinsky, Jessica. 2005. Small minds think alike. *Discover: Science, Technology and the Future* 26: 12.

MIX
Papier aus verantwortungsvollen Quellen
Paper from responsible sources
FSC® C105338

If you have any concerns about our products,
you can contact us on
ProductSafety@springernature.com

In case Publisher is established outside the EU,
the EU authorized representative is:
Springer Nature Customer Service Center GmbH
Europaplatz 3, 69115 Heidelberg, Germany

Printed by Libri Plureos GmbH
in Hamburg, Germany